LIST OF ABBREVIATIONS

Handbook Series in Organic Electrochemistry
Vol. IV

ALL OTHER (cont.)

IU	double potential-step chronoamperometry	MG	1,2-dimethoxyethane (monoglyme)	**Q**	
IV	polarographic chrono-amperometry	Mn	minimum		
i vs. E	log i vs. E	MP	mercury-pool electrode	Q	incision quotient
i_A	anodic current	MPDE	mercury-plated disc electrode	QE	controlled-potential coulometry
i_a	adsorption current	MSE	mercury-mercurous sulfate electrode	QP	controlled-current coulometry without a reagent precursor
I_{ac}	diffusion current constant for ac polarography (mmho m$\underline{M}^{-1}$)	Mx	maximum		
i_C	cathodic current	**N**		**R**	
i_c	capacity current			R	reversible
i_{cat}	catalytic current	NCE	normal calomel electrode	r	reliable
i_d	diffusion current	NHE	normal hydrogen electrode	RCCE	rotating cylindrical carbon electrode
i_k	kinetic current	NMR	nuclear magnetic resonance	RCGE	rotating glassy carbon cylinder
i_ℓ	limiting current (μA)	ns	not stated	rms	root mean square
i_{max}	current at the maximum	n_a	number of electrons transferred through rate-determining step of a cathodic process	RP	radio-frequency polarography
i_{min}	current at the minimum				
i_p	peak current (μA)	n_b	number of electrons transferred through rate-determining step of an anodic process	RPDE	rotating platinum-disc electrode
$i_{p,A}$	anodic peak current			RPE	rotating platinum electrode
$i_{p,C}$	cathodic peak current			rxn	reaction
i_{su}	summit current (μA)	**O**			
J		o	organic	**S**	
		O	no wave observed	S	merging with succeeding wave
j_p	peak current density (μA cm^{-2})	**P**		satd	saturated
				SCE	saturated calomel electrode
$j_T^{\frac{1}{2}}/C$	chronopotentiometric constant (μA s$^{1/2}$cm^{-2}m$\underline{M}^{-1}$)			Se	energy sufficient
				SME	streaming mercury electrode
		P	merging with previous wave	SSCE	mercury—mercurous chloride—saturated sodium chloride electrode
		p	preliminary		
		p=	number of protons consumed through rate-determining step	sttd	stated
K				supp elect	supporting electrolyte
		PA	triangular-wave polarography		
k	heterogeneous rate constant (cm s^{-1})	PbDE	lead-disc electrode	**T**	
$k_{f,h}$	forward heterogeneous rate constant (cm s^{-1})	PBE	platinum-bead electrode		
		PbSE	lead sulfide electrode		
		PCA	propylene carbonate		
$k_{f,h}^o$	the value of $k_{f,h}$ at 0 V vs. NHE	PD	derivative polarography	T	temperature
		PDE	platinum-disc electrode	t	time (s)
		PE	pretreated electrode	Tc	temperature coefficient of the limiting current (% deg^{-1})
L		PG	square-wave polarography		
		PGE	pyrolytic graphite electrode	THF	tetrahydrofuran
ℓp	limited proof	PHEN	phenol	THFA	tetrahydrofurfuryl alcohol
LUMO	lowest unoccupied molecular orbital	PHOS	phosphate	TLC	thin-layer chromatography
		PHTH	phthalate	TLIR	thin-layer chronoamperometry
		pK_a	-log(acidic dissociation constant)		
M		Pl	plateau	TLQE	thin-layer coulometry
		PO	oscillopolarography	Tomeš	$E_{\frac{3}{4}} - E_{\frac{1}{4}}$ (mV)
		Po	postwave	TX100	Triton X-100
		PQ	polarographic coulometry		
M	merges with final current rise	Pr	prewave		
m	rate of flow of Hg (mg s^{-1})	PrCN	butyronitrile	**U**	
MAS	mass spectrometry	PRW	Prideaux and Ward		
MB	McIlvaine	PT	Tast polarography		
Mc( )	mistake corrected (column)	PV	total ac polarography		
MeCN	acetonitrile	PVI	in-phase ac polarography	UB	universal buffer
		PY	polarography	UVS	ultraviolet spectroscopy
		PYR	pyridine		

LIST OF ABBREVIATIONS

Handbook Series in Organic Electrochemistry

Vol. IV

APPARATUS:

Cell		Oxygen removal		Data acquisition	
0	two-electrode	A	inert gas	0	recorder
1	iR correction	B	air	1	manual
2	three-electrode	C	hydrogen	2	oscillographic
3	agar bridge	D	sulfite	3	digital
4	liquid junction (Kalousek)	E	hydrazine	4	other
5	porous separator	F	other		
6	other				

ALL OTHER:

A

A	anodic
$A=$	area (cm^2)
ACET	acetate
AE	abraded electrode
AM	ammonia
Ap	approximate value
AuBE	gold-bead electrode
AuDE	gold-disc electrode
Aux	auxiliary (counter) electrode
Av(n)	average of n values

B

BARB	barbital
BOR	borate or borax
BR	Britton and Robinson

C

C	cathodic
$C=$	concentration
$C=x-x$	concentration range
$C\rightarrow$	concentration up to
CARB	carbonate
CHA	carbon-hydrogen analysis
CITR	citrate
CL	Clark and Lubs
ClACET	chloroacetic acid
CP	controlled-potential electrolysis
CPDO	carbon-paste electrode containing dissolved organic substances
CPE	carbon-paste electrode
C_O	concentration of compound oxidized
C_R	concentration of compound reduced

D

D	diffusion coefficient ($10^{-6} cm^2 s^{-1}$)
d	diameter (cm)
De	not energy sufficient
DI	differential pulse polarography
DME	dropping mercury electrode
DMF	N,N-dimethylformamide
DMSO	dimethyl sulfoxide
dr	drawn out

E

E	exponential (e.g., 1E4=1x10^4)
$E=$	potential
EC	programmed-current chronopotentiometry
$EC=$	electrode constant $m^{2/3}t^{1/6}$ ($mg^{2/3}s^{-1/2}$)
ece	an electron-transfer step followed by a chemical reaction, then a second electron-transfer step
ecl	electrochemiluminescence
EE	current-step chronopotentiometry
El	electrode
Elog	E vs. $\log[(i_\ell-i)/i]$
Elogi^a	E vs. $\log[(i_\ell-i)/i^a]$ (a = some numerical constant)
Elogv	E_p vs. $\log v$
Elog$_\tau$	E vs. $\log[(\tau^{\frac{1}{2}}-t^{\frac{1}{2}})/t^{\frac{1}{2}}]$
Elog$_\omega$	$E_{\frac{1}{2}}$ vs. $\log \omega$
ER	chronopotentiometry
ESR	electron-spin resonance
E_{app}	applied emf (V)
E_{disc}	potential of the disc (for a ring-disc electrode)
E_{max}	potential of a maximum
E_{min}	potential of a minimum
E_p	peak potential
$E_{p,A}$	anodic peak potential
$E_{p,C}$	cathodic peak potential
$E_{p/2}$	half-peak potential
E_{rf}	radio-frequency potential at which current changes sign
E_{su}	summit potential
$E_{\pi max}$	potential of peak emission (on scanning)
$E_{\tau/4}$	quarter-transition-time potential
$E_{\frac{1}{2}}$	half-wave potential
$E_{\frac{1}{2}}^{\circ}$	half-wave potential at pH=0
$E_{0.22}$	0.22-transition-time potential
$E_{0.85}$	0.85-peak potential

F

F	from a figure
f	frequency of a sinusoidal signal (Hz)
f( )	functional dependence, not given

G

GCE	glassy carbon electrode
GE	graphite electrode
GEL	gelatin
GLC	gas-liquid chromatography

H

H	reduction of hydrogen ion
h	height of the mercury column (cm)
HL	high-level faradaic rectification
HMDE	hanging mercury-drop electrode
HOMO	highest occupied molecular orbit

I

I	diffusion current constant ($\mu A \ s^{1/2} mg^{-2/3} m\underline{M}^{-1}$)
i	current (μA)
I:I	comparison of diffusion current constants
i:i	comparison of (limiting) currents
IL	chronoamperometry with linear potential sweep
Ilk	from Ilkovič equation
IP	Izmailov and Pivneva
IR	chronoamperometry
iR	ohmic potential as in iR drop
IRS	infrared spectroscopy
i-t	potentiostatic measurement

LIST OF ABBREVIATIONS

Handbook Series in Organic Electrochemistry

Vol. IV

ALL OTHER (cont.)

IU	double potential-step chronoamperometry	MG	1,2-dimethoxyethane (monoglyme)		
IV	polarographic chrono-amperometry	Mn	minimum		
i vs. E	log i vs. E	MP	mercury-pool electrode		
		MPDE	mercury-plated disc electrode		
i_A	anodic current	MSE	mercury-mercurous sulfate electrode		
i_a	adsorption current	Mx	maximum		
i_{ac}	diffusion current constant for ac polarography (mmho m$\underline{M}^{-1}$)				

Q

		Q	incision quotient
i_C	cathodic current	QE	controlled-potential coulometry
i_c	capacity current	QP	controlled-current coulometry without a reagent precursor

N

NCE	normal calomel electrode
NHE	normal hydrogen electrode
NMR	nuclear magnetic resonance
ns	not stated
n_a	number of electrons transferred through rate-determining step of a cathodic process
n_b	number of electrons transferred through rate-determining step of an anodic process

i_{cat}	catalytic current
i_d	diffusion current
i_k	kinetic current
i_ℓ	limiting current (μA)
i_{max}	current at the maximum
i_{min}	current at the minimum
i_p	peak current (μA)
$i_{p,A}$	anodic peak current
$i_{p,C}$	cathodic peak current
i_{su}	summit current (μA)

R

R	reversible
r	reliable
RCCE	rotating cylindrical carbon electrode
RCGE	rotating glassy carbon cylinder
rms	root mean square
RP	radio-frequency polarography
RPDE	rotating platinum-disc electrode
RPE	rotating platinum electrode
rxn	reaction

J

j_p	peak current density (μA cm^{-2})
$j_T^{1/2}/C$	chronopotentiometric constant (μA s$^{1/2}$cm^{-2}m$\underline{M}^{-1}$)

O

o	organic
0	no wave observed

P

P	merging with previous wave
p	preliminary
p=	number of protons consumed through rate-determining step
PA	triangular-wave polarography
PbDE	lead-disc electrode
PBE	platinum-bead electrode
PbSE	lead sulfide electrode
PCA	propylene carbonate
PD	derivative polarography
PDE	platinum-disc electrode
PE	pretreated electrode
PG	square-wave polarography
PGE	pyrolytic graphite electrode
PHEN	phenol
PHOS	phosphate
PHTH	phthalate
pK_a	-log(acidic dissociation constant)
Pl	plateau
PO	oscillopolarography
Po	postwave
PQ	polarographic coulometry
Pr	prewave
PrCN	butyronitrile
PRW	Prideaux and Ward
PT	Tast polarography
PV	total ac polarography
PVI	in-phase ac polarography
PY	polarography
PYR	pyridine

S

S	merging with succeeding wave
satd	saturated
SCE	saturated calomel electrode
Se	energy sufficient
SME	streaming mercury electrode
SSCE	mercury—mercurous chloride—saturated sodium chloride electrode
sttd	stated
supp elect	supporting electrolyte

K

k	heterogeneous rate constant (cm s^{-1})
$k_{f,h}$	forward heterogeneous rate constant (cm s^{-1})
$k_{f,h}^o$	the value of $k_{f,h}$ at 0 V vs. NHE

L

ℓp	limited proof
LUMO	lowest unoccupied molecular orbital

T

T	temperature
t	time (s)
Tc	temperature coefficient of the limiting current (% deg^{-1})
THF	tetrahydrofuran
THFA	tetrahydrofurfuryl alcohol
TLC	thin-layer chromatography
TLIR	thin-layer chronoamperometry
TLQE	thin-layer coulometry
Tomeš	$E_{\frac{3}{4}} - E_{\frac{1}{4}}$ (mV)
TX100	Triton X-100

M

M	merges with final current rise
m	rate of flow of Hg (mg s^{-1})
MAS	mass spectrometry
MB	McIlvaine
Mc( )	mistake corrected (column)
MeCN	acetonitrile

U

UB	universal buffer
UVS	ultraviolet spectroscopy

LIST OF ABBREVIATIONS

Handbook Series in Organic Electrochemistry
Vol. IV

ALL OTHER (cont.)

V

V	volts
v	scan rate (mV s^{-1})
VA	triangular-wave voltammetry
VIS	visible spectroscopy
VR	cyclic triangular-wave voltammetry
VV	ac voltammetry
VY	hydrodynamic voltammetry

W

W	well defined

X

X	ill defined
X preceding another symbol, see Introduction to Table I	
x→x(units)	continuous variation
x-x(units)	range

Symbols:

⇌̸	irreversible
≠	unequal
∝̸	unproportional
∅	quantum yield efficiency
□	square wave
∧	triangular wave

α	cathodic charge-transfer coefficient
β	Hückel integral energy unit
Γ	surface excess (10^{-10} mol cm^{-2})
ΔE	pulse amplitude (mV)
Δe	amplitude of sinusoidal potential (mV)
ΔE_p	$E_{p,A} - E_{p,C}$
$\Delta E_{su/2}$	half-summit width
λ	wavelength (nm)
λ_{max}	wavelength of the maximum
π	intensity of luminescence
ρ	Hammett reaction constant
σ, σ_x	Hammett substituent constant
τ	transition time (s)
τ_A	anodic transition time (s)
τ_C	cathodic transition time (s)
τ_f	forward transition time (s)
τ_r	reverse transition time (s)
ω	rotation rate (s^{-1})

CRC Handbook Series in Organic Electrochemistry

Volume IV

Authors:

Louis Meites
Petr Zuman

Professors of Chemistry
Department of Chemistry
Clarkson College of Technology

Elinore B. Rupp

Research Associate in Chemistry
Department of Chemistry
Clarkson College of Technology

and

Theodore L. Fenner
Lawrence Spritzer

CRC PRESS, Inc.
Boca Raton, Florida 33431

Library of Congress Cataloging in Publication Data

Meites, Louis.
　　Handbook of organic electrochemistry.

　　(CRC handbook series in organic electrochemistry;
v.4).
　　Bibliography: p.
　　Includes indexes.
　　1. Electrochemical analysis — Handbooks, manuals,
etc. 2. Chemistry, Organic — Handbooks, manuals,
etc. I. Zuman, Petr, joint author. II. Title.
III. Series.
QD272.E4M44　　　　547'.1'37　　　　77-24273
0-8493-7220-8 (Complete Set)
ISBN 0-8493-7224-0(v.4)

　　This Handbook was prepared with the support of the National Science Foundation Grants No. CHE74-19761 and CHE76-17435. However, any opinions, findings, conclusions, or recommendations expressed herein are those of the editors and do not necessarily reflect the view of NSF.

　　This book represents information obtained from authentic and highly regarded sources. Reprinted material is quoted with permission, and sources are indicated. A wide variety of references are listed. Every reasonable effort has been made to give reliable data and information, but the author and the publisher cannot assume responsibility for the validity of all materials or for the consequences of their use.

　　All rights reserved. This book, or any parts thereof, may not be reproduced in any form without written consent from the publisher.

© Clarkson College of Technology

International Standard Book Number 0-84903-7220-8 (Complete Set)
International Standard Book Number 0-8493-7224-0 (Volume IV)

Library of Congress Card Number 77-24273
Printed in the United States

PREFACE

Almost 60 years have gone by since Professor Jaroslav Heyrovský conducted the preliminary investigations that led to the development of polarography. In that time, nurtured by the efforts of a host of brilliant and dedicated scientists, it has grown from promising infancy to vigorous maturity. Mathematicians, electrochemists, analytical chemists, hydrodynamicists, biochemists, clinical chemists and analysts, inorganic chemists, physical organic and synthetic organic chemists, and others have contributed to its development and have conferred a wide spectrum of abilities on it. Analysts use it for detection, identification, and determination; organic chemists use it for studying the rates and equilibria of organic reactions, elucidating their mechanisms, making structural assignments, and designing or improving synthetic procedures; inorganic chemists use it for deducing the compositions of complexes and studying the thermodynamics and kinetics of their formations and dissociations; clinicians use it for diagnosis and for following the treatment of diseases. Many other kinds of uses have already been found for it, and many more can be foreseen for the future. Without doubt it is one of the most powerful and most widely applicable techniques of scientific measurement and investigation.

As polarography itself has grown, other techniques related to it have been revived or invented and have shared in its growth. Some of these, like ac polarography and oscillopolarography, involve changes of the natures of the exciting signal and the measured response. Some, like voltammetry, involve different kinds of mass-transfer processes and different indicator electrodes. Others, like controlled-potential electrolysis and coulometry, involve the same fundamental relationships among current, potential, composition, and time, but employ them in different ways and for different purposes.

The development and growth of this family of interrelated techniques has been recorded in a large number of publications. About 125 research papers dealing with polarography had been published by 1930, about 1000 by 1940, about 3500 by 1950, and about 10,000 by 1960. Professor Heyrovský's annual *Bibliography of Publications Dealing with the Polarographic Method* included a total of 17,306 citations through 1964, while the *Bibliografica Polarografica* published by the Consiglio Nazionale delle Richerche lists 21,798 through 1967. No more recent reliable figure is available, but an estimate of 50,000 through 1978 can hardly be very far from the mark.

Ready access to the information previously gathered is of course essential to further progress throughout the development of any field of science. The analytical chemist and the clinical chemist need data on the behaviors of the substances they wish to determine and of other substances that might accompany these so that they can design and use analytical methods that yield accurate and reliable results. Data on the behaviors of individual substances at different pH-values, ionic strengths, and temperatures and in different buffers and solvents are needed in correlating these with their chemical and biochemical properties; data obtained with different techniques and indicator electrodes are needed in clarifying electrochemical phenomena; data on the behaviors of different but related substances provide not only structure-reactivity correlations but also correlations with many other kinds of data from different fields, ranging from ultraviolet, infrared, and nuclear magnetic resonance spectroscopy to the abilities of polycyclic hydrocarbons to act as semiconductors; data on the behaviors of starting materials often enable the synthetic electrochemist to select conditions that will give the best possible yield of the purest possible product in the least possible time. Data compilations can also indicate which compounds or groups of compounds need further study, aid in the selection of systems that will repay study by newly devised techniques, and guide efforts to achieve uniformity in reporting the results of research.

A number of authors, including two of ourselves, strove to achieve these goals in compilations of electroanalytical data published between about 1950 and 1965. These differed widely in scope, length, and arrangement, and in the amounts of critical judgment they embodied. Some were limited to inorganic substances and others to organic ones. Some were intended only as representative samples of the available information; others were attempts to present it much more fully. Some were limited to half-wave potentials, to polarography alone, to data obtained in aqueous media, or in other ways; others, especially the more recent ones, attempted greater diversity. It is not our purpose here to evaluate them in detail, for some of them were our own; all served a useful and important purpose, and we have striven to learn from the merits and defects of each. It is more appropriate to lament their increasing age, which renders them all less satisfactory and less useful than they were when they were published. The simple fact is that the volume of data available is already so large and is growing so rapidly that their collection, evaluation, and selection is no longer a feasible task for any single electrochemist.

That the need for such work was becoming more acute as the possibility of doing it decreased has been a matter of concern to Commission V.5 on Electroanalytical Chemistry of the International Union of Pure and Applied Chemistry for over a decade. A subcommission charged with responsibility for electroanalytical data and nomenclature, established as early as 1963 and having first one and then another of us as chairman, has attempted for years to encourage, advise, and cooperate with everyone engaged in data compilation in this area. During this time there have been several groups, in as many different countires, that seriously considered engaging in such work, but the magnitudes of the task and of the expenditures of money and time required by work on a worthwhile scale have prevented some of these from actually beginning work and rendered significant achievement impossible for those who made the attempt.

When two of us came together on the staff of Clarkson College of Technology in 1970, we had published 12 different compilations of polarographic data: one of us had been involved in this work since 1950 and the other since 1955. In our common concern for continuing it we were fortunate enough to be able to secure financial support from the National Library of Medicine of the National Institutes of Health. That support enabled us to publish, in 1974, a volume containing data on 2015 organic compounds having 1 to 11 carbon atoms per molecule.

Since mid-1974, the work has been supported by the National Science Foundation, and a new publisher had to be selected for reasons beyond our control. The first two volumes of the present *Handbook* were published in 1977. They included the data contained in the 1974 volume, with various changes and corrections, and also gave data on 1848 new compounds having 12 or more carbon atoms per molecule. The third volume was published in 1978 and contained data on 1309 compounds including 228 that had appeared in the first two volumes, and for which additional data (obtained by different techniques, under different conditions, or in different solvents and supporting electrolytes) were given in Volume III. The first three volumes thus dealt with a total of 4944 compounds.

This volume extends the series still further. It follows the same pattern as Volume III and includes 731 compounds. There are newer or additional data for 190 compounds that also appeared in previous volumes.

Although these four volumes contain the largest collection of organic electrochemical data ever made, it is essential to describe their limitations because they include only a fraction of the data that have appeared in the original literature. We have excluded electrochemical techniques, such as conductometry, high-frequency conductometry, and dielectrometry*, in which neither the electrical double layer nor any electrode reaction need be considered. These furnish information so widely different from, and so rarely used in conjunction with, that provided by polarography and its congeners that combining them would be difficult and of little use. We have also excluded a number of other techniques, such as electrography and potentiometric, amperometric, and other titration techniques that, although of great analytical utility, do not in general provide fundamental data of lasting importance. Finally, we have excluded potentiometry itself to avoid duplicating the efforts of others who have prepared or are preparing compilations of standard and formal potentials. These exclusions leave a group of approximately 50 closely interrelated techniques, which include polarography, voltammetry, amperometry, controlled-potential coulometry, chronopotentiometry, chronoamperometry, and chronocoulometry, and a host of variants on these such as ac and square-wave polarography, programmed-current chronopotentiometry, and double potential-step chronocoulometry. Table VIII lists the techniques that were used to obtain the data that appear in this volume.

This series of volumes is devoted to data on the behaviors of organic substances and organometallic compounds in which the metal is bound to a carbon atom. The importance of porphyrin and its derivatives to scientists in health-related fields led us to include hemes and related compounds in these volumes, and we also included substitution-inert complexes in which the ligand is electroactive so that data on such complexes and ligands would not be divided between separate series. A companion series, to be entitled CRC *Handbook Series of Inorganic Electrochemistry*, will deal with inorganic substances, including the complexes of metal ions with organic ligands that are not electroactive, and data on complex compounds included in this volume will be cross-referenced there.

Volumes I and II of the CRC *Handbook Series of Organic Electrochemistry* were still further restricted to information published in the 12-year period from 1960 through 1971. Volume III contained additional material from the same period, and also covered the 4-year period from 1972 through 1975. Volume IV follows the same pattern, covering additional information published between 1960 and 1975 and a limited amount of information published in 1976. Subsequent volumes will extend this coverage in both directions.

Within these restrictions the coverage of the literature in this volume and its predecessors is comprehensive but by no means complete. According to the estimates we gave above, there must have been roughly 30,000 to 35,000 publications in the field between the beginning of 1960 and the end of 1976, and we have not been able to inspect all of these. We have never relied on secondary sources such as *Chemical Abstracts*, and our coverage of the original publications has naturally been influenced by the availability of journals and reprints. Although we have occasionally allowed our desire to record the fact that a compound has been found to be amenable to investigation by polarographic and related techniques, and our conviction that further study of its behavior would be worthwhile,

---

* The names used in this compilation are taken from Meites, L., Nurnberg, H. W., and Zuman, P., The classification and nomenclature of electroanalytical techniques, *Pure and Applied Chemistry*, 45, 81, 1976.

to persuade us to include preliminary or dubious results (all of which we have scrupulously attempted to identify as such), we have rejected many more such results than we have included. Our attempt has been to select the best of the available data, the most useful, and the ones most likely to withstand the test of time.

To this end we and our collaborators have repeatedly scrutinized every bit of information recorded here. We have separately evaluated the conformity of the apparatus and experimental techniques to the best modern standards, the accuracy of the handling of the data obtained, the certainty with which products and intermediates were identified, the strength of the evidence given for each suggested reaction course or mechanism, and every other point that the published description of each research allowed us to check. In an effort to minimize inconsistency we have compared the data obtained under different conditions, by different techniques, and for different but related compounds. We have assayed alternative courses and mechanisms proposed by different authors and for related compounds. Despite all these precautions it would of course be idle to pretend that the result is wholly free from errors or contradictions. Some of these are in the original literature and have survived our best attempts to detect and expunge them; others have been introduced in transcription and typing. We can only hope that the merits and utility of this collection are sufficient to induce its users to forgive its defects, and urge those who find errors in the pages that follow to call them to our attention so that others may be apprised of them.

There is, however, one sort of defect for which we must disclaim responsibility: this is the result of the woeful lack of standardization in reporting data that runs rampant through the original literature and that we have been powerless to correct. In some circumstances this is trivial enough to be merely unaesthetic; in others it is an ineradicable blot on the meaningfulness or utility of the work that the data represent. In polarography some authors deduce ratios of n-values for processes that give rise to successive waves on a polarogram by comparing the heights of the waves, while others apply a correction for the dependence of $m^{2/3}t^{1/6}$ on potential by comparing diffusion current constants instead. In Table I, which constitutes the major portion of this volume, these two practices are represented by the entries "i:i" and "I:I", respectively, in the 21st column. Though the second of them is unarguably superior for diffusion-controlled waves, it would be so unlikely that the first would lead to incorrect values of n that the universal adoption of the better expedient would merely improve the beauty and legibility of a tabulation like this one. At the other extreme, there are still authors who report potentials that are referred to mercury-pool "reference" electrodes even though every reputable textbook or monograph published in the field for over a quarter of a century has inveighed against this archaic practice. There are actually even some who fail to describe the compositions of the solvents or supporting electrolytes they have used. Such defects becloud the significance of the data reported and render them far less useful than they might have been for identification, devising analytical methods, and many other purposes. In the present state of the art it has not been possible to exclude all of the data that are thus or otherwise blemished, but we strongly hope that these volumes will help to promote standardization and uniformity so that such blemishes will be less numerous in the raw material gathered for future ones.

We said above that this work owes its existence to the financial support we have received. That of the National Science Foundation is acknowledged in official terms on the title page. But in addition we are glad to have this opportunity to record our personal gratitude to Dr. A. F. Findeis and Dr. Janet Osteryoung at the Foundation: their sympathy with our aspirations and patience with our problems have given us moral support that has meant as much to us as the Foundation's financial support.

This work would also have been impossible without the advice, encouragement, and cooperation that we have received from many of our electrochemical colleagues. First among these are the ones whose names appear on the title page. Dr. Elinore B. Rupp joined the project early in 1975 and played a leading role in every stage of the preparation of this volume. Dr. Theodore L. Fenner joined us at the same time, and his care, diligence, and accuracy as an editor have contributed much to its quality. Dr. John J. Rupp assisted us in checking the ACS Registry Numbers, which appear in the present volume for the first time in this series as a further aid in identifying the compounds listed. We also owe a continuing debt of gratitude to those who have worked on the project in prior years. These include Dr. William J. Scott, who shared in its birth pangs; Dr. Bruce H. Campbell, who has continued to bear the responsibility for the WLN code designations given in Table I; and Dr. Alex M. Kardos, who helped to prepare Volumes I and II and to develop most of the procedures employed in preparing this volume. It is a pleasure to be able to express our thanks to all of these for the talents, devotion, and enthusiasm that they brought to tasks that are both tedious and difficult. The bulk of of this volume is a reproduction of a typescript prepared by Mrs. Helen Tyler. Her contributions to its appearance and legibility and to its accuracy and reliability are visible whereas her care, dedication, and patient good cheer are not, but all have been equally indispensable throughout our work. We are grateful for the devoted cooperation of Mrs. Sharon Whittier, who prepared and typed the indices. We are grateful to Professor Ernest I. Becker for his continued generous advice and help with problems of organic nomenclature. We have also had the benefit of consultation and help from our fellow members of Commission V.5 on Electroanalytical Chemistry of IUPAC, whose encouragement, support, and experience in data compilation in many diverse fields have been invaluable to us. Finally, it is a pleasure to acknowledge the advice and assistance that we have received at various stages from many eminent leaders of the electrochemical community.

Louis Meites
Petr Zuman
Elinore B. Rupp
Potsdam, New York
September, 1978

TABLE OF CONTENTS

General Introduction 1

Table I.   Electrochemical Data 4

Table II.   Structural Formulas 265

Table III.   Courses and Mechanisms of Half-Reactions 295

Table IV.   Compounds Included in Table I 361

Table V.   Functional-Group Index 391

Table VI.   Chemical Abstracts Registry Numbers 437

Table VII.   Index of Solvents Employed 445

Table VIII. Index of Techniques Employed 451

Table IX.   Index of Indicator Electrodes Employed 457

Table X.   Key to Literature Citations 461

Table XI.   Author Index 471

Corrigenda   477

GENERAL INTRODUCTION

We have attempted in these volumes to give as much information as possible in as little space as possible. The information is arranged in ways that are often arbitrary, and it is presented with the aid of symbols and abbreviations whose meanings are often not obvious. This General Introduction, and the introductions to the individual tables that follow, are provided in the hope that they will enable the reader to make full use of the tables in the shortest possible time.

This volume of the CRC *Handbook Series in Organic Electrochemistry* comprises 11 tables. Of these, the longest and most important is Table I, which gives information about the electrochemical behaviors of 731 organic and organometallic compounds and about the experimental techniques and conditions employed in studying them. The other ten tables serve to supplement, cross-index, and interpret the contents of Table I.

Despite the physical bulk of this volume and its predecessors in this series and the stringency with which we applied the criteria, described in the Preface, for restricting their scopes, ruthless abbreviation and condensation have been necessary in compressing the information that is given here into the space available for it. The list of abbreviations and symbols is therefore essential to the correct decoding and interpretation of entries that would be unintelligible without it. One copy of this list appears inside the front and back covers of this volume, and there is a separate copy of it on unbound sheets that can be removed and used to follow the rows of data across the wide pages of Table I. In preparing this volume we added a few abbreviations to the lists that appeared in Volumes I, II, and III. Hence the current list can be used in conjunction with those volumes, whereas the earlier lists will occasionally fail to permit the complete interpretation of an entry appearing in this volume.

Each table is preceded by a description of its contents, its organization, and the manner in which it can be used. The present introductory discussion gives a more general description of the volume as a whole, stressing the purpose of each table and the ways in which different tables are related and supplement each other.

In this volume, Table I is divided into two parts. Its first, and very much its longer, part contains the purely electrochemical data that have been obtained by polarography and the other techniques listed in Table VIII. This is followed by a much shorter second part that gives data and qualitative observations pertaining to electrochemiluminescence. Both parts of Table I contain so many columns that they extend across two facing pages. In each case the division between these pages is very nearly such that the left-hand pages identify the compounds for which data are given, the techniques by which the data were obtained, and the electrodes used, and also describe the solvent and supporting electrolyte, apparatus, and experimental conditions; the right-hand pages give the data and other information obtained and provide cross-references to additional information contained in other tables.

The compounds listed in the first part of Table I are arranged according to their empirical formulas, and this of course provides one mode of access to the information on any particular compound. Empirical formulas are written in the sequence C, H, and all other atoms in the alphabetical order of their symbols. These formulas are then arranged alphabetically and according to increasing number of atoms; thus, for example, $C_6H_5NO_2$ precedes $C_6H_5NO_4$, and this in turn precedes C_6H_6. To permit the rapid location of any empirical formula, the empirical formula of the first compound that appears on each pair of facing pages is given in the upper left-hand corner of the left-hand page, and that of the last compound appearing on that pair of pages is given in the upper right-hand corner of the right-hand page. By virtue of these facts, dimethyl maleate (*cis*-$CH_3OOCCH:CHCOOCH_3$ or $C_6H_8O_4$) can be located quickly on the pair of pages that contain information for compounds having formulas between $C_6H_8AsNO_3$ and $C_6H_{10}O$, and then further identified as the fourth compound on this pair of pages by means of either the empirical formulas listed in the second column or the names listed in the third. Different compounds having the same empirical formulas are listed in the alphabetical order of the names assigned to them.

Another mode of access to the same information is provided by Table IV. This is an index of names and a few common synonyms. It contains the entry "Dimethyl maleate, DB22". The code number DB22 appears in both the first and last columns of Table I. The code number and empirical formula of the first compound appearing on a pair of facing pages are given in the upper left-hand corner of the left-hand page, while those of the last compound appearing on that pair of pages are given in the upper right-hand corner of the right-hand page. Each code number consists of two letters followed by two digits; in Table I the code numbers appear in the order DA00, DA01, ..., DA99, DB00, DB01, ..., DH64. The code numbers serve not only to abbreviate cross-references between tables and in the tightly packed columns of Table I but also to distinguish between different compounds that have identical empirical formulas. The first letter of the code number identifies the volume of this series: the letters A and B refer to compounds given in Table I of Volumes I and II, C to those in Volume III, and D to those in the present volume.

Table V is another index that makes it possible to locate compounds of interest in Table I: it classifies the compounds according to the electroactive (or possibly electroactive) functional groups they contain. One frequently wants to examine data on unsaturated compounds, on carboxylic acids, or on another more or less specific class of substances, and the purpose of Table V is to simplify searches of this sort. For example, maleic acid appears among the α,β-unsaturated acids and esters in Table V, and inspection of the compounds listed under this heading would quickly reveal any other closely related compound that appears in this volume. To minimize false leads in such searches, this index is selective: maleic acid does not appear among the aliphatic carboxylic acids, because its behavior, so far as it is known, is not representative of compounds bearing a carboxylic group. On the other hand, 4-nitroazobenzene would appear both among the nitro compounds and among the azo compounds because both the nitro group and the azo group are known (or can be assumed) to be involved in its electrochemical behavior.

Table VI lists the compounds under American Chemical Society Registry Numbers, arranged sequentially. These Numbers, which also appear in Column 3 of Table I, should enable the user to find a compound in Table I without having to be concerned with nomenclature.

Table VII is a list of the solvents that appear in the individual entries of Table I. It permits comparing data on different substances in a single solvent or solvent mixture. The typical entry

Benzonitrile, DG81, DG87, DG90, DH01, DH13

provides rapid and easy access to all of the information that this volume includes about electrochemistry in this solvent.

Table VIII is an index to the techniques employed. For the sake of brevity classical dc polarography is excluded from it, as the dropping mercury electrode is from Table IX, but for each of the other 50-odd techniques that are included in these volumes an entry like,

Chronopotentiometry (ER), DA49, DF39, DG73, DH12, DH19, DH21, DH22

provides not only access to all of the chronopotentiometric information that these volumes contain, but also an indication of how much chronopotentiometry has contributed to our knowledge of the electrochemical properties of organic substances during the years that these volumes cover. Since Table VIII lists some techniques for which no citations are given, it may be useful to state the policy we adopted in dealing with data obtained by rarely used techniques. There were a few of these that were rejected because they, or the interpretations based on them, disagreed radically with other information, but such cases were very few and we much more often found ourselves deciding to include an entry because it was obtained by a technique little represented in the data file. Such a selection procedure tends to yield overestimates of the utility and importance of rarely used techniques. The overwhelming preponderance of classical dc polarography is reflected by the absence of any entry for it from Table VIII because it would have consumed more space than we thought reasonable. If there were an entry for it, its length would, if anything, have understated the predominating importance that dc polarography still has in obtaining fundamental electrochemical information.

Table IX is an index to the indicator electrodes employed. It provides access to all of the data obtained with mercury-pool, carbon paste, rotating disc, and each of the numerous other electrode materials and configurations represented in Table I, with the single exception of the dropping mercury electrode, which is omitted here for the same reason that polarography is omitted from Table VIII.

Tables IV to IX are thus indexes to Table I. Tables II, III, X, and XI, on the other hand, are supplements to Table I, providing additional information and used with Table I as a starting point. Inspection of these Tables is often necessary for the proper use of Table I. Table II gives, in the order of their code numbers, the structural formulas that could not conveniently be compressed into line form for inclusion in Column 4 of Table I. Table III gives equations that represent the courses and mechanisms of half-reactions, together with values of the rate and equilibrium constants for many of the homogeneous chemical steps involved in these. The equations are given in numerical order and keyed to entries in the fifteenth column (headed "C/M") of Table I; the appearance of a number in that column of Table I in the entry for any compound signifies that information about the course or mechanism of the half-reaction is given opposite the same number in Table III. Table X provides full literature references and is used together with the condensed citations that appear in the fourteenth column (headed "Ref.") of Table I. For each such citation, Table X gives the name of the journal, the volume number, the year of publication, the page number, and the names of the authors. Table XI is an author index which gives, in alphabetical order, the names of all of the authors listed in Table X and, for each, a reference to each citation of his work in Table X. Finally, there is a list of corrigenda giving all of the previously uncorrected errors in prior volumes of which we are aware.

This volume is dedicated to all those whose names appear in Table XI, in recognition of the labor that they have expended on assembling and bringing to its present state the enormous body of knowledge and understanding that is summarized here. Much remains to be discovered, and much of our understanding is still very fragmentary. We hand the volume over to its users in the hope that it will both ease and help to guide their work in the future, as well as the work of those who will follow them.

TABLE I.
ELECTROCHEMICAL DATA

This table is divided into two parts. The first, which is very much the longer of the two, contains the purely electrochemical data that have been obtained by polarography and the other techniques listed in Table VIII. The arrangement of this part is virtually identical with that of Table I in Volume I of this series. The second part, which is much shorter, gives data and qualitative observations pertaining to electrochemiluminescence and follows the pattern introduced in Volume III of this series.

The first part of the table consists of 28 columns divided between two facing pages. Here we shall first describe the contents, significance, and arrangement of these columns. A brief description will then be given of the ways in which the second part of the table differs from the first.

Columns 1 and 28: "Code No."

The contents of these columns are identical so that the user can more easily follow the lines of this very wide table across two pages. The columns give the four-character code number (see explanation given in the General Introduction) assigned to each compound and also indicates whether or not more information about a compound will be found on the following pair of pages.

In many instances the data for a single compound had to be divided between two, and occasionally even among three, pairs of facing pages. To avoid overlooking part of an entry thus divided, the user should observe the following points:

1a. When the last line is blank in Columns 1 and 28, the entry does *not* continue onto the next pair of pages, whereas
1b. An entry that *is* continued overleaf is identified by the symbol "CONT" on the last line in both Column 1 and Column 28.
2a. When the index in the upper left-hand corner of a left-hand page consists solely of a code number and an empirical formula (for example, "DB12, $C_6H_7AsO_4$") the first entry on that page is *not* continued from the preceding pair of pages, whereas
2b. An entry that *is* being continued from the preceding pair of pages is identified by the appearance of "(CONT.)" between the code number and the empirical formula in the upper left-hand corner of the left-hand page (for example, "DB08 (CONT.) $C_6H_6O_2$"), and in addition the WLN designation does not appear beneath the name of the compound on the first line in Column 3.

For a number of these compounds additional data will be found in an earlier volume of this series. Each compound that also appears in an earlier volume is identified by the appearance, in both Column 1 and Column 28, of the code number assigned to it in each previous volume in which it appeared, but with small letters in place of the capitals used elsewhere. For dimethyl maleate the entry in these columns,

DB22
cc03

shows that additional data appear in Volume III under the code number CC03. As was stated in the General Introduction, the first letter of a code number identifies the volume to which reference is made. The second letter and the two digits of the code number have only an ordinal significance.

Column 2: "Empirical Formula"

The empirical formulas in this column are given in the customary way, with carbon first, hydrogen second, and then the other elements in the alphabetical order of their chemical symbols. Deuterated compounds appear immediately after the corresponding protonated ones.

Column 3: "Name and WLN CODE"

This column gives the name and, below it, the American Chemical Society Registry Number, introduced by the abbreviation C.A. for *Chemical Abstracts,* where available. On the following line the Wiswesser Line Notation (WLN) designation of each compound is given. Some of the names are the simplest and shortest trivial names for the compounds they denote, but most follow IUPAC or *Chemical Abstracts* nomenclature. The Registry Number and WLN designation are given only once for each compound; when an entry continues onto the following pair of pages the name of the compound is repeated but the Registry Number and the WLN designation are not. Some compounds have not been assigned Registry Numbers or WLN designations, either because their structures are unknown, because Registry Numbers have not yet been published by *Chemical Abstracts,* or because the WLN rules applicable to them have not yet been accepted.

TABLE I. Electrochemical Data

Column 4: "Structural Formula"

Only line structural formulas are given here. For many of these compounds, however, the structural formula cannot conveniently be given in line form, and is therefore shown in Table II. The entry "Table II" in Column 4 means that the structural formula may be found in Table II of this volume opposite the appropriate code number taken from Column 1. To avoid duplicating structural formulas in different volumes, the entry "Table II-2", which appears in Column 4, provides a cross-reference to Table II in Volume II, while the entry "Table II-3" is a cross-reference to Table II in Volume III. For the compound for which the entries in Columns 1 and 3 are

$$\begin{array}{ll} \text{DB67} & \text{thianthrene} \\ \text{ba42} & \\ \text{cg97} & \end{array}$$

the reference "Table II-2" means that the structural formula may be found in Table II of Volume II with the code number (BA42) given for thianthrene in Column 1.

Column 5: "Solvent"

The symbols used in this column to denote non-aqueous solvents are given in the list of abbreviations. The entry "H_2O" denotes an aqueous solution; an entry like "MeCN" denotes a solution in the nominally pure (and anhydrous) solvent given; in an entry like "MeCN 50" the number denotes the percentage (by volume, unless otherwise stated) of the specified non-aqueous solvent in a solvent mixture, of which the balance is understood to be water. An occasional entry like "EtOH(aq)" reflects our inability to deduce the composition of the solvent mixture from the published information given; such entries appear only when the importance of the data seemed to us to override the laxity of the reporting. Of the entries

$$\begin{array}{cc} \text{MeOH} \quad 50 \\ C_6H_6 \quad 50 \end{array} \text{ and } \begin{array}{cc} \text{EtOH} \quad 50 \\ Me_2CO \quad 25 \end{array}$$

the left-hand one denotes a nominally anhydrous mixture containing equal parts by volume of methanol and benzene. The right-hand one denotes a mixture containing 50% (V/V) of ethanol, 25% (V/V) of acetone, and 25% (V/V) of water; when the sum of the percentages given for the constituents of any solvent mixture is less than 100, the balance is always understood to be water. Aqueous solutions are listed first, then alcohols, and finally other solvents in the order in which they appear on the list of abbreviations. Entries for mixtures of a non-aqueous solvent with water immediately precede those for the corresponding anhydrous solvent. A few data that seemed worth including were taken from papers that did not fully describe the composition of the solvent; for these data the symbol "ns" ("not stated") is given in this column.

Column 6: "Technique"

The technique is identified by a two-letter symbol that can usually be interpreted by means of the list of abbreviations. Polarography ("PY") appears first in the data for any solvent, and the other techniques follow in the alphabetical order of the symbols assigned to them. Some of the information in this volume has been obtained by techniques as yet so rarely used that we have hesitated to assign symbols to them until further evidence of their utility appears. Each such technique is identified by the letters "XT" in Column 6, and Column 27 always gives its name on the same line in the form "XT = second-derivative chronoamperometry with linear potential sweep". In assigning such names we have followed the principles embodied in the 1975 IUPAC recommendations regarding the nomenclature of electroanalytical techniques (*Pure and Applied Chemistry*, 45, 81, 1976).

Column 7: "Medium"

This column describes the composition of the supporting electrolyte. Concentrations are given in moles per cubic decimeter wherever possible: the entry "KOH 0.1" denotes 0.1 M potassium hydroxide. Britton-Robinson, McIlvaine, and other commonly used buffers, both mixed and simple, are identified by means of abbreviations. The word "buffer" means that the solution was said to be buffered at the pH quoted in Column 9, but that no information was given about the composition of the buffer employed. Maximum suppressors are identified in this column and their concentrations are given in weight / volume % (grams per 100 cm³).

Column 8: "μ, M"

This column gives the ionic strength of the solution, in moles per cubic decimeter, except where this is fully specified by the entry in Column 7. A dash in this column means that no information about the ionic strength is available in addition to that given in Column 7.

Column 9: "pH"

Values of the pH are generally given to the nearest 0.1 unit. An entry like "5.7→6.2" would signify that the pH-value changed over the range given during the course of a measurement; one like "5-7" means that all of the data in the columns that follow are independent of pH over the stated range. The entries for any given medium are usually arranged in order of increasing pH. A dash in this column means that no information about the acidity of the solution is available in addition to that in Column 7.

Column 10: "T,°C"

This column gives the temperature at which the data were obtained and, where possible, the precision with which it was maintained. A dash in this column means that no information about the temperature of the solution was given in the original publication.

Column 11: "Electrodes"

The indicator or working electrode is given first; then, after a solidus or virgule ("/") or on a second line, the reference electrode. In three-electrode configurations the nature of the auxiliary or counter electrode is given in Column 13.

Reference electrodes are divided into two groups. One comprises the saturated calomel electrode, its variants (such as the "lithium S.C.E.", $Hg/Hg_2Cl_2(s)$, LiCl(s), and others of the same ilk), and the normal hydrogen electrode. These are almost invariably prepared with water, so that their use with a non-aqueous solution entails a liquid-junction potential between the non-aqueous solution of the compound being studied and the aqueous solution in the reference electrode. Some workers have sought to circumvent this by preparing similar electrodes in the same solvents or solvent mixtures that contain the compounds they study; when this has been done, the symbol "(o)" (for "organic") follows the abbreviation that would denote the ordinary aqueous form of the reference electrode.

The other group comprises silver-silver halide electrodes, mercury pools, metal-metal-ion electrodes, and others normally prepared in the solvent used for the compound being studied (and often, indeed, employed as internal "reference" electrodes). For such an electrode, the abbreviation alone signifies that the solvent was the same throughout the cell, while the symbol "(w)" (for "water") following the abbreviation signifies that the reference electrode was prepared with water and used as an external reference electrode.

Some authors have used "pilot ions", whose half-wave potentials in non-aqueous solvents have been reliably established in prior work by themselves or others, to provide internal reference potentials and minimize problems like those that arise from variations of liquid-junction potentials with time. Potassium ion, rubidium ion, ferrocene, ferricinium ion, and several others have been used. This practice gives rise to entries of the form

$$\text{DME/MP} \\ \text{K}^+$$

which denotes the use of a dropping mercury electrode as the indicator electrode, of a mercury pool as the counter electrode, and of potassium ion as the added pilot ion.

A dash in place of the symbol for one electrode (e.g., "DME/-") means that the nature of that electrode was not given in the original paper. A dash unaccompanied by anything else ("-") means that neither electrode was identified.

Column 12: "App."

Some salient features of the apparatus are summarized in this column. The basic code is a three-character alphanumeric, of which the first character is a digit that describes the cell, the second is a letter that indicates whether and how dissolved oxygen was removed, and the third is a digit that identifies the technique of data acquisition. By means of the list of abbreviations (in which the abbreviations used in this column appear at the very beginning), the typical entry "0A0" may be decoded to find that a two-electrode cell was used, that deaeration was accomplished with a stream of inert gas, and that the data were recorded on a pen-and-ink recorder. Inert gases are not identified because that would, in our judgment, consume more space than it would be worth, but they do not include hydrogen because hydrogen is sometimes not inert. More elaborate entries, such as "23CD3", are occasionally necessary; this one denotes the use of a three-electrode cell with an agar bridge, deaeration by both hydrogen and sulfite, and digital data-acquisition. When insufficient information was given to permit completing any part of this three-character code, a dash always appears in the appropriate position, as in the entry "0-0". This permits easy differentiation between, for example, "--0" and "0--".

TABLE I. Electrochemical Data

Column 13: "Experimental Parameters"

Some of the information given in this column, such as the concentration C of the compound studied (which is always given in millimoles per cubic decimeter), would be significant no matter what technique was employed; some of it is governed by the nature of that technique. For polarographic data we give, wherever possible, values of the drop time t (in seconds), the rate of the flow mercury m (in milligrams per second), and the height h (in centimeters) of the column of mercury above the capillary tip.

The values of t and m should be regarded only as approximations, for there is little agreement about the conditions under which they are measured and the ways in which they are reported. An author chiefly interested in mass-transfer phenomena is apt to measure t at a potential on the plateau of the wave, one chiefly interested in charge-transfer phenomena is apt to measure it at the half-wave potential, and one studying a number of different compounds having different half-wave potentials is apt to measure it at the potential of the electrocapillary maximum (open circuit) or at the potential of the reference electrode employed (short circuit). Similarly, the value of m may be measured at the same potential as that of t, but often it is measured at the potential of the electrocapillary maximum while t is measured at a different potential. Sometimes m is measured with the tip of the capillary immersed in mercury. In view of this diversity, exact specification would have been difficult and wasteful of space, and the difficulty would have been compounded in dealing with papers that report values of t and m without specifying how they were measured. Consequently we have usually foregone any attempt at greater exactness; the user who needs more information must seek it in the original literature. The value of h serves as a crude but useful indication of whether the capillary was a conventional one or not: an abnormally high or low value would signify that there was something unusual about the capillary, and the original literature should again be consulted for details.

Sometimes the value of $m^{2/3}t^{1/6}$ (abbreviated "EC") is given, always in $mg^{2/3}s^{-1/2}$, when this was stated in the paper cited in lieu of the value of t or m or both. What was said above about the difficulty of quoting, and often even of deducing, the potential at which a value of m or t was measured applies equally to the potential at which $m^{2/3}t^{1/6}$ was evaluated.

For other techniques other kinds of data are given instead. For chronopotentiometry these include the area of the indicator electrode and the current or current density; for stationary-electrode voltammetry they include the area of the indicator electrode and the scan rate; for cyclic voltammetry they include the starting and reversal potentials and the area of the indicator electrode, and so on. The list of abbreviations must always be consulted regarding the units of the quantities given in this column. In the space that was available for these purposes it is quite impossible to give a full description of the experimental conditions, but an attempt has been made to give an accurate idea of their nature.

This column is also used to give the identity of the auxiliary or counter electrode when a three-electrode configuration was used, and to indicate whether the indicator or working electrode was subjected to chemical, mechanical, or electrolytic pretreatment.

The symbol "ns" (for "not stated") is often used in this column to mean that information was not available about the concentration of the compound studied ("C = ns") or about important experimental parameters.

Column 14: "Ref."

Each entry in this column is a nine-character alphanumeric of which two letters and three digits appear on the first line while the remaining four digits appear on the second line. The first two letters denote the journal; these abbreviations do not appear in the list of abbreviations, but may be deciphered with the aid of Table X. The next three digits give the volume, packed to three digits by the addition of leading zeros if necessary. The four digits on the second line give the page number, again with leading zeros added if necessary to yield a total of four digits. Since the letters "AA" denote *"Analytica Chimica Acta"*, the typical entries

AA072   and   AA080
0169   0017

would refer to papers published in *Anal. Chim. Acta*, **72**, 169 and *Anal. Chim. Acta*, **80**, 17, respectively. The year is not given here but may be obtained by consulting Table X, which gives the names of the authors as well. In a reference to a journal that does not employ volume numbers, the two leading digits "19" are dropped from the number of the year, and the two remaining digits (e.g., 66 for a paper published in 1966) are followed by a letter. This is "O" if, like the *Bulletin de la Societe Chimique de France*, the journal has no subsections; otherwise it denotes the subsection of a journal like the *Journal of the Chemical Society (London)*. Thus the entries

$$\begin{array}{ccc} \text{JL66B} & & \text{BF63O} \\ & \text{and} & \\ 0103 & & 2252 \end{array}$$

would denote *J. Chem. Soc.*, Sec. B, 103, 1966 and *Bull. Soc. Chim. Fr.*, 2252, 1963, respectively.

To the shame chiefly of the authors, but also to some extent of the referees, editors, and journals involved, we have found a number of cases in which nearly, and sometimes exactly, identical papers have been published in two or three different journals. This practice has been repeatedly criticized even in the formerly common situation where publication in a local orregional journal in the language of a small country was followed by publication in an international journal having a much wider circulation and in a more widely known language. Repeated publication of the same material in journals of comparable circulation and in the same language seems to us to defraud the scientific community. In such cases we have simply suppressed all of the citations save one; we prefer the risk of being criticized for having covered the literature a little less thoroughly than we have actually done to that of allowing the offenders to profit from actions that we must hope to have been ill-judged rather than something more reprehensible.

Columns 15 through 26

Information that belongs in any of these columns but that was not given in the original paper is represented by a dash on the first line of an entry to aid the user in following separate entries across the page.

Column 15: "C/M"

A number that appears in this column is a cross-reference to Table III. If the number is followed by parentheses enclosing the number of an earlier volume (e.g., "221(2)") the cross-reference is to Table III in that volume; otherwise the cross-reference is to Table III in the present volume. Opposite the same number in that table, equations are given that describe the course or mechanism of the half-reaction. Table III may also contain other information, such as the rate constants of homogeneous chemical steps in the overall mechanism; its introduction should be consulted for further details.

The numbers that appear in Column 15 are not in numerical order because virtually identical equations may often be written for the reductions or oxidations of a number of compounds that have widely different empirical formulas and are therefore widely scattered in Table I. Cross-references in Table III list the code numbers of other compounds to which the same mechanism, or a closely related one, is applicable. For further details see the introduction to Table III.

Columns 16—18: "Characteristic Potential"

This term denotes a potential whose nature depends on the technique used. Typical characteristic potentials are the half-wave potential in polarography, the quarter-transition-time potential in chronopotentiometry, and the peak or half-peak potential in linear-sweep (stationary-electrode) voltammetry. Regardless of its nature, the characteristic potential always depends on the identity of the electroactive substance, on the kinetics or thermodynamics of the electron-transfer process, and of course on the experimental conditions; for any particular technique and under any completely defined set of experimental conditions the value of any characteristic potential is a reproducible property of the electroactive substance.

Column 16 gives the symbol of the characteristic potential whose numerical value (in volts) appears in Column 17. Column 18 identifies the reference electrode to which that value is referred; this is not necessarily the same as the reference electrode used for the experimental measurements, which was identified in Column 11.

As is also true of the techniques employed, some of the characteristic potentials appearing here have not yet been assigned individual symbols. Each of these is identified by the letters "XE" in Column 16, and Column 27 always defines it on the same line in the form "XE = potential where $d^2i/dE^2 = 0$".

Some investigators report values of the half-wave potential obtained by triangular-wave voltammetry and other techniques that do not yield experimental values of the half-wave potential. These values are of dubious significance and are identified by the symbol "$E_{1/2}(?)$" in Column 16.

Columns 19—20: "Response Const."

This generic term may denote either the measured value of the independent variable under the experimental conditions employed or some function of that value: as with the characteristic potential, tte nature of the response constant depends on the technique used, and in addition it depends on the behavior of the system being studied. For a diffusion-controlled process the preferred polarographic response constant is the diffusion current constant $I = i_d/Cm^{2/3}t^{1/6}$, but the ratio i_d/C is often given instead when a value of $m^{2/3}t^{1/6}$ could not be obtained from the original, and even values

TABLE I. Electrochemical Data

of the limiting current i_l alone are sometimes quoted. Values of i_d/C and i_l provide at least some indication of the relative heights of different waves. Wherever possible, values of i_l are accompanied (in Column 27) by an indication of whether the current is controlled by diffusion or some other process. For diffusion-controlled processes the preferred response constants in chronopotentiometry and stationary-electrode voltammetry are $i\tau^{1/2}/AC$ and $i_p/v^{1/2} AC$, respectively, given here in terms of current densities $j\tau^{1/2}/C$ and $j_p/v^{1/2}C$ for compactness. We have attempted to give all currents in microamperes, concentrations in millimoles per cubic decimeter, and areas in square centimeters, but as currents are sometimes reported in arbitrary units such as millimeters of recorder deflection we have occasionally had to quote these, using the symbol "(u)" to denote their limited significance.

For processes that are not diffusion-controlled, and even for diffusion-controlled processes when some techniques are used, closed-form descriptions of the preferred response constant may not be available, and this is another reason why currents, current densities, current- or current density-concentration ratios, and other incomplete response "constants" appear in these columns. There are even some authors who have reported diffusion coefficients in place of the response constants obtained experimentally; the justification for doing so is meager, but we have sometimes been forced to quote these values for want of anything with a more solid basis.

Column 19 identifies the response constant whose value is given in Column 20. Units are given in the list of abbreviations: a polarographic diffusion current constant identified by an "I" in Column 19 is always in $\mu A\ mmol^{-1}dm^{-3}mg^{-2/3}s^{1/2}$, a calculated diffusion coefficient identified by a "D" in Column 20 is always 10^6 times the reported value in cm^2s^{-1}, and so on.

Some rarely used response constants have not been assigned individual symbols but are identified by the letters "XI" in Column 19 and defined on the same line in Column 27.

Columns 21 and 22: "n"

Usually the value of n appearing in Column 22 is the total number of electrons involved in the overall half-reaction that consumes one molecule or ion of the electroactive substance, and the abbreviation appearing in Column 21 identifies the technique by which that value was obtained. Values obtained by controlled-potential coulometry ("QE" in Column 21), however, generally pertain to the experimentally determined ratio of the number of faradays consumed to the number of moles of electroactive substance taken or consumed; in the original literature this ratio is often denoted by a symbol like "n_{app}" to emphasize its origin. Nonintegral values of n obtained by controlled-potential coulometry and other techniques reflect the occurrence of coupled chemical reactions that produce or consume electroactive material during a measurement.

Columns 23—25: "Electrokinetic Data"

Information about charge-transfer kinetics is given in these columns in various ways whose diversity again reflects the lack of standardization in the literature. Some authors calculate and report values of the symmetry parameter αn_a (usually abbreviated here as αn), the heterogeneous rate constant at 0 V vs. N.H.E. ($k_{f,h}$) or at the standard or formal potential of the couple ($k_{s,h}$), or of other parameters similar to these; others report values of more directly accessible quantities, such as (in polarography) the slope of a plot of E vs. $\log[i/(i_d-i)]$ or $E_{3/4} - E_{1/4}$. Some calculate and report the number of hydrogen or hydroxide ions consumed in the rate-determining electron-transfer step and the steps that precede it; some report the slope of a plot of the characteristic potential against the pH. Column 23 gives the symbol of the parameter whose value is given in Column 24, while Column 25 identifies the experimental data from which that value was deduced.

Column 26: "Products and Identification"

This column indicates what final products were obtained and what techniques were used to isolate and identify them. Products are usually identified by names; only for unusual structures (e.g., some radical ions) are their formulas given. The percent yield of the product, when available, follows its name. Techniques used for isolation and identification are indicated by abbreviations given in the list of abbreviations. Additional information, such as the lifetime of a radical, is sometimes included.

Column 27: "Description and Remarks"

This column, in which abbreviations are very dense, indicates whether the wave or other signal is cathodic (reduction) or anodic (oxidation), whether it is well- or ill-defined, whether or not it shows a maximum or some other reproducible deviation from the idealized simple form, whether the response is controlled by mass-transfer or some other process such as adsorption, and gives other information not provided in the preceding columns, including definitions of the special symbols "XT", "XE", and "XI" where these appear in Columns 6, 16, or 19 respectively. Correlations of measured quantities, such as the half-wave potential, with structural parameters (such as the exchange integral β, the Hammett substituent constant σ, or the Taft polar constant σ^*) are often indicated here. Relationships between measured quantities, such as the current, and experimental variables, such as the concentration, are indicated together with the limits of their validities. Sometimes the nature of the electrochemical process is indicated by entries such as "H redn." (which means that the process consists of the direct or catalytic reduction of hydrogen ion rather than the reduction of the organic species) or "Maleic acid redn.", which for a maleate salt means that it is not the organic cation but the maleate anion that is responsible for the behavior observed. In addition, this column gives, very briefly, our assessment of the reliability of the data and interpretation. Usually we have confined ourselves to stating whether we believe the information and conclusions to be reliable and reasonably accurate and precise, of limited precision, or in need of further confirmation, but occasionally we have indicated doubts concerning individual components of the entry. The symbol "(?)" is used to signify that we doubt the correctness of the information it follows, while the symbol "(!)" usually means either that the information reported cannot be validly obtained in the manner specified, as for example when the slope of a "log plot" is adduced as evidence for an overall n-value, or that we are dubious of its significance.

When data reported in one of the papers abstracted for this volume were thought to be inferior to data reported in a previous volume for the same compound, only the name of the compound and the experimental conditions are given, and a cross-reference (in the form "See AG07") to the earlier data appears in this column.

The second part of the tables begins with the compound "Wurster's Blue" (code number DH32) and deals with data concerning electrochemiluminescence. Its arrangement is generally similar to that of the first part, but there are some differences that arise from the special nature of experiments in this area of electrochemistry.

Columns 1 and 23: "Code No."

The four-character code numbers used to identify the compounds involved in electrochemiluminescence follow the same principle as those in Columns 1 and 28 in the first section of this table.

Since two different compounds are often involved in experiments on electrochemiluminescence, cross-references are more complex here than in the main portion of this table. Thus, the sixth entry in this column on page 257 beginning with code number DH32 is

$$\begin{array}{c} \text{DH37} \\ \text{dd67} \\ \text{de35} \end{array}$$

"DH37" is the code number assigned to this combination of electroactive substances: thianthrene as the one that is oxidized (which is denoted as "R" and identified in Column 2), and 2,5-diphenyl-1,3,4-oxadiazole as the one that is reduced (which is denoted as "O" and identified in Column 3). The "dd67" on the second line pertains to the first of these and signifies that electrochemical data on the behavior of R (when it is present alone) may be found under code number DD67 in this volume. The "de35" on the third line pertains to the second compound and signifies that electrochemical data on the behavior of O may be found under code number DE35 in the present volume. On consulting code number DE35 the user will find that additional data on 2,5-diphenyl-1,3,4-oxadiazole appear in Volumes I and II under code number BE90 and in Volume III under code number CI55.

Column 2: "Compound oxidized (= R), empirical formula"

For the compound that is oxidized in the experiment, this column gives the information contained in Columns 2 and 3 ("Empirical Formula" and "Name and WLN Code") of the first part of this table.

Column 3: "Compound reduced (= O), empirical formula"

This column gives the same information for the compound that is reduced in the experiment. The structural formulas of both compounds may be obtained through the cross-references provided in Columns 1 and 23.

Many investigations of electrochemiluminescence involve only a single compound, which undergoes both reduction at the cathode and oxidation at the anode. In such cases the entry in this column is "same".

Column 4: "Solvent"

The same way of presentation and the same abbreviations are used in this column as in Column 5: "Solvent" in the first part of this table.

TABLE I. Electrochemical Data

Column 5: "Medium" through Column 13: "C/M"

Data in these columns are arranged in the same way, using the same units, symbols, and abbreviations as in Column 7: "Medium" through Column 15: "C/M" in the first part. Column 11: "Exptl. parameters" gives, wherever possible, both the concentration of the substance that is oxidized (C_R) and the concentration of that which is reduced (C_o). Information that we could not obtain from the original literature is represented by dashes in columns 6, 7, 8, 10, and 12, and by the symbol "ns" ("not stated") in the other columns of this group.

Columns 14—18: "Excitation"

These columns give information about the electrical signal employed. Column 14 shows whether a square wave ("□"), a square wave superimposed on a ramp ("□ + /"), or a triangular wave ("∧") was used, and Column 15 gives the frequency, in hertz, of the square or triangular component. Columns 16 and 17 give the more positive (E_1) and the more negative (E_2) limits, respectively, of the range of potentials employed, and Column 18 specifies the reference electrode or value to which E_1 and E_2 are referred. Missing information is indicated by dashes in these columns.

Columns 19—21: "Emission"

These columns give information about the luminescence obtained. Column 19 identifies the species responsible for it, if this is known, using symbols like "^{1}R*" to denote an excited singlet species of R (the compound oxidized). Column 20 tells whether the process is energy-sufficient ("Se") or -deficient ("De"), and Column 21 gives information on the wavelength of maximum emission, the excitation frequency at which the intensity of emission is a maximum, the quantum efficiency, and other quantities of interest. Missing information is indicated by a dash in Column 20, or by a blank in Column 21.

Column 22: "Remarks"

This column contains an "O" if luminescence was not observed under the conditions stated, gives other information not contained in the preceding columns, and includes our assessment of the reliability of the data and interpretation.

Code No.	Empirical Formula	Name and WLN Code	Structural Formula	Solvent	Tech.	Medium		μ, M	pH	T, °C	Electrodes	App.	Experimental Parameters
DA00 aa04 ca00	CCl_4	carbontetrachloride C.A. 56-23-5 GXGGG	CCl_4	EtOH 75	PD	Et_4NOH	0.05	-	-	-	DME/MP	-	-
DA01 aa38	CN_4O_8	tetranitromethane C.A. 509-14-8 WN 4X	$C(NO_2)_4$	i-PrOH 65	PY	HNO_3 NH_4NO_3	0.1 0.1	-	-	-	DME/MSE	0-0	C=0.5, m=0.605, t=0.35
DA02 aa09	CHN_3O_6	trinitromethane C.A. 517-25-9 WNYNWNW	$CH(NO_2)_3$	i-PrOH 65	PY	HNO_3 NH_4NO_3	0.1 0.1	-	-	-	DME/MSE	0-0	C=0.5, m=0.605, t=0.35
DA03 aa21	CH_3ClHg	methylmercuric chloride C.A. 115-09-3 1-HG- &G	CH_3HgCl	EtOH 50	PY	KI	0.1	-	-	25	DME/SCE	13A1	C=0.656
						Et_4NI	0.1	-	-	25	DME/SCE	13A1	C=0.5
DA04 aa28	CH_3NaO_3S	formaldehyde sodium sulfoxylate C.A. 149-44-0 Q1OSO &-NA- &9/5	$HOCH_2SO_2Na$	H_2O	VY	MB		1.0	4.0	20	RPE/SCE	OB0	$0 \to 1.2$ V, C=0.2, ω=10, v=3.3, A=1.1E-4
DA05 aa49	$C_2H_2O_4$	oxalic acid C.A. 144-62-7 QVVQ	$(\cdot COOH)_2$	H_2O	IL	H_2SO_4	0.5	0.5	0	25±0.1	PGE/SCE	235A0	C=0.504, A=0.125, v=5, Pt Aux
						ACET	1		2.3				
						ACET	?		4.7				
						BOR	?		8.1				
									11.1				
DA06	$C_2H_4N_2O_2$	oxamide C.A. 471-46-5 ZVVZ	$[H_2NC(O)\cdot]_2$	H_2O	PY	UB	?	0.5	5.6	25±0.1	DME/SCE	2A0	t=2, Pt Aux
						BR			8.1				
						BM			11.1F				
					VA	UB	?	0.5	5.6	25±0.1	HMDE SCE	2A0	C=0.5, v=5, Pt Aux
													v > 5E4
									11.6				v=5
													v > 5E4
CONT													

TABLE I. Electrochemical Data $C_2H_4N_2O_2$ (CONT.)   DA06

| Ref. | C/M | Charact. Potential | | Response | Const. | n | | Electrokinetic Data | | | Products and | Description and | Code |
		Value	vs.		Value	Tech.		Parameter	Value	From	Identification	Remarks	No.
RL032 0344	-	-	-	-	-	-	-	-	-	-	-	see AA04, CA00	DA00 aa04 ca00
ER002 0108	4b	$E_{\frac{1}{2}}$ -0.1F -0.58F	SCE	-	- 1.0F -	i:i	2 -	-	-	-	-	$C,M,i_d,E_{\frac{1}{2}}\uparrow,i_{\ell}=kC,r$ C C,X	DA01 aa38
ER002 0108	-	$E_{\frac{1}{2}}$ -0.1F -0.58F	SCE	-	-	-	-	-	-	-	-	C,p C	DA02 aa09
JE006 0034	9a	$E_{\frac{1}{2}}$ -0.594	SCE	D	5.90	-	-	αn_a	1.1	Elog	-	C,i_d,R,r	DA03 aa21
JE006 0034	9a	$E_{\frac{1}{2}}$ -0.596	SCE	D	5.53	-	-	αn_a $dE_{\frac{1}{2}}/d\log C$	1.0 +30	Elog sttd	-	C,i_d,R,r	
		-1.324			4.67			αn_a $dE_{\frac{1}{2}}/d\log C$	0.6 -80	Elog sttd		C,i_d	
AN100 0735	-	$E_{\frac{1}{2}}$ 0.68F	SCE	i_{ℓ}	100F	-	-	-	-	-	-	$A, i \propto C$ for $C=0.02\rightarrow$ 0.2,r	DA04 aa28
AA072 0209	-	E_p	-	-	-	-	-	-	-	-	-	O,p	DA05 aa49
	19c	1.01	SCE	$j_p/Cv^{\frac{1}{2}}$	31.59	QE	2.02 ±0.2	αn_a	0.36 ±0.04	E_p- $E_{p/2}$	CO_2, gravimetrically as $BaCO_3$	$A,E_p\neq f(pH),E_p=f(v),r$	
		0.98			29.12		2.02 ±0.2		0.36 ±0.04		CO_2, gravimetrically as $BaCO_3$	$A,E_p\neq f(pH),E_p=f(v),r$	
		0.99			36.56		2.02 ±0.2		0.36 ±0.04		CO_2, gravimetrically as $BaCO_3$	$A,E_p\neq f(pH),E_p=f(v),r$	
		0.92			32.67		2.02 ±0.2		-		CO_2, gravimetrically as $BaCO_3$	$A,E_p\neq f(pH),E_p=f(v),r$	
AA067 0415	359a	$E_{\frac{1}{2}}$ -1.59 ±0.01	SCE	I	7.00	QE	4	$dE_{\frac{1}{2}}/dpH$	0	-	-	C,i_d,ece,r	DA06
		-1.59 ±0.01			7.00F		3.9		0	-	QE at -1.65 → NH_3, 100±20%, Nessler's reagent, glycolamide, >90%, MAS, NMR, IRS, MPT	$C,i_d,Tc=0.4,ece,r$	
		-1.59 ±0.01			5.32F		4		0	-	QE at -1.65 → NH_3, 100±20%, Nessler's reagent, glycolamide, >90%, MAS, NMR, IRS, MPT	$C,i_d,I\propto -pH,ece,r$	
AA067 0415	359a	E_p -1.63 ±0.2	SCE	-	-	-	-	-	-	-	-	$C,\neq,ece,E_p=f(v^{\frac{1}{2}}),r$	
		- -						-	-	-		$C,\neq,r$	
								ΔE_p	200	sttd		$A,i_{p,C}\gg i_{p,A}$	
		-1.63 ±0.2						-	-	-		$C,=,ece,E_p=f(v^{\frac{1}{2}}),r$	
		- -						-	-	-		$C,\neq,r$	
								ΔE_p	200	sttd		$A,i_{p,C}\gg i_{p,A}$	

CONT

13

DA06 (CONT.) $C_2H_4N_2O_2$

Code No.	Empirical Formula	Name and WLN Code	Structural Formula	Solvent	Tech.	Medium		μ, M	pH	T, °C	Electrodes	App.	Experimental Parameters
DA06	$C_2H_4N_2O_2$	oxamide	$[H_2NC(O)\cdot]_2$	MeCN 50	VA	CARB	?	-	10.0	25± 0.1	HMDE SCE	2AO	C=0.5, v=1E5, Pt Aux
DA07 aa78	$C_2H_4O_2S$	thioglycolic acid C.A. 68-11-1 SHIVQ	$HSCH_2COOH$	H_2O	PY	BOR	0.6	-	9.2	-	DME/SCE	OA-	C=0.01
						BOR H_3BO_3 $CoCl_2$	0.06 0.48 4E-4		7.7				C=0.1
DA08	$C_2H_5N_3O_2$	N-methyl-N-nitroso-urea C.A. 684-93-5 ZVN1&NO	$NH_2CON(CH_3)NO$	H_2O	DI	BR		-	2.0	-	DME/SCE	O4AO	C=0.05, m=1.73, t=1, h=80, v=10, ΔE=100
					PT	BR		-	2.0	-	DME/SCE	O4AO	C=0.05, m=1.73, t=1, h=80, v=10
									4.0				
									6.0				
DA09	$C_2H_5N_5O_3$	N-methyl-N-nitroso-N'-nitroguanidine C.A. 70-25-7 WNMYZN1&NO	$NO_2NHC(:NH)N-(CH_3)NO$	H_2O	DI	BR		-	2.0	-	DME/SCE	O4AO	C=0.05, m=1.73, t=1, h=80, v=10, ΔE=100
									10.0				
DA10 aa95	$C_2H_6N_2O$	dimethyl-N-nitroso-amine C.A. 62-75-9 ONN1&1	$(CH_3)_2NNO$	H_2O	DI	BR		-	2.0	-	DME/SCE	O4AO	C=0.05, m=1.73, t=1, h=80, v=10, ΔE=100
									10.0				
					PT	BR		-	2.0	-	DME/SCE	O4AO	C=0.05, m=1.73, t=1, h=80, v=10
DA11 ab34 ca23	C_3H_3N	acrylonitrile C.A. 107-13-1 NC1U1	$CH_2:CHCN$	DMF	PY	Bu_4NI	0.05?	-	-	-	DME/MP?	-	-
DA12 ab36	$C_3H_3N_3$	1,3,5-triazine C.A. 290-87-9 T6N CN ENJ	TABLE II-1.	MeCN	PY	Et_4NClO_4	0.1	-	-	-	DME Ag/Ag^+	235AO	C=0.54, h=38
						Et_4NClO_4 ClACET	0.1 3.0E-4						
						Et_4NClO_4 ClACET	0.1 3.9E-4						
						Et_4NClO_4 ClACET	0.1 6.0E-4						
						Et_4NClO_4 ClACET	0.1 8.1E-4						
CONT													

TABLE I. Electrochemical Data $C_3H_3N_3$ (CONT.) DA12

Ref.	C/M	Charact. Potential Value	vs.	Response Const. Value		n Tech.	Electrokinetic Data Parameter	Value	From	Products and Identification	Description and Remarks	Code No.
AA067 0415	359 a	E_p	-1.6F -2.0F -1.5F	SCE	i_p	100F 290F 80F	- -	- -	- -	- -	-	$C, \ne, r$ C, i_a A	DA06
C0036 1035	-	$E_{\frac{1}{2}}$	-1.56F	SCE	-	-	-	-	-	-	-	-	C, i_{cat}, Mx, p	DA07 aa78
			-1.07F -1.26F -1.65F		- - -	- - -	- - -	- - -	- - -	- - -	- - -	- - -	C, i_{cat}, Mx, p C, i_{cat}, Mx C, i_{cat}	
AA078 0081	-	E_{su}	-0.65	SCE	i_{su}	1.5	-	-	-	-	-	-	C, r	DA08
AA078 0081	-	$E_{\frac{1}{2}}$	-0.65F	SCE	i_ℓ	40F	-	-	-	-	-	-	C, r	
			-0.90F			40F	-	-	-	-	-	-	C, r	
			-0.90F -1.02F -1.43F			18F 14F 15F	- - -	- - -	- - -	- - -	- - -	- - -	C, r C C	
AA078 0081	-	E_{su}	-0.28 -0.40 -0.65	SCE	i_{su}	1.0 2.0 2.5	- - -	- - -	- - -	- - -	- - -	- - -	C, r C C	DA09
			-1.1 -1.28			0.8 1.2	- -	- -	- -	- -	- -	- -	C, r C	
AA078 0081	82b	E_{su}	-0.87	SCE	i_{su}	2.0	QE	4	-	-	-	-	C, r	DA10 aa95
			-1.55			0.75		2					C, i_d, r	
AA078 0081	82b	$E_{\frac{1}{2}}$	-0.90	SCE	-	-	QE	4	-	-	-	-	C, i_d, r	
ER003 0494	-	-	-	-	-	-	-	-	-	-	-	-	see AB34, CA23	DA11 ab34 ca23
JA094 7941	75c	$E_{\frac{1}{2}}$	-	SCE	i_ℓ	0.23F	-	-	-	-	-	-	C, Pr, r	DA12 ab36
			-2.074		i_ℓ	2.35F 3.40			Tomeš	50	sttd		$C, i_\ell \propto h^{0.49}$	
			-		i_ℓ	1.41F 0.45F 1.41F	- - -	- - -	- - -	- - -	- - -	- - -	C, p C, Pr C	
			-			1.88F 0.45F 1.41F	- - -	- - -	- - -	- - -	- - -	- - -	C, p C, Pr C	
			-			2.82F 0.59F 1.10F	- - -	- - -	- - -	- - -	- - -	- - -	C, p C, Pr C	
			-		i_ℓ/C	3.65F 0.54F 0.70F	- - -	- - -	- - -	- - -	- - -	- - -	C, p C, Pr C	
														CONT

DA12 (CONT.) $C_3H_3N_3$

Code No.	Empirical Formula	Name and WLN Code	Structural Formula	Solvent	Tech.	Medium		μ, M	pH	T, °C	Electrodes	App.	Experimental Parameters
DA12 ab36	$C_3H_3N_3$	1,3,5-triazine		MeCN	PY	Et_4NClO_4 H_2O	0.1 0.01	-	-	-	DME Ag/Ag^+	235A0	C=0.54,h=38
						Et_4NClO_4 H_2O	0.1 0.2						
						Et_4NClO_4 H_2O	0.1 0.5						
DA13	C_3H_4HgIN	2-cyanoethylmercuric iodide C.A. 2517-78-4 NC2-HG- &I	$IHgCH_2CH_2CN$	H_2O	PY	Me_4NI	0.1	-	-	-	DME/SCE	O-O	C=satd,EC= 1.263
						LiCl	0.1						
DA14 ab48 ca27	C_3H_4IN	3-iodopropanonitrile C.A. 2517-76-2 NC2I	ICH_2CH_2CN	H_2O	PY	LiCl	0.1	-	-	-	DME/SCE	O-O	C=2.5,EC= 1.263
				EtOH 50	PY	Me_4NI	0.1	-	-	-	DME/SCE	O-O	C=2.5,EC= 1.263
DA15 ca28	C_3H_4O	acrolein C.A. 107-02-8 VH1U1	$CH_2{:}CHCHO$	EtOH 48	PY	BR,KCl		0.15	1.55 to 4.09	25± 0.1	DME/SCE	O14A0	C=1
DA16	$C_3H_5Br_2Cl$	1,3-dibromo-1-chloropropane G2YEE	TABLE II.	MeOH 40	PY	Cs_2CO_3	0.5	-	-	20± 0.1	DME/SCE	O-O	C=1
				MeOH 60	PY	Cs_2CO_3	0.5	-	-	20± 0.1	DME/SCE	O-O	C=1
				MeOH 70	PY	Cs_2CO_3	0.05	-	-	10± 0.1	DME/SCE	O-O	C=ns
						Cs_2CO_3	0.5			10± 0.1			
										20± 0.1			C=1
						KCl	0.05			10± 0.1			C=ns
						LiCl	0.05						
						Rb_2CO_3	0.05						
				MeOH 80	PY	Cs_2CO_3	0.5	-	-	20± 0.1	DME/SCE	O-O	C=1
				EtOH 40	PY	KCl	0.1	-	-	20± 0.1	DME/SCE	O-O	C=ns,t= 1(controlled), m=1.5
DA17	$C_3H_5KOS_2$	potassium ethylxanthate C.A. 104-89-6 SUCS&O2 &-KA-	$C_2H_5OC(S)SK$	H_2O	VA	BOR	0.1	-	-	-	PbDE SCE	OAO	C=10,A=6, v=10
											PbSE SCE		C=20,A=6, v=40
DA18 ab63	C_3H_5NO	acrylamide C.A. 79-06-1 ZV1U1	$CH_2{:}CHCONH_2$	H_2O	PO	Me_4NI	satd	-	-	-	XEI	---	C=2.59, ω=600
DA19 CONT	$C_3H_5NO_5$	2-nitratopropanoic acid C.A. 1860-18-0 WNOY1&VQ	$CH_3CH(ONO_2)COOH$	H_2O	PY	HCl KCl	0.1 1	-	1.0	25± 0.1	DME/SCE	O-O	EC=2.10

TABLE I. Electrochemical Data $C_3H_5NO_5$ (CONT.) DA19

Ref.	C/M	Charact. Potential		Response Const.		n Tech.	Electrokinetic Data			Products and Identification	Description and Remarks	Code No.
		Value	vs.		Value		Parameter	Value	From			
JA094 7941	75c	$E_{\frac{1}{2}}$	-	SCE	i_ℓ/C	4.5F	-	-	-	-	-	-	C,p	DA12 ab36
		-2.000F			4.4F	-	-	-	-	-	-	C,p	
		-1.971F			4.6F	-	-	-	-	-	-	C,p	
VZ010 0101	8d	$E_{\frac{1}{2}}$	-0.30	SCE	i_ℓ	0.43	Ilk	1	-	-	-	-	C,i_d,r	DA13
		-0.885			0.80		1					C	
		-0.39			0.586		1	-	-	-	-	C,i_d,r	
		-1.13			0.586		1					C	
VZ010 0101	8e	$E_{\frac{1}{2}}$	-1.03	SCE	i_ℓ	4.42	Ilk	1	-	-	-	-	C,i_d,r	DA14 ab48 ca27
		-1.42			5.27		1					C	
VZ010 0101	8e	$E_{\frac{1}{2}}$	-0.895	SCE	i_ℓ	4.07	Ilk	1	-	-	-	-	C,i_d,r	
		-1.14			3.22		1					C	
GA096 0578	92c	$E_{\frac{1}{2}}^o$	-0.860	SCE	i_ℓ/C	3.38	-	-	$dE_{\frac{1}{2}}/dpH$	66	sttd	-	$C,Av(5),\ell p$	DA15 ca28
RR020 1167	-	$E_{\frac{1}{2}}$	-1.21F	SCE	i_ℓ	6F	-	-	-	-	-	-	C,X,p	DA16
RR020 1167	-	$E_{\frac{1}{2}}$	-1.25F	SCE	i_ℓ	6.0F	-	-	-	-	-	-	C,X,p	
RR020 1167	-	$E_{\frac{1}{2}}$	-1.55	SCE	i_ℓ	7.6	-	-	-	-	-	-	$C,X,iR?,p$	
		-1.36			8.0	-	-	-	-	-	-	C,X,p	
		-1.31F			6.3F	-	-	-	-	-	-	C,X,p	
		-1.63			6.9	-	-	-	-	-	-	$C,X,iR?,p$	
		-1.80			4.8	-	-	-	-	-	-	$C,X,iR?,p$	
		-1.60			6.5	-	-	-	-	-	-	$C,X,iR?,p$	
RR020 1167	-	$E_{\frac{1}{2}}$	-1.36F	SCE	i_ℓ	6.8F	-	-	-	-	-	-	C,X,p	
RR020 1167	-	$E_{\frac{1}{2}}$	-1.19F	SCE	i_ℓ	8.0F	-	-	-	-	-	-	C,X,p	
		-1.45F			2.0F							C	
AJ025 2329	-	E_p	-0.39F	NHE	j_p	1400F	-	-	-	-	-	lead ethylxanthate, IRS, elemental analysis	A,R,r	DA17
		0.014F			228F	-	-	-	-	-	lead ethylxanthate, diethyl dixanthogen, IRS	A,r	
		-0.40F			200F							C	
RU032 0244	-	Q	-1.68	-	-	-	-	-	-	-	-	-	$C,i_a?,XEl$=rotating Hg-plated Ag electrode,p	DA18 ab63
ER001 0759	-	$E_{\frac{1}{2}}$	-0.50	SCE	i_ℓ/C	7.7	-	-	-	-	-	-	C,i_d,r	DA19
		-1.00			14.3							C,i_d	CONT

Code No.	Empirical Formula	Name and WLN Code	Structural Formula	Solvent	Tech.	Medium		μ, M	pH	T, °C	Electrodes	App.	Experimental Parameters
DA19	$C_3H_5NO_5$	2-nitratopropanoic acid	$CH_3CH(ONO_2)COOH$	H_2O	PY	UB KCl	1	-	2.9	25± 0.1	DME/SCE	O-O	EC=2.10
									3.8				
									4.2				
									4.4				
									6.0				
									7.0				
									11.0				
DA20	$C_3H_6NNaS_2$	sodium dimethyl-dithiocarbamate C.A. 128-04-1 SUCS&N1&1 &-NA-	$(CH_3)_2NC(S)SNa$	H_2O	PY	KOH KCl TX100	0.01 0.3 0.008	-	-	25± 0.2	DME/SCE	O1AO	C=1.0,m= 2.52,t=3.00, h=40
					IL	KOH KCl	0.01 0.3	-	-	25± 0.2	HMDE SCE	O1AO	C=1.0,v=20
						KOH KCl TX100	0.01 0.3 0.008						
DA21	$C_3H_6O_2S$	methyl vinyl sulfone C.A. 3680-02-2 1U1SW1	$CH_3SO_2CH:CH_2$	MeOH 20	PY	LiCl	0.07	-	-	25	DME/SCE	O-O	C=0.5,m= 1.63,t=2.5
DA22 ab90	C_3H_7BrMg	isopropylmagnesium bromide C.A. 920-39-8 E-MG-Y	TABLE II-2.	MG	PY	Bu_4NClO_4	0.1	-	-	-	DME Ag/AgClO$_4$	-	-
DA23 ac06	$C_3H_7NO_2S$	cysteine C.A. 52-90-4 SH1YZVQ	$HSCH_2CH(NH_2)-COOH$	H_2O	PY	BOR $CoCl_2$	0.6 4E-4	-	-	-	DME/SCE	OA-	C=0.5
DA24	$C_3H_7N_3O_2$	N-nitrososarcosinamide C.A. 39935-46-1 ZV1N1&NO	$ONN(CH_3)CH_2CO-NH_2$	H_2O	DI	BR		-	10.0	-	DME/SCE	O4AO	C=0.05,m= 1.73,t=1,h= 80,v=10, ΔE=100
					PT	BR		-	2.0	-	DME/SCE	O4AO	C=0.05,m= 1.73,t=1,h= 80,v=10

TABLE I. Electrochemical Data $C_3H_7N_3O_2$ DA24

Ref.	C/M	Charact. Potential		Response Const.		n Tech.	n	Electrokinetic Data			Products and Identification	Description and Remarks	Code No.
		Value	vs.		Value			Parameter	Value	From			
ER001 0759	-	$E_{\frac{1}{2}}$	-0.63	SCE	i_ℓ/C	6.2	TLQE llk	2	-	-	-	$CP \rightarrow NO_2^-$, PY, rxn. with sulfanilic acid and α-naphthylamine; lactic acid, CP to oxalic acid	$C, i_d, Mc(product), r$	DA19
			-1.05			14.4		-					C, i_d	
			-0.65			5.8		2	-	-	-	-	C, i_d, r	
			-1.20			12.8		-					C, i_d	
			-1.50			7.7		-					C, i_d	
			-0.70			6.2		2	-	-	-	-	C, i_d, r	
			-1.24			11.2		-					C, i_d	
			-1.55			8.9		-					C, i_d	
			-0.70			6.2		2	-	-	-	-	C, i_d, r	
			-1.27			11.6		-					C, i_d	
			-1.55			9.7		-					C, i_d	
			-0.69			7.0		2	-	-	-	-	C, i_d, r	
			-			-		-					0	
			-1.58			9.3		-					C, i_d	
			-0.69			7.0		2	-	-	-	-	C, i_d, r	
			-			-		-					0	
			-			-		-					0	
			-0.69			7.0		2	-	-	-	-	C, i_d, r	
			-			-		-					0	
			-			-		-					0	
AJ028 0021	-	$E_{\frac{1}{2}}$	-0.285 -0.475 -0.544	SCE	i_ℓ	3.7F - -		-	-	-	-	-	A, r A A, i_a, Pr	DA20
AJ028 0021	-	E_p	-0.395	SCE	-	-	-	-	$E_p - E_{p/2}$	4	-	-	A, r	
			-0.245	-	-	-	-	-		73	-	-	A, r	
DS171 0860	43a	$E_{\frac{1}{2}}$	-1.76	SCE	I	3.10	llk	≈2	αn_a	0.5	Elog	ethene	$C, i_d, \neq, r$	DA21
JA088 5132	-	-	-	-	-	-	-	-	-	-	-	-	see AB90	DA22 ab90
C0036 1035	-	$E_{\frac{1}{2}}$	-1.47	SCE	-	-	-	-	-	-	-	-	C, i_{cat}, Mx, p	DA23 ac06
AA078 0081	82c	E_{su}	-1.35	SCE	i_{su}	0.50	QE	2	-	-	-	-	C, r	DA24
AA078 0081	82a	$E_{\frac{1}{2}}$	-0.81	SCE	-	-	QE	4	-	-	-	-	C, r	

Code No.	Empirical Formula	Name and WLN Code	Structural Formula	Solvent	Tech.	Medium		μ, M	pH	T, °C	Electrodes	App.	Experimental Parameters
DA25	$C_3H_9AsO_3$	1-propanearsonic acid C.A. 107-34-6 Q-AS-QO&3	$CH_3CH_2CH_2AsO(OH)_2$	H_2O	PY	HCl	0.1	-	-	-	DME/SCE	O4AO	C=0.1-1.0, t=2.4, v=1.7
					IV	HCl	0.1	-	-	-	DME/SCE	O4AO	C=0.2, t=2.4, E_{app}=-0.900
													E_{app}=-1.000
													E_{app}=-1.100
													E_{app}=-1.200
DA26	$C_3H_{10}N_6$	2,5-dimethyl-1,2,3,-4,5,6-hexaazabicyclo[2.2.1]heptane C.A. L55 A BNMNNMNJ D1 G1	TABLE II.	MeCN	VR	$NaClO_4$	0.1	-	-	-	AuBE SCE	2AO	C=0.418, d=0.15, v=99, PE, MP Aux
DA27	C_4Cl_6	hexachlorobuta-1,3-diene C.A. 87-68-3 GYGUYGYGUYGG	$Cl_2C:CClCCl:CCl_2$	EtOH 75	PD	Et_4NOH	0.05	-	-	-	DME/MP	OAO	t=1.4
DA28	C_4Cl_6Hg	bis(trichlorovinyl)-mercury(II) C.A. 10507-38-7 GYGUYG-HG-YGUYGG	$Hg(CCl:CCl_2)_2$	MeOH 50	PY	$LiClO_4$	0.1	-	-	2.5	DME/SCE	O-O	m=0.62, t=3.00
DA29 ac26	$C_4H_2Br_2S$	2,3-dibromothiophene C.A. 3140-93-0 T5SJ BE CE	TABLE II-2.	EtOH	PY	Et_4NI	0.05	-	-	25±0.2	DME/MP (satd Et_4NI)	-AO	m=1.07, t=0.26
DA30 ac38	$C_4H_3NO_3$	2-nitrofuran C.A. 609-39-2 T5OJ BNW	TABLE II-2.	H_2O	PY	BR		0.45	2.0	25±0.1	DME/SCE	OAO	C=0.2, m=1.18, t=0.26
									3.0				
									4.0				
									6.0				
									8.0				
									9.0				
									11.0				
									11.9				
				EtOH 10	PY	BR		0.45	2.0	25±0.1	DME/SCE	OAO	C=0.2, m=1.18, t=0.26
									3.0				
									4.0				
									5.0				
									6.0				
									7.0				
									8.0				
									9.0				
									12				

CONT

TABLE I. Electrochemical Data — $C_4H_3NO_3$ (CONT.) DA30

Ref.	C/M	Charact. Potential		Response Const.		n Tech.	n	Electrokinetic Data			Products and Identification	Description and Remarks	Code No.
		Value	vs.		Value			Parameter	Value	From			
AN100 0489	-	$E_{\frac{1}{2}}$	-1.11	SCE	-	-	i:i	4	-	-	-	-	$C, i_d, i \propto C, r$	DA25
AN100 0489	-	-	-	-	dlogi/dlogt	0.39	-	-	-	-	-	-	C,r	
		-	-	-		0.30	-	-	-	-	-	-	C,r	
		-	-	-		0.25	-	-	-	-	-	-	C,r	
		-	-	-		0.24	-	-	-	-	-	-	C,r	
JA094 7108	-	E_p	1.07	SCE	-	-	-	-	-	-	-	-	$A, \frac{1}{2}, p$	DA26
RL032 0344	-	E_{su}	-1.45 -2.55	MP	-	-	QE	13 -	-	-	-	butane(?)	$C, \frac{1}{2}, p$ C	DA27
ER002 0587	23d	$E_{\frac{1}{2}}$	-0.63	SCE	-	-	sttd	2	α	0.44	sttd	-	$C, \frac{1}{2}, r$	DA28
ER001 0060	14b	$E_{\frac{1}{2}}$	-1.41 -2.20	SCE	-	-	sttd	2 2	-	-	-	-	$C, i_d, \frac{1}{2}, E_{\frac{1}{2}}=f[EtOH], r$ $C, i_d, \frac{1}{2}, E_{\frac{1}{2}}=f[EtOH]$	DA29 ac26
LV670 0023	-	$E_{\frac{1}{2}}$	-0.18	SCE	i_ℓ	0.89	llk	4	-	-	-	2-furylhydroxylamine	C,r	DA30 ac38
			-0.36			0.36		2				2-aminofuran	C	
			-0.23 -0.50			0.92 0.13		4 2	-	-	-	-	C,r C	
			-0.28			0.93		4	-	-	-	-	C,r	
			-0.42			0.90		4	-	-	-	-	C,r	
			-0.55			0.88		4	-	-	-	-	C,r	
			-0.58			0.88		4	-	-	-	-	C,r	
			-0.62			0.90		4	-	-	-	-	C,r	
			-0.62			0.90		4	-	-	-	-	C,r	
LV670 0023	-	$E_{\frac{1}{2}}$	-0.18 -0.36	SCE	i_ℓ	0.89 0.36	llk ns	4 2	-	-	-	-	C,r C	
			-0.28 -0.54			0.82 0.13	llk ns	4 2	-	-	-	-	C,r C	
			-0.43			0.80	llk	4	-	-	-	-	C,r	
			-0.45			0.76		-	-	-	-	-	C,r	
			-0.54 -0.99			0.64 0.19		-	-	-	-	-	C,r C	
			-0.58 -0.92			0.38 0.41		-	-	-	-	-	C,r C	
			-0.62 -0.92			0.28 0.48		-	-	-	-	-	C,r C	
			-0.64 -0.92			0.25 0.55		-	-	-	-	-	C,r C	
			-0.64 -0.92			0.20 0.56		-	-	-	-	-	C,r C	CONT

DA30 (CONT.) $C_4H_3NO_3$ *CRC Handbook Series in Organic Electrochemistry*

Code No.	Empirical Formula	Name and WLN Code	Structural Formula	Solvent	Tech.	Medium	μ, M	pH	T, °C	Electrodes	App.	Experimental Parameters
DA30 ac38	$C_4H_3NO_3$	2-nitrofuran		EtOH 20	PY	BR	0.45	2.0	25±0.1	DME/SCE	OAO	C=0.2,m=1.18,t=0.26
								3.0				
								4.0				
								5.0				
								6.0				
								7.0				
								8.0				
								9.0				
								10.0				
								11.0				
								12.0				
				EtOH 30	PY	BR	0.45	2.0	25±0.1	DME/SCE	OAO	C=0.2,m=1.18,t=0.26
								3.0				
								4.0				
								5.0				
								6.0				
								7.0				
								8.0				
								10.0				
								12.0				
				EtOH 50	PY	BR	0.45	2.0	25±0.1	DME/SCE	OAO	C=0.2,m=1.18,t=0.26
								3.0				
								4.0				
								5.0				
								6.0				
								7.0				
								8.0				
								12.0				
DA30 ac38	$C_4H_3NO_3$	2-nitrofuran										

TABLE I. Electrochemical Data $C_4H_3NO_3$ DA30

Ref.	C/M	Charact. Potential		Response Const.		n Tech.	n	Electrokinetic Data			Products and Identification	Description and Remarks	Code No.
		Value	vs.		Value			Parameter	Value	From			
LV670 0023	-	$E_{\frac{1}{2}}$	-0.28	SCE	i_ℓ	1.85	IIk	4	-	-	-	-	c,r	DA30 ac38
			-0.35			1.85		4	-	-	-	-	c,r	
			-0.44			1.80		4	-	-	-	-	c,r	
			-0.50			1.60		-	-	-	-	-	c,r	
			-0.58 -0.95			1.10 0.72		-	-	-	-	-	c,r c	
			-0.60 -0.95			0.70 1.10		-	-	-	-	-	c,r c	
			-0.61 -0.95			0.50 1.22		-	-	-	-	-	c,r c	
			-0.62 -0.95			0.50 1.19		-	-	-	-	-	c,r c	
			-0.62 -0.95			0.50 1.10		-	-	-	-	-	c,r c	
			-0.62 -0.95			0.48 1.00		-	-	-	-	-	c,r c	
			-0.62 -0.95			0.46 0.80		-	-	-	-	-	c,r c	
LV670 0023	-	$E_{\frac{1}{2}}$	-0.32	SCE	i_ℓ	1.5	IIk	4	-	-	-	-	c,r	
			-0.45			1.55		4	-	-	-	-	c,r	
			-0.53			1.47		4	-	-	-	-	c,r	
			-0.58 -1.02			1.40 0.30		-	-	-	-	-	c,r c	
			-0.60 -1.02			0.80 0.80		-	-	-	-	-	c,r c	
			-0.61 -1.01			0.40 0.98		-	-	-	-	-	c,r c	
			-0.61 -1.01			0.40 1.10		-	-	-	-	-	c,r c	
			-0.61 -1.01			0.36 1.10		-	-	-	-	-	c,r c	
			-0.62 -1.01			0.37 1.10		-	-	-	-	-	c,r c	
LV670 0023	-	$E_{\frac{1}{2}}$	-0.38	SCE	i_ℓ	1.58	IIk	4	-	-	-	-	c,r	
			-0.51			1.58		4	-	-	-	-	c,r	
			-0.58			1.46		4	-	-	-	-	c,r	
			-0.60 -1.07			0.92 0.68		-	-	-	-	-	c,r c	
			-0.61 -1.06			0.66 0.84		-	-	-	-	-	c,r c	
			-0.61 -1.07			0.50 0.90		-	-	-	-	-	c,r c	
			-0.62 -1.07			0.40 0.94		-	-	-	-	-	c,r c	
			-0.62 -1.07			0.40 0.96		-	-	-	-	-	c,r c	

DA31 C$_4$H$_4$N$_2$

Code No.	Empirical Formula	Name and WLN Code	Structural Formula	Solvent	Tech.	Medium	μ, M	pH	T, °C	Electrodes	App.	Experimental Parameters
DA31 ac42 ca54	C$_4$H$_4$N$_2$	pyrazine C.A. 290-37-9 T6N DNJ	TABLE 11-2.	H$_2$O	PY	HClO$_4$ NaClO$_4$	1.0	0.01	25± 21	DME Ag/AgCl, satd NaCl	0-0	C=0.129
												C=4.13
								0.1				C=1.29
								0.60				
								1.01				
DA31 CONT	C$_4$H$_4$N$_2$	pyrazine C.A. 290-37-9	TABLE 11-2.	H$_2$O	PY	HClO$_4$ NaClO$_4$	1.0	0.01	25± 21	DME Ag/AgCl, satd NaCl	0-0	C=0.129

TABLE I. Electrochemical Data

$C_4H_4N_2$ (CONT.) DA31

Ref.	C/M	Charact. Potential		Response Const.		Tech.	n	Electrokinetic Data			Products and Identification	Description and Remarks	Code No.
		Value	vs.		Value			Parameter	Value	From			
JA094 7295	74d	$E_{\frac{1}{2}}$	-	Ag/AgCl, satd NaCl	-	-	sttd	1	-	-	-	1,4-dihydropyrazine radical cation	$C, i_\ell \propto C, i_\ell \neq f(pH)$, $i_\ell \propto h^{\frac{1}{2}}, i_d, S, R, r$	DA31 ac42 ca54
		-					1	-	-	-	1,4-dihydropyrazine	$C, i_\ell \propto C, i_\ell \neq f(pH)$, $i_\ell \propto h^{\frac{1}{2}}, i_d, P, R$	
		-0.62						-	$dE_{\frac{1}{2}}/dlogC$	7	sttd		$C, i_\ell \propto C, i_\ell = f(pH)$,	
								Tomeš	37.2	sttd		$i_\ell \propto h^o, E_{\frac{1}{2}}=0.048pH$	
								$dE_{\frac{1}{2}}/dpH$	48			-0.619 for pH < 2, $Elogi^{\frac{2}{3}}$ nonlinear, i_c	
		-		-	-		1	-	-	-	1,4-dihydropyrazine radical cation	$C, i_\ell \propto C, i_\ell \neq f(pH)$, $i_\ell \propto h^{\frac{1}{2}}, i_d, S, R, r$	
		-					1	-	-	-	1,4-dihydropyrazine	$C, i_\ell \propto C, i_\ell \neq f(pH)$, $i_\ell \propto h^{\frac{1}{2}}, i_d, P, R$	
		-0.62						-	$dE_{\frac{1}{2}}/dlogC$	7	sttd		$C, i_\ell \propto C, i_\ell = f(pH)$,	
								Tomeš	54.3	sttd		$i_\ell \propto h^o, E_{\frac{1}{2}}=-0.048pH$	
								$dE_{\frac{1}{2}}/dpH$	48			-0.619 for pH < 2, $Elogi^{\frac{2}{3}}$ nonlinear, i_c	
		-		-	-	sttd	1	-	-	-	1,4-dihydropyrazine radical cation	$C, i_\ell \propto C, i_\ell \neq f(pH)$, $i_\ell \propto h^{\frac{1}{2}}, i_d, S, R, r$	
		-					1	-	-	-	1,4-dihydropyrazine	$C, i_\ell \propto C, i_\ell \neq f(pH)$, $i_\ell \propto h^{\frac{1}{2}}, i_d, P, R$	
		-0.625						-	$dE_{\frac{1}{2}}/dlogC$	7	sttd		$C, i_\ell \propto C, i_\ell = f(pH)$,	
								Tomeš	43.2	sttd		$i_\ell \propto h^o, E_{\frac{1}{2}}=-0.048pH$	
								$dE_{\frac{1}{2}}/dpH$	48			-0.619 for pH < 2, $Elogi^{\frac{2}{3}}$ nonlinear, i_c	
		-		-	-	sttd	1	-	-	-	1,4-dihydropyrazine radical cation	$C, i_\ell \propto C, i_\ell \neq f(pH)$, $i_\ell \propto h^{\frac{1}{2}}, i_d, S, R, r$	
		-					1	-	-	-	1,4-dihydropyrazine	$C, i_\ell \propto C, i_\ell \neq f(pH)$, $i_\ell \propto h^{\frac{1}{2}}, i_d, P, R$	
		-0.65						-	$dE_{\frac{1}{2}}/dlogC$	7	sttd		$C, i_\ell \propto C, i_\ell = f(pH)$,	
								Tomeš	71.5	sttd		$i_\ell \propto h^o, E_{\frac{1}{2}}=-0.048pH$	
								$dE_{\frac{1}{2}}/dpH$	48			-0.619 for pH < 2, $Elogi^{\frac{2}{3}}$ nonlinear, i_c	
		-0.24F		i_ℓ	5.6F	sttd	1	-	-	-	1,4-dihydropyrazine radical cation	$C, i_\ell \propto C, i_\ell \neq f(pH)$, $i_\ell \propto h^{\frac{1}{2}}, i_d, S, R, r$	
		-0.36F			4.4F		1	-	-	-	1,4-dihydropyrazine	$C, i_\ell \propto C, i_\ell \neq f(pH)$, $i_\ell \propto h^{\frac{1}{2}}, i_d, P, R$	
		-0.67			3.2F			-	$dE_{\frac{1}{2}}/dlogC$	7	sttd		$C, i_\ell \propto C, i_\ell = f(pH)$,	
								Tomeš	94.0	sttd		$i_\ell \propto h^o, E_{\frac{1}{2}}=-0.048pH$	
								$dE_{\frac{1}{2}}/dpH$	48			-0.619 for pH < 2, $Elogi^{\frac{2}{3}}$ nonlinear, i_c	
JA094 7295	74d	$E_{\frac{1}{2}}$	Ag/AgCl, satd NaCl										DA31 CONT

DA31 (CONT.) $C_4H_4N_2$

Code No.	Empirical Formula	Name and WLN Code	Structural Formula	Solvent	Tech.	Medium	μ, M	pH	T, °C	Electrodes	App.	Experimental Parameters
DA31 ac42 ca54	$C_4H_4N_2$	pyrazine		H_2O	PY	PHOS $HClO_4$ $NaClO_4$	?	1.0	2	-	DME Ag/AgCl, satd KCl	0-0	
									7				
					IU	$HClO_4$	1.0	1.0	-	-	DME Ag/AgCl, satd NaCl	0-0	$C=1.3$, $(t-\tau)/\tau=0.1$
													$(t-\tau)/\tau=0.3$
													$(t-\tau)/\tau=0.6$
					PR	$HClO_4$	?	1.0	-0.04	25.0 ±0.1	DME Ag/AgCl, satd NaCl	0-0	$C=1.29, v=50$
													$v=100$
													$v=1E4$
													$v=1E5$
													$C=1.0, v=20$
									0.96				$C=1.29, v=50$
													$v=1E4$
									1.41				$C=1.29, v=50$
													$v=1E4$
									2.94				$C=1.29, v=50$
													$v=1E3$
													$v=1E5$
						PHOS			4.70				$C=1.29, v=50$
													$v=1E3$
													$v=1E5$
DA31 CONT	$C_4H_4N_2$	pyrazine		H_2O	PY	PHOS $HClO_4$ $NaClO_4$	?	1.0					

TABLE I. Electrochemical Data $C_4H_4N_2$ (CONT.) DA31

Ref.	C/M	Charact. Potential		Response Const.		n Tech.	Electrokinetic Data			Products and Identification	Description and Remarks	Code No.
		Value	vs.	Value			Parameter	Value	From			
JA094 7295	-	$E_{\frac{1}{2}}$	0.094	Ag/AgCl, satd KCl	-	-	sttd 2	$dE_{\frac{1}{2}}/dpH$	86	sttd	-	C,r	DA31 ac42 ca54
	74e		-0.336				2		86			$C, i_d, i_\ell = f(t)$ for $pH > 7, r$	
JA094 7295	74d	-	-	-	i_A/i_C	0.552	-	-	-	-	-	$C, k\tau = 0.34, k_{decomp} = 0.47\ s^{-1}, r$	
					0.332	-	-	-	-	-	-	$C, k\tau = 0.47, r$	
					0.217	-	-	-	-	-	-	$C, k\tau = 0.46, r$	
JA094 7295	74d	E_p	-0.230	Ag/AgCl, satd NaCl	$i_p/v^{\frac{1}{2}}$	75.4	sttd 1	dE_p/dpH	104	sttd	-	C,r	
			-0.347			69.6	1		60			C	
			-0.292			12.0	1		-			A	
			-0.164			24.1	1					A	
			-0.232			69.1	1		104			C,r	
			-0.342			65.2	1		60			C	
			-0.295			28.3	1		-			A	
			-0.167			39.6	1					A	
			-0.230			69.8	1		104			C,r	
			-0.353			71.4	1		60			C	
			-0.297			36.6	1		-			A	
			-0.176			42.8	1					A	
			-0.228			67.0	1		104			C,r	
			-0.360			79.0	1		60			C	
			-0.300			34.0	1		-			A	
			-0.176			51.0						A	
			-0.24F			21F	-	-	104			C,r	
			-0.36F			12F			60			C	
			-0.67F			20F			-			C	
			-0.59F			15F						A	
			-0.29F			4F						A	
			-0.17F			8F						A	
			-0.334		-	-	-	1	104			C,r	
			-0.428					1	60			C	
			-0.360					1	-			A	
			-0.267					1				A	
			-0.338		-	-	-	1	104			C,r	
			-0.432					1	60			C	
			-0.362					1	-			A	
			-0.270					1				A	
			-0.375		-	-	-	1	104			C,r	
			-0.442					1	60			C	
			-0.378					1	-			A	
			-0.300					1				A	
			-0.375		-	-	-	1	104			C,r	
			-0.446					1	60			C	
			-0.383					1	-			A	
			-0.308					1				A	
	74e		-0.545		-	-	-	2	90		-	C,R,r	
			-0.496					2	-			A,R	
			-0.550		-	-	-	2	90			C,R,r	
			-0.494					2	-			A,R	
			-0.578		-	-	-	2	90			C,R,r	
			-0.470					2	-			A,R	
			-0.700		-	-	-	2	90			C,R,r	
			-0.641					2	-			A,R	
			-0.707		-	-	-	2	90			C,R,r	
			-0.651					2	-			A,R	
			-0.823		-	-	-	2	90			C,R,r	
			-0.614					2	-			A,R	

CONT

DA31 (CONT.) $C_4H_4N_2$ (1)   *CRC Handbook Series in Organic Electrochemistry*

Code No.	Empirical Formula	Name and WLN Code	Structural Formula	Solvent	Tech.	Medium		μ, M	pH	T, °C	Electrodes	App.	Experimental Parameters
DA31 ac42 ca54	$C_4H_4N_2$	pyrazine		H_2O	PR	PHOS	?	1.0	5.25	25.0 ±0.1	DME Ag/AgCl, satd NaCl	0-0	C=1.29, v=50
													v=1E3
													v=1E5
									6.98				C=1.29, v=50
													v=1E3
													v=1E5
					QE	$HClO_4$ PHOS	?	–	<2	25± 0.1	– Ag/AgCl, satd KCl	0-0	C=0.65-20, E_{app}=-0.550
				MeCN	PY	Et_4NClO_4 0.1		–	–	–	DME Ag/AgI	235A 0	ns
						Et_4NClO_4 0.1 H_2O 0.010							C=0.68, h=38
						Et_4NClO_4 0.1 H_2O 0.02							
						Et_4NClO_4 0.1 H_2O 0.3							
						Et_4NClO_4 0.1 H_2O 0.7							
						Et_4NClO_4 0.1 H_2O 1							
						Et_4NClO_4 0.1 ClACET 3.0E-4							
						Et_4NClO_4 0.1 ClACET 8E-4							
						Et_4NClO_4 0.1 ClACET 1.1E-3							
					VA	Et_4NClO_4 0.1		–	–	–	MPDE Ag/AgI	235A 0	C=0.40, v=400
													C=0.076, v=5.4E4
													C=0.76
													C=1.35, v=4.8
													v=20.5
													v=43
													v=420
													v=5.6E3
													v=2.7E4
DA31 CONT													

TABLE I. Electrochemical Data $C_4H_4N_2$ (CONT.) DA31

Ref.	C/M	Charact. Potential Value	vs.	Response Const.	Value	Tech.	n	Electrokinetic Data Parameter	Value	From	Products and Identification	Description and Remarks	Code No.
JA094 7295	74e	E_p -0.736	Ag/AgCl, satd NaCl	-	-	sttd	1	-	-	-	-	C,r	DA31 ac42 ca54
		-0.678					1					A	
		-0.748		-	-		1	-	-	-	-	C,r	
		-0.682					1					A	
		-0.880		-	-		1	-	-	-	-	C,r	
		-0.649					1					A	
		-0.916		-	-		1	-	-	-	-	C,⇌,r	
		-0.776					1					A,⇌	
		-0.925		-	-		1	-	-	-	-	C,⇌,r	
		0.773					1					A,⇌	
		-1.006		-	-		1	-	-	-	-	C,⇌,r	
		-0.738					1					A,⇌	
JA094 7295	74d	-	-	-	-	QE	2.00 ± 0.01	-	-	-	blue intermediate, ESR, final product light amber	C, log i=k_1-k_2t, r	
JA094 7941	74c	$E_{\frac{1}{2}}$ -2.158	SCE	i I	5.18 5.72	QE	4-4.5	Tomeš	56	sttd	-	C, $i_\ell \propto h^{0.50}$, i_d, logi-t curve is segmented, ece, $n_{t \to 0}$=2.2, r	
		-		i_ℓ/C	7.7F	-	-	-	-	-	-	C,p	
		-2.147F			8.3	-	-	-	-	-	-	C,p	
		-2.005F			10.6F	-	-	-	-	-	-	C,p	
		-2.108F			8.6F	-	-	-	-	-	-	C,p	
		-1.801F			11.1F	-	-	-	-	-	-	C,p	
		-	-		1.18F 0.09F 3.53F	-	-	-	-	-	-	C,p C,Pr C	
		-	-		1.88F 0.35F 3.29F	-	-	-	-	-	-	C,p C,Pr C	
		-	-	i_ℓ	4.47F 0.47F 0.71F	-	-	-	-	-	-	C,p C,Pr C	
JA094 7941	74c	E_p -2.26F -2.15F -0.6 to -1.9	SCE	i_p	1.32F 0.2F -	-	-	-	-	-	-	C,p A A,series of 4-6 peaks	
		-2.29 -2.21		i_A/i_C	0.72 -	-	-	-	-	-	-	C,p A	
		-2.36 -2.17			0.86 -	-	-	-	-	-	-	C,p A	
		-	-	$i_p/ACv^{\frac{1}{2}}$	3.1F	-	-	-	-	-	-	C,p	
		-	-		46.9F	-	-	-	-	-	-	C,p	
		-	-		27.5F <0.6F 0.6F	-	-	-	-	-	-	C,p A A	
		-	-		20.8F 2.5F 1.9F	-	-	-	-	-	-	C,p A A	
		-	-		15F 5.7F 1.3F	-	-	-	-	-	-	C,p A A	
		-	-		13F 7.2F 0.6F	-	-	-	-	-	-	C,p A A	

CONT

DA31 (CONT.) $C_4H_4N_2$ (2) *CRC Handbook Series in Organic Electrochemistry*

Code No.	Empirical Formula	Name and WLN Code	Structural Formula	Solvent	Tech.	Medium	μ, M	pH	T, °C	Electrodes	App.	Experimental Parameters
DA31 ac42 ca54	$C_4H_4N_2$	pyrazine		MeCN	VA	Et_4NClO_4 0.1	-	-	-	MPDE Ag/Ag+	235A 0	C=1.35, v= 1.3E5
												C=1.52
												C=7.59
DA32 ac43	$C_4H_4N_2$	pyridazine C.A. 289-80-5 T6NNJ	TABLE II-2.	MeCN	PY	Et_4NClO_4 0.1	-	-	-	DME Ag/Ag+	235A 0	ns
						Et_4NClO_4 0.1 ClACET 5.4E-5						C=0.46, h=38
						Et_4NClO_4 0.1 H_2O 0.013						C=0.92
						Et_4NClO_4 0.1 H_2O 0.065						
						Et_4NClO_4 0.1 H_2O 0.30						
						Et_4NClO_4 0.1 H_2O 0.50						
DA33 ac44 ca55	$C_4H_4N_2$	pyrimidine C.A. 289-95-2 T6N CNJ	TABLE II-2.	MeCN	PY	Et_4NClO_4 0.1	-	-	-	DME Ag/Ag+	235A 0	ns
						Et_4NClO_4 0.1 H_2O 0.01						h=38
						Et_4NClO_4 0.1 H_2O 0.040						C=1.47
						Et_4NClO_4 0.1 H_2O 0.18						
						Et_4NClO_4 0.1 H_2O 0.45						
						Et_4NClO_4 0.1 H_2O 0.7						
						Et_4NClO_4 0.1 H_2O 1						
						Et_4NClO_4 0.1 ClACET 6.9E-4						C=1.03
						Et_4NClO_4 0.1 ClACET 1.42E-3						
DA34	$C_4H_4O_2$	methyl propiolate C.A. 922-67-8 1UU1V01	$CH{\equiv}CCO_2CH_3$	MeCN	PY	Bu_4NClO_4 0.1	-	-	23	DME Ag/0.01 Ag+	2A0	C=2, Pt Aux
DA35	C_4H_5N	crotonitrile C.A. 627-26-9 NC1U2 -T	trans-$CH_3CH{:}CHCN$	DMF	PY	Bu_4NBF_4 0.50	-	-	25	DME/SCE	23-0	C=0.7-1.5, Pt Aux
DA36 ac66 CONT	$C_4H_5N_3O$	cytosine C.A. 71-30-7 T6N CNJ BQ DZ	TABLE II-2.	H_2O	PV	ACET ?	0.5	4.2	25.0 ±0.1	DME/SCE	235A 03	C=1, m=1.0, t=6.7, f=50, Pt Aux
					PVI	ACET ?	-	-	-	DME/SCE	235A 03	C=1, m=1.0, t=6.7, f=50, Pt Aux

TABLE I. Electrochemical Data — $C_4H_5N_3O$ (CONT.) DA36

Ref.	C/M	Charact. Potential		Response Const.		n Tech.	n	Electrokinetic Data			Products and Identification	Description and Remarks	Code No.
		Value	vs.		Value			Parameter	Value	From			
JA094 7941	74c	-	-	$i_p/ACv^{\frac{1}{2}}$	12.5F 7.2F <0.6F	-	-	-	-	-	-	C,p A A	DA31 ac42 ca54
	E_p	-2.42 -2.14	SCE	i_A/i_C	0.88 -	-	-	-	-	-	-	C,p A	
		-2.69 -1.99			0.84 -	-	-	-	-	-	-	C,p A	
JA094 7941	75c	$E_{\frac{1}{2}}$ -2.203	SCE	i_ℓ I	3.88F 6.50	-	-	Tomeš	58	sttd	-	$C, i_\ell \propto h^{0.41}, r$	DA32 ac43
		-1.464 -2.30	Ag/AgI	i	3.41F 0.23F	-	-	- $dE_{\frac{1}{2}}/dpH$	- <0	- sttd	-	C,r C,Pr,i_{cat},H	
		-2.494			0.23F	-	-	-	-	-	-	C	
		-2.190F		i_ℓ/C	8.33F	-	-	-	-	-	-	C,r	
		-2.136F			9.8F	-	-	-	-	-	-	C,r	
		-2.020F			10.25F	-	-	-	-	-	-	C,r	
		-1.940F			10.3F	-	-	-	-	-	-	C,r	
JA094 7941	75c	$E_{\frac{1}{2}}$ -2.337	SCE	i_ℓ I	3.29F 3.39	-	-	Tomeš	54	sttd	-	$C, i_\ell \propto h^{0.54}, \log i = k_1 - k_2 t, r$	DA33 ac44 ca55
		-		i_ℓ/C	3.8F	-	-	-	-	-	-	C,r	
		-2.330F			4.0F	-	-	-	-	-	-	C,r	
		-2.297F			4.5F	-	-	-	-	-	-	C,r	
		-2.260F			6.3F	-	-	-	-	-	-	C,r	
		-2.224F			7.0F	-	-	-	-	-	-	C,r	
		-2.188F			7.1F	-	-	-	-	-	-	C,r	
		-		i_ℓ	1.88F 1.53F	-	-	-	-	-	-	C,p C	
		-			4.0F	-	-	-	-	-	-	C,p	
JA095 1703	-	$E_{\frac{1}{2}}$ -2.36	Ag/0.01 Ag$^+$	I	2.9	sttd	1.08	αn_a	0.30	Tomeš	-	C,r	DA34
JA094 8471	-	$E_{\frac{1}{2}}$ -2.37	SCE	-	-	-	-	αn_a	0.6	Elog	-	C,r	DA35
JA095 0991	89c	E_{su} -5.11F	SCE	i_{su}	0.53F	-	-	-	-	-	-	C,r	DA36 ac66
JA095 0991	89c	E_{su} -1.50	SCE	i_{su}	0.60	-	-	-	-	-	-	C,r	

CONT

DA36 (CONT.) $C_4H_5N_3O$

Code No.	Empirical Formula	Name and WLN Code	Structural Formula	Solvent	Tech.	Medium		μ, M	pH	T, °C	Electrodes	App.	Experimental Parameters
DA36 ac66	$C_4H_5N_3O$	cytosine		H_2O	PVI	ACET	?	0.5	4.2	25	DME/SCE	235A 03	C=0.97, f=500, Δe=3.5 rms, Pt Aux
													f=1250
						MB		0.13	5.0	0			C=0.1, f=50, Δe=3.5 rms
					VR	ACET	?	0.5	4.1	25.0 ±0.1	HMDE SCE	235A 03	C=0.97, v=19, Pt Aux
													v=100
													v=1E3
													v=4.5E3
													v=1E4
DA37	$C_4H_6HgO_4$	bis(methoxycarbonyl)-mercury(II) C.A. 10507-39-8 10V-HG-V01	$Hg(COOCH_3)_2$	MeOH 50	PY	$LiClO_4$		0.1	-	-	25	DME/SCE	0-0	m=0.62, t=3.00
DA38 ac83	C_4H_6O	crotonaldehyde C.A. 123-73-9 VH1U2 -T	trans-CH_3CH:CHCHO	DMF	PY	Bu_4NBF_4		0.50	-	-	25	DME/SCE	23-0	C=1.4-1.7, Pt Aux
DA39	C_4H_6OS	divinyl sulfoxide C.A. 1115-15-7 1U1SO&1U1	$SO(CH:CH_2)_2$	MeOH 20	PY	LiCl		0.07	-	-	25	DME/SCE	0-0	C=0.5, m=1.63, t=2.5
DA40 ca68	$C_4H_6O_2$	methyl 2-propenoate C.A. 96-33-3 1U1VO1	CH_2:CHCOOCH$_3$	MeOH 92	PY	Et_4NI		0.02	-	-	25±1	DME/SCE	---	t=6.8, m=0.733, h=64
				MeCN	PY	Bu_4NClO_4		0.1	-	-	23	DME Ag/0.01 $AgNO_3$	2A0	C=2, Pt Aux
DA41	$C_4H_6O_2S$	divinyl sulfone C.A. 77-77-0 1U1SW1U1	$SO_2(CH:CH_2)_2$	MeOH 20	PY	LiCl		0.07	-	-	25	DME/SCE	0-0	C=0.5, m=1.63, t=2.5
DA42	$C_4H_7NO_4$	2-methyl-2-nitrato-propanal C.A. WNOX1&1&VH	$(CH_3)_2C(ONO_2)$-CHO	H_2O	PY	HCl KCl		0.1 1	-	1	25± 0.1	DME/SCE	0-0	EC=2.10
						UB KCl		1		2.9				
										3.8				
										4.2				
										4.4				
CONT										6.0				

TABLE I. Electrochemical Data $C_4H_7NO_4$ (CONT.) DA42

Ref.	C/M	Charact. Potential Value	vs.	Response Const. Value		n Tech.	n	Electrokinetic Data Parameter	Value	From	Products and Identification	Description and Remarks	Code No.
JA095 0991	89c	E_{su} -1.53	SCE	i_{su}	0.60	-	-	-	-	-	-	c,r	DA36 ac66
		-1.54			0.62	-	-	-	-	-	-	c,r	
		-	-	-	-	-	-	-	-	-	-	o,p	
JA095 0991	89c	-	-	$i_p/ACv^{\frac{1}{2}}$	46.3	-	-	-	-	-	-	c,D=11,r	
		E_p -1.48F	SCE		31.8F	-	-	-	-	-	-	c,r	
		-1.54F			29.2F	-	-	-	-	-	-	c,r	
		-1.57F			-	-	-	-	-	-	-	c,r	
		-			26.0F	-	-	-	-	-	-	c,r	
ER002 0587	23d	$E_{\frac{1}{2}}$ -1.31	SCE	-	-	sttd	2	α	0.42	sttd	-	c,≠,r	DA37
JA094 8471	-	$E_{\frac{1}{2}}$ -1.91	SCE	-	-	-	-	αn_a	0.8	Elog	-	c,r	DA38 ac83
DS171 0860	-	$E_{\frac{1}{2}}$ -2.05	SCE	-	-	llk	≈2	αn_a	0.5	Elog	ethene	c,i_d,≠,r	DA39
UK031 0392	-	$E_{\frac{1}{2}}$ -1.87	SCE	-	-	-	-	-	-	-	-	c,p	DA40 ca68
JA095 1703	-	$E_{\frac{1}{2}}$ -2.46	Ag/0.01 AgNO$_3$	I	1.95	sttd	0.73	αn_a	0.48	Tomeš	-	c,r	
DS171 0860	43a	$E_{\frac{1}{2}}$ -1.55	SCE	I	3.04	llk	≈2	αn_a	0.5	Elog	ethene	c,i_d,≠,r	DA41
ER001 0759	-	$E_{\frac{1}{2}}$ -0.49	SCE	i_ℓ /C	5.6	TLQE llk	2	-	-	-	CP→NO$_2^-$,PY	c,i_d,r	DA42
		-0.97			10.8		-					c,i_d,NO$_2^-$ redn.	
		-0.50			5.3		2	-	-	-	-	c,i_d,r	
		-1.08			9.9		-					c,i_d,NO$_2^-$ redn.	
		-0.50			4.9		2	-	-	-	-	c,i_d,r	
		-1.22			9.5		-					c,i_d,NO$_2^-$ redn.	
		-1.51			7.9		-					c,i_d,NO$_2^-$ redn.	
		-0.48			4.9		2	-	-	-	-	c,i_d,r	
		-1.23			8.9		-					c,i_d,NO$_2^-$ redn.	
		-1.53			7.9		-					c,i_d,NO$_2^-$ redn.	
		-0.47			5.2		2	-	-	-	-	c,i_d,r	
		-1.25			9.2		-					c,i_d,NO$_2^-$ redn.	
		-1.54			8.2		-					c,i_d,NO$_2^-$ redn.	
		-0.49			4.9		2	-	-	-	-	c,i_d,r	
		-			-		-					o	
		-1.60			11.1							c,i_d	CONT

DA42 (CONT.) $C_4H_7NO_4$

Code No.	Empirical Formula	Name and WLN Code	Structural Formula	Solvent	Tech.	Medium		μ, M	pH	T, °C	Electrodes	App.	Experimental Parameters
DA42	$C_4H_7NO_4$	2-methyl-2-nitrato-propanal	$(CH_3)_2C(ONO_2)-CHO$	H_2O	PY	UB KCl	? 1	-	11.0	25± 0.1	DME/SCE	0-0	EC=2.10
DA43	$C_4H_7NO_5$	2-methyl-2-nitrato-propanoic acid C.A. 1617-35-2 WNOX1&1&VQ	$(CH_3)_2C(ONO_2)-COOH$	H_2O	PY	HCl KCl	0.1 1	-	1	25± 0.1	DME/SCE	0-0	EC=2.10
						UB KCl	? 1		2.9				
									3.8				
									4.2				
									4.4				
									6.0				
									7.0				
									11.0				
DA44	$C_4H_8N_2O_3$	ethyl N-methyl-N-nitrosocarbamate C.A. 615-53-2 ONN1&VO2	$C_2H_5OCON(CH_3)NO$	H_2O	PT	BR		-	2.0	-	DME/SCE	O4AO	C=0.05, m= 1.73, t=1, h=80, v=10
									4.0				
									7.0				
DA45	C_4H_8O	ethyl vinyl ether C.A. 109-92-2 2O1U1	$H_2C:CHOC_2H_5$	MeOH	VY	$NaClO_4$ Na tosylate	?	-	-	25	RCCE SCE	06-0	C=1, A=0.8
				MeCN	VY	$NaClO_4$	?	-	-	25	RCCE SCE	06-0	C=1, v=20, ω=ns, A=0.8
											RPE/SCE		A=ns
DA46	$C_4H_8O_2S$	ethyl vinyl sulfone C.A. 1889-59-4 2SW1U1	$C_2H_5SO_2CH:CH_2$	MeOH 20	PY	LiCl	0.07	-	-	25	DME/SCE	0-0	C=0.5, m= 1.63, t=2.5
DA47	This Code no. not used												

TABLE I. Electrochemical Data

Ref.	C/M	Charact. Potential		Response Const.		n Tech.	n	Electrokinetic Data			Products and Identification	Description and Remarks	Code No.
		Value	vs.		Value			Parameter	Value	From			
ER001 0759	-	$E_{\frac{1}{2}}$	-1.65	SCE	i_ℓ/C	2.3	TLQE llk	2	-	-	-	-	C, i_d, r	DA42
ER001 0759	-	$E_{\frac{1}{2}}$	-0.57	SCE	i_ℓ/C	5.8	QE llk	2	-	-	-	$CP \to NO_2^-, PY$	C, i_d, r	DA43
			-0.96			12.0		-					C, i_d	
			-0.67			4.2		2					C, i_d, r	
			-1.10			11.3		-					C, i_d	
			-0.77			4.2		2					C, i_d, r	
			-1.18			11.7		-					C, i_d	
			-			-							C, X	
			-0.83			4.9		2					C, i_d, r	
			-1.23			12.1		-					C, i_d	
			-1.53			7.3							C, i_d	
			-0.84			5.8		2					C, i_d, r	
			-1.25			11.6		-					C, i_d	
			-1.53			9.7							C, i_d	
			-0.87			5.8		2					C, i_d, r	
			-			-							0	
			-1.56			10.6							C, i_d	
			-0.85			6.0		2					C, i_d, r	
			-			-							0	
			-0.85			6.2		2					C, i_d, r	
AA078 0081	-	$E_{\frac{1}{2}}$	-0.58F	SCE	$i_\ell(u)$	98F	-	-	-	-	-	-	C, r	DA44
			-0.74F			68F	-	-					C, r	
			-1.0F			48F							C	
			-1.4F			46F							C	
			-0.80F			50F	-	-					C, r	
			-1.03F			35F							C	
			-1.45F			30F							C	
EA017 1595	372 b	$E_{\frac{1}{2}}$	1.460	SCE	-	-	i:i	2.1	-	-	-	-	A, p	DA45
			1.510					2.1	-	-	-	-	A, p	
EA017 1595	372 a	$E_{\frac{1}{2}}$	1.520	SCE	-	-	i:i	2	-	-	-	-	A, p	
			1.900		-	-		2	-	-	-	-	A, p	
DS171 0860	43a	$E_{\frac{1}{2}}$	-1.79	SCE	I	3.04	llk	≈2	αn_a	0.5	Elog	ethene	$C, i_d, \neq, r$	DA46
													This Code no. not used	DA47

Code No.	Empirical Formula	Name and WLN Code	Structural Formula	Solvent	Tech.	Medium	μ, M	pH	T, °C	Electrodes	App.	Experimental Parameters
DA48	C_4H_9NO	morpholine C.A. 110-91-8 T6M DOTJ	TABLE II.	m-cresol	PY	Et_4NClO_4 0.1	-	-	-	DME Ag/AgCl (0.1 Me_4NCl)	216AO	C=0.5-1.5, m=2.87, t=2.70, h=56, MP Aux
DA49 ad45	$C_4H_{11}N$	1-aminobutane C.A. 109-73-9 Z4	$CH_3(CH_2)_3NH_2$	m-cresol	PY	Et_4NClO_4 0.1	-	-	20±0.01	DME Ag/AgCl (0.1 Me_4NCl)	216AO	C=0.5-1.5, m=2.87, t=2.70, h=56, MP Aux
					ER	Et_4NClO_4 0.1	-	-	-	MP Ag/AgCl (0.1 Me_4NCl)	2156AO	C=0.5, A=3.87, i_a=43.3, Pt Aux
												C=1
												C=2
												C=0.5, i_a=26.4
												C=1.5
												C=0.3, i_a=16.8
												C=1.5
					VY	Et_4NClO_4 0.1	-	-	-	RPDE Ag/AgCl (0.1 Me_4NCl)	216AO	C=1, ω=30, A=0.08, MP Aux
DA50	$C_4H_{12}N_2$	tetramethylhydrazine C.A. 6415-12-9 1N1&N1&1	$(CH_3)_2NN(CH_3)_2$	MeCN	VR	$NaClO_4$ 0.1	-	-	-	AuBe SCE	2AO	C=0.2-0.9, d=0.15, v=53, MP Aux, PE
DA51	$C_5H_2N_2O_3$	2-cyano-5-nitrofuran C.A. 59-82-5 T5OJ BNW ECN	TABLE II.	H_2O	PY	BR	0.45	2.0	25±0.1	DME/SCE	OAO	C=0.2, m=1.18, t=0.26
								5.0				
								9.0				
								10.0				
								11.9				
				EtOH 10	PY	BR	0.45	2.0	25±0.1	DME/SCE	OAO	C=0.2, m=1.18, t=0.26
								4.0				
								5.0				
								7.0				
								9.0				
CONT								10.0				

TABLE I. Electrochemical Data $C_5H_2N_2O_3$ (CONT.) DA51

Ref.	C/M	Charact. Potential		Response	Const. Value	n Tech.	n	Electrokinetic Data			Products and Identification	Description and Remarks	Code No.
		Value	vs.					Parameter	Value	From			
AA072 0169	-	$E_{\frac{1}{2}}$	0.40	Ag/AgCl	I	0.408	-	-	-	-	-	-	$A, \neq, i_d, i \propto h^{\frac{1}{2}}, \underline{m}$-cresolate oxidn., $pK_a = 12.1, r$	DA48
AA072 0169	-	$E_{\frac{1}{2}}$	0.41	Ag/AgCl	I	0.419	QE	1	-	-	-	-	$A, \neq, \underline{m}$-cresolate oxidn., r	DA49 ad45
AA072 0169	-	$E_{\tau/4}$	0.350	Ag/AgCl	τ	8.8	-	-	-	-	-	-	$A, \tau_r/\tau_f = 0.32, i\tau^{\frac{1}{2}} =$ const., $\underline{m}$-cresolate oxidn., r	
			0.349			25.4	-	-	-	-	-	-	$A, \tau_r/\tau_f = 0.12, i\tau^{\frac{1}{2}} =$ const., $\underline{m}$-cresolate oxidn., r	
			0.336			95.0	-	-	-	-	-	-	$A, \tau_r/\tau_f = 0.05, i\tau^{\frac{1}{2}} =$ const., $\underline{m}$-cresolate oxidn., r	
			0.352			20.0	-	-	-	-	-	-	$A, \tau_r/\tau_f = 0.26, i\tau^{\frac{1}{2}} =$ const., $\underline{m}$-cresolate oxidn., r	
			0.337			139	-	-	-	-	-	-	$A, \tau_r/\tau_f = 0.06, i\tau^{\frac{1}{2}} =$ const., $\underline{m}$-cresolate oxidn., r	
			0.352			70.0	-	-	-	-	-	-	$A, \tau_r/\tau_f = 0.1, i\tau^{\frac{1}{2}} =$ const., $\underline{m}$-cresolate oxidn., r	
			0.338			313	-	-	-	-	-	-	$A, \tau_r/\tau_f = 0.04, i\tau^{\frac{1}{2}} =$ const., $\underline{m}$cresolate oxidn., r	
AA072 0169	-	$E_{\frac{1}{2}}$	0.45F	Ag/AgCl	i_ℓ	5.0F	-	-	-	-	-	-	$A, \neq, \underline{m}$-cresolate oxidn., $i \propto \omega^{\frac{1}{2}}, r$	
JA094 7108	-	XE	0.22	SCE	-	-	sttd	1	ΔE_p	90	sttd	radical cation, ESR	$A, R, XE = \frac{1}{2}(E_{p,A} + E_{p,C})$, r	DA50
			-					1		-			c	
LV670 0023	-	$E_{\frac{1}{2}}$	-0.08	SCE	i_ℓ	1.22	IIk	4	-	-	-	2-cyano-5-hydroxyl-aminofuran	C, r	DA51
			-0.21			1.04		4	-	-	-	-	C, r	
			-0.41			0.97		4	-	-	-	-	C, r	
			-0.45			0.92		4	-	-	-	-	C, r	
			-0.48			0.90		4	-	-	-	-	C, r	
LV670 0023	-	$E_{\frac{1}{2}}$	-0.10	SCE	i_ℓ	1.10	IIk	4	-	-	-	-	C, r	
			-0.20			1.17		4	-	-	-	-	-	
			-0.25 -1.17			1.17 0.58		-	-	-	-	-	C, r c	
			-0.36 -1.16			1.18 0.52		-	-	-	-	-	C, r c	
			-0.48 -1.15			1.10 0.52		-	-	-	-	-	C, r c	
			-0.50 -1.12			1.00 0.40		-	-	-	-	-	C, r c	CONT

DA51 (CONT.) $C_5H_2N_2O_3$

Code No.	Empirical Formula	Name and WLN Code	Structural Formula	Solvent	Tech.	Medium	μ, M	pH	T, °C	Electrodes	App.	Experimental Parameters
DA51	$C_5H_2N_2O_3$	2-cyano-5-nitrofuran		EtOH 10	PY	BR	0.45	12.0	25±0.1	DME/SCE	OAO	C=0.2, m=1.18, t=0.26
				EtOH 20	PY	BR	0.45	2.0	25±0.1	DME/SCE	OAO	C=0.2, m=1.18, t=0.26
								3.0				
								5.0				
								6.0				
								7.0				
								8.0				
								10.0				
								12.0				
				EtOH 30	PY	BR	0.45	2.0	25±0.1	DME/SCE	OAO	C=0.2, m=1.18, t=0.26
								3.0				
								5.0				
								6.0				
								7.0				
								8.0				
								9.0				
				EtOH 50	PY	BR	0.45	2.0	25±0.1	DME/SCE	OAO	C=0.2, m=1.18, t=0.26
								3.0				
								5.0				
								6.0				
								7.0				
								9.0				
								11.0				
								12.0				
DA52	$C_5H_3NO_3S$	4-nitro-2-thienyl-carboxaldehyde C.A. 57500-53-5 T5SJ BVH DNW	TABLE II.	H_2O	PY	PRW	-	10.21	-	DME/-	---	C=1
DA53 ad67	$C_5H_3NO_3S$	5-nitro-2-thienyl-carboxaldehyde C.A. 4521-33-9 T5SJ BVH ENW	TABLE II-2.	H_2O	PY	PRW	-	10.21	-	DME/-	---	C=1
DA54 ad68	$C_5H_3NO_4$	5-nitro-2-furaldehyde C.A. 698-63-5 T5OJ BVH ENW	TABLE II-2.	H_2O	PY	BR	0.45	2.0	25±0.1	DME/SCE	OAO	C=0.2, m=1.18, t=0.26
								3.0				

CONT

TABLE I. Electrochemical Data $C_5H_3NO_4$ (CONT.) DA54

Ref.	C/M	Charact. Potential		Response Const.		n Tech.		Electrokinetic Data			Products and Identification	Description and Remarks	Code No.
		Value	vs.		Value			Parameter	Value	From			
LV670 0023	–	$E_{\frac{1}{2}}$	-0.52 -1.08	SCE	i_ℓ	0.85 0.31	–	–	–	–	–	–	C,r C	DA51
LV670 0023	–	$E_{\frac{1}{2}}$	-0.13	SCE	i_ℓ	1.53	IIk	4	–	–	–	–	C,r	
			-0.18			1.60		4	–	–	–	–	C,Mc($E_{\frac{1}{2}}$),r	
			-0.25			1.54	–	–	–	–	–	–	C,r	
			-0.32			1.10	–	–	–	–	–	–	C,r	
			-0.38			0.70	–	–	–	–	–	–	C,Mc($E_{\frac{1}{2}}$),r	
			-0.45			0.50	–	–	–	–	–	–	C,r	
			-0.49			0.40	–	–	–	–	–	–	C,r	
			-0.50			0.30	–	–	–	–	–	–	C,r	
LV670 0023	–	$E_{\frac{1}{2}}$	-0.13	SCE	i_ℓ	1.30	IIk	4	–	–	–	–	C,r	
			-0.18			1.40	–	–	–	–	–	–	C,r	
			-0.31			1.40	–	–	–	–	–	–	C,r	
			-0.38			1.10	–	–	–	–	–	2-cyano-5-hydroxyl-aminofuran	C,r	
			-0.70			0.30						2-cyano-5-aminofuran	C	
			-0.41 -0.66			0.70 0.88	–	–	–	–	–	–	C,r C	
			-0.44 -0.66			0.44 0.90	–	–	–	–	–	–	C,r C	
			-0.44 -0.66			0.30 0.92	–	–	–	–	–	–	C,r C	
LV670 0023	–	$E_{\frac{1}{2}}$	-0.18	SCE	i_ℓ	1.15	IIk	4	–	–	–	–	C,r	
			-0.23			1.14		4	–	–	–	–	C,r	
			-0.39			1.10	–	–	–	–	–	–	C,r	
			-0.42 -0.71			0.60 0.30	–	–	–	–	–	–	C,r C	
			-0.45 -0.71			0.40 0.60	–	–	–	–	–	–	C,r C	
			-0.48 -0.69			0.30 0.80	–	–	–	–	–	–	C,r C	
			-0.49 -0.70			0.25 0.75	–	–	–	–	–	–	C,r C	
			-0.50 -0.71			0.23 0.15	–	–	–	–	–	–	C,r C	
BF630 0479	–	$E_{\frac{1}{2}}$	-0.59F	–	i_ℓ	14.6F	sttd	4	–	–	–	–	C,W,r	DA52
BF630 0479	–	$E_{\frac{1}{2}}$	-0.36F	–	i_ℓ	13.4F	sttd	4	–	–	–	–	C,W,r	DA53 ad67
LV670 0023	–	$E_{\frac{1}{2}}$	–	SCE	i_ℓ	–	–	–	–	–	–	–	O,r	DA54 ad68
			-0.15 -0.63 -1.14			0.32 0.46 0.67							C C C	
			-0.09			0.42	–	–	–	–	–	5-hydroxylamino-furaldehyde	C,r	
			-0.19			0.35						5-hydroxylamino-furaldehyde geminal diol	C	
			-0.70			0.40						5-aminofuraldehyde	C	
			-1.20			0.65						5-amino-2-hydroxy-furan	C	

CONT

DA54 (CONT.) $C_5H_3NO_4$

Code No.	Empirical Formula	Name and WLN Code	Structural Formula	Solvent	Tech.	Medium	μ, M	pH	T, °C	Electrodes	App.	Experimental Parameters
DA54 ad68	$C_5H_3NO_4$	5-nitro-2-furaldehyde		H_2O	PY	BR	0.45	4.0	25±0.1	DME/SCE	OAO	c=0.2, m=1.18, t=0.26
								5.0				
								6.0				
								8.0				
								10.0				
								11.0				
								11.9				
				EtOH 10	PY	BR	0.45	2.0	25±0.1	DME/SCE	OAO	c=0.2, m=1.18, t=0.26
								3.0				
								4.0				
								5.0				
								6.0				
								7.0				
								9.0				
								10.0				
								11.0				
								12.0				
DA55 ad71	$C_5H_3NO_5$	5-nitro-2-furoic acid C.A. 645-12-5 T5OJ BVQ ENW	TABLE II-2.	H_2O	PY	BR	0.45	2.0	25±0.1	DME/SCE	OAO	c=0.2, m=1.18, t=0.26
								4.0				

TABLE I. Electrochemical Data $C_5H_3NO_5$ (CONT.) DA55

Ref.	C/M	Charact. Potential Value	vs.	Response Const. Value		n Tech.		Electrokinetic Data Parameter	Value	From	Products and Identification	Description and Remarks	Code No.
LV670 0023	-	$E_{\frac{1}{2}}$	-0.12 -0.22 -0.80 -1.25	SCE	i_ℓ	0.45 0.36 0.37 0.60	-	-	-	-	-	-	C,r C C C	DA54 ad68
			-0.15 -0.27 -0.88 -1.31			0.47 0.38 0.40 0.50	-	-	-	-	-	-	C,r C C,Mc($E_{\frac{1}{2}}$) C	
			-0.18 -0.34 -0.98 -1.36			0.47 0.42 0.43 0.35	-	-	-	-	-	-	C,r C C C	
			-0.24 -0.45 -1.08 -1.52			0.43 0.42 0.48 0.35	-	-	-	-	-	-	C,i_ℓ=f(t) for pH>8,r C C C	
			-0.32 -0.55 -1.20 -1.63			0.41 0.39 0.47 0.30	-	-	-	-	-	-	C,i_ℓ=f(t) for pH>8,r C C C	
			-0.34 -0.56 - -			0.35 0.28 0.40 0.20	-	-	-	-	-	-	C,i_ℓ=f(t) for pH>8.r C C C	
			-			-	-	-	-	-	-	-	O,r	
LV670 0023	-	$E_{\frac{1}{2}}$	- -0.16 -0.72 -1.20	SCE	i_ℓ	- 0.47 0.65 0.40	-	-	-	-	-	-	O,r C C,Mc(i_ℓ) C	
			-0.10 -0.22 -0.80 -1.23			0.37 0.56 0.62 0.34	-	-	-	-	-	-	C,r C C C	
			-0.13 -0.28 -0.86 -1.33			0.42 0.55 0.60 0.33	-	-	-	-	-	-	C,r C C C	
			-0.17 -0.33 -0.92 -1.40			0.45 0.54 0.53 0.33	-	-	-	-	-	-	C,r C C C	
			-0.21 -0.41 -0.98 -1.43			0.46 0.51 0.41 0.35	-	-	-	-	-	-	C,r C C C	
			-0.24 -0.44 -1.04 -1.50			0.45 0.42 0.33 0.48	-	-	-	-	-	-	C,r C C C	
			-0.30 -0.55 -1.13 -1.60			0.47 0.32 0.31 0.52	-	-	-	-	-	-	C,i_ℓ=f(t) for pH>8,r C C C	
			-0.34 -0.56 -1.18 -1.70			0.45 0.27 0.31 0.40	-	-	-	-	-	-	C,i_ℓ=f(t) for pH>8,r C C C	
			-0.36 -0.56 -1.20 -1.40			0.40 0.20 0.24 0.18	-	-	-	-	-	-	C,i_ℓ=f(t) for pH>8,r C C C	
			-			-	-	-	-	-	-	-	O,r	
LV670 0023	-	$E_{\frac{1}{2}}$	-0.09	SCE	i_ℓ	0.76	llk	4	-	-	-	-	C,r	DA55 ad71
			-0.20			0.84		4	-	-	-	-	C,r	

Code No.	Empirical Formula	Name and WLN Code	Structural Formula	Solvent	Tech.	Medium	μ, M	pH	T, °C	Electrodes	App.	Experimental Parameters
DA55 ad71	$C_5H_3NO_5$	5-nitro-2-furoic acid		H_2O	PY	BR	0.45	5.0	25±0.1	DME/SCE	OAO	C=0.2, m=1.18, t=0.26
								6.0				
								7.0				
								8.0				
								10.0				
								12.0				
				EtOH 10	PY	BR	0.45	2.0	25±0.1	DME/SCE	OAO	C=0.2, m=1.18, t=0.26
								3.0				
								5.0				
								6.0				
								7.0				
								8.0				
								10.0				
								12.0				
DA56	$C_5H_4N_2O_3$	2-hydroxy-3-nitro-pyridine C.A. 6332-56-5 T6NJ BQ CNW	TABLE II.	H_2O	PY	BR	-	2.4	-	DME/MP?	OAO	C=0.2, m=1.64, t=5.54, h=35
								3.0				
								4.0				
								6.2				
								8.2				
								10.0				
DA57	$C_5H_4N_2O_3$	2-hydroxy-5-nitro-pyridine C.A. 6191-11-3 T6NJ BQ ENW	TABLE II.	H_2O	PY	BR	-	2.4	-	DME/MP?	OAO	C=0.2, m=1.64, t=5.54, h=35
								3.0				
								4.0				
								6.2				
								8.2				
								10.0				

TABLE I. Electrochemical Data $C_5H_4N_2O_3$ DA57

Ref.	C/M	Charact. Potential Value	vs.	Response Const. Value		Tech.	n	Electrokinetic Data Parameter	Value	From	Products and Identification	Description and Remarks	Code No.
LV670 0023	-	$E_{\frac{1}{2}}$ -0.25	SCE	i_ℓ	0.86	llk	4	-	-	-	-	C,r	DA55 ad71
		-0.30			0.89		4	-	-	-	5-hydroxylamino-2-furoic acid	C,r	
		-1.55			0.22		-	-	-	-	5-aminofuroic acid	C	
		0.36			0.99		4	-	-	-	-	C,r	
		-1.55			0.21		-	-	-	-		C	
		-0.41			0.88		4	-	-	-	-	C,r	
		-1.54			0.28		-	-	-	-		C	
		-0.53			0.87		4	-	-	-	-	C,r	
		-1.55			0.28		2	-	-	-		C	
		-0.56			0.85		4	-	-	-	-	C,r	
		-1.55			0.28		2	-	-	-		C	
LV670 0023	-	$E_{\frac{1}{2}}$ -0.10	SCE	i_ℓ	0.81	llk	4	-	-	-	-	C,r	
		-0.17			0.90		4	-	-	-	-	C,r	
		-0.30			0.82		-	-	-	-	-	C,r	
		-0.38			0.62		-	-	-	-	-	C,r	
		-0.86			0.36			-	-	-	-	C	
		-0.45			0.28		-	-	-	-	-	C,r	
		-0.86			0.55			-	-	-	-	C	
		-0.51			0.12		-	-	-	-	-	C,r	
		-0.86			0.60			-	-	-	-	C	
		-0.54			0.1		-	-	-	-	-	C,r	
		-0.86			0.58			-	-	-	-	C	
		-0.54			0.07		-	-	-	-	-	C,r	
		-0.88			0.57			-	-	-	-	C	
BJ048 2176	-	$E_{\frac{1}{2}}$ -0.20F	SCE	-	-	-	-	$dE_{\frac{1}{2}}/dpH$	78	plot	-	C, $i \propto C$ for C=0.05-2, $i_d, i_\ell \neq f(pH), W, r$	DA56
		-0.90							150			C,X	
		-0.26F		-	-	-	-		78		-	C, $i \propto C$ for C=0.05-2, $i_d, i_\ell \neq f(pH), W, r$	
		-1.04F							150			C,X	
		-0.34F		-	-	-	-		78		-	C, $i \propto C$ for C=0.05-2, $i_d, i_\ell \neq f(pH), W, r$	
		-1.20F							150			C,X,0 for pH > 5	
		-0.50F		-	-	-	-		68		-	C, $i \propto C$ for C=0.05-2, $i_d, i_\ell \neq f(pH), W, r$	
		-0.63F		-	-	-	-		-		-	C, $i \propto C$ for C=0.05-2, $i_d, i_\ell \neq f(pH), W, r$	
		-0.76F		-	-	-	-		68		-	C, $i \propto C$ for C=0.05-2, $i_d, i_\ell \neq f(pH), W, r$	
BJ048 2176	-	$E_{\frac{1}{2}}$ -0.30F	SCE	-	-	-	-	$dE_{\frac{1}{2}}/dpH$	78	plot	-	C, i_d, W, r	DA57
		-0.36F		-	-	-	-				-	C, i_d, W, r	
		-0.44F		-	-	-	-				-	C, i_d, W, r	
		-0.60F		-	-	-	-	$dE_{\frac{1}{2}}/dpH$	53	plot	-	C, i_d, W, r	
		-0.70F		-	-	-	-		53		-	C, W, r	
		-1.0F							0			C, W, $\Sigma i_\ell \neq f(pH)$	
		-0.81F		-	-	llk	-		53			C, W, r	
		-1.0F							0			C, W, $\Sigma i_\ell \neq f(pH)$	

DA58 $C_5H_4N_2O_4$

Code No.	Empirical Formula	Name and WLN Code	Structural Formula	Solvent	Tech.	Medium	μ, M	pH	T, °C	Electrodes	App.	Experimental Parameters
DA58 ad94	$C_5H_4N_2O_4$	uracil-6-carboxylic acid C.A. 65-86-1 T6N CNJ BQ DQ FVQ	TABLE II-2.	H_2O	PY	BR	-	2.93	25± 0.1	DME/SCE	2-0	C=0.5, m=2.23, t=3.38, h=70, Pt Aux
								4.36				
								5.83				
								7.56				
								10.20				
DA59 ad95 ca92	$C_5H_4N_4$	purine C.A. 120-73-0 T56 BM DN FN HNJ	TABLE II-2.	MeCN	PY	Et_4NClO_4 0.1	-	-	25± 0.1	DME Ag/0.1 $AgClO_4$, 0.1 Et_4NClO_4	25A0	C=0→2, EC= 3.14, Pt Aux
						Et_4NClO_4 0.1 $HClO_4$						
						Et_4NClO_4 0.1 ClACET						
						Et_4NClO_4 0.1 C_6H_5COOH						
					VR	Et_4NClO_4 0.1	-	-	25± 0.1	HMDE Ag/0.1 $AgClO_4$, 0.1 Et_4NClO_4	25A0	C=?, v=20, Pt Aux
				DMF	PY	Et_4NClO_4 0.1	-	-	25± 0.1	DME Ag/0.1 $AgClO_4$, 0.1 Et_4NClO_4	25A0	C=0→2, EC= 3.14, Pt Aux
					VR	Et_4NClO_4 0.1	-	-	25± 0.1	HMDE Ag/0.1 $AgClO_4$, 0.1 Et_4NClO_4	25A0	C=?, v=20, Pt Aux
DA60 ca93	$C_5H_4N_4O$	allopurinol C.A. 315-30-0 T56 BMN GN INJ FQ	TABLE II-3.	H_2O	PY	H_2SO_4 1	1.0	0	25± 0.1	DME/SCE	2A0	C=0.4-1.0, Pt Aux
						ACET ?	0.5	4.7				
CONT				H_2O		MB		6.0				

TABLE I. Electrochemical Data $C_5H_4N_4O$ (CONT.) DA60

Ref.	C/M	Charact. Potential		Response Const.		n Tech.	Electrokinetic Data			Products and Identification	Description and Remarks	Code No.
		Value	vs.		Value		Parameter	Value	From			
BJ048 0435	-	$E_{\frac{1}{2}}$	-0.92F	SCE	i_ℓ	3.7F	QE 2.17	Tomeš αn_a } $p=$	40 1.29 2.36	- Tomeš $dE_{\frac{1}{2}}$/dpH	-	$C,i_d,\frac{1}{4},r$	DA58 ad94
		-1.11F			2.0F	2.17	}	39 1.33 2.43		-	$C,i_d,\frac{1}{4},r$	
		-1.32F			1.4F	1.83	Tomeš αn_a	38 1.36	- Tomeš	-	$C,i_d,\frac{1}{4}$	
		- -1.32F			- 2.4F	- 1.83	- Tomeš αn_a }	- 32 1.61	- Tomeš	-	O,r $C,i_d,\frac{1}{4}$	
		-1.56F			1.4F	1.86	}	50 1.03			$C,i_d,\frac{1}{4}$	
		-1.32F			0.2F	1.83		-		-	$C,i_d,\frac{1}{4},r$	
		-1.56F			3.2F	1.86	Tomeš αn_a	57 0.90	- Tomeš		$C,i_d,\frac{1}{4}$	
		- -1.56F			- 2.0	- -	- Tomeš αn_a	- 51 1.01	- Tomeš	-	O,r $C,i_d,\frac{1}{4}$	
BJ047 2650	113 j	$E_{\frac{1}{2}}$	-2.16	Ag/Ag$^+$	I	1.60	- -	Tomeš	80	-	-	$C, i \propto C, i_d, E_{\frac{1}{2}} = -1.64 + 0.16 \log[CH_3COOH], E_{\frac{1}{2}} = -2.11 + 0.09 \log [H_2O], r$	DA59 ad95 ca92
	113 l		-1.07 -1.24 -1.84		-	-	- -	-	-	-	-	C,r $C, i \propto [HClO_4]$ $C, i \propto [HClO_4]$	
	113 k		-1.89		-	-	- -	-	-	-	-	$C, i \propto [acid], r$	
	113 k		-2.01		-	-	- -	-	-	-	-	$C, i \propto [acid], r$	
BJ047 2650	113 j	E_p	-2.05 -0.56 -0.52 -0.44	Ag/AgClO$_4$	-	-	- -	-	-	-	-	C,r A C A	
BJ047 2650	113 j	$E_{\frac{1}{2}}$	-2.38	Ag/AgClO$_4$	I	1.03	- -	Tomeš	110	-	radical anion; ESR	$C, i \propto C, i_d, E_{\frac{1}{2}} = -1.86 + 0.15 \log[HClO_4], E_{\frac{1}{2}} = -1.92 + 0.10 \log [CH_2ClCOOH], E_{\frac{1}{2}} = -2.02 + 0.10 \log [C_6H_5COOH], E_{\frac{1}{2}} = -2.29 + 0.03 \log [CH_3COOH], r$	
			-2.66			0.22					-	$C, i \propto C, i_d$	
BJ047 2650		E_p	-2.38 -2.52 -2.46 -0.72 -0.63 -0.52	Ag/AgClO$_4$	-	-	- -	-	-	-	-	C,r C A A C A	
AA058 0183	-	$E_{\frac{1}{2}}$	-1.118	SCE	i/C	10.3F	sttd 4	$dE_{\frac{1}{2}}$/dpH	67	sttd	-	$C,r?$	DA60 ca93
			-1.423			6.67F	2		67		-	$C,r?$	
			-1.520			5.58F	2		67		-	$C,r?$	

CONT

DA60 (CONT.) $C_5H_4N_4O$

Code No.	Empirical Formula	Name and WLN Code	Structural Formula	Solvent	Tech.	Medium		μ, M	pH	T, °C	Electrodes	App.	Experimental Parameters
DA60 ca93	$C_5H_4N_4O$	allopurinol		H_2O	IL	H_2SO_4	1	1.0	0	25± 0.1	PGE/SCE	2AO	Pt Aux
						ACET	?	0.5	4.7				
						MB			6.0				
DA61 ad99 ca95	$C_5H_4N_4O_3$	uric acid C.A. 69-93-2 T56 BM DN FN HNJ CQ GQ IQ	TABLE II-2.	H_2O	IL	H_2SO_4	1	1.0	0	25± 0.1	PGE/SCE	2AO	Pt Aux
						ACET	?	0.5	4.7				
						MB			6.0				
					VV	ACET	?	0.5	4.7	25± 0.1	PGE/SCE	2AO	C=0.5,Δe= 10,f=100, Pt Aux
						MB			6.0				
DA62	C_5H_5N	2,4-pentadiene-nitrile C.A. 1615-70-9 NC1U2U1	CH_2:CHCH:CHCN	DMF 80.0	PY	Bu_4NI	0.05	-	-	20	DME/MP?	OAO	C=1.114,EC= 3.08,t=1.77, h=35
				DMF 95.0	PY	Bu_4NI	0.05	-	-	20	DME/MP?	OAO	C=1.114,EC= 3.08,t=1.77, h=35
				DMF	PY	Bu_4NI	0.05	-	-	20	DME/MP?	OAO	C=1.114,EC= 3.08,t=1.77, h=35
DA63 ae06 cb04	C_5H_5N	pyridine C.A. 110-86-1 T6NJ	TABLE II-2.	MeCN	PY	Et_4NClO_4	0.1	-	-	-	DME Ag/AgI	235A O	C=0.87,h=38
						Et_4NClO_4 0.1 C_6H_5COOH 1.48E-3							
						Et_4NClO_4 0.1 C_6H_5COOH 7.6E-4							
					VR	Et_4NClO_4	0.1	-	-	-	MPDE Ag/AgI	235A O	ns
DA64	$C_5H_5NO_3$	2-methyl-5-nitro-furan C.A. 823-74-5 T5OJ BNW E1	TABLE II.	H_2O	PY	BR		0.45	2.0	25± 0.1	DME/SCE	OAO	C=0.2,m= 1.18,t=0.26
									3.0				
									4.0				
									6.0				
									8.0				
									10.0				
									11.9				
				EtOH 10	PY	BR		0.45	2.0	25± 0.1	DME/SCE	OAO	C=0.2,m= 1.18,t=0.26
									4.0				
									5.0				
									6.0				
CONT													

TABLE I. Electrochemical Data $C_5H_5NO_3$ (CONT.) DA64

Ref.	C/M	Charact. Potential Value	vs.	Response Const.	Value	Tech.	n	Electrokinetic Data Parameter	Value	From	Products and Identification	Description and Remarks	Code No.
AA058 0183	-	E_p 1.405	SCE	-	-	-	-	dE_p/dpH	51	sttd	-	A,r?	DA60 ca93
		1.165		-	-	-	-		51		-ª	A,r?	
		1.099		-	-	-	-		51		-	A,r?	
AA058 0183	-	E_p 0.71	SCE	-	-	-	-	-	-	-	-	A,$E_p=f(pH)$,r?	DA61 ad99 ca95
		0.57		-	-	-	-	-	-	-	-	A,$E_p=f(pH)$,r?	
		0.40		-	-	-	-	-	-	-	-	A,$E_p=f(pH)$,r?	
AA058 0183	-	E_p 0.44F	SCE	i_p	12.1F	-	-	-	-	-	-	A,r?	
		0.338			17.5F	-	-	-	-	-	-	A,r?	
ER003 0494	71f	-	-	l	1.6F	-	-	-	-	-	QP at i=0.6 amp→ polymer,78%;hydro-dimer,trace	C,p	DA62
		-			0.7F							C	
ER003 0494	71f	$E_{\frac{1}{2}}$ -	-	l	1.2F	sttd	1	-	-	-	-	C,p	
		-			1.0F		1					C	
ER003 0494	71e	-	-	l	1.0F	-	1	-	-	-	-	C,p	
		-			1.0F		1					C	
JA094 7941	7b	$E_{\frac{1}{2}}$ -2.66	SCE	l i_l	6.07 5.65F	-	-	Tomeš	67	sttd	-	C,r	DA63 ae06 cb04
		-			4.94F 0.70F	-	-	-	-	-	-	C,r C	
		-			2.70F 1.88F	-	-	-	-	-	-	C,r C	
JA094 7941	7b	E_p -3.05 -1.39 -1.4	Ag/AgI	-	-	-	-	-	-	-	-	C,r A C,0 on first scan	
LV670 0023	-	$E_{\frac{1}{2}}$ -0.22	SCE	i_l	1.30	IIk	4	-	-	-	5-methyl-2-hydroxyl-aminofuran	C,r	DA64
		-0.40			0.24		2				5-methyl-2-amino-furan	C	
		-0.26 -0.54			1.00 0.10		4 2				-	C,r C	
		-0.31			1.0		4	-	-	-	-	C,r	
		-0.46			1.00		-	-	-	-	-	C,r	
		-0.59			0.94		4	-	-	-	-	C,r	
		-0.65			0.90		4	-	-	-	-	C,r	
		-0.65			0.82		4	-	-	-	-	C,r	
LV670 0023		$E_{\frac{1}{2}}$ -0.24	SCE	i_l	0.80	IIk	4	-	-	-	-	C,r	
		-0.35			0.72	-	-	-	-	-	-	C,r	
		-0.44			0.70	-	-	-	-	-	-	C,r	
		-0.54 -0.90			0.69 0.16	sttd	6	-	-	-	-	C,r C,Mc($E_{\frac{1}{2}}$,i_l)	
													CONT

DA64 (CONT.) $C_5H_5NO_3$

Code No.	Empirical Formula	Name and WLN Code	Structural Formula	Solvent	Tech.	Medium	μ, M	pH	T, °C	Electrodes	App.	Experimental Parameters
DA64	$C_5H_5NO_3$	2-methyl-5-nitro-furan		EtOH 10	PY	BR	0.45	7.0	25±0.1	DME/SCE	OAO	C=0.2, m=1.18, t=0.26
								9.0				
								11.0				
								12.0				
				EtOH 20	PY	BR	0.45	2.0	25±0.1	DME/SCE	OAO	C=0.2, m=1.18, t=0.26
								4.0				
								5.0				
								6.0				
								7.0				
								8.0				
								10.0				
								11.0				
								12.0				
				EtOH 30	PY	BR	0.45	2.0	25±0.1	DME/SCE	OAO	C=0.2, m=1.18, t=0.26
								4.0				
								5.0				
								6.0				
								7.0				
								8.0				
								10.0				
								12.0				
				EtOH 50	PY	BR	0.45	2.0	25±0.1	DME/SCE	OAO	C=0.2, m=1.18, t=0.26
								4.0				
								5.0				
								6.0				
								7.0				
								9.0				
								11.0				
								12.0				
DA64	$C_5H_5NO_3$	2-methyl-5-nitro-furan		EtOH 10								

TABLE I. Electrochemical Data $C_5H_5NO_3$ DA64

Ref.	C/M	Charact. Potential		Response Const.		n Tech.	n	Electrokinetic Data			Products and Identification	Description and Remarks	Code No.
		Value	vs.		Value			Parameter	Value	From			
LV670 0023	-	$E_{\frac{1}{2}}$	-0.60 -0.94	SCE	i_ℓ	0.55 0.50	-	-	-	-	-	-	c,r c	DA64
			-0.63 -0.95			0.30 0.64	-	-	-	-	-	-	c,r c	
			-0.66 -0.94			0.20 0.70	-	-	-	-	-	-	c,r c	
			-0.56 -0.93			0.19 0.56	sttd	4	-	-	-	-	c,r c	
LV670 0023	-	$E_{\frac{1}{2}}$	-0.28	SCE	i_ℓ	1.80	-	-	-	-	-	-	c,r	
			-0.42			1.74	-	-	-	-	-	-	c,r	
			-0.53			1.60	-	-	-	-	-	-	c,r	
			-0.60 -1.00			1.40 0.70	-	-	-	-	-	-	c,r c	
			-0.64 -1.02			0.82 1.18	-	-	-	-	-	-	c,r c	
			-0.66 -1.03			0.50 1.53	-	-	-	-	-	-	c,r c	
			-0.66 -1.04			0.44 1.55	-	-	-	-	-	-	c,r c	
			-0.66 -1.03			0.44 1.46	-	-	-	-	-	-	c,r c	
			-0.66 -1.02			0.40 1.30	-	-	-	-	-	-	c,r c	
LV670 0023	-	$E_{\frac{1}{2}}$	-0.34	SCE	i_ℓ	1.48	llk	4	-	-	-	-	c,r	
			-0.52			1.48		4	-	-	-	-	c,r	
			-0.60			1.36	-	-	-	-	-	-	c,r	
			-0.64 -1.08			0.98 0.80	-	-	-	-	-	-	c,r c	
			-0.67 -1.09			0.58 1.11	-	-	-	-	-	-	c,r c	
			-0.67 -1.10			0.39 1.24	-	-	-	-	-	-	c,r c	
			-0.68 -1.08			0.38 1.30	-	-	-	-	-	-	c,r c	
			-0.68 -1.08			0.36 1.10	-	-	-	-	-	-	c,r c	
LV670 0023	-	$E_{\frac{1}{2}}$	-0.42	SCE	i_ℓ	1.40	llk	4	-	-	-	-	c,r	
			-0.61			1.35		4	-	-	-	-	c,r	
			-0.65 -1.10			1.20 0.35	-	-	-	-	-	-	c,r c	
			-0.68 -1.10			0.75 0.67	-	-	-	-	-	-	c,r c	
			-0.69 -1.10			0.43 0.82	-	-	-	-	-	-	c,r c	
			-0.70 -1.11			0.35 1.10	-	-	-	-	-	-	c,r c	
			-0.70 -1.11			0.36 1.00	-	-	-	-	-	-	c,r c	
			-0.70 -1.11			0.30 0.99	-	-	-	-	-	-	c,r c	

DA65 $C_5H_5NO_4$ CRC Handbook Series in Organic Electrochemistry

Code No.	Empirical Formula	Name and WLN Code	Structural Formula	Solvent	Tech.	Medium	μ, M	pH	T, °C	Electrodes	App.	Experimental Parameters
DA65 a111	$C_5H_5NO_4$	5-nitrofurfuryl alcohol C.A. 2493-04-1 T5OJ BNW E1Q	TABLE II-2.	EtOH 10	PY	BR	0.45	2.0	25± 0.1	DME/SCE	OAO	C=0.2, m=1.18, t=0.26
								4.0				
								5.0				
								6.0				
								7.0				
								8.0				
								11.0				
								12.0				
DA66	$C_5H_5N_3O_2$	2-amino-3-nitro-pyridine C.A. 4214-75-9 T6NJ BZ CNW	TABLE II.	H_2O	PY	BR	–	2.4	–	DME/MP?	OAO	C=0.2, m=1.64, t=5.54, h=35
								3.0				
								4.0				
								6.2				
								8.0				
								10.0				
DA67	$C_5H_5N_3O_2$	2-amino-5-nitro-pyridine C.A. 4214-76-0 T6NJ BZ ENW	TABLE II.	H_2O	PY	BR	–	2.4	–	DME/MP?	OAO	C=0.2, m=1.64, t=5.54, h=35
								4.0				
								6.0				
								8.0				
								10.0				
DA68	C_5H_6O	2-cyclopentenone C.A. 930-30-3 L5V BUTJ	TABLE II.	DMF	PY	Bu_4NBF_4 0.50	–	–	25	DME/SCE	23-0	C=2.2-2.6, Pt Aux
DA69	$C_5H_7ClO_4S_2$	3,5-dimethyl-1,2-dithiolylium perchlorate C.A. 12275-31-9 T5SSTJ C1 E1 & G-O4	TABLE II.	CH_2Cl_2	VR	Bu_4NBF_4 0.1	–	–	–	PBE/SCE	26A2	v=150

TABLE I. Electrochemical Data $C_5H_7ClO_4S_2$ DA69

Ref.	C/M	Charact. Potential		Response Const.		n		Electrokinetic Data			Products and Identification	Description and Remarks	Code No.
		Value	vs.		Value	Tech.		Parameter	Value	From			
LV670 0023	-	$E_{\frac{1}{2}}$	-0.21	SCE	i_ℓ	1.13	IIk	4	-	-	-	-	C,r	DA65 al11
		-0.31			1.13		4	-	-	-	-	C,r	
		-0.40			1.02	-	-	-	-	-	-	C,r	
		-0.51			0.68	-	-	-	-	-	-	C,r	
		-0.57 -1.07			0.47 1.04	-	-	-	-	-	-	C,r C	
		-0.58 -1.05			0.25 1.24	-	-	-	-	-	-	C,r C	
		-0.63 -0.77 -1.20			0.25 0.64 0.74	-	-	-	-	-	-	C,r C C	
		-0.66 -0.75 -1.24			0.22 0.44 0.64	-	-	-	-	-	-	C,r C C	
BJ048 2176	-	$E_{\frac{1}{2}}$	-0.14F	SCE	-	-	-	-	$dE_{\frac{1}{2}}/dpH$	97	plot	-	C,i∝C,i_d,i=f(pH), W,r	DA66
		-0.99F								47			C,X	
		-0.18F		-	-	-	-		97		-	C,i∝C,i_d,i=f(pH), W,r	
		-1.04								47			C,X	
		-0.30F		-	-	-	-		97		-	C,i∝C,i_d,i=f(pH), W,r	
		-1.12F								47			C,X	
		-0.51F		-	-	-	-		65		-	C,i∝C,i_d,i=f(pH), W,r	
		-1.17F								95			C,X	
		-0.64F		-	-	-	-		65		-	C,i∝C,i_d,i_ℓ=f(pH), W,r	
		-1.36F								95			C,X,0 for pH>9	
		-0.76F		-	-	-	-		65		-	C,i∝C,i_d,i_ℓ=f(pH), W,r	
BJ048 2176	-	$E_{\frac{1}{2}}$	-0.21F	SCE	-	-	-	-	-	-	-	-	C,i∝C,i_d,i_ℓ=f(pH), W,r	DA67
		-0.86F											C,X	
		-0.30F		-	-	-	-	-	-	-	-	C,i∝C,i_d,i_ℓ=f(pH), W,r	
		-1.11F											C,X	
		-0.49F		-	-	-	-	$dE_{\frac{1}{2}}/dpH$	69	plot	-	C,i∝C,i_d,i_ℓ=f(pH), W,r	
		-0.62F		-	-	-	-		-		-	C,i∝C,i_d,i_ℓ=f(pH), W,r	
		-0.76F		-	-	-	-		69		-	C,i∝C,i_d,i_ℓ=f(pH), W,r	
JA094 8471	-	$E_{\frac{1}{2}}$	-2.16	SCE	-	-	-	-	αn_a	0.7	Elog	-	C,r	DA68
JA095 4373	-	E_p	-0.53	SCE	-	-	-	-	-	-	-	-	C,p	DA69

Code No.	Empirical Formula	Name and WLN Code	Structural Formula	Solvent	Tech.	Medium		μ, M	pH	T, °C	Electrodes	App.	Experimental Parameters
DA70 ae46	$C_5H_8NNaS_2$	sodium 1-pyrrolidine-dithiocarboxylate C.A. 872-71-9 T5NTJ ACS&US &-NA-	TABLE II.	H_2O	PY	KOH KCl TX100	0.01 0.3 0.008	-	-	25± 0.2	DME/SCE	01A0	C=1.0,m=2.52, t=3.00,h=40
					IL	KOH KCl	0.01 0.3	-	-	25± 0.2	HMDE SCE	01A0	C=1.0,v=20
DA71	C_5H_8O	trans-3-penten-2-one C.A. 625-33-2 2U1V1 -T	trans-CH_3CH: $COCH_3$	DMF	PY	Bu_4NBF_4	0.50	-	-	25	DME/SCE	23-0	C=2.7-11.7, Pt Aux
DA72	$C_5H_8O_2$	methyl 3-butenoate C.A. 3724-55-8 1U2V01	$CH_3O_2CCH_2CH:CH_2$	MeCN	PY	Bu_4NClO_4	0.1	-	-	23	DME Ag/0.01 Ag^+	2A0	C=2,Pt Aux
DA73 cb18	$C_5H_8O_2$	methyl crotonate C.A. 623-43-8 1YU1V01 -T	trans-CH_3CH: $CHCO_2CH_3$	DMF	PY	Bu_4NBF_4	0.50	-	-	25	DME/SCE	23-0	C=8.8, Pt Aux
DA74 ae65 cb28	$C_5H_{10}NNaS_2$	sodium diethyldi-thiocarbamate C.A. 148-18-5 SUYS&N2&2 &-NA-	$(C_2H_5)_2NC(:S)SNa$	H_2O	IL	KOH KCl KOH KCl TX100	0.01 0.3 0.01 0.3 0.008	-	-	25± 0.1	HMDE SCE	01A0	C=1.0,v=20
DA75 ae67	$C_5H_{10}N_2O$	N-nitrosopiperidine C.A. 100-75-4 T6NTJ ANO	TABLE II-2.	H_2O	DI	BR		-	2.0	-	DME/SCE	04A0	C=0.05,m=1.73,t=1,h=80,v=10,ΔE=100
									10.0				
					PT	BR		-	2.0	-	DME/SCE	04A0	C=0.05,m=1.73,t=1,h=80,v=10
DA76	$C_5H_{10}N_2O_3S$	3-(N-nitrosomethyl-amino)thiolane C.A. 13256-21-8 T5SWTJ CN1&NO	TABLE II.	H_2O	PT	BR		-	2.0	-	DME/SCE	04A0	C=0.05,m=1.73,t=1,h=80,v=10
DA77 ae80	$C_5H_{10}O_5$	D-xylose C.A. 58-86-6 T6OTJ BQ CQ DQ EQ	TABLE II-2.	H_2O	PY	AM	0.2	-	8.59	25± 0.05	DME/SCE	0A0	C=60.0,m=1.306,t=5.65,h=52.5
DA78	$C_5H_{12}N_2$	1,2-dimethylpyrazol-idine C.A. 38704-89-1 T5NNTJ A1 B1	TABLE II.	MeCN	VR	$NaClO_4$	0.1	-	-	-	AuBE SCE	2A0	C=0.2-0.9, d=0.15,v=50,MP Aux, PE
DA79	$C_5H_{12}N_2S_2$	ammonium pyrrol-idine-1-dithiocarb-oxylate C.A. 50732-69-9 T5NTJ ACS&US &K	TABLE II.	H_2O	PY	NaOH	0.1	-	-	25± 0.1	DME/SCE	25A0	C=0.5,m=1.32,t=5.57,h=65,Pt Aux
					PV	NaOH	0.1	-	-	25± 0.1	DME/SCE	25A0	C=0.3,m=1.32,t=5.57,h=65,Pt Aux

TABLE I. Electrochemical Data $C_5H_{12}N_2S_2$ DA79

Ref.	C/M	Charact. Potential			Response Const.		n Tech.	n	Electrokinetic Data			Products and Identification	Description and Remarks	Code No.
			Value	vs.		Value			Parameter	Value	From			
AJ028 0021	-	$E_{\frac{1}{2}}$	-0.438 -0.593	SCE	i_ℓ	20F -	-	-	-	-	-	-	A,r A,i_a,Pr	DA70 ae46
AJ028 0021	-	E_p	-0.462	SCE	-	-	-	-	$E_p/E_{p/2}$	9	-	-	A,r	
JA094 8471	-	$E_{\frac{1}{2}}$	-2.08	SCE	-	-	-	-	αn_a	1.0	Elog	-	C,p	DA71
JA095 1703	-	-	-	-	-	-	-	-	-	-	-	-	O,p	DA72
JA094 8471	-	$E_{\frac{1}{2}}$	-2.33	SCE	-	-	-	-	αn_a	1.2	Elog	-	C,p	DA73 cb18
AJ028 0021	-	E_p	-0.460 -0.448	SCE	-	-	-	-	$E_p-E_{p/2}$	12 27	sttd	-	A,r A,r	DA74 ae65 cb28
AA078 0081	82d	E_{su}	-0.75	SCE	i_{su}	1.5	QE	4	-	-	-	-	C,i_d,r	DA75 ae67
	82b		-1.35			0.75		2	-	-	-	-	C,i_d,r	
AA078 0081	82d	$E_{\frac{1}{2}}$	-0.74	SCE	-	-	QE	4	-	-	-	-	C,i_d,r	
AA078 0081	82 b,d	$E_{\frac{1}{2}}$	-0.75	SCE	-	-	QE	4	-	-	-	-	C,i_d,r	DA76
BJ046 2107	-	$E_{\frac{1}{2}}$	1.53F	SCE	i_ℓ	2.8F	-	-	Elog	100	sttd	-	C,r	DA77 ae80
JA094 7108	-	E_p	0.06 -	SCE	-	-	sttd	1 1	ΔE_p	75 -	sttd	radical cation,ESR	A,R,XE=$\frac{1}{2}(E_{p,A}+E_{p,C})$, r C,R	DA78
BJ046 2151	-	$E_{\frac{1}{2}}$	-0.42 -0.70	SCE	i_ℓ	0.71F 0.14F	-	-	-	-	-	Hg salt	A,$i_\ell \propto h^{\frac{1}{2}}$,$i_d$,$E_{\frac{1}{2}}=k_1+k_2$(pH) for pH<4.2, $E_{\frac{1}{2}} \neq f$(pH) for pH>4.2, $T\tilde{c}$=1.3,Γ=6.9,0 for C<0.15,r A,Pr,$i_\ell \propto h$,i_a,$i_\ell \propto$ C for C<0.15	DA79
BJ046 2151	-	E_{su}	-0.64F -0.44	SCE	i_{su}	0.66F 0.42F	-	-	-	-	-	Hg salt	A,p A	

Code No.	Empirical Formula	Name and WLN Code	Structural Formula	Solvent	Tech.	Medium		μ, M	pH	T, °C	Electrodes	App.	Experimental Parameters
DA80	$C_5H_{13}N_3$	s-tetramethyl-guanidine C.A. 50854-51-8 1N1&YUMN1&1	$(CH_3)_2NC(:NH)N-(CH_3)_2$	m-cresol	PY	Et_4NClO_4	0.1	–	–	–	DME Ag/AgCl, 0.1 Me_4NCl	216AO	C=0.5-1.5, m=2.87, t=2.70, h=56, MP Aux
DA81	$C_5H_{14}N_2$	N,N,N',N'-tetra-methylmethylene-diamine C.A. 51-80-9 1N1&1N1&1	$[(CH_3)_2N]_2CH_2$	MeCN	VR	$NaClO_4$	0.1	–	–	–	AuBE SCE	2AO	C=3.26, d=0.15, v=71, MP Aux, PE
DA82	C_6HCl_5S	pentachlorothio-phenol C.A. 133-49-3 SHR BG CG DG EG FG	TABLE II.	H_2O	PY	BOR $CoCl_2$	4E-4	–	–	–	DME/SCE	OA-	ns
DA83	$C_6H_4N_2O_3S$	5-nitro-2-(thio-cyanatomethyl) furan C.A. 4063-38-1 T5OJ BNW E1SCN	TABLE II.	EtOH 86	PY	BR		0.45	1.9 2.9 3.9 5.9 8.0 8.9 9.8 10.8	25.0 ±0.1	DME/SCE	OCO	C=0.2, m=1.56, t=0.22
DA84 ag00	$C_6H_4N_2O_4$	1,3-dinitrobenzene C.A. 99-65-0 WNR CNW	$3-NO_2C_6H_4NO_2$	MeCN	VR	Bu_4NClO_4		–	–	–	HMDE Ag/AgNO_3	2-O	v=33.3?, Pt Aux
DA85 ag07	$C_6H_4O_2$	1,2-benzoquinone C.A. 583-63-1 L6VVJ	TABLE II-2.	MeCN	PY	Et_4NClO	0.1	–	–	–	DME/MP	–	–

TABLE I. Electrochemical Data $C_6H_4O_2$ DA85

Ref.	C/M	Charact. Potential Value	vs.	Response Const.	Value	Tech.	n	Electrokinetic Data Parameter	Value	From	Products and Identification	Description and Remarks	Code No.
AA072 0169	-	$E_{1/2}$ 0.38	Ag/AgCl	I	0.389	-	-	-	-	-	-	A,≠,m-cresolate oxidn.,pK_a=14.5,r	DA80
JA094 7114	-	E_p 0.87	SCE	-	-	-	-	$E_p-E_{p/2}$	120	sttd	-	A,p	DA81
C0036 1035	-	-	-	-	-	-	-	-	-	-	-	C,i_{cat}, i reaches max. value at pH 3, pK=3.5,p	DA82
		-			-				-			C,i_{cat}, $i\uparrow$ as $T\downarrow$ for pH < 4	
		-			-				-			C,i_{cat}	
RG035 0773	130 k	$E_{1/2}$ -0.15 -0.20 -0.50 -0.74	SCE	i_d	1.26 0.49 0.73 0.85	IIk	4.08 -	-	-	-	-	C,i_d,R,r C,i_d C,i_a C	DA83
		-0.18 -0.35 -0.59 -0.88 -1.35			1.33 0.64 0.64 1.04 0.58		4.08 -					C,i_d,R,r C,i_d C,i_a C C	
		-0.19 -0.45 -0.64 -0.93 -1.45			1.26 0.67 0.58 1.42 0.24		4.08 -					C,i_d,R,r C,i_a C,i_a C C	
		-0.23 -0.48 - -1.12 -1.71			0.96 0.33 - 1.45 1.18		4.08 -					C,i_d,R,r C,i_d 0 C,i_d C	
		-0.26 -0.61 -1.12 -1.64			0.89 1.27 1.17 1.58		4.08 -					C,i_d,R,r C,i_d C,i_d C,i_d	
		-0.22 -0.64 -1.17 -1.69			0.69 0.27 1.08 1.34		4.08 -					C,i_d,R,r C,i_d C,i_d C,i_d	
		- - -1.18 -1.63			- - 0.91 0.95		- -					$0,r$ 0 C,i_d C,i_d	
		-1.31 -			0.69 -		-					C,i_d,p 0	
AJ027 2495	-	$E_{p/2}$ -1.10 -1.45	Ag/AgNO₃	-	-	-	-	ΔE_p αn_a?	60 60 0.9	- $E_p-E_{p/2}$	-	$C,A;R,i_{p,C}/i_{p,A}=1,r?$ $C,A;i_{p,C}/i_{p,A}=1$	DA84 ag00
RY002 1096	-	-	-	-	-	-	-	-	-	-	-	see AG07	DA85 ag07

Code No.	Empirical Formula	Name and WLN Code	Structural Formula	Solvent	Tech.	Medium		μ, M	pH	T, °C	Elec-trodes	App.	Experimental Parameters
DA86 ag08 cb70	$C_6H_4O_2$	1,4-benzoquinone C.A. 106-51-4 L6V DVJ	TABLE II-2.	DMF	VR	Bu_4NClO_4	0.1	-	-	-	PDE/SCE	3F-	C=0.1-1,v= 248
					PY	Et_4NClO_4	0.1				DME/MP		
DA87	$C_6H_4O_3S_3$	2,6-dimercapto-1,4-thiopyrone-3-carb-oxylic acid C.A. 1738-10-9 T6V DSJ BVQ CSH ESH	TABLE II.	H_2O	PY	NaOH	0.1	-	-	-	DME/ -	O-O	C=0.15
					PV	NaOH	0.1	-	-	-	DME/ -	O-O	C=0.1,f=50, Δe=8
													C=0.2
													C=0.39
													C=0.57
DA88	C_6H_5ClHgO	2-hydroxyphenyl-mercury chloride C.A. 90-03-9 QR B-HG-G	2-HOC_6H_4HgCl	H_2O	PY	HNO_3 GEL	0.05 0.005	0.50	-	25± 0.1	DME/SCE	O15A 1	C=1.0,EC= 1.810,h=100
						ACET GEL	0.005		4.38				
DA89 ag41	C_6H_5NO	nitrosobenzene C.A. 586-96-9 ONR	C_6H_5NO	EtOH 10	PY	HCl	0.01	0.5	-	30± 0.1	DME/SCE	-CO	m=1.17 & 1.24
						AM			9.0				
				EtOH 20	PY	HCl	0.1	0.5	-	30± 0.1	DME/SCE	-CO	m=1.17 & 1.24
				EtOH 40	PY	HCl	0.1	0.5	-	30± 0.1	DME/SCE	-CO	m=1.17 & 1.24
						HCl	0.01						
						AM			9.0				
				EtOH 60	PY	HCl	0.01	0.5	-	30± 0.1	DME/SCE	-CO	m=1.17 & 1.24
				EtOH 70	PY	HCl	0.1	0.5	-	30± 0.1	DME/SCE	-CO	m=1.17 & 1.24
						AM			9.0				
				Me_2CO 10	PY	HCl	0.01	0.5	-	30± 0.1	DME/SCE	-CO	m=1.17 & 1.24
				Me_2CO 20	PY	HCl	0.1	0.5	-	30± 0.1	DME/SCE	-CO	m=1.17 & 1.24
				Me_2CO 40	PY	HCl	0.1	0.5	-	30± 0.1	DME/SCE	-CO	m=1.17 & 1.24
						HCl	0.01						
				Me_2CO 60	PY	HCl	0.1	0.5	-	30± 0.1	DME/SCE	-CO	m=1.17 & 1.24
						HCl	0.01						
CONT				dioxane 10	PY	AM	?	0.5	9.0	30± 0.1	DME/SCE	-CO	m=1.17 & 1.24

TABLE I. Electrochemical Data C_6H_5NO (CONT.) DA89

Ref.	C/M	Charact. Potential		Response Const.		n Tech.		Electrokinetic Data			Products and Identification	Description and Remarks	Code No.
		Value	vs.		Value			Parameter	Value	From			
JA094 0691	-	E_p	-0.52	SCE	-	-	-	-	-	-	-	-	C,R,p	DA86 ag08
RY002 1096	-	-	-	-	-	-	-	-	-	-	-	-	see AG08,CB70	cb70
ER003 0623	-	$E_{\frac{1}{2}}$	-0.35F -0.7F	NCE	i_ℓ	0.12F 0.38F	-	-	-	-	-	-	A,i_d,M,r A,i_a,X	DA87
ER003 0623	-	E_{su}	-0.58F -0.7F -0.95F	NCE	i_{su}	0.36F 0.16F 0.02F	-	-	-	-	-	-	A,r A A	
			-0.58F -0.7F -1.0F			0.56F 0.44F 0.02F	-	-	-	-	-	-	A,r A A	
			-0.45F -0.78F -1.0			1.66F 0.46F 0.04F	-	-	-	-	-	-	A,r A A	
			-0.43F -0.78F -1.0F			2.6F 0.6F 0.09F	-	-	-	-	-	-	A,r A A	
BJ046 2129	8f	$E_{\frac{1}{2}}$	0.2 0.05F -0.6	SCE	i_ℓ	- 3.1F 2.7F	-	-	Elog	167	sttd	-	C,Pr,Hg^{2+} redn.,i_d,r C,$E_{\frac{1}{2}}=k_1+k_2C$,i_d,⊁ C,i_a	DA88
			0.02F -0.63F			2.7F 2.9F	-	-	-	-	-	-	C,$i_\ell \propto C$ C	
EA017 0471	-	$E_{\frac{1}{2}}$	- -0.680 -0.206	SCE	I	3.33? 2.42 2.03	-	-	-	-	-	-	C,R,r C C,R,r	DA89 ag41
EA017 0471	-	$E_{\frac{1}{2}}$	- -0.633	SCE	I	2.88 1.67	-	-	-	-	-	-	C,R,r C	
EA017 0471	-	$E_{\frac{1}{2}}$	- -0.675	SCE	I	2.57 1.82	-	-	-	-	-	-	C,R,r C	
			- -0.750			2.67? 2.12	-	-	-	-	-	-	C,R,r C	
			-0.215			2.15	-	-	-	-	-	-	C,R,r	
EA017 0471	-	$E_{\frac{1}{2}}$	- -0.792	SCE	I	2.88? 2.57	-	-	-	-	-	-	C,R,r C	
EA017 0471	-	$E_{\frac{1}{2}}$	- -0.720	SCE	I	3.57 2.42	-	-	-	-	-	-	C,R,r C	
			-0.208			3.06	-	-	-	-	-	-	C,R,r	
EA017 0471	-	$E_{\frac{1}{2}}$	- -0.898	SCE	I	1.80? 1.29	-	-	-	-	-	-	C,R,r C	
EA017 0471	-	$E_{\frac{1}{2}}$	- -0.839	SCE	I	2.08? 1.60	-	-	-	-	-	-	C,R,r C	
EA017 0471	-	$E_{\frac{1}{2}}$	- -0.866	SCE	I	3.53? 2.94	-	-	-	-	-	-	C,R,r C	
			- -0.934			3.64? 2.88	-	-	-	-	-	-	C,R,r C	
EA017 0471	-	$E_{\frac{1}{2}}$	- -0.854	SCE	I	5.14? 4.0?	-	-	-	-	-	-	C,R,r C	
			- -0.950			4.24? 2.57	-	-	-	-	-	-	C,R,r C	
EA017 0471	-	$E_{\frac{1}{2}}$	-0.197	SCE	I	2.66	-	-	-	-	-	-	C,R,r	

CONT

Code No.	Empirical Formula	Name and WLN Code	Structural Formula	Solvent	Tech.	Medium		μ, M	pH	T, °C	Electrodes	App.	Experimental Parameters
DA89 ag41	C_6H_5NO	nitrosobenzene		dioxane 40	PY	AM	?	0.5	9.0	30±0.1	DME/SCE	-CO	m=1.17 & 1.24
				dioxane 70	PY	AM	?	0.5	9.0	30±0.1	DME/SCE	-CO	m=1.17 & 1.243
DA90 ag46	$C_6H_5NO_2$	nicotinic acid C.A. 59-67-6 T6NJ CVQ	TABLE II-2.	H_2O	DI	BOR	0.1	-	8.7	25±0.1	DME/SCE	2AO	C=0.09,t=2, v=1,ΔE=5, Pt Aux
DA91 ag47 cb73	$C_6H_5NO_2$	nitrobenzene C.A. 98-95-3 WNR	$C_6H_5NO_2$	H_2O	PY	HCl NaF GEL	0.004 0.1 0.005	-	-	25±0.1	DME/SCE	O3AO	C=0.4,m=1.85, t=4.9,h=30
						UB GEL	0.005		2.40				
									5.40				
									7.80				
									11.40				
				i-PrOH 6.5w/v	PY	LiCl	0.1	-	-	-	DME/SCE	O13AO	C=0.4,m=1.85, t=4.95,h=30
				i-PrOH 22.6w/v	PY	LiCl	0.1	-	-	-	DME/SCE	O13AO	C=0.4,m=1.85, t=4.95,h=30
						LiCl PHEN	0.1 0.004						
						LiCl PHEN	0.1 0.04						
						LiCl PHEN	0.1 0.4						
						LiCl C_6H_5COOH	0.1 4E-4						
						LiCl C_6H_5COOH	0.1 8E-4						
						UB			5.5				
									7.74				
									9.5				
									11.2				
				i-PrOH 69.4w/v	PY	LiCl	0.1	-	-	-	DME/SCE	O13AO	C=0.4,m=1.85, t=4.95,h=30
DA92 ag49	$C_6H_5NO_2$	4-nitrosophenol C.A. 104-91-6 QR DNO	$4-ONC_6H_4OH$	H_2O	DI	BR		-	2.0	-	DME/SCE	O4AO	C=0.05,m= 1.73,t=1,h= 80,v=10,ΔE= 100
									10.0				
DA93	$C_6H_5NO_3$	2-hydroxynicotinic acid C.A. 609-71-2 T6NJ BQ CVQ		EtOH ≈ 40	DI	PHOS	≈ 0.1	-	5.0	-	DME/SCE	2A-	-1.108→ -1.80 V,C= 0.14,t=0.5, EC=1.561,v=5, Δe=50,Pt Aux

TABLE I. Electrochemical Data $C_6H_5NO_3$ DA93

Ref.	C/M	Charact. Potential Value	vs.	Response Const.	Value	n Tech.	n	Electrokinetic Data Parameter	Value	From	Products and Identification	Description and Remarks	Code No.
EA017 0471	-	$E_{\frac{1}{2}}$	-0.185	SCE	I	3.23	-	-	-	-	-	-	C,R,r	DA89 ag41
EA017 0471	-	$E_{\frac{1}{2}}$	-0.177	SCE	I	3.01	-	-	-	-	-	-	C,R,r	
AN100 0377	-	E_{su}	-1.72F	SCE	i_{su}	0.273F	-	-	-	-	-	-	C,$i_{su} \propto$ C for C= 0.005 - 0.09, r	DA90 ag46
EA017 0511	-	$E_{\frac{1}{2}}$	-0.33F	SCE	i_ℓ	5.4F	-	-	-	-	-	-	C,r	DA91 ag47 cb73
			-0.21			-	-	-	αn_a p=	1.07 0.97	Elog $dE_{\frac{1}{2}}$/dpH	-	C,r	
			-0.435		-	-	-	-		0.80 0.97		-	C,r	
			-0.590		-	-	-	-		0.84 0.97		-	C,r	
			-0.730		-	-	-	-		1.00 0.97		-	C,r	
EA017 2065	-	$E_{\frac{1}{2}}$	-0.72F	SCE	i_ℓ	4.8F	-	-	-	-	-	-	C,r	
EA017 2065	-	$E_{\frac{1}{2}}$	-0.79F -0.98F	SCE	i_ℓ	1.2F 2.7F	-	-	-	-	-	-	C,r C	
			-0.76F -0.97F			1.6F 2.7F	-	-	-	-	-	-	C,r C	
			-0.74F -1.02F			3.2F 1.1F	-	-	-	-	-	-	C,r C	
			-0.72F -1.02F			3.36F 0.8F	-	-	-	-	-	-	C,r C	
			-0.50F -0.82F -1.01F			1.3F 0.8F 1.8F	-	-	-	-	-	-	C,r C C	
			-0.50F -0.82F -1.04F			2.4F 0.6F 0.8F	-	-	-	-	-	-	C,r C C	
			-0.56F			4.13F	-	-	-	-	-	-	C,r	
			-0.66F -			3.6F 0.7F	-	-	-	-	-	-	C,r	
			-0.79F -1.32F			2.4F 3.0F	-	-	-	-	-	-	C,r C	
			-0.79F -1.32F			1.0F 3.3F	-	-	-	-	-	-	C,r C	
EA017 2065	-	$E_{\frac{1}{2}}$	-0.93F	SCE	i_ℓ	4.5F	-	-	-	-	-	-	C,r	
AA078 0081	-	E_{su}	0.13 -1.0	SCE	i_{su}	0.5 0.5	-	-	-	-	-	-	C,r	DA92 ag49
			-0.39 -0.89 -1.08			3.5 2.0 1.0	-	-	-	-	-	-	C,r C C	
AA066 0023	-	E_{su}	-1.50	SCE	i_{su}	2.7	-	-	-	-	-	2,3-dihydro-2-hydroxynicotinic acid	C,p	DA93

Code No.	Empirical Formula	Name and WLN Code	Structural Formula	Solvent	Tech.	Medium	μ, M	pH	T, °C	Electrodes	App.	Experimental Parameters
DA94 ag54	$C_6H_5NO_3$	4-hydroxynitrobenzene C.A. 100-02-7 WNR DQ	$4-O_2NC_6H_4OH$	H_2O	DI	BR	-	2.0	-	DME/SCE	O4AO	C=0.05,m=1.73,t=1,h=80,v=10,Λe=100
								10.0				
DA95 ag57	$C_6H_5NO_4$	2-acetyl-5-nitrofuran C.A. 5275-69-4 T5OJ BVI ENW	TABLE II-2.	EtOH 10	PY	BR	0.45	2.0	25±0.1	DME/SCE	OAO	C=0.2,m=1.18,t=0.26
								3.0				
								6.0				
								8.0				
								10.0				
								11.0				
								12.0				
				EtOH 20	PY	BR	0.45	2.0	25±0.1	DME/SCE	OAO	C=0.2,m=1.18,t=0.26
								4.0				
								5.0				
								7.0				
								9.0				
								11.0				
								12.0				
				EtOH 30	PY	BR	0.45	2.0	25±0.1	DME/SCE	OAO	C=0.2,m=1.18,t=0.26
								3.0				
								5.0				
								7.0				
								8.0				
								10.0				
								11.0				
								12.0				
				EtOH 50	PY	BR	0.45	2.0	25±0.1	DME/SCE	OAO	C=0.2,m=1.18,t=0.26
								4.0				
								6.0				
DA94 CONT		4-hydroxynitrobenzene WNR DQ	$4-O_2NC_6H_4OH$					8.0				
								10.0				

TABLE I. Electrochemical Data $C_6H_5NO_4$ (CONT.) DA95

| Ref. | C/M | Charact. Potential | | Response Const. | | n | Electrokinetic Data | | | Products and | Description and | Code |
		Value	vs.		Value	Tech.	Parameter	Value	From	Identification	Remarks	No.
AA078 0081	-	E_{su}	-0.03 -0.22 -1.0	SCE	i_{su}	0.5 3.0 0.5	-	-	-	-	-	-	c,r c c	DA94 ag54
			-0.90 -1.08			2.0 1.5	-	-	-	-	-	-	c,r c	
LV670 0023	-	$E_{\frac{1}{2}}$	-0.06	SCE	i_ℓ	0.90	-	-	-	-	-	2-acetyl-5-hydroxyl-aminofuran	c,r	DA95 ag57
			-0.79			0.58						2-acetyl-5-amino-furan	c	
			-1.23			0.55						-	c,carbonyl redn.	
			-0.10 -0.84 -1.35			0.93 0.62 0.79	-	-	-	-	-	-	c,r c c	
			-0.22 1.06 1.55			0.92 0.65 0.50	-	-	-	-	-	-	c,r c c	
			-0.30 -1.10 -1.64			0.88 0.60 0.52	-	-	-	-	-	-	c,r c c	
			-0.37 -1.12 -1.66			0.85 0.54 0.40	-	-	-	-	-	-	c,r c c	
			-0.40 -1.12 -			0.80 0.53 -	-	-	-	-	-	-	c,r c 0	
			-0.42 -1.12 -			0.67 0.34 -	-	-	-	-	-	-	c,r c 0	
LV670 0023	-	$E_{\frac{1}{2}}$	-0.11	SCE	i_ℓ	1.6	IIk	4	-	-	-	-	c,r	
			-0.19			1.70	-	-	-	-	-	-	c,r	
			-0.23			1.70	-	-	-	-	-	-	c,r	
			-0.36			1.60	-	-	-	-	-	-	c,r	
			-0.44			1.50	-	-	-	-	-	-	c,r	
			-0.46			1.30	-	-	-	-	-	-	c,r	
			-0.46			0.70	-	-	-	-	-	-	c,r	
LV670 0023	-	$E_{\frac{1}{2}}$	-0.14	SCE	i_ℓ	1.36	IIk	4	-	-	-	-	c,r	
			-0.16			1.44		4	-	-	-	-	c,r	
			-0.28			1.44	-	-	-	-	-	-	c,r	
			-0.43			1.40	-	-	-	-	-	-	c,r	
			-0.46			1.30	-	-	-	-	-	-	c,r	
			-0.47			1.20	-	-	-	-	-	-	c,r	
			-0.46			1.10	-	-	-	-	-	-	c,r	
			-0.47			0.70	-	-	-	-	-	-	c,r	
LV670 0023	-	$E_{\frac{1}{2}}$	-0.18	SCE	i_ℓ	1.15	IIk	4	-	-	-	-	c,r	
			-0.27			1.25	-	-	-	-	-	-	c,r	
			-0.43			1.30	-	-	-	-	-	-	c,r	
			-0.47			1.30	-	-	-	-	-	-	c,r	
			-0.49			1.20	-	-	-	-	-	-	c,r	CONT

DA95 (CONT.) $C_6H_5NO_4$

Code No.	Empirical Formula	Name and WLN Code	Structural Formula	Solvent	Tech.	Medium	μ, M	pH	T, °C	Electrodes	App.	Experimental Parameters
DA95 ag57	$C_6H_5NO_4$	2-acetyl-5-nitro-furan		EtOH 50	PY	BR	0.45	11.0 12.0	25±0.1	DME/SCE	OAO	$C=0.2, m=1.18, t=0.26$
DA96 ag66	C_6H_6	benzene C.A. 71-43-2 RH	C_6H_6	NH_3	VR	KI	0.1	-	-50	AuDE Ag/0.01 $AgNO_3$, 0.1KI	125F2	$C=1, A=0.5$, Pt Aux
DA97	$C_6H_6AsClO_3$	4-chlorophenyl-arsonic acid C.A. 5540-04-0 Q-AS-QO&R DG	$4-ClC_6H_4AsO(OH)_2$	H_2O	PY	HCl	0.1	-	-	DME/SCE	O4AO	$C=0.01-1.0, t=2.4, v=1.7$
					IV	HCl	0.1	-	-	DME/SCE	O4AO	$C=0.2, t=2.4, E_{app}=-0.700$
												$E_{app}=-0.800$
												$E_{app}=-0.900$
												$E_{app}=-0.950$
												$E_{app}=-1.000$
DA98 ag67	$C_6H_6AsNO_5$	2-nitrophenyl-arsonic acid C.A. 5410-29-7 WNR B-AS-QQO	$2-O_2NC_6H_4AsO(OH)_2$	H_2O	PY	HCl	0.1	-	-	DME/SCE	O4AO	$C=0.1-1.0, t=2.4, v=1.7$
					IV	HCl	0.1	-	-	DME/SCE	O4AO	$C=0.2, t=2.4, E_{app}=-0.760$
												$E_{app}=-0.800$
												$E_{app}=-0.900$
												$E_{app}=-0.950$
												$E_{app}=-1.050$
DA99 ag68	$C_6H_6AsNO_5$	3-nitrophenyl-arsonic acid C.A. 618-07-5 WNR C-AS-QQO	$3-O_2NC_6H_4AsO(OH)_2$	H_2O	PY	HCl	0.1	-	-	DME/SCE	O4AO	$C=0.1-1.0, t=2.4, v=1.7$
					IV	HCl	0.1	-	-	DME/SCE	O4AO	$C=0.2, t=2.4, E_{app}=-0.750$
												$E_{app}=-0.800$
												$E_{app}=-0.900$
												$E_{app}=-1.000$
												$E_{app}=-1.100$
DB00	$C_6H_6AsNO_5$	4-nitrophenyl-arsonic acid C.A. 98-72-6 WNR D-AS-QQO	$4-O_2NC_6H_4AsO(OH)_2$	H_2O	PY	HCl	0.1	-	-	DME/SCE	O4AO	$C=0.1-1.0, t=2.4, v=1.7$
CONT					IV	HCl	0.1	-	-	DME/SCE	O4AO	$C=0.2, t=2.4, E_{app}=-0.800$
												$E_{app}=-0.850$

TABLE I. Electrochemical Data $C_6H_6AsNO_5$ (CONT.) DB00

Ref.	C/M	Charact. Potential		Response Const.		n		Electrokinetic Data			Products and Identification	Description and Remarks	Code No.
		Value	vs.		Value	Tech.		Parameter	Value	From			
LV670 0023	-	$E_{\frac{1}{2}}$ -0.428	SCE	i_ℓ	1.15	-	-	-	-	-	-	C,r	DA95 ag57
		-0.49			1.00	-	-	-	-	-	-	C,r	
JA095 3495	-	-	-	-	-	-	-	-	-	-	-	O,p	DA96 ag66
AN100 0489	367 ab	$E_{\frac{1}{2}}$ -0.905	SCE	-	-	i:i	4	$dE_{\frac{1}{2}}/dpH$ p=	155 1	- Elog	-	C,⊁,i_d,$i \propto C$,$E_{\frac{1}{2}}=k_1+$ $k_2(pH)$,$\Delta E_{\frac{1}{2}}=0.173\sigma$,r	DA97
AN100 0489	367 ab	-	SCE	dlogi/ dlogt	0.31	-	-	-	-	-	-	C,r	
		-			0.28	-	-	-	-	-	-	C,r	
		-			0.24	-	-	-	-	-	-	C,r	
		-			0.21	-	-	-	-	-	-	C,r	
		-			0.24	-	-	-	-	-	-	C,r	
AN100 0489	367 ab	$E_{\frac{1}{2}}$ -0.860	SCE	-	-	i:i	4	$dE_{\frac{1}{2}}/dpH$ p=	86 1	- Elog	-	C,⊁,i_d,$i \propto C$,$E_{\frac{1}{2}}=k_1+$ $k_2(pH)$,r	DA98 ag67
		-					4				2-hydroxylaminophen-ylarsonic acid,sttd	C	
AN100 0489	367 ab	-	-	dlogi/ dlogt	0.29	-	-	-	-	-	-	C,r	
		-			-0.27	-	-	-	-	-	-	C,r	
		-			0.20	-	-	-	-	-	-	C,r	
		-			0.21	-	-	-	-	-	-	C,r	
		-			0.23	-	-	-	-	-	-	C,r	
AN100 0489	367 ab	$E_{\frac{1}{2}}$ -0.950	SCE	-	-	i:i	4	$dE_{\frac{1}{2}}/dpH$ p=	110 1	- Elog	-	C,⊁,i_d,$i \propto C$,$E_{\frac{1}{2}}=k_1+$ $k_2(pH)$,$\Delta E_{\frac{1}{2}}=0.173\sigma$,r	DA99 ag68
		-					4		-		3-hydroxylaminophen-ylarsonic acid,sttd	C	
AN100 0489	367 ab	-	-	dlogi/ dlogt	0.34	-	-	-	-	-	-	C,r	
		-			0.29	-	-	-	-	-	-	C,r	
		-			0.24	-	-	-	-	-	-	C,r	
		-			0.22	-	-	-	-	-	-	C,r	
		-			0.24	-	-	-	-	-	-	C,r	
AN100 0489	367 ab	$E_{\frac{1}{2}}$ -1.000	SCE	-	-	i:i	4	$dE_{\frac{1}{2}}/dpH$ p=	106 1	- Elog	-	C,⊁,i_d,$i \propto C$,$E_{\frac{1}{2}}=k_1+$ $k_2(pH)$,$\Delta E_{\frac{1}{2}}=0.173\sigma$,r	DB00
		-0.05F					4		-		4-hydroxylaminophen-ylarsonic acid,sttd	C	
AN100 0489	367 ab	-		dlogi/ dlogt	0.33	-	-	-	-	-	-	C,r	
		-			0.30	-	-	-	-	-	-	C,r	CONT

Code No.	Empirical Formula	Name and WLN Code	Structural Formula	Solvent	Tech.	Medium		μ, M	pH	T, °C	Electrodes	App.	Experimental Parameters
DB00	$C_6H_6AsNO_5$	4-nitrophenyl-arsonic acid	$4-O_2NC_6H_4AsO(OH)_2$	H_2O	IV	HCl	0.1	-	-	-	DME/SCE	04A0	$C=0.2, t=2.4$, $E_{app}=-0.900$
													$E_{app}=-1.000$
													$E_{app}=-1.100$
DB01	$C_6H_6AsNO_6$	4-hydroxy-3-nitro-phenylarsonic acid C.A. 121-19-7 WNR BQ E-AS-QQO	TABLE II.	H_2O	PY	HCl	0.1	-	-	-	DME/SCE	04A0	$C=0.1-1.0$, $t=2.4, v=1.7$
					IV	HCl	0.1	-	-	-	DME/SCE	04A0	$C=0.2, t=2.4$, $E_{app}=-0.850$
													$E_{app}=-0.900$
													$E_{app}=-1.000$
													$E_{app}=-1.100$
													$E_{app}=-1.200$
DB02	$C_6H_6AsNO_7$	2,4-dihydroxy-5-nitrophenylarsonic acid C.A. 57178-74-2 WNR BQ DQ E-AS-QQO	TABLE II.	H_2O	PY	HCl	0.1	-	-	-	DME/SCE	04A0	$C=0.1-1.0$, $t=2.4, v=1.7$
					IV	HCl	0.1	-	-	-	DME/SCE	04A0	$C=0.2, t=2.4$, $E_{app}=-0.850$
													$E_{app}=-0.900$
													$E_{app}=-1.000$
													$E_{app}=-1.100$
													$E_{app}=-1.200$
DB03 ag74 cb75	C_6H_6ClN	4-chloroaniline C.A. 106-47-8 ZR DG	$4-ClC_6H_4NH_2$	MeCN	VR	Bu_4NClO_4		-	-	-	HMDE Ag/AgNO$_3$	2-0	$v=33.3?$, Pt Aux
DB04 ag85 cb79	$C_6H_6N_2O$	nicotinamide C.A. 98-92-0 T6NJ CVZ	TABLE II-2.	H_2O	PY	H_2SO_4 Benax	4 0.0025	-	-	25± 0.1	DME Ag/AgCl, satd KCl	2A0	$C=0.2, m=2.313, t=3.790, h=58.3$, Pt Aux
						CITR Benax	0.0025		1.17				
						CITR			7.60				
						AM			9.15				
						NaOH			13.0				
					DI	NaOH		-	-	25± 0.1	DME Ag/AgCl, satd KCl	2A0	$C=1, t=0.5$, $\Delta E=50, v=2$, Pt Aux
CONT					PV?	H_2SO_4 Benax	4 0.0025	-	-	25± 0.1	DME Ag/AgCl, satd KCl	2A0	$C=0.2, m=2.313, t=3.790, h=58.3, \Delta e=10$, Pt Aux

TABLE I. Electrochemical Data $C_6H_6N_2O$ (CONT.) DB04

| Ref. | C/M | Charact. Potential | | Response | Const. | n | | Electrokinetic Data | | | Products and | Description and | Code |
		Value	vs.		Value	Tech.		Parameter	Value	From	Identification	Remarks	No.
AN100 0489	367 ab	-	-	-	dlogi/dlogt	0.29	-	-	-	-	-	-	C,r	DB00
				-		0.25	-	-	-	-	-	-	C,r	
				-		0.24	-	-	-	-	-	-	C,r	
AN100 0489	367 ab	$E_{\frac{1}{2}}$	-1.015	SCE	-	-	i:i	4	$dE_{\frac{1}{2}}/dpH$ $p=$	96 1	- Elog	4-hydroxy-3-hydroxyl-aminophenylarsonic acid,sttd	$C,\neq,i_d,i \propto C,E_{\frac{1}{2}}=k_1+k_2(pH),\Delta E_{\frac{1}{2}}=0.173\sigma,r$ C	DB01
			-					4						
AN100 0489	367 ab	-	-	-	dlogi/dlogt	0.32	-	-	-	-	-	-	C,r	
				-		0.29	-	-	-	-	-	-	C,r	
				-		0.25	-	-	-	-	-	-	C,r	
				-		0.23	-	-	-	-	-	-	C,r	
				-		0.24	-	-	-	-	-	-	C,r	
AN100 0489	367 ab	$E_{\frac{1}{2}}$	-1.030	SCE	-	-	i:i	4	$dE_{\frac{1}{2}}/dpH$ $p=$	104 1	- Elog	2,4-dihydroxy-5-hydroxylaminophenylarsonic acid,sttd	$C,\neq,i_d,i \propto C,E_{\frac{1}{2}}=k_1+k_2(pH),r$ C	DB02
			-					4						
AN100 0489	367 ab	-	-	-	dlogi/dlogt	0.33	-	-	-	-	-	-	C,r	
				-		0.29	-	-	-	-	-	-	C,r	
				-		0.27	-	-	-	-	-	-	C,r	
				-		0.24	-	-	-	-	-	-	C,r	
				-		0.24	-	-	-	-	-	-	C,r	
AJ027 2495	-	E_p	-2.66	-	-	-	-	-	αn_a	0.43	$E_p-E_{p/2}$	-	$C,\neq,r?$	DB03 ag74 cb75
AA071 0175	277 d	$E_{\frac{1}{2}}$	-0.868	Ag/AgCl	i_ℓ	1.33	QE	1.89	Tomeš αn_a	42 1.37	- Tomeš	-	$C,i_d,Tc=1.54,\neq,r$	DB04 ag85 cb79
			-1.010			3.26		1.89		104 1.37		-	$C,i_d,Tc=1.54,\neq,r$	
	277 e		-1.492			1.60		1.89		76 1.37		-	$C,i_d,Tc=1.54,\neq,r$	
			-1.578			1.21		1.89		50 1.37		-	$C,i_d,Tc=1.54,\neq,r$	
			-1.730			1.32		1.89		38 1.37		-	$C,i_d,Tc=1.54,\neq,r$	
AA071 0175	277 e	E_{su}	-1.70F	Ag/AgCl	i_{su}	89.50	-	-	-	-	-	-	$C,\neq,i_{su} \propto C$ for $C=$ 5E-3 → 1,r	
AA071 0175	277 d	E_{su}	-0.916	Ag/AgCl	i_{su}	0.32	-	-	$\Delta E_{su/2}$	100	-	-	$C,i_d+i_k,\neq,Tc=1.05,r$	

CONT

DB04 (CONT.) $C_6H_6N_2O$

Code No.	Empirical Formula	Name and WLN Code	Structural Formula	Solvent	Tech.	Medium		μ, M	pH	T, °C	Electrodes	App.	Experimental Parameters
DB04 ag85 cb79	$C_6H_6N_2O$	nicotinamide		H_2O	PV	CITR Benax	? 0.0025	-	1.17	25± 0.1	DME Ag/AgCl, satd KCl	2AO	C=0.2, m= 2.31, t= 3.79, h= 58.3, Δe=10, Pt Aux
						CITR	?		7.60				
						AM	?		9.15				
						NaOH			13.0				
					VA	Et_4NOH	0.1	-	-	25± 0.1	HMDE Ag/AgCl, satd KCl	2AO	C=1, v=50, Pt Aux
DB05 ag92	$C_6H_6N_2O_2$	3-nitroaniline C.A. 99-09-2 ZR CNW	$3-O_2NC_6H_4NH_2$	MeCN	VR	Bu_4NClO_4	?	-	-	-	HMDE Ag/AgNO_3	2-0	v=33.3, Pt Aux
DB06 ag93	$C_6H_6N_2O_2$	4-nitroaniline C.A. 100-01-6 ZR DNW	$4-O_2NC_6H_4NH_2$	MeCN	VR	Bu_4NClO_4	?	-	-	-	HMDE Ag/AgNO_3	2-0	v=33.3?, Pt Aux
DB07 ah06 cb85	$C_6H_6O_2$	hydroquinone C.A. 123-31-9 QR DQ	$4-HOC_6H_4OH$	DMF	VY	PHEN	?	-	-	-	Pt/ Ag/AgCl (aq), satd KCl, 0.8 $KClO_4$	26--	PE, $E_{ferrocene}$= 0.456 V vs. Ag/AgCl(aq)
						$HClO_4$	?		1				PE
						$Cl_2ACETOH$	?		7.1				
						Chlorobenzoic acid	?		10.2				
						AcOH	?		12.9				
DB08 ah07 cb86	$C_6H_6O_2$	resorcinol C.A. 108-46-3 QR CQ	$3-HOC_6H_4OH$	H_2O	VY	KOH	2	-	-	30	RPDE/ -	0--	C=0.1, A=1.8, v=7.2, ω= 7.75, PE
													C=0.20
													C=0.30
													C=0.40
											40		C=0.10
													C=0.20
													C=0.30
													C=0.40
											50		C=0.10
													C=0.20
CONT													C=0.30

TABLE I. Electrochemical Data $C_6H_6O_2$ (CONT.) DB08

Ref.	C/M	Charact. Potential		Response Const.		n Tech.	Electrokinetic Data			Products and Identification	Description and Remarks	Code No.
		Value	vs.		Value		Parameter	Value	From			
AA071 0175	277 d	E_{su}	-1.062	Ag/AgCl	i_{su}	0.45	-	-	$\Delta E_{su/2}$	104	-	-	C,i_d+i_k,⇌,Tc=1.05,r	DB04 ag85 cb79
	277 e		-1.531			0.50	-	-		107	-	-	C,i_d+i_k,⇌,Tc=1.05,r	
			-1.702			0.36	-	-		102	-	-	C,i_d+i_k,Tc=1.05,r	
			-1.765			0.36	-	-		88	-	-	C,i_d+i_k,Tc=1.05,r	
AA071 0175	277 e	E_p	-1.74	Ag/AgCl	i_p	15.0	-	-	-	-	-	-	C,⇌,r	
AJ027 2495	-	$E_{p/2}$	-1.40	Ag/AgNO$_3$	i_p	36F	-	-	ΔE_p	70	-	-	C,R,$i_{p,C}/i_{p,A}$=1,r?	DB05 ag92
		E_p	-2.08			-	-	-	αn_a	0.27	$E_p - E_{p/2}$		C,⇌	
AJ027 2495	-	$E_{p/2}$	-1.55	Ag/AgNO$_3$	-	-	-	-	ΔE_p	65	-	-	C,R,$i_{p,C}/i_{p,A}$=1,r?	DB06 ag93
		E_p	-2.40				-	-	αn_a	0.30	$E_p - E_{p/2}$		C,⇌	
AL002 0123	-	$E_{\frac{1}{2}}$	-0.39	Ag/AgCl (aq)	-	-	-	-	$dE_{\frac{1}{2}}/dlogi_\ell$	70	sttd	-	A, ref. electrode connected by 5% methylcellulose bridge in satd Et$_4$NClO$_4$, $E_{\frac{1}{2}}$ uncorrected for liquid junction potential, r	DB07 ah06 cb85
			0.73			-	-	sttd	1.92	160		-	A,r	
			0.62			-	-	-	-	160		-	A,r	
			0.47			-	-	-	-	230		-	A,r	
			0.05			-	-	-	-	210		-	A,r	
LU021 0150	-	$E_{\frac{1}{2}}$	0.360 ±0.005	SCE	i_d	4.0	-	-	-	-	-	-	A,i_d,$i_\ell=f(\omega^{\frac{1}{2}})$,r	DB08 ah07 cb86
			0.385			5.5	-	-	-	-	-	-	A,i_d,$i_\ell=f(\omega^{\frac{1}{2}})$,r	
			0.400			8.0	-	-	-	-	-	-	A,i_d,$i_\ell=f(\omega^{\frac{1}{2}})$,r	
			0.410			10.5	-	-	-	-	-	-	A,i_d,$i_\ell=f(\omega^{\frac{1}{2}})$,r	
			0.350			4.0	-	-	-	-	-	-	A,i_d,$i_\ell=f(\omega^{\frac{1}{2}})$,r	
			0.375			8.0	-	-	-	-	-	-	A,i_d,$i_\ell=f(\omega^{\frac{1}{2}})$,r	
			0.385			12.0	-	-	-	-	-	-	A,i_d,$i_\ell=f(\omega^{\frac{1}{2}})$,r	
			0.400			15.5	-	-	-	-	-	-	A,i_d,$i_\ell=f(\omega^{\frac{1}{2}})$,r	
			0.330			5.0	-	-	-	-	-	-	A,i_d,$i_\ell=f(\omega^{\frac{1}{2}})$,r	
			0.365			9.5	-	-	-	-	-	-	A,i_d,$i_\ell=f(\omega^{\frac{1}{2}})$,r	
			0.370			14.0	-	-	-	-	-	-	A,i_d,$i_\ell=f(\omega^{\frac{1}{2}})$,r	CONT

DB08 (CONT.) $C_6H_6O_2$

Code No.	Empirical Formula	Name and WLN Code	Structural Formula	Solvent	Tech.	Medium		μ, M	pH	T, °C	Electrodes	App.	Experimental Parameters
DB08 ah07 cb86	$C_6H_6O_2$	resorcinol	$3\text{-}HOC_6H_4OH$	H_2O	VY	KOH	2	-	-	50	RPDE/ -	O--	C=0.40
										60			C=0.10
													C=0.20
													C=0.30
													C=0.40
DB09	C_6H_6S	benzenethiol C.A. 108-98-5 SHR	C_6H_5SH	H_2O	PY	NaOH	0.1	-	-	-	DME/ -	O-O	C=0.1
													C=0.2
													C=0.38
													C=0.57
													C=0.92
						BOR $CoCl_2$	4E-4	-	-	-	DME/SCE	OA-	
					PV	NaOH	0.1	-	-	-	DME/ -	O-O	C=0.05, f=50
													C=0.2
													C=0.57
													C=0.92
				EtOH	PY	NaOH	1	-	-	-	DME/ -	O-O	C=0.1
													C=0.2
													C=0.58
					PV	NaOH	1	-	-	-	DME/ -	O-O	C=0.1
													C=0.2
													C=0.58

TABLE I. Electrochemical Data — C_6H_6S DB09

Ref.	C/M	Charact. Potential Value	vs.	Response Const.	Value	n Tech.	Electrokinetic Data Parameter	Value	From	Products and Identification	Description and Remarks	Code No.
LU021 0150	–	$E_{\frac{1}{2}}$	0.385	SCE	i_d	18.0	–	–	–	–	–	–	$A, i_d, i_\ell = f(\omega^{\frac{1}{2}}), r$	DB08 ah07 cb86
		0.330			6.0	–	–	–	–	–	–	$A, i_d, i_\ell = f(\omega^{\frac{1}{2}}), r$	
		0.360			11.0	–	–	–	–	–	–	$A, i_d, i_\ell = f(\omega^{\frac{1}{2}}), r$	
		0.370			16.5	–	–	–	–	–	–	$A, i_d, i_\ell = f(\omega^{\frac{1}{2}}), r$	
		0.380			21.0	–	–	–	–	–	–	$A, i_d, i_\ell = f(\omega^{\frac{1}{2}}), r$	
ER003 0623	141 ij	$E_{\frac{1}{2}}$	-0.63F	NCE	i_ℓ	0.14F	–	–	–	–	–	–	A, i_a, p	DB09
		-0.63F			0.20F	–	–	–	–	–	–	A, i_a, p	
		-0.4F			0.18F	–	–	–	–	–	–	A, i_a	
		-0.63F			0.23F	–	–	–	–	–	–	A, i_a, X, p	
		-0.4F			0.66F	–	–	–	–	–	–	A, i_a	
		-0.63			0.26F	–	–	–	–	–	–	A, i_a, p	
		-0.45			0.79F	–	–	–	–	–	–	$A, i_a, i_\ell \propto h$ for $C > 0.5$	
		-0.63			0.29F	–	–	–	–	–	–	A, i_a, p	
		-0.45			1.00F	–	–	–	–	–	–	A, i_a, Mx	
C0036 1035	–	–	–	–	–	–	–	–	–	–	–	C, i_{cat}, i reaches max. value at pH=6, p	
		–			–							C, i_{cat}, i ↑ as T ↓ for pH < 7	
		–			–							C, i_{cat}	
ER003 0623	141 ij	E_{su}	-0.7F	NCE	i_{su}	0.11F	–	–	–	–	–	–	A, p	
		-0.7F			0.06F	–	–	–	–	–	–	A, p	
		-0.6F			0.04F	–	–	–	–	–	–	A	
		-0.45F			0.22F	–	–	–	–	–	–	A	
		-0.8F			0.44F	–	–	–	–	–	–	A, p	
		-0.65F			0.18F	–	–	–	–	–	–	A	
		-0.45F			0.22F	–	–	–	–	–	–	A	
		-0.85F			0.80F	–	–	–	–	–	–	A, p	
		-0.65F			0.13F	–	–	–	–	–	–	A	
		-0.45F			0.18F	–	–	–	–	–	–	A	
ER003 0623	141 ij	$E_{\frac{1}{2}}$	-0.67	NCE	i_ℓ	0.12F	–	–	–	–	–	–	A, i_a, p	
		-0.60			0.24F	–	–	–	–	–	–	A, i_d, p	
		–			–							A, i_a	
		-0.56			0.68F	–	–	–	–	–	–	A, i_d, p	
		-0.67			–							A, i_a	
ER003 0623	141 ij	E_{su}	-0.67F	NCE	i_{su}	0.08F	–	–	–	–	–	–	A, p	
		-0.67F			0.3F	–	–	–	–	–	–	A, p	
		-0.67F			0.6F	–	–	–	–	–	–	A, p	

DB10 $C_6H_7AsO_2$

Code No.	Empirical Formula	Name and WLN Code	Structural Formula	Solvent	Tech.	Medium		μ, M	pH	T, °C	Electrodes	App.	Experimental Parameters
DB10	$C_6H_7AsO_2$	phenylarsonous acid C.A. 25400-22-0 Q-AS-QR	$C_6H_5As(OH)_2$	H_2O	PY	HCl	0.1	-	-	-	DME/SCE	O4AO	$C=0.05, t=2.4, v=1.7$
													$C=0.1$
													$C=0.15$
													$C=0.4$
													$C=0.75$
					IV	HCl	0.1	-	-	-	DME/SCE	O4AO	$C=0.1, t=2.4, E_{app}=0.05 \rightarrow 0.15$
													$E_{app}=0.4 \rightarrow 0.55$
DB11	$C_6H_7AsO_3$	phenylarsonic acid C.A. 98-05-5 Q-AS-QO&R	$C_6H_5AsO(OH)_2$	H_2O	PY	HCl	0.1	-	-	-	DME/SCE	O4AO	$C=0.01-1.0, t=2.4, v=1.7$
					IV	HCl	0.1	-	-	-	DME/SCE	O4AO	$C=0.2, t=2.4, E_{app}=0.750$
													$E_{app}=-0.800$
													$E_{app}=-0.900$
													$E_{app}=-1.000$
													$E_{app}=-1.100$
DB12	$C_6H_7AsO_4$	4-hydroxyphenyl-arsonic acid C.A. 98-14-6 Q-AS-QO&R DQ	$4\text{-}HOC_6H_4AsO(OH)_2$	H_2O	PY	HCl	0.1	-	-	-	DME/SCE	O4AO	$C=0.01-1.0, t=2.4, v=1.7$
					IV	HCl	0.1	-	-	-	DME/SCE	O4AO	$C=0.2, t=2.4, E_{app}=-0.750$
													$E_{app}=-0.800$
													$E_{app}=-0.900$
CONT													$E_{app}=-1.000$

TABLE I. Electrochemical Data $C_6H_7AsO_4$ (CONT.) DB12

Ref.	C/M	Charact. Potential		Response Const.		n Tech.		Electrokinetic Data			Products and Identification	Description and Remarks	Code No.
		Value	vs.		Value			Parameter	Value	From			
AN100 0573	367 cd	$E_{\frac{1}{2}}$	-0.09F	SCE	i_ℓ	0.36F	QE	1.30 ± 0.06	$E\log i^0$	44F	sttd	-	$C,R,i_\ell \propto C$ for C=0.01-0.1, $i_\ell \propto h^{\frac{1}{2}}$, i_ℓ = $f(pH)$, $E_{\frac{1}{2}}=f(pH)$ for pH = 1-2.2, -0.61 + 0.44log(i_d-i),r	DB10
							$E\log i$!	1.25 to 1.35						
			-0.48F			0.24F			$dE_{\frac{1}{2}}/dpH$	59		-	$C, i_\ell \propto C$ for C=0.01-0.1, $i_\ell \propto h^{\frac{1}{2}}$, $i_\ell \neq$ $f(pH)$, $E_{\frac{1}{2}}=f(pH)$ for pH = 1-2.2	
			-	-	-	-	-	-	$dE_{\frac{1}{2}}/dlogC$	43	sttd	-	C,i_a,wave split into 3 parts,r	
			-0.46F -0.56F							0 160F			C C	
			-		-	-	-	-	-	-		-	C,i_a,wave split into 3 parts,r	
			-0.46F -0.59F							0 160F			C C	
			-		-	-	-	-	-	-		-	$C,i_a,i_\ell \neq f(T)$,wave split into 3 parts,r	
			-0.63F -0.70F							0 160F			C C	
			-0.08F		i_ℓ	0.8F	-	-	-	-		-	C,i_a,r	
			-0.26F			0.8F							C,i-t curves complex,$i_\ell \neq f(C)$,i_a	
			-0.43F			0.8F							C,i-t curves complex,$i_\ell \downarrow$ as C$\uparrow$,i_a	
			-0.72F -0.78F			3.2F 2.0F				160F 320F			C C,Mx	
AN100 0573	367 cd	-	-	SCE	$dlogi_\ell/dlogt$	≈0.19F	-	-	-	-	-	-	C,i_d,r	
		-	-	-		≈0.19F	-	-	-	-	-	-	C,$dlogi_\ell/dlogt\neq 0.19$ on rising part of wave,r	
AN100 0489	367 ab	$E_{\frac{1}{2}}$	-0.945	SCE	-	-	QE	4.05 ± 0.15	$dE_{\frac{1}{2}}/dpH$ p= αn_a	112 1 0.49	- Elog	-	$C,\neq,i_d,i \propto C, E_{\frac{1}{2}}=$ $f(pH), \Delta E_{\frac{1}{2}}=0.44pK_1$	DB11
AN100 0489	367 ab	-	-	-	$dlogi_\ell/dlogt$	0.37	-	-	-	-	-	-	C,r	
		-	-	-		0.33	-	-	-	-	-	-	C,r	
		-	-	-		0.27	-	-	-	-	-	-	C,r	
		-	-	-		0.24	-	-	-	-	-	-	C,r	
		-	-	-		0.26	-	-	-	-	-	-	C,r	
AN100 0489	367 ab	$E_{\frac{1}{2}}$	-0.9625	SCE	-	-	QE	4	$dE_{\frac{1}{2}}/dpH$ p=	115 1	- Elog	-	$C,\neq,i_d,i \propto C, E_{\frac{1}{2}}=$ $f(pH), \Delta E_{\frac{1}{2}}=0.056\sigma,r$	DB12
AN100 0489	367 ab	-	-	-	$dlogi_\ell/dlogt$	0.41	-	-	-	-	-	-	C,r	
		-	-	-		0.35	-	-	-	-	-	-	C,r	
		-	-	-		0.27	-	-	-	-	-	-	C,r	
		-	-	-		0.20	-	-	-	-	-	-	C,r	CONT

DB12 (CONT.) $C_6H_7AsO_4$

Code No.	Empirical Formula	Name and WLN Code	Structural Formula	Solvent	Tech.	Medium		μ, M	pH	T, °C	Electrodes	App.	Experimental Parameters
DB12	$C_6H_7AsO_4$	4-hydroxyphenyl-arsonic acid	$4\text{-HOC}_6H_4AsO(OH)_2$	H_2O	IV	HCl	0.1	-	-	-	DME/SCE	04A0	$C=0.2, t=2.4,$ $E_{app}=-1.100$
DB13	$C_6H_7AsO_5$	2,4-dihydroxyphenyl-arsonic acid C.A. 6269-96-1 Q-AS-QO&R BQ DQ	TABLE II.	H_2O	PY	HCl	0.1	-	-	-	DME/SCE	04A0	$C=0.1-1.0,$ $t=2.4, v=1.7$
					IV	HCl	0.1	-	-	-	DME/SCE	04A0	$C=0.2, t=2.4,$ $E_{app}=-0.750$
													$E_{app}=-0.800$
													$E_{app}=-0.900$
													$E_{app}=-1.000$
													$E_{app}=-1.100$
DB14	$C_6H_7LiO_2$	lithium 2-hexynoate C.A. 37881-11-1 OV1UU4 &-LI-	$CH_3CH_2CH_2C{\equiv}C\text{-}COOLi$	DMF	PY	Bu_4NBF_4	0.50	-	-	25	DME/SCE	23-0	$C=5.0-5.1,$ Pt Aux
DB15 ah21 cb89	C_6H_7NO	4-aminophenol C.A. 123-30-8 ZR DQ	$4\text{-HOC}_6H_4NH_2$	H_2O	VY	H_2SO_4	0.20	-	-	-	RRDE/-	---	E_{disc} on oxidn. Pl, $\omega=1.5$ $\omega=40$
DB16 ah34	C_6H_7NS	2-mercaptoaniline C.A. 137-07-5 ZR BSH	$2\text{-HSC}_6H_4NH_2$	H_2O	PY	BOR $CoCl_2$	? 4E-4	-	-	-	DME/SCE	0A-	ns
DB17	C_6H_7NS	4-mercaptoaniline C.A. 1193-02-8 ZR DSH	$4\text{-HSC}_6H_4NH_2$	H_2O	PY	BOR $CoCl_2$	? 4E-4	-	-	-	DME/SCE	0A-	ns
DB18	$C_6H_8AsNO_3$	2-aminophenyl-arsonic acid C.A. 2045-00-3 ZR B-AS-QQO	$2\text{-}H_2NC_6H_4AsO(OH)_2$	H_2O	PY	HCl	0.1	-	-	-	DME/SCE	04A0	$C=0.01-1.0,$ $t=2.4, v=1.7$
					IV	HCl	0.1	-	-	-	DME/SCE	04A0	$E_{app}=-0.750$
													$E_{app}=-0.800$
													$E_{app}=-0.900$
													$E_{app}=-1.000$
													$E_{app}=-1.100$
DB19 CONT	$C_6H_8AsNO_3$	4-aminophenyl-arsonic acid C.A. 98-50-0 ZR D-AS-QQO	$4\text{-}H_2NC_6H_4AsO(OH)_2$	H_2O	PY	HCl	0.1	-	-	-	DME/SCE	04A0	$C=0.01-1.0,$ $t=2.4, v=1.7$
					IV	HCl	0.1	-	-	-	DME/SCE	04A0	$C=0.2, t=2.4,$ $E_{app}=-0.750$
													$E_{app}=-0.800$

TABLE I. Electrochemical Data $C_6H_8AsNO_3$ (CONT.) DB19

Ref.	C/M	Charact. Potential Value	vs.	Response Const.	Value	n	Tech.	Electrokinetic Data Parameter	Value	From	Products and Identification	Description and Remarks	Code No.
AN100 0489	367 ab	$E_{\frac{1}{2}}$	-	-	$dlogi_\ell/dlogt$	0.25	-	-	-	-	-	-	C,r	DB12
AN100 0489	367 ab	$E_{\frac{1}{2}}$	-0.990	SCE	-	-	i:i	4	$dE_{\frac{1}{2}}/dpH$ p=	113 1	- Elog	-	C,≠,i_d,$i_\ell \propto C$,$E_{\frac{1}{2}}=$ f(pH),$\Delta E_{\frac{1}{2}} = -0.44 pK_1$	DB13
AN100 0489	367 ab	-	-	-	$dlogi_\ell/dlogt$	0.45	-	-	-	-	-	-	C,r	
		-	-	-		0.38	-	-	-	-	-	-	C,r	
		-	-	-		0.31	-	-	-	-	-	-	C,r	
		-	-	-		0.21	-	-	-	-	-	-	C,r	
		-	-	-		0.23	-	-	-	-	-	-	C,r	
JA094 8471	-	$E_{\frac{1}{2}}$	-2.31	SCE	-	-	-	-	αn_a	0.9	Elog	-	C,p	DB14
JE016 0041	105 c	-	-	-	N_0	0.424	-	-	-	-	-	-	A,r	DB15 ah21 cb89
		-	-	-		0.460	-	-	-	-	-	-	A,r	
C0036 1035	-	-	-	-	-	-	-	-	-	-	-	-	C,i_{cat}, i reaches max. value at pH=5, $pK_1=2.9$, $pK_2=5.9$, p	DB16 ah34
		-	-	-	-	-	-	-	-	-	-	-	C,i_{cat}, $i\uparrow$ as $T\downarrow$ for pH<6	
		-	-	-	-	-	-	-	-	-	-	-	C,i_{cat}	
C0036 1035	-	-	-	-	-	-	-	-	-	-	-	-	C,i_{cat}, i reaches max. value at pH=7, pK=7.95, p	DB17
		-	-	-	-	-	-	-	-	-	-	-	C,i_{cat}, $i\uparrow$ as $T\downarrow$ for pH<8	
		-	-	-	-	-	-	-	-	-	-	-	C,i_{cat}	
AN100 0489	367 ab	$E_{\frac{1}{2}}$	-0.937	SCE	-	-	i:i	4	$dE_{\frac{1}{2}}/dpH$ p=	123 1	- Elog	-	C,≠,i_d,$i_\ell \propto C$,$E_{\frac{1}{2}}=$ f(pH),r	DB18
		-	-	-	$dlogi_\ell/dlogt$	0.35	-	-	-	-	-	-	C,r	
		-	-	-		0.32	-	-	-	-	-	-	C,r	
		-	-	-		0.26	-	-	-	-	-	-	C,r	
		-	-	-		0.23	-	-	-	-	-	-	C,r	
		-	-	-		0.28	-	-	-	-	-	-	C,r	
AN100 0489	367 ab	$E_{\frac{1}{2}}$	-0.980	SCE	-	-	i:i	4	$dE_{\frac{1}{2}}/dpH$ p=	123 1	- Elog	-	C,≠,i_d,$i_\ell \propto C$,$E_{\frac{1}{2}}=$ f(pH),$\Delta E_{\frac{1}{2}} = 0.056\sigma$, $\Delta E_{\frac{1}{2}} = -0.44 pK_1$,r	DB19
AN100 0489	367 ab	-	-	-	$dlogi_\ell/dlogt$	0.43	-	-	-	-	-	-	C,r	
		-	-	-		0.36	-	-	-	-	-	-	C,r	

CONT

DB19 (CONT.) $C_6H_8AsNO_3$

Code No.	Empirical Formula	Name and WLN Code	Structural Formula	Solvent	Tech.	Medium		μ, M	pH	T, °C	Electrodes	App.	Experimental Parameters
DB19	$C_6H_8AsNO_3$	4-aminophenyl-arsonic acid	$4-H_2NC_6H_4AsO(OH)_2$	H_2O	IV	HCl	0.1	-	-	-	DME/SCE	04A0	$C=0.2, t=2.4$, $E_{app}=-0.900$
													$E_{app}=-1.000$
													$E_{app}=-1.100$
DB20	$C_6H_8AsNO_4$	3-amino-4-hydroxy-phenylarsonic acid C.A. 2163-77-1 ZR BQ E-AS-QQO	TABLE II.	H_2O	PY	HCl	0.1	-	-	-	DME/SCE	04A0	$C=0.01-1.0$, $t=2.4, v=1.7$
					IV	HCl	0.1	-	-	-	DME/SCE	04A0	$C=0.2, t=2.4$, $E_{app}=-0.750$
													$E_{app}=-0.800$
													$E_{app}=-0.900$
													$E_{app}=-1.000$
													$E_{app}=-1.100$
DB21 cc02	$C_6H_8O_4$	dimethyl fumarate C.A. 624-49-7 1OV1U1VO1 -T	$CH_3O_2CCH:CH-COOCH_3$	MeCN	PY	Bu_4NClO_4	0.1	-	-	23	DME Ag/0.01 $AgNO_3$	2A0	$C=2$, Pt Aux
DB22 cc03	$C_6H_8O_4$	dimethyl maleate C.A. 624-48-6 1OV1U1VO1 -C	cis-$CH_3O_2CH:CH-COOCH_3$	MeCN	PY	Bu_4NClO_4	0.1	-	-	23	DME Ag/0.01 $AgNO_3$	2A0	$C=2$, Pt Aux
DB23 ah54	$C_6H_8O_6$	ascorbic acid C.A. 10504-35-5 T5OV EHJ CQ DQ EYQ1Q	TABLE II-2.	H_2O	IL	HCl	0.1	-	1	25±0.1	XEI/SCE	26B0	$C=0.01-1.0$, $v=10, A=0.28$, Pt Aux
						BR			2				
									3				
									5.6				
DB24	C_6H_9NO	1-vinyl-2-pyrrolid-one C.A. 88-12-0 T5NVTJ A1U1	TABLE II.	DMF	PY	Et_4NI	0.05	-	-	-	DME/MP	0A0	$C=0.728, EC=0.77, t=17.5$
DB25	$C_6H_{10}HgO_2$	diacetonylmercury C.A. 6704-33-2 1OV1-HG-1VO1	$(CH_3COCH_2)_2Hg$	MeOH 50	PY	KNO_3	0.1	-	-	25	DME/SCE	0-0	$t=3.00, m=0.67$
DB26	$C_6H_{10}NNaS_2$	sodium 1-piperidine-dithiocarboxylate C.A. 873-57-4 T6NTJ AYS&US &-NA-	TABLE II.	H_2O	PY	KOH KCl TX100	0.01 0.3 0.008	-	-	25±0.2	DME/SCE	01A0	$C=1.0, m=2.52, t=3.00, h=40$
					IL	KOH KCl	0.01 0.3	-	-	25±0.2	HMDE SCE	01A0	$C=1.0, v=20$
DB27	$C_6H_{10}O$	3-methyl-3-penten-2-one C.A. 565-62-8 2UY1&V1	$CH_3CH:C(CH_3)CO-CH_3$	DMF	PY	Bu_4NBF_4	0.50	-	-	25	DME/SCE	23-0	$C=3.0-7.5$, Pt Aux

TABLE I. Electrochemical Data $C_6H_{10}O$ DB27

Ref.	C/M	Charact. Potential			Response Const.		n Tech.	n	Electrokinetic Data			Products and Identification	Description and Remarks	Code No.
			Value	vs.		Value			Parameter	Value	From			
AN100 0489	367 ab	-	-	-	$dlogi_\ell/dlogt$	0.30	-	-	-	-	-	-	C,r	DB19
			-			0.19	-	-	-	-	-	-	C,r	
			-			0.25	-	-	-	-	-	-	C,r	
AN100 0489	367 ab	$E_{\frac{1}{2}}$	-0.975	SCE	-	-	i:i p=	4	$dE_{\frac{1}{2}}/dpH$	122 1	- Elog	-	$C,\neq,i_d,i_\ell \propto C, E_{\frac{1}{2}} = f(pH), \Delta E_{\frac{1}{2}} = 0.056\sigma, r$	DB20
AN100 0489	367 ab	-	-	-	$dlogi_\ell/dlogt$	0.40	-	-	-	-	-	-	C,r	
			-			0.35	-	-	-	-	-	-	C,r	
			-			0.28	-	-	-	-	-	-	C,r	
			-			0.21	-	-	-	-	-	-	C,r	
			-			0.27	-	-	-	-	-	-	C,r	
JA095 1703	-	$E_{\frac{1}{2}}$	-1.685	Ag/0.01 AgNO$_3$	I	1.89	sttd	0.76	αn_a	1.1	Tomeš	-	C,r	DB21 cc02
JA095 1703	-	$E_{\frac{1}{2}}$	-1.82	Ag/0.01 AgNO$_3$	I	1.89	sttd	0.75	αn_a	0.67	Tomeš	-	C,r	DB22 cc03
AN100 0339	-	$E_{p/2}$	0.42F	SCE	-	-	-	-	$dE_{p/2}/dpH$	73	sttd	-	$A, i_p \propto C, XEI=CPE$ with wax and silicone oil, r	DB23 ah54
			0.35F		-	-	-	-		73		-	$A, i_p \propto C, r$	
			0.30F		-	-	-	-		73		-	$A, i_p \propto C, r$	
			0.14		-	-	-	-	-	-	-	-	$A, i_p \propto C, r$	
RR020 1274	-	-	-	-	-	-	-	-	-	-	-	-	O,p	DB24
ER002 0587	23 d	$E_{\frac{1}{2}}$	-0.62	SCE	-	-	sttd	2	α	0.44	sttd	-	$C,\neq,r?$	DB25
AJ028 0021	-	$E_{\frac{1}{2}}$	-0.456 -0.623	SCE	i_ℓ	2.4F	-	-	-	-	-	-	A,r	DB26
AJ028 0021	-	E_p	-0.470	SCE	-	-	-	-	$E_p - E_{p/2}$	19	sttd	-	A,r	
JA094 8471	-	$E_{\frac{1}{2}}$	-2.18	SCE	-	-	-	-	αn_a	0.9	Elog	-	C,p	DB27

DB28 C₆H₁₀OS₂   CRC Handbook Series in Organic Electrochemistry

Code No.	Empirical Formula	Name and WLN Code	Structural Formula	Solvent	Tech.	Medium		μ, M	pH	T, °C	Electrodes	App.	Experimental Parameters
DB28 cc11	$C_6H_{10}OS_2$	O-ethyl thioaceto-thioacetate C.A. 34827-66-2 SHYU1YUS&O2	HSC(CH₃):CHC-(:S)OC₂H₅	Me₂CO	PY	Et₄NClO₄	0.1	-	-	20	DME Ag/AgCl, 0.1LiCl	12A-	m=1.48,t=4.70,h=40, EC=1.68, W Aux
DB29	$C_6H_{10}O_2$	methyl 4-pentenoate C.A. 818-57-5 1U3VO1	CH₂:CHCH₂CH₂-COOCH₃	MeCN	PY	Bu₄NClO₄	0.1	-	-	23	DME Ag/0.01 AgNO₃	2AO	C=2, Pt Aux
DB30 ah85	$C_6H_{11}NO_2$	nitrocyclohexane C.A. 1122-60-7 L6TJ ANW	TABLE II-2.	ns	PY	ns		-	-	-	DME/SCE	-	-
DB31	$C_6H_{12}N_2$	1,4-diazabicyclo-[2.2.2]octane C.A. 280-57-9 T66 A B CN FNTJ	TABLE II.	MeCN	VR	NaClO₄	0.1	-	-	-	AuBE SCE	2AO	C=3.11,d=0.15,v=56, MP Aux,PE v=110
DB32	$C_6H_{12}N_2$	1,5-diazabicyclo-[3.2.1]octane C.A. 280-28-4 T56 A BN ENTJ	TABLE II.	MeCN	VR	NaClO₄	0.1	-	-	-	AuBE SCE	2AO	C=4.64,d=0.15,v=124, MP Aux,PE
DB33	$C_6H_{12}N_2$	1,5-diazabicyclo-[3.3.0]octane T55N ENTJ	TABLE II.	MeCN	VR	NaClO₄	0.1	-	-	-	AuBE SCE	2AO	C=0.2-0.9, d=0.15,v=120,MP Aux, PE
DB34	$C_6H_{12}N_2$	1,2-dimethyl-1,2,3,6-tetrahydropyridazine C.A. 26163-36-0 T6NN DUTJ A1 B1	TABLE II.	MeCN	VR	NaClO₄	0.1	-	-	-	AuBE SCE	2AO	C=0.2-0.9, d=0.15,v=57, MP Aux,PE
DB35 ah94 cc20	$C_6H_{12}N_2O_4S_2$	L-cystine C.A. 56-89-3 QVYZ15 2	[HOOCCH(NH₂)CH₂S·]₂	H₂O	PY	HCl	0.1	-	-	-	DME/SCE	05AO	C=1,m=1.75, t=4.66,h=45
						ACET	?		1.99				
									4.4				
						BOR PHOS	?		7.2				
									8.8				
						PHOS	?		9.95				
						HCl 0.1 dodecylamine ClO₄ 0.0015			-				
						ACET ? dodecylamine ClO₄ 0.0015			3.18				
									4.3				
CONT						PHOS BOR dodecylamine ClO₄ 0.0015	?		7.2				

TABLE I. Electrochemical Data $C_6H_{12}N_2O_4S_2$ (CONT.) DB35

Ref.	C/M	Charact. Potential		Response Const.		n		Electrokinetic Data			Products and Identification	Description and Remarks	Code No.
		Value	vs.		Value	Tech.		Parameter	Value	From			
JA095 1449	335 c	$E_{\frac{1}{2}}$	0.26	Ag/AgCl, 0.1LiCl	-	-	-	-	Tomeš Elog	60 63	sttd -	-	$A,E_{\frac{1}{2}}=f(C),E_{\frac{1}{2}}$ at t = 1.78,r $C,E_{\frac{1}{2}}$ at t = 1.85	DB28 cc11
			-1.428						Tomeš	68	-			
JA095 1703	-	-	-	-	-	-	-	-	-	-	-	-	O,p	DB29
ER001 1345	-	-	-	-	-	-	-	-	-	-	-	-	see AH85	DB30 ah85
JA094 7114	-	E_p	0.60 -	SCE	-	-	-	-	$E_p-E_{p/2}$	50 -	sttd	-	A,p C,O	DB31
		XE	0.57 -		-	-	-	-	ΔE_p	85 -	sttd	-	$A,XE=\frac{1}{2}(E_{p,A}+E_{p,C})$, r C	
JA094 7114	-	E_p	1.20 -	SCE	-	-	-	-	$E_p-E_{p/2}$	140	sttd	-	A,p	DB32
JA094 7108	-	XE	0.02 -	SCE	-	-	sttd	1 1	ΔE_p	90 -	sttd	radical cation, ESR	$A,R,XE=\frac{1}{2}(E_{p,A}+E_{p,C})$, r C	DB33
JA094 7108	-	XE	0.28 -	SCE	-	-	sttd	1 1	ΔE_p	70 -	sttd	radical cation, ESR	$A,R,XE=\frac{1}{2}(E_{p,A}+E_{p,C})$, r C	DB34
EA017 1615	141 g	$E_{\frac{1}{2}}$	- -0.40F	SCE	-	-	-	-	-	-	-	-	C,r C,i ∝ C,Mx	DB35 ah94 cc20
			- -0.45F		-	-	-	-	-	-	-	-	C,r C,i ∝ C,Mx	
			- -0.62F		-	-	-	-	-	-	-	-	C,r C,i ∝ C	
	141 h		- -0.75		-	-	-	-	-	-	-	-	C,r C,i ∝ C	
			- -1.0F		-	-	-	-	-	-	-	-	C,r C,i ∝ C	
			- -1.30F		-	-	-	-	-	-	-	-	C,r C,i ∝ C	
	-		- -0.87F		-	-	-	-	-	-	-	-	C,r C,i ∝ C	
			- -0.97F		-	-	-	-	-	-	-	-	C,r C,i ∝ C,i = f[surfactant]	
			- -1.06F		-	-	-	-	-	-	-	-	C,r C,i ∝ C,i = f[surfactant]	
			- -1.13F		-	-	-	-	-	-	-	-	C,r C,i ∝ C,i = f[surfactant]	

CONT

DB35 (CONT.) $C_6H_{12}N_2O_4S_2$

Code No.	Empirical Formula	Name and WLN Code	Structural Formula	Solvent	Tech.	Medium	μ, M	pH	T, °C	Electrodes	App.	Experimental Parameters
DB35 ah94 cc20	$C_6H_{12}N_2O_4S_2$	L-cystine	[HOOCCH(NH_2) $CH_2S\cdot$]$_2$	H_2O	PY	PHOS ? dodecylamine ClO_4 0.0015	-	10.0	-	DME/SCE	O5AO	C=1,m=1.75, t=4.66,h=45
						HCl 0.1 dodecylben- zene SO_3Na 0.004		-				
						ACET ? dodecylben- zene SO_3Na 0.004		3.4				
								4.6				
						BOR ? PHOS dodecylben- zene SO_3Na 0.004		6.6				
						PHOS ? dodecylben- zene SO_3Na 0.004		10.0				
						HCl 0.1 TX100 0.004		-				
						ACET ? TX100 0.004		2.0				
								3.0				
								4.0				
						PHOS ? TX100 0.004		6.8				
						BOR ? TX100 0.004		9.9				
						NH_4Cl 0.1 NH_3 0.1 $CoCl_2$ 0.001	-	?	-	DME/NCE	OAO	C=4E-3
DB36	$C_6H_{12}N_4$	1,3,5,7-tetraaza- tricyclo[3.3.1.1^{3,7}]- decane C.A. 100-97-0 T666/BI/DJ/HJ A J 2BF J BN DN FN HNTJ	TABLE II.	MeCN	VR	$NaClO_4$ 0.1	-	-	-	AuBE SCE	2AO	C=1.14,d= 0.15,v=114, MP Aux,PE
DB37 ai09	$C_6H_{12}O_6$	D-glucose C.A. 50-99-7 VH/YQ 41Q	TABLE II-2.	H_2O	PY	KCl 0.1	-	-	-	DME/SCE	-	-
DB38 ai13	$C_6H_{13}NO$	cyclohexylhydroxyl- amine C.A. 2211-64-5 L6TJ AMO	TABLE II.	H_2O?	PY	buffer?	-	-	-	DME/SCE	-	-
DB39	$C_6H_{14}N_2$	1,4-dimethylhexa- hydropyrazine C.A. 106-58-1 TGN DNTJ A1 D1	TABLE II.	MeCN	VR	$NaClO_4$ 0.1	-	-	-	AuBE SCE	2AO	C=3.16,d= 0.15,v=99, MP Aux,PE

TABLE I. Electrochemical Data $C_6H_{14}N_2$ DB39

Ref.	C/M	Charact. Potential Value	vs.	Response Const.	Value	n Tech.		Electrokinetic Data Parameter	Value	From	Products and Identification	Description and Remarks	Code No.
EA017 1615	-	$E_{\frac{1}{2}}$	-1.25F	SCE	-	-	-	-	-	-	-	-	C,r $C, i \propto C, i = f[\text{surfactant}]$	DB35 ah94 cc20
			-0.45		-	-	-	-	-	-	-	-	C,r $C, i \propto C, i = f[\text{surfactant}]$	
			-0.72F		-	-	-	-	-	-	-	-	C,r $C, i \propto C, i = f[\text{surfactant}]$	
			-1.00F		-	-	-	-	-	-	-	-	C,r $C, i \propto C, i = f[\text{surfactant}]$	
			-1.23F		-	-	-	-	-	-	-	-	C,r $C, i \propto C, i = f[\text{surfactant}]$	
			-1.42F		-	-	-	-	-	-	-	-	C,r $C, i \propto C, i = f[\text{surfactant}]$	
	141 g		-1.08F		-	-	-	-	-	-	-	-	C,r $C, i \propto C, i = f[\text{surfactant}]$	
			-1.23F		-	-	-	-	-	-	-	-	C,r $C, i \propto C, i = f[\text{surfactant}]$	
			-1.31F		-	-	-	-	-	-	-	-	C,r $C, i \propto C, i = f[\text{surfactant}]$	
			-1.50F		-	-	-	-	-	-	-	-	C,r $C, i \propto C, i = f[\text{surfactant}]$	
			-1.65F		-	-	-	-	-	-	-	-	C,r $C, i \propto C, i = f[\text{surfactant}]$	
	141 h		-1.72		-	-	-	-	-	-	-	-	C,r $C, i \propto C, i = f[\text{surfactant}]$	
C0032 0246	-	-	-	-	i_{Mx}	50	-	-	-	-	-	-	$C, i_{cat}, Mx, i_\ell = f(pH)$ and reaches max value at pH = 10.3, $i_\ell = 9350C/(1+55C)$, r	
JA094 7114	-	E_p	1.37	SCE	-	-	-	-	$E_p - E_{p/2}$	130	sttd	-	A,p	DB36
BJ046 2107	-	-	-	-	-	-	-	-	-	-	-	-	see A109	DB37 ai09
ER001 1345	-	-	-	-	-	-	-	-	-	-	-	-	see A113	DB38 ai13
JA094 7114	-	E_p	0.75	SCE	-	-	-	-	$E_p - E_{p/2}$	115	sttd	-	A,p	DB39

DB40 $C_6H_{14}N_2$

Code No.	Empirical Formula	Name and WLN Code	Structural Formula	Solvent	Tech.	Medium		μ, M	pH	T, °C	Electrodes	App.	Experimental Parameters
DB40	$C_6H_{14}N_2$	1,2-dimethylhexahydropyridazine C.A. 26163-37-1 T6NNTJ A1 B1	TABLE II.	MeCN	VR	$NaClO_4$	0.1	-	-	-	AuBE SCE	2A0	C=0.2-0.9, d=0.15,v=61, MP Aux,PE
DB41	$C_6H_{14}N_2O$	3,4-diethyl-1,3,4-oxadiazolidine C.A. 38705-00-9 T5NN DOTJ A2 B2	TABLE II.	MeCN	VR	$NaClO_4$	0.1	-	-	-	AuBE SCE	2A0	C=0.2-0.9, d=0.15,v=8, MP Aux,PE
DB42 ai31	$C_6H_{14}N_2O$	diisopropyl-N-nitrosoamine C.A. 601-77-4 OUNNY1&1&Y1&1	$[(CH_3)_2CH]_2NNO$	H_2O	PT	BR		-	2.0	-	DME/SCE	04A0	C=0.05,m=1.73,t=1,h=80,v=10
DB43 ai38	$C_6H_{15}N$	triethylamine C.A. 121-44-8 2N2&2	$(CH_3CH_2)_3N$	m-cresol	PY	Et_4NClO_4 0.1 Me_4NCl 0.1		-	-	-	DME Ag/AgCl,	216A0	C=0.5-1.5, m=2.87,t=2.70,h=56, MP Aux
				MeCN	VR	$NaClO_4$	0.1	-	-	-	AuBE SCE	2A0	C=2.77,d=0.15,v=63, MP Aux,PE
DB44	$C_6H_{15}N_3$	1,3,5-trimethylperhydro-1,3,5-triazine C.A. 108-74-7 T6N CN ENTJ A1 C1 E1	TABLE II.	MeCN	VR	$NaClO_4$	0.1	-	-	-	AuBE SCE	2A0	C=2.32,d=0.15,v=104, MP Aux,PE
DB45	$C_6H_{15}N_3O_2S_2$	diammonium proline-N-dithiocarboxylate C.A. 49540-66-1 T5NTJ AS&US BVO &K &K	TABLE II.	H_2O	PY	NaOH	0.1	-	-	25±0.1	DME/SCE	25A0	C=0.5,m=1.32,t=5.57,h=65,Pt Aux
					PV	NaOH	0.1	-	-	25±0.1	DME/SCE	25A0	C=0.5,m=1.32,t=5.57,h=65,Pt Aux
DB46	$C_6H_{16}N_2$	N,N,N',N'-tetramethylethylenediamine C.A. 110-18-9 1N1&2N1&1	$[(CH_3)_2NCH_2\cdot]_2$	MeCN	VR	$NaClO_4$	0.1	-	-	-	AuBE SCE	2A0	C=4.20,d=0.15,v=54, MP Aux,PE
DB47	$C_6H_{16}N_4$	1,2,4,5-tetramethylperhydro-s-tetrazine C.A. 20717-38-8 T6NN DNNTJ A1 B1 D1 E1	TABLE II.	MeCN	VR	$NaClO_4$	0.1	-	-	-	AuBE SCE	2A0	C=0.2-0.9, d=0.15,v=64, MP Aux,PE
DB48	C_7H_4BrN	2-bromobenzonitrile C.A. 2042-37-7 NCR BE	$2\text{-}BrC_6H_4CN$	DMF	VR	Et_4NClO_4	0.1	-	-	-	PBE/SCE	125A F02	v=80.6, Pt Aux
DB49 ai59	C_7H_4BrN	3-bromobenzonitrile C.A. 6952-59-6 NCR CE	$3\text{-}BrC_6H_4CN$	DMF	VR	Et_4NClO_4	0.1	-	-	-	PBE/SCE	125A F02	v=80.6, Pt Aux
DB50 ai60	C_7H_4BrN	4-bromobenzonitrile C.A. 623-00-7 NCR DE	$4\text{-}BrC_6H_4CN$	DMF	VR	Et_4NClO_4	0.1	-	-	-	PBE/SCE	125A F02	v=80.6, Pt Aux

TABLE I. Electrochemical Data C_7H_4BrN DB50

Ref.	C/M	Charact. Potential		Response Const.		n		Electrokinetic Data			Products and Identification	Description and Remarks	Code No.
		Value	vs.		Value	Tech.		Parameter	Value	From			
JA094 7108	-	XE 0.18 -	SCE	-	-	sttd	1 1	ΔE_p	60 -	sttd	radical cation, ESR	A,R,$XE=\frac{1}{2}(E_{p,A}+E_{p,C})$, r C	DB40
JA094 7108	-	XE 0.56 -	SCE	-	-	sttd	1 1	ΔE_p	70 -	sttd	radical cation, ESR	A,R,$XE=\frac{1}{2}(E_{p,A}+E_{p,C})$, r C	DB41
AA078 0081	82 bd	$E_\frac{1}{2}$ -0.84	SCE	-	-	QE	4	-	-	-	-	C,i_d,r	DB42 ai31
AA072 0169	-	$E_\frac{1}{2}$ 0.39	Ag/AgCl	I	0.395	-	-	-	-	-	-	A,$\frac{1}{4}$,m-cresolate oxidn., pK_a=14.2,r	DB43 ai38
JA094 7114	-	E_p 0.78	SCE	-	-	-	-	$E_p-E_{p/2}$	110	sttd	-	A,p	
JA094 7114	-	E_p 0.94	SCE	-	-	-	-	$E_p-E_{p/2}$	95	sttd	-	A,p	DB44
BJ046 2151	-	$E_\frac{1}{2}$ -0.65 -	SCE	i_ℓ	0.15F -	-	-	-	-	-	Hg salt	A,i_a,i$\propto$C for C< 0.1,r A,M,Γ=4.8	DB45
BJ046 2151	-	E_{su} -0.61 -	SCE	i_{su}	0.66F -	-	-	-	-	-	Hg salt	A,r A	
JA094 7114	-	E_p 0.67	SCE	-	-	-	-	$E_p-E_{p/2}$	100	sttd	-	A,p	DB46
JA094 7108	-	XE 0.32 -	SCE	-	-	sttd	1 1	ΔE_p	55 -	sttd	radical cation, ESR	A,R,$XE=\frac{1}{2}(E_{p,A}+E_{p,C})$, r C	DB47
JA094 7526	72 c	E_p -1.87 -2.32	SCE	-	-	QE	1.99 -	$E_p-E_{p/2}$	80 -	sttd	benzonitrile,88%, GLC,VR	C,r C	DB48
JA094 7526	72 c	E_p -1.95 -2.32	SCE	-	-	QE	2.09 -	$E_p-E_{p/2}$	100 -	sttd	benzonitrile,90%, GLC,VR	C,r C	DB49 ai59
JA094 7526	72 c	E_p -1.92 -2.32	SCE	-	-	QE	1.92 -	$E_p-E_{p/2}$	90 -	calc	benzonitrile,92-100%,GLC,VR	C,r C	DB50 ai60

Code No.	Empirical Formula	Name and WLN Code	Structural Formula	Solvent	Tech.	Medium		μ, M	pH	T, °C	Electrodes	App.	Experimental Parameters
DB51	C_7H_4ClN	2-chlorobenzonitrile C.A. 873-32-5 NCR BG	$2\text{-}ClC_6H_4CN$	DMF	VR	Et_4NClO_4	0.1	-	-	-	PBE/SCE	125A F02	v=80.6, Pt Aux
DB52	C_7H_4ClN	3-chlorobenzonitrile C.A. 766-84-7 NCR CG	$3\text{-}ClC_6H_4CN$	DMF	VR	Et_4NClO_4	0.1	-	-	-	PBE/SCE	125A F02	v=80.6, Pt Aux
DB53 ai64	C_7H_4ClN	4-chlorobenzonitrile C.A. 623-03-0 NCR DG	$4\text{-}ClC_6H_4CN$	DMF	VA	Et_4NClO_4	0.1	-	-	-	PBE/SCE	125A F02	C=2.15, v=80.6, Pt Aux
					VR	Et_4NCN	0.2	-	-	-	PBE/SCE	125A F02	0→-2.2 (30 s)→ -0.8→-2.2, C=3.75, v=80.6
					IR	Et_4NClO_4	0.1	-	-	-	PBE/SCE	125A F02	E_{app}=-2.60, t=0.004-8 s
DB54 ai77	$C_7H_4F_3NO_2$	2-nitro(trifluoro-methyl)benzene C.A. 384-22-5 WNR BXFFF	$2\text{-}CF_3C_6H_4NO_2$	DMF	VR	Pr_4NClO_4	0.1	-	-	-	PDE/SCE	235A 0	C=1, A=0.8, v=283.17, Pt Aux
DB55 ai78	$C_7H_4F_3NO_2$	3-nitro(trifluoro-methyl)benzene C.A. 98-46-4 WNR CXFFF	$3\text{-}CF_3C_6H_4NO_2$	DMF	VR	Pr_4NClO_4	0.1	-	-	-	PDE/SCE	235A 0	C=1, A=0.8, v=295.17, Pt Aux
DB56	$C_7H_4F_3NO_2$	4-nitro(trifluoro-methyl)benzene C.A. 402-54-0 WNR DCFFF	$4\text{-}CF_3C_6H_4NO_2$	DMF	VR	Pr_4NClO_4	0.1	-	-	-	PDE/SCE	235A 0	C=1, A=0.8, v=267.66, Pt Aux
DB57 ai80	C_7H_4IN	4-iodobenzonitrile C.A. 3058-39-7 NCR DI	$4\text{-}IC_6H_4CN$	DMF	IR	Et_4NI	0.1	-	-	-	PBE/SCE	125A F0	E_{app}=-2.60
					VR	Et_4NI	0.1	-	-	-	PBE/SCE	125A F02	v=80.6, Pt Aux
DB58	$C_7H_4N_2O_2$	benzimidazole-4,7-quinone C.A. 7711-39-9 T56 BM DN FV IVJ	TABLE II.	DMF	PY	Et_4NClO_4	0.1	-	-	-	DME/MP	0-0	EC=1.23
DB59 ai87 cc40	$C_7H_4N_2O_2$	4-nitrobenzonitrile C.A. 619-72-7 WNR DCN	$4\text{-}O_2NC_6H_4CN$	DMF	VR	Et_4NClO_4	0.1	-	-	-	PBE/SCE	125A F02	v=80.6, Pt Aux

TABLE I. Electrochemical Data $C_7H_4N_2O_2$ DB59

Ref.	C/M	Charact. Potential		Response Const.		n Tech.	n	Electrokinetic Data			Products and Identification	Description and Remarks	Code No.
		Value	vs.		Value			Parameter	Value	From			
JA094 7526	72 c	E_p	-1.91 -2.32	SCE	-	-	QE	2.20 -	$E_p-E_{p/2}$	110 -	calc	benzonitrile,90%, GLC,VR	C,r C	DB51
JA094 7526	72 c	E_p	-1.99 -2.32	SCE	-	-	QE	2.10 -	$E_p-E_{p/2}$	80 -	calc	benzonitrile,89% GLC,VR	C,r C	DB52
JA094 7526	72 c	E_p	-1.96 -2.32 -2.26F	SCE	i_p	166F 73F 100F	QE	2.02 -	$E_p-E_{p/2}$	80 -	calc	benzonitrile,85-95%,GLC,VR	C,ΔE_p=75 for v = 810,r C A	DB53 ai64
JA094 7526	72 c	E_p	-2.04F -1.46F -1.56F	SCE	i_p	290F 40F 40F	-	-	-	-	-	-	C,r A,R C,R	
JA094 7526	72 c	-	-	-	$it^{\frac{1}{2}}/C$	120	sttd	3	-	-	-	-	C,r	
AA071 0433	-	E_p	-1.02 -2.10	SCE	$i_p/v^{\frac{1}{2}}$	6.54 12.04	- IR	- 1.8	ΔE_p	70 -	sttd	- 2-$F_3CC_6H_4NO^-$,$E_{p,c}$= -0.68 V vs. SCE, $i_p/v^{\frac{1}{2}}$=0.22 for v= 100-180,ΔE_p=59, $i_{p,A}/i_{p,C}$=1.0	C,R,$i_{p,A}/i_{p,C}$=1.0, $i_{p,C}/v^{\frac{1}{2}} \neq f(v^{\frac{1}{2}})$,r C,ece	DB54 ai77
AA071 0433	-	E_p	-0.99 -2.21	SCE	$i_p/v^{\frac{1}{2}}$	12.04 18.19	- IR	- 1.9	ΔE_p	70 -	sttd	- 3-$F_3CC_6H_4NO^-$,$E_{p,c}$= -0.72 V vs. SCE, $i_p/v^{\frac{1}{2}}$=0.09 for v= 100-180,ΔE_p=67	C,R,$i_{p,A}/i_{p,C}$=1.0, $i_{p,C}/v^{\frac{1}{2}} \neq f(v^{\frac{1}{2}})$,r C,ece	DB55 ai78
AA071 0433	-	E_p	-0.88 -1.95	SCE	$i_p/v^{\frac{1}{2}}$	8.22 19.24	- IR	- 2.1	ΔE_p	60 -	-	-	C,R,$i_{p,A}/i_{p,C}$=1.0, $i_{p,C}/v^{\frac{1}{2}} \neq f(v^{\frac{1}{2}})$,r C,ece	DB56
JA094 7526	72 c	-	-	-	$i\tau^{\frac{1}{2}}/C$	110	sttd	3	-	-	-	-	C,i_d,r	DB57 ai80
JA094 7526	72 c	E_p	-1.35 -1.84 -2.32	SCE	-	-	-	-	$E_p-E_{p/2}$	130 -	calc	-	C,Pr dependent on pretreatment of electrode,i_a,r C C	
RY002 1096	-	$E_{\frac{1}{2}}$	-0.44 -	SCE	-	-	-	-	-	-	-	-	C,p C,X	DB58
JA094 7526	-	E_p	-0.83 -1.64	SCE	-	-	sttd	1 1	$E_p-E_{p/2}$	60 -	calc	radical anion	C,R,r C	DB59 ai87 cc40

Code No.	Empirical Formula	Name and WLN Code	Structural Formula	Solvent	Tech.	Medium		μ, M	pH	T, °C	Electrodes	App.	Experimental Parameters
DB60 ai93 cc44	C_7H_5ClO	4-chlorobenzaldehyde C.A. 104-88-1 VHR DG	$4\text{-}ClC_6H_4CHO$	MeOH 98.8	PY	ACET GEL	0.003	-	≈1.3?	-	DME/MP	OAO	C=3.18
				EtOH 96.0	PY	ACET GEL	0.003	-	≈1.3?	-	DME/MP	OAO	C=3.18
				PrOH 96	PY	ACET GEL	0.003	-	≈1.3?	-	DME/MP	OAO	C=3.18
				BuOH 96	PY	ACET GEL	0.003	-	≈1.3?	-	DME/MP	OAO	C=3.18
DB61 aj09 cc45	C_7H_5N	benzonitrile C.A. 100-47-0 NCR	C_6H_5CN	DMF	VR	Et_4NClO_4	0.1	-	-	-	PBE/SCE	125A F02	C=1.44, v= 80.6, Pt Aux
DB62 aj14 cc47	$C_7H_5NO_3$	3-nitrobenzaldehyde C.A. 99-61-6 WNR CVH	$3\text{-}O_2NC_6H_4CHO$	DMF	PY	CsI	0.1	-	-	20± 0.1	DME/SCE	3CO	C=1, m=2.02, t=4.0, h=49
						LiCl	0.1						
						$NaClO_4$	0.1						
						NaI	0.1						
						Et_4NI	0.1						
DB63 aj15 cc48	$C_7H_5NO_3$	4-nitrobenzaldehyde C.A. 555-16-8 WNR DVH	$4\text{-}O_2NC_6H_4CHO$	DMF	PY	CsI	0.1	-	-	20± 0.1	DME/SCE	3CO	C=1, m=2.02, t=4.0, h=49
						LiCl	0.1						
						$NaClO_4$	0.1						
						NaI	0.1						
						Et_4NI	0.1						
DB64 aj? cc?	$C_7H_5NaO_2Ru$	sodium cyclopentadienyldicarbonyl-ruthenate(0) C.A. 42802-20-0 L70 AHJ Ø-RU-CO&CO &-NA-	$(C_5H_5)Ru(CO)_2^-$ Na^+	MG	VY	Bu_4NClO_4	0.1	-	-	25	RPE Ag/Ag^+	12AO	C=2, ω=10, MP Aux
DB65 aj55 cc54	$C_7H_6ClNO_2$	4-nitrobenzyl chloride C.A. 100-14-1 WNR D1G	$4\text{-}O_2NC_6H_4CH_2Cl$	MeCN	IL	Et_4NClO_4	0.1	-	-	-	Pt/SCE	12F-	ns
DB66 aj80 cc59	C_7H_6O	benzaldehyde C.A. 100-52-7 VHR	C_6H_5CHO	EtOH 10	PY	ACET GEL	0.003	-	≈1.3	-	DME/MP	OAO	C=1.59, EC_1=1.99, t_1=4.7, EC_2=2.04, t_2=4.6 C=3.18
				EtOH 20	PY	ACET GEL	0.003	-	≈1.3	-	DME/MP	OAO	C=3.18, EC_1=1.99, t_1=4.7, EC_2=2.04, t_2=4.6
				EtOH 40	PY	ACET GEL	0.003	-	≈1.3	-	DME/MP	OAO	C=3.18, EC_1=1.99, t_1=4.7, EC_2=2.04, t_2=4.6
				EtOH 60	PY	ACET GEL	0.003	-	≈1.3	-	DME/MP	OAO	C=3.18, EC_1=1.99, t_1=4.7, EC_2=2.04, t_2=4.6
CONT													

TABLE I. Electrochemical Data C_7H_8O (CONT.) DB66

Ref.	C/M	Charact. Potential		Response Const.		n		Electrokinetic Data			Products and Identification	Description and Remarks	Code No.
		Value	vs.		Value	Tech.		Parameter	Value	From			
ER002 0042	-	$E_{\frac{1}{2}}$	-0.87?F	-	i_ℓ	1.8F	-	-	-	-	-	-	$C, i_\ell=f([ROH]),p$	DB60 ai93 cc44
ER002 0042	-	$E_{\frac{1}{2}}$	-0.74?F	-	i_ℓ I	7F 1.37	-	-	-	-	-	-	C,p	
ER002 0042	-	$E_{\frac{1}{2}}$	-0.75?F	-	i_ℓ	5.0F	-	-	-	-	-	-	C,p	
ER002 0042	-	$E_{\frac{1}{2}}$	-0.75?F	-	i_ℓ	8F	-	-	-	-	-	-	C,p	
JA094 7526	-	E_p	-2.32 -2.27F	SCE	i_p	90F 80F	sttd	1	$E_p-E_{p/2}$	60 -	calc	radical anion	C,R,r A	DB61 aj09 cc45
JE004 0321	130 g	$E_{\frac{1}{2}}$	-0.935	SCE	i_ℓ	3.4	-	-	-	-	-	-	C,r	DB62 aj14 cc47
			-0.915			3.6	-	-	-	-	-	-	C,r	
			-0.915			3.5	-	-	-	-	-	-	C,r	
			-0.910			3.5	-	-	-	-	-	-	C,r	
			-0.980			3.4	-	-	-	-	-	-	C,r	
JE004 0321	130 b	$E_{\frac{1}{2}}$	-0.790	SCE	i_ℓ	3.5	-	-	-	-	-	-	C,r	DB63 aj15 cc48
			-0.765			4.2	-	-	-	-	-	-	C,r	
			-0.775			4.1	-	-	-	-	-	-	C,r	
			-0.775			3.8	-	-	-	-	-	-	C,r	
			-0.800			-	-	-	-	-	-	-	C,r	
JA088 5121	116 a	$E_{\frac{1}{2}}$	-1.5	Ag/Ag$^+$	-	-	-	-	-	-	-	-	$A,E_{\frac{1}{2}}=f(\log k_2'),p$	DB64
JA094 0640	69 b	$E_{\frac{1}{2}}$	-1.09 -1.20	SCE	-	-	IU	2.0	-	-	-	-	C,p C	DB65 aj55 cc54
ER002 0042	-	-	-	-	I	1.50	-	-	-	-	-	-	C,p	DB66 aj80 cc59
	-	-	-	-		1.47	-	-	-	-	-	-	C,p	
ER002 0042	-	-	-	-	I	1.54	-	-	-	-	-	-	C,p	
ER002 0042	-	-	-	-	I	1.36	-	-	-	-	-	-	C,p	
ER002 0042	-	-	-	-	I	1.32	-	-	-	-	-	-	C,p	
		$E_{\frac{1}{2}}$	-0.87?F		i_ℓ									
		$E_{\frac{1}{2}}$	-0.74?F			7F								
														CONT

DB66 (CONT.) C_7H_6O

Code No.	Empirical Formula	Name and WLN Code	Structural Formula	Solvent	Tech.	Medium		μ, M	pH	T, °C	Electrodes	App.	Experimental Parameters
DB66 aj80 cc59	C_7H_6O	benzaldehyde	C_6H_5CHO	EtOH 80	PY	ACET GEL	? 0.003	–	≈1.3?	–	DME/MP	OAO	$C=3.18, EC_1=1.99, t_1=4.7, EC_2=2.04, t_2=4.6$
				EtOH 90	PY	ACET GEL	? 0.003	–	≈1.3?	–	DME/MP	OAO	$C=3.18, EC_1=1.99, t_1=4.7, EC_2=2.04, t_2=4.6$
				EtOH 96	PY	ACET GEL	? 0.003	–	≈1.3?	–	DME/MP	OAO	$C=3.18, EC_1=1.99, t_1=4.7, EC_2=2.04, t_2=4.6$
DB67	C_7H_6OS	thiobenzoic acid C.A. 98-91-9 SHVR	$C_6H_5C(O)SH$	EtOH 50	PY	$HClO_4$	?	–	1.30	30±0.2	DME/SCE	OA1	$C=0.6, m=1.31, t=5.91, EC=1.610, h=68.4$
						CITR PHOS	?		3.50				
						ACET	?		5.30				
						PHOS	?		7.45				
						AM	?		9.10				
						BOR	?		10.50				
						NaOH	?		13.70				
DB68 aj81	$C_7H_6O_2$	benzoic acid C.A. 65-85-0 QVR	C_6H_5COOH	MeCN	IL	$LiClO_4$	0.5	–	–	30	Pt/SCE	245A1	$C=0.5, A=0.289$, MP Aux
													$C=5$
													$C=50$
DB69 cc60	$C_7H_6O_2$	methyl-1,4-benzoquinone C.A. 553-97-9 L6V DVJ B	TABLE II-3.	EtOH 75	PY	ACET	?	0.05	6.4	25±0.1	DME/SCE	2AO	$C=0.2 \rightarrow 1.0$
				i-PrOH 40	PY	PHOS	?	–	7.6	25	DME/NCE	---	$C=0.4$
				MeCN	PY	Et_4NClO_4	0.1	–	–	25	DME Ag/Ag$^+$	3AO	ns
					PV	Et_4NClO_4	0.1	–	–	25	DME/MP	OAO	$t=4.2, \wedge e=26.5, f=35.0$

TABLE I. Electrochemical Data $C_7H_6O_2$ DB69

Ref.	C/M	Charact. Potential Value	vs.	Response	Const. Value	Tech.	n	Electrokinetic Data Parameter	Value	From	Products and Identification	Description and Remarks	Code No.
ER002 0042	-	-	-	-	I	1.41	-	-	-	-	-	-	C,p	DB66 aj80 cc59
ER002 0042	-	-	-	-	I	1.42	-	-	-	-	-	-	C,p	
ER002 0042	-	-	-	-	I	1.37	-	-	-	-	-	-	C,p	
EA017 2085	-	$E_{\frac{1}{2}}$	-0.090 -0.265	SCE	i_ℓ	1.217 0.405	-	-	-	-	-	-	$A, i_d + i_k, i_\ell \propto C, R, r$ $A, i_d + i_a$	DB67
			-0.220 -0.405			1.030 0.582	-	-	-	-	-	-	$A, i_d + i_k, i_\ell \propto C, R, r$ $A, i_d + i_a$	
			-0.265 -0.450			1.050 0.582	-	-	-	-	-	-	$A, i_d + i_k, i_\ell \propto C, R, r$ $A, i_d + i_a$	
			-0.270 -0.445 -0.595			1.030 0.582 0.104	-	-	-	-	-	-	$A, i_d + i_k, i_\ell \propto C, R, r$ $A, i_d + i_a$ A, i_a	
			-0.270 -0.450 -0.645			1.008 0.582 0.125	-	-	-	-	-	-	$A, i_d + i_k, i_\ell \propto C, R, r$ $A, i_d + i_a$ A, i_a	
			-0.270 -0.450 -0.680			1.030 0.572 0.104	-	-	-	-	-	-	$A, i_d + i_k, i_\ell \propto C, R, r$ $A, i_d + i_a$ A, i_a	
			-0.270 -0.450 -0.770			0.988 0.582 0.218	-	-	-	-	-	-	$A, i_d + i_k, i_\ell \propto C, R, r$ $A, i_d + i_a$ A, i_a	
BJ046 0430	-	$E_{\frac{1}{2}}$	2.33F	SCE	j_ℓ	0.57F	-	-	-	-	-	-	A,p	DB68 aj81
			2.30F			3.92F	-	-	-	-	-	-	A,p	
			2.23F			8.52F	-	-	-	-	-	-	A,p	
JL62A 4558	-	$E_{\frac{1}{2}}^\circ$	-0.64	NHE	I	2.30	Elog!	2	-	-	-	-	$C, i_\ell \propto h^{\frac{1}{2}}, p$	DB69 cc60
BP068 0003	-	$E_{\frac{1}{2}}$	-0.085	NCE	-	-	-	-	-	-	-	-	$C, E_{\frac{1}{2}} = f(\lambda_{max})$	
TF061 1516	-	$E_{\frac{1}{2}}$	-0.58	SCE	-	-	sttd	1	-	-	-	-	$C, W, R, E_{\frac{1}{2}} = f(\sigma$ and λ_{max} charge transfer), r	
			-1.12					1					C,R	
JL62B 4540	-	E_{su}	-0.58 -1.10	SCE	$\dfrac{i_{su} \cos\varnothing}{t^{\frac{1}{2}} i_\ell}$	1.53 1.50	-	-	-	-	-	-	C,r C	

Code No.	Empirical Formula	Name and WLN Code	Structural Formula	Solvent	Tech.	Medium		μ, M	pH	T, °C	Electrodes	App.	Experimental Parameters
DB70 cc61	$C_7H_6O_2$	salicylaldehyde C.A. 90-02-8 VHR BQ	$2\text{-}HOC_6H_4CHO$	EtOH 10	PY	ACET GEL	? 0.003	-	≈1.3	-	DME/MP	OAO	C=0.64
				EtOH 20	PY	ACET GEL	? 0.003	-	≈1.3	-	DME/MP	OAO	C=1.59
				EtOH 40	PY	ACET GEL	? 0.003	-	≈1.3	-	DME/MP	OAO	C=1.59
				EtOH 60	PY	ACET GEL	? 0.003	-	≈1.3	-	DME/MP	OAO	C=3.18
				EtOH 80	PY	ACET GEL	? 0.003	-	≈1.3	-	DME/MP	OAO	C=3.18
				EtOH 90	PY	ACET GEL	? 0.003	-	≈1.3	-	DME/MP	OAO	C=3.18
				EtOH 96	PY	ACET GEL	? 0.003	-	≈1.3	-	DME/MP	OAO	C=3.18
DB71 ak32 cc66	$C_7H_7NO_2$	4-nitrotoluene C.A. 99-99-0 WNR D	$4\text{-}O_2NC_6H_4CH_3$	MeCN	IL	Et_4NClO_4	0.1	-	-	-	Pt/SCE	12F-	ns
DB72 ak38	$C_7H_7NO_5$	ethyl 5-nitro-2-furoate C.A. 943-37-3 T5OJ BNW EVO2	TABLE II-2.	EtOH 10	PY	BR		0.45	2.0	25±0.1	DME/SCE	OAO	C=0.2,m=1.18,t=0.26
									4.0				
									6.0				
									8.0				
									9.0				
									10.0				
									11.0				
									12.0				
DB73	C_7H_7NaO	sodium 2-methylphenoxide C.A. 4549-72-8 OR B1 &-NA-	$2CH_3C_6H_4O^-\ Na^+$	m-cresol	PY	Et_4NClO_4	0.1	-	-	-	DME Ag/AgCl, 0.1 Me_4NCl	216AO	C=0.5-1.5, m=2.87,t=2.70,h=56, MP Aux
DB74	$C_7H_8ClN_3O_4S_2$	6-chloro-3,4-dihydro-7-sulfamoyl-2H-1,2,4-benzothiadiazine 1,1-dioxide C.A. 58-93-5 T66 BSWM EM DHJ HG ISWZ	TABLE II.	H_2O	PY	BOR Me_4NBr	? 0.1	-	9.9 10.7	21±0.2	DME/SCE	OAO	C=0.52,m=1.58,t=3.3 C=1.04
DB75	$C_7H_8N_2O$	N-methyl-N-nitrosoaniline C.A. 614-00-6 ONN1&R	$C_6H_5N(CH_3)NO$	H_2O	DI	BR		-	2.0 10.0	-	DME/SCE	O4AO	C=0.05,m=1.73,t=1,h=80,v=10,ΔE=100
					PT	BR		-	2.0	-	DME/SCE	O4AO	C=0.05,m=1.73,t=1,h=80,v=10

TABLE I. Electrochemical Data $C_7H_8N_2O$ DB75

Ref.	C/M	Charact. Potential Value	vs.	Response Const. Value		n Tech.		Electrokinetic Data Parameter	Value	From	Products and Identification	Description and Remarks	Code No.
ER002 0042	-	-	-	I	1.83	-	-	-	-	-	-	C,p	DB70 cc61
ER002 0042	-	-	-	I	1.60	-	-	-	-	-	-	C,p	
ER002 0042	-	-	-	I	1.32	-	-	-	-	-	-	C,p	
ER002 0042	-	-	-	I	1.30	-	-	-	-	-	-	C,p	
ER002 0042	-	-	-	I	1.50	-	-	-	-	-	-	C,p	
ER002 0042	-	-	-	I	1.61	-	-	-	-	-	-	C,p	
ER002 0042	-	-	-	I	1.69	-	-	-	-	-	-	C,p	
JA094 0640	69 b	$E_{\frac{1}{2}}$	-1.20	SCE	-	-	1U at -1.5	1.0	-	-	-	-	C,p	DB71 ak32 cc66
LV670 0023	-	$E_{\frac{1}{2}}$	-0.09	SCE	i_d	1.05	-	-	-	-	-	ethyl 5-hydroxyl-amino-2-furoate	C,p	DB72 ak38
			-1.02			0.56						ethyl 5-amino-2-furoate	C	
			-0.15			1.00	-	-	-	-	-	-	C	
			-1.08			0.55							C	
			-0.26			1.08	-	-	-	-	-	-	C	
			-1.07			0.56							C	
			-0.37			1.04	-	-	-	-	-	-	C	
			-1.04			0.60							C	
			-0.40			0.86	-	-	-	-	-	-	C	
			-1.04			0.58							C	
			-0.47			0.76	-	-	-	-	-	-	C	
			-1.04			0.50							C	
			-0.53			0.18	-	-	-	-	-	-	C	
			-1.08			0.23							C	
			-0.54			0.15	-	-	-	-	-	-	C	
			-			-							O	
			-1.04			0.21							C	
			-			-							O	
AA072 0169	-	$E_{\frac{1}{2}}$	0.38	Ag/AgCl	I	0.398	-	-	-	-	-	-	A,≠,r	DB73
AA080 0017	-	$E_{\frac{1}{2}}$	-1.675	SCE	-	-	-	-	-	-	-	-	C,r?	DB74
			-1.85F		-	-	-	-	-	-	-	-	C,r?	
AA078 0081	82 a	E_{su}	-0.65	SCE	i_{su}	3.0	QE	4	-	-	-	-	C,r	DB75
	82 e		-1.20			1.00		2	-	-	-	-	C,r	
AA078 0081	82 a	$E_{\frac{1}{2}}$	-0.38	SCE	-	-	QE	4	-	-	-	-	C,r	

Code No.	Empirical Formula	Name and WLN Code	Structural Formula	Solvent	Tech.	Medium	μ, M	pH	T, °C	Electrodes	App.	Experimental Parameters
DB76	$C_7H_8N_2O_3$	4-methyl-2-nitro-aniline C.A. 89-62-3 ZR D1 FNW	4-CH_3-2-NO_2-$C_6H_3NH_2$	MeCN	VR	Bu_4NClO_4	?	-	-	HMDE Ag/$AgNO_3$	2-O	v=33.3?, Pt Aux
DB77	$C_7H_8O_3S$	methyl 5-(mercapto-methyl)-2-furoate T5OJ BVO1 E1SH	TABLE II.	EtOH 86	PY	BR	0.45	2.3 3.0 5.0 6.9 8.0 10.8	25.0 ±0.1	DME/SCE	OCO	C=0.2, m=1.15, t=0.1
DB78	$C_7H_9AsO_3$	4-methylphenyl-arsonic acid C.A. 3969-54-8 Q-AS-QO&R D1	4-$H_3CC_6H_4$-AsO(OH)$_2$	H_2O	PY	HCl	0.1	-	-	DME/SCE	O4AO	C=0.1-1.0, t=2.4, v=1.7
					IV	HCl	0.1	-	-	DME/SCE	O4AO	C=0.2, t=2.4, E_{app}=-0.750 E_{app}=-0.800 E_{app}=-0.900 E_{app}=-1.000 E_{app}=-1.100
DB79	$C_7H_9NO_3S$	2-amino-4-methyl-benzenesulfonic acid ZR C FSWQ	TABLE II.	H_2O	PY	KCl	0.1	-	20	DME/SCE	OA-	C=4
DB80	$C_7H_9NO_3S$	2-amino-5-methyl-benzenesulfonic acid C.A. 88-44-8 ZR D BSWQ	TABLE II.	H_2O	PY	KCl	0.1	-	20	DME/SCE	OA-	C=4
DB81	$C_7H_9NO_3S$	5-amino-2-methyl-benzenesulfonic acid C.A. 118-88-7 ZR D ESWQ	TABLE II.	H_2O	PY	KCl	0.1	-	20	DME/SCE	OA-	C=4
DB82 CONT	$C_9H_9NO_7$	5-nitro-2-furaldehyde diacetate T5OJ BYOV1&OV1 ENW	TABLE II.	EtOH 10	PY	BR	0.45	2.0 4.0 6.0 8.0 9.0	25± 0.1	DME/SCE	OAO	C=0.2, m=1.18, t=0.26

TABLE I. Electrochemical Data $C_7H_9NO_3S$ (CONT.) DB82

Ref.	C/M	Charact. Potential		Response Const.		Tech.	n	Electrokinetic Data			Products and Identification	Description and Remarks	Code No.
		Value	vs.		Value			Parameter	Value	From			
AJ027 2495	–	$E_{p/2}$	-1.40	Ag/AgNO$_3$	–	–	–	–	ΔE_p	70	–	–	$C,R,i_{p,c}/i_{p,A}=1,r?$	DB76
		E_p	-2.25						αn_a	0.29	$E_p - E_{p/2}$		$C,\neq$	
RG035 0773	–	$E_{\frac{1}{2}}$	-1.23	–	i_ℓ	2.42	llk	2.2	–	–	–	–	C,R,r	DB77
			-1.16			2.31	–	–	–	–	–	–	C,R,r	
			-1.14			1.45	–	–	–	–	–	–	C,i_d,R,r	
			-1.13			1.24	–	–	–	–	–	–	C,i_d,R,r	
			-1.13			1.04	–	–	–	–	–	–	C,i_d,R,r	
			-1.13			1.24	–	–	–	–	–	–	C,i_d,R,r	
AN100 0489	367 ab	$E_{\frac{1}{2}}$	-0.955	SCE	–	–	i:i	4	$dE_{\frac{1}{2}}/dpH$ p=	110 1	– Elog	–	$C,\neq,i_d,i_\ell \propto C, E_{\frac{1}{2}}=f(pH), \Delta E_{\frac{1}{2}}=0.056\sigma, \Delta E_{\frac{1}{2}}=-0.438 pK_1, r$	DB78
AN100 0489	367 ab	–	–	–	dlogi/dlogt	0.39	–	–	–	–	–	–	C,r	
			–			0.35	–	–	–	–	–	–	C,r	
			–			0.28	–	–	–	–	–	–	C,r	
			–			0.22	–	–	–	–	–	–	C,r	
			–			0.27	–	–	–	–	–	–	C,r	
EA011 1189	–	$E_{\frac{1}{2}}$	-0.86 -1.61	SCE	i_ℓ	21.17 138.54	i:i	0.97 6.30	–	–	–	–	C,i_d,r C	DB79
EA011 1189	–	$E_{\frac{1}{2}}$	-0.93 -1.67	SCE	i_ℓ	32.71 65.42	i:i	1.50 3.00	–	–	–	–	C,i_d,r C	DB80
EA011 1189	–	$E_{\frac{1}{2}}$	-0.87 -1.62	SCE	i_ℓ	21.17 181.83	i:i	0.97 8.30	–	–	–	–	C,i_d,r C	DB81
LV670 0023	–	$E_{\frac{1}{2}}$	0.16 -0.94	SCE	i_d	0.70 0.45	–	–	–	–	–	–	C,r C	DB82
			-0.25 -1.03			0.71 0.42	–	–	–	–	–	–	C,r C	
			-0.36 -1.05			0.70 0.39	–	–	–	–	–	–	C,r C	
			-0.49 -1.06			0.65 0.39	–	–	–	–	–	–	C,r C	
			-0.53 -1.10			0.63 0.29	–	–	–	–	–	–	C,r C	

CONT

Code No.	Empirical Formula	Name and WLN Code	Structural Formula	Solvent	Tech.	Medium		μ, M	pH	T, °C	Electrodes	App.	Experimental Parameters
DB82	$C_9H_9NO_7$	5-nitro-2-furaldehyde diacetate	TABLE II.	EtOH 10	PY	BR		0.45	10.0	25±0.1	DME/SCE	OAO	C=0.2, m=1.18, t=0.26
									11.0				
									12.0				
DB83	$C_7H_{10}BrCl$	2-exo-bromo-2-endo-chloronorbornane C.A. 21690-94-8 L55 A CXTJ CG CE -AB	TABLE II.	DMF	PY	Et_4NBr	0.1	-	-	-	DME Cd(Hg)	O4A-	C=1, t=0.5
DB84	$C_7H_{10}BrCl$	2-exo-chloro-2-endo-bromonorbornane C.A. 21690-95-9 L55 A CXTJ CG CE -BA	TABLE II.	DMF	PY	Et_4NBr	0.1	-	-	-	DME Cd(Hg)	O4A-	C=1, t=0.5
DB85	$C_7H_{10}Cl_2$	2,2-dichloronorbornane C.A. 19916-65-5 L55 A CXTJ CG CG	TABLE II.	DMF	PY	Et_4NBr	0.1	-	-	-	DME Cd(Hg)	O4A-	C=1, t=0.5
DB86	$C_7H_{10}O$	3-formylcyclohexene C.A. 1321-16-0 L6UTJ CVH	TABLE II.	DMF	PY	Bu_4NBF_4	0.50	-	-	25	DME/SCE	23-0	C=2.0-5.0, Pt Aux
DB87	$C_7H_{10}O$	3-heptyn-2-one C.A. 26059-43-8 4UU1V1	$CH_3CH_2CH_2C\equiv CCO-CH_3$	DMF	PY	Bu_4NBF_4	0.50	-	-	25	DME/SCE	23-0	C=4.9-6.4, Pt Aux
DB88	$C_7H_{10}O$	5-methyl-2-cyclohexenone C.A. 7214-50-8 L6V BUTJ E1	TABLE II.	DMF	PY	Bu_4NBF_4	0.50	-	-	25	DME/SCE	23-0	C=1.4-1.6, Pt Aux
DB89	$C_7H_{10}O_2$	2-ethoxy-2-cyclopentenone C.A. 40006-64-2 L5V BUTJ BO2	TABLE II.	DMF	PY	Bu_4NBF_4	0.50	-	-	25	DME/SCE	23-0	C=2.0-2.4, Pt Aux
DB90	$C_7H_{10}O_2$	methyl 2-hexynoate C.A. 18937-79-6 4UU1VO1	$C_3H_7C\equiv CCO_2CH_3$	DMF	PY	Bu_4NBF_4	0.50	-	-	25	DME/SCE	23-0	C=2.9-6.5, Pt Aux
DB91	$C_7H_{11}BrO_3$	methyl 4-bromo-5-oxohexanoate 1VYE2OV1	$CH_3COCHBrCH_2-CH_2OCOCH_3$	H_2O	PY	PHOS	?	-	2.2	-	DME/SCE	0-0	C=2.5, EC=2.24(0.4V)
									3.9				
									5.7				
									7.2				
									9.7?				
						ACET	?		5.6				

TABLE I. Electrochemical Data $C_7H_{11}BrO_3$ DB91

Ref.	C/M	Charact. Potential		Response Const.		n Tech.		Electrokinetic Data			Products and Identification	Description and Remarks	Code No.
			vs.		Value			Parameter	Value	From			
LV670 0023	-	$E_{\frac{1}{2}}$	SCE	i_d	0.53	-	-	-	-	-	-	C,r	DB82
		-0.52			0.15							C	
		-1.10											
		-0.48			0.40	-	-	-	-	-	-	C,r	
		-			-							O	
		-										O	
JA094 8475	362 c	$E_{\frac{1}{2}}$	-	-	-	QE	2.00 ± 0.05	-	-	-	nortricyclene; endo-2-chloronorbornane	C,≠,r	DB83
JA094 8475	362 cd	$E_{\frac{1}{2}}$	-	-	-	QE	2.00 ± 0.05	-	-	-	nortricyclene; endo-2-chloronorbornane	C,≠,r	DB84
JA094 8475	362 ab	$E_{\frac{1}{2}}$	-1.54	-	-	QE	2.00 ± 0.05	-	-	-	nortricyclene; endo-2-chloronorbornane	C,≠,r	DB85
JA094 8471	-	$E_{\frac{1}{2}}$	-2.03	SCE	-	-	-	-	αn_a	1.0	Elog	-	C,p	DB86
JA094 8471	-	$E_{\frac{1}{2}}$	-1.99	SCE	-	-	-	-	αn_a	0.8	Elog	-	C,p	DB87
JA094 8471	-	$E_{\frac{1}{2}}$	-2.07	SCE	-	-	-	-	αn_a	1.1	Elog	-	C,p	DB88
JA094 8471	-	$E_{\frac{1}{2}}$	-2.23	SCE	-	-	-	-	αn_a	0.8	Elog	-	C,p	DB89
JA094 8471	-	$E_{\frac{1}{2}}$	-2.26	SCE	-	-	-	-	αn_a	0.7	Elog	-	C,p	DB90
RR022 0263	14a	$E_{\frac{1}{2}}$	-0.29F	SCE	i_ℓ	15F	IIk	2	-	-	-	-	C,i_d,X,p	DB91
		E_{Mx}	-0.56			-		-					C,Mx	
		$E_{\frac{1}{2}}$	-0.29F			16F		2	-	-	-	-	C,i_d,X,p	
		E_{Mx}	-0.55F			-		-					C,Mx	
		$E_{\frac{1}{2}}$	-0.2F			7F		2	-	-	-	-	C,i_d,X,p	
		E_{Mx}	-0.5F			-		-					C,Mx	
		$E_{\frac{1}{2}}$	-0.15F			6F		2	-	-	-	-	C,i_d,X,p	
			-0.2F			2F		2	-	-	-	-	C,i_d,O at pH=10.5,p	
			-0.1		i_ℓ	-		-	-	-	-	-	C,$i_\ell=kC$, $i_\ell \propto h^{\frac{1}{2}}$, i_d, Mx for pH=3.5-5.0,p	

Code No.	Empirical Formula	Name and WLN Code	Structural Formula	Solvent	Tech.	Medium		μ, M	pH	T, °C	Electrodes	App.	Experimental Parameters
DB92	$C_7H_{12}Br_2O$	2,4-dibromo-2,4-dimethyl-3-pentanone C.A. 17346-16-6 EX1&1&VXE1&1	$[(CH_3)_2CBr]_2CO$	MeCN	PY	$LiClO_4$	0.1	-	-	-	DME/SCE	-A-	ns
				DMF	QE	NaACET	0.2	-	-	14	MP/SCE	235A	E_{app} = -1.3, GE Aux
DB93	$C_7H_{12}N_2$	2,3-dimethyl-2,3-diazabicyclo[2.2.1]-hept-5-ene C.A. 14288-15-4 T55 A CNN FUTJ C1 D1	TABLE II.	MeCN	VR	$NaClO_4$	0.1	-	-	-	AuBE SCE	2AO	C=0.2-0.9, d=0.15,v=55,MP Aux, PE
DB94	$C_7H_{12}O_3$	4-oxo-1-pentyl acetate C.A. 5185-97-7 1VO3V1	$CH_3COCH_2CH_2CH_2$-$OCOCH_3$	H_2O	PY	PHOS		-	-	-	DME/SCE	O-O	C=2.5,EC= 2.24(-0.4 V)
DB95	$C_7H_{13}N$	1-azabicyclo[2.2.2]-octane C.A. 100-76-5 T66 A B CNTJ	TABLE II.	MeCN	VR	$NaClO_4$	0.1	-	-	-	AuBE SCE	2AO	C=4.50,d= 0.15,v=96, MP Aux,PE
DB96	$C_7H_{14}NNaS_2$	sodium dipropyl-dithiocarbamate C.A. 4143-50-4 SUCS&N3&3 &-NA-	$(CH_3CH_2CH_2)_2N$-$C(S)SNa$	H_2O	PY	KOH KCl TX100	0.01 0.3 0.008	-	-	25± 0.2	DME/SCE	O1AO	C=1.0,m=2.52, t=3.00,h=40
					IL	KOH KCl	0.01 0.3	-	-	25± 0.2	HMDE SCE	O1AO	C=1.0,v=20
						KOH KCl TX100	0.01 0.3 0.008						
DB97	$C_7H_{14}N_2$	1,5-diazabicyclo-[3.2.2]nonane C.A. 283-47-6 T67 A B CN FNTJ	TABLE II.	MeCN	VR	$NaClO_4$	0.1	-	-	-	AuBE SCE	2AO	C=1.90,d= 0.15,v=66, MP Aux,PE
DB98	$C_7H_{14}N_2$	1,5-diazabicyclo-[3.3.1]nonane C.A. 281-17-4 T66 A BN FNTJ	TABLE II.	MeCN	VR	$NaClO_4$	0.1	-	-	-	AuBE SCE	2AO	C=1.79,d= 0.15,v=60, MP Aux,PE
DB99	$C_7H_{14}N_2$	2,3-dimethyl-2,3-diazabicyclo[2.2.1]-heptane C.A. 14287-89-9 T55 A CNNTJ C1 D1	TABLE II.	MeCN	VR	$NaClO_4$	0.1	-	-	-	AuBE SCE	2AO	C=0.2-0.9, d=0.15,v= 54,MP Aux, PE
DC00	$C_7H_{14}N_2$	1,2,3-trimethyl-1,2,3,6-tetrahydro-pyridazine C.A. 38704-94-8 T6NN DUTJ A1 B1 C1	TABLE II.	MeCN	VR	$NaClO_4$	0.1	-	-	-	AuBE SCE	2AO	C=0.2-0.9, d=0.15,v= 50,MP Aux, PE
DC01	$C_7H_{16}N_2$	1,2-diethylpyrazol-idine C.A. 22825-58-7 T5NNTJ A2 B2	TABLE II.	MeCN	VR	$NaClO_4$	0.1	-	-	-	AuBE SCE	2AO	C=0.2-0.9, d=0.15,v= 50,MP Aux, PE

TABLE I. Electrochemical Data $C_7H_{16}N_2$ DC01

Ref.	C/M	Charact. Potential		Response Const.		n Tech.		Electrokinetic Data			Products and Identification	Description and Remarks	Code No.
		Value	vs.		Value			Parameter	Value	From			
JA094 0240	357 a	$E_{\frac{1}{2}}$	-0.26	SCE	-	-	-	-	-	-	-	-	C,r	DB92
JA094 0240	357 a	-	-	-	-	-	QE	2.0	-	-	-	2-acetoxy-2,4-di-methyl-3-pentanone, IRS,NMR;85%$(CH_3)_2$-$CHC(:O)C(CH_3)_2O$ $C(:O)CH_3$ 2-hydroxy-2,4-di-methyl-3-pentanone, IRS,NMR,15%$(CH_3)_2$-$CHC(:O)C(CH_3)_2OH$	C,r	
JA094 7108	-	XE	0.24 -	SCE	-	-	sttd	1 1	ΔE_p	70 -	sttd	radical cation,ESR	A,R,XE=$\frac{1}{2}(E_{p,A}+E_{p,C})$, r C	DB93
RR022 0263	-	-	-	-	-	-	-	-	-	-	-	-	O,p	DB94
JA094 7114	-	E_p	1.10	SCE	-	-	-	-	$E_p-E_{p/2}$	260	sttd	-	A,p	DB95
AJ028 0021	-	$E_{\frac{1}{2}}$	-0.645 -0.501	SCE	i_ℓ	- 3.0F	-	-	-	-	-	-	A,i_a,Pr,r A	DB96
AJ028 0021	-	E_p	-0.500 -0.497	SCE	-	-	-	-	$E_p-E_{p/2}$	20 22	sttd	-	A,r A,r	
JA094 7114	-	E_p	0.56	SCE	-	-	-	-	$E_p-E_{p/2}$	90	sttd	-	A,p	DB97
JA094 7114	-	E_p	0.69	SCE	-	-	-	-	$E_p-E_{p/2}$	85	sttd	-	A,p	DB98
JA094 7108	-	XE	0.10 -	SCE	-	-	sttd	1 1	ΔE_p	80 -	sttd	radical cation,ESR	A,R,XE=$\frac{1}{2}(E_{p,A}+E_{p,C})$, r C	DB99
JA094 7108	-	XE	0.24 -	SCE	-	-	sttd	1 1	ΔE_p	110 -	sttd	radical cation,ESR	A,R,XE=$\frac{1}{2}(E_{p,A}+E_{p,C})$, r C,R	DC00
JA094 7108	-	XE	0.06 -	SCE	-	-	sttd	1	ΔE_p	85	sttd	radical cation,ESR	A,R,XE=$\frac{1}{2}(E_{p,A}+E_{p,C})$, r C,R	DC01

Code No.	Empirical Formula	Name and WLN Code	Structural Formula	Solvent	Tech.	Medium		μ, M	pH	T, °C	Electrodes	App.	Experimental Parameters
DC02	$C_7H_{16}N_2$	1,2,3-trimethyl-perhydropyridazine C.A. 38704-92-6 T6NNTJ A1 B1 C1 A&C	TABLE II.	MeCN	VR	$NaClO_4$	0.1	-	-	-	AuBE SCE	2A0	C=2.81, d=0.15, v=55, MP Aux, PE
DC03 a146 cd11	$C_8H_4N_2$	1,4-dicyanobenzene C.A. 623-26-7 NCR DCN	$4-NCC_6H_4CN$	DMF	VR	Et_4NClO_4	0.1	-	-	-	PBE/SCE	125A F02	C=1.80, v=80.6, Pt Aux
DC04	$C_8H_5ClN_2O_2$	1-methyl-2-chloro-benzimidazole-4,7-quinone T56 BN DN FV IVJ B1 CG	TABLE II.	DMF	PY	Et_4NClO_4	0.1	-	-	-	DME/MP	0-0	EC=1.23
DC05	$C_8H_5ClN_2O_2$	1-methyl-2-chloro-benzimidazole-6,7-quinone T56 BN DN FVVJ CG D1	TABLE II.	DMF	PY	Et_4NClO_4	0.1	-	-	-	DME/MP	0-0	EC=1.23
DC06	$C_8H_5CrNaO_3$	sodium (cyclopentadienyl)tricarbonyl-chromate(0) C.A. 12203-12-2 L7Ø AHJ Ø-CR-/CO 3 &-NA-	$(\pi-C_5H_5)Cr(CO)_3^-$ Na^+	MG	VY	Bu_4NClO_4	0.1	-	-	25	RPE Ag/Ag$^+$	12A0	C=2, ω=10, MP Aux
DC07	$C_8H_5MoNaO_3$	sodium cyclopentadienetricarbonyl-molybdate(0) C.A. 12107-35-6 L5ØJ Ø-MO-OCW &-NA-	$(\pi-C_5H_5)Mo(CO)_3^-$ Na^+	MG	VY	Bu_4NClO_4	0.1	-	-	25	RPE Ag/Ag$^+$	12A0	C=2, ω=10, MP Aux
DC08	C_8H_6	phenylacetylene C.A. 536-74-3 1UU1R	$C_6H_5C{\equiv}CH$	MeCN	VY	$NaClO_4$	?	-	-	25	RPE/SCE	06-0	C=1, v=20, A=?, ω=ns
DC09	$C_8H_6Br_2$	1,2-dibromobenzocyclobutene C.A. 22250-72-2 T46T&J BE CE	TABLE II.	MeCN	PY	Et_4NClO_4	0.1	-	-	-	DME/SCE	6A0	C=1
					VR	Et_4NClO_4	0.1	-	-	-	HMDE SCE	6A02	C=1
				PrCN	VR	Et_4NClO_4	0.1	-	-	-	PBE/SCE	6A02	C=1
DC10	$C_8H_6Cl_3NO$	trichloroacetanilide C.A. 2563-97-5 GXGGVMR	$C_6H_5NHCOCCl_3$	EtOH 50	PY	HMe_3NI	0.05	-	-	-	DME/SCE	0-0	C=1
DC11	$C_8H_6N_2O_2$	1-methylbenzimidazole-4,7-quinone C.A. 7711-63-9 T56 BN DN FV IVJ B1	TABLE II.	DMF	PY	Et_4NClO_4	0.1	-	-	-	DME/MP	0-0	EC=1.23
DC12	$C_8H_6N_2O_2$	1-methylbenzimidazole-6,7-quinone T56 BN DN FVVJ D1	TABLE II.	DMF	PY	Et_4NClO_4	0.1	-	-	-	DME/MP	0-0	EC=1.23

TABLE I. Electrochemical Data $C_8H_6N_2O_2$ DC12

Ref.	C/M	Charact. Potential		Response Const.		n Tech.		Electrokinetic Data			Products and Identification	Description and Remarks	Code No.
		Value	vs.		Value			Parameter	Value	From			
JA094 7108	-	XE	0.14	SCE	-	-	sttd	1	ΔE_p	95	sttd	radical cation, ESR	A,R,XE=$\frac{1}{2}$($E_{p,A}$+$E_{p,C}$), r	DC02
			-					1		-			C,R	
JA094 7526	-	E_p	-1.57	SCE	i_p	100F	sttd	1	E_p-$E_{p/2}$	60	calc	radical anion benzonitrile	C,R,r	DC03
			-2.39					1		-			C	a146
			-1.51F			100F		-					A,R	cd11
RY002 1096	-	$E_\frac{1}{2}$	-0.60	SCE	-	-	-	-	-	-	-	-	C,p	DC04
			-										C,X	
RY002 1096	-	$E_\frac{1}{2}$	-0.54	SCE	-	-	-	-	-	-	-	-	C,p	DC05
			-										C,X	
JA088 5121	116 a	$E_\frac{1}{2}$	-0.80	Ag/Ag$^+$	-	-	-	-	-	-	-	-	A,$E_\frac{1}{2}$=f(logk_2'),p	DC06
JA088 5121	116 a	$E_\frac{1}{2}$	-0.55	Ag/Ag$^+$	-	-	-	-	-	-	-	-	A,$E_\frac{1}{2}$=f(logk_2'),p	DC07
EA017 1595	-	$E_\frac{1}{2}$	2.250	SCE	-	-	i:i	2	-	-	-	-	A,p	DC08
JA095 2646	-	$E_\frac{1}{2}$	-1.04	SCE	i(u)	2	QE	1.96	-	-	-	QE at -1.3 V → benzo-[a]biphenylene, 90%	C,r	DC09
			-1.92			1		-					C	
JA095 2646	-	E_p	-1.23	SCE	-	-	-	-	-	-	-	-	C,⊀,p	
			-2.04										C,⊀ for v < 1E5	
JA095 2646	-	E_p	-1.32	SCE	-	-	-	-	-	-	-	-	C,⊀	
			-2.10										C,⊀ for v < 1E5	
RG035 0843	14c	$E_\frac{1}{2}$	-0.48	SCE	-	-	-	-	-	-	-	-	C,⊀,p	DC10
			-1.00										C	
			-1.52										C	
RY002 1096	-	$E_\frac{1}{2}$	-0.67	SCE	-	-	-	-	-	-	-	-	C,p	DC11
			-										C,X	
RY002 1096	-	$E_\frac{1}{2}$	-0.54	SCE	-	-	-	-	-	-	-	-	C,p	DC12
			-										C,X	

DC13 $C_8H_6N_2O_2$

Code No.	Empirical Formula	Name and WLN Code	Structural Formula	Solvent	Tech.	Medium	μ, M	pH	T, °C	Electrodes	App.	Experimental Parameters
DC13	$C_8H_6N_2O_2$	4-nitrophenylacetonitrile C.A. 555-21-5 WNR D1CN	$4\text{-}O_2NC_6H_4CH_2CN$	MeCN	VA	Et_4NClO_4 0.1	–	–	–	PBE/SCE	12F-	$+0.4 \rightarrow -1.5$, C=1.15, v=100
						Et_4NClO_4 0.1 Et_4NOH 0.00148						
DC14	$C_8H_6N_2O_2S$	4-nitrobenzyl thiocyanate C.A. 13287-49-5 WNR D1SCN	$4\text{-}O_2NC_6H_4CH_2SCN$	MeCN	VR	Et_4NClO_4 0.1	–	–	22.5 ±0.5	Pt/SCE	15A0	$-0.6 \rightarrow -1.4$ V, C=2.31, A=0.25, v=81, Pt Aux
DC15 am01 cd27	$C_8H_6O_3$	phenylglyoxalic acid C.A. 611-73-4 QVVR	$C_6H_5COCOOH$	H_2O	PY	ACET ? PHOS	–	4.0	25	DME/SCE	0-0	C=2, t=2.7
						ACET ? PHOS Camphor 0.03%						
						ACET ? PHOS GEL 0.003%						
						ACET ? PHOS PV alcohol 0.01%						
						ACET ? PHOS Bu_4NI 4E-5						
						ACET ? PHOS		6.6				t=3
						ACET ? PHOS PV alcohol 0.006%						
						ACET ? PHOS PV alcohol 0.01%						
DC16	$C_8H_7BrClNO$	3'-bromo-2-chloroacetanilide G1VMR CE	$3\text{-}BrC_6H_4NHCO\text{-}CH_2Cl$	EtOH 50	PY	HMe_3NI 0.05	–	–	–	DME/SCE	0-0	C=1
DC17	$C_8H_7BrClNO$	4'-bromo-2-chloroacetanilide C.A. 2564-02-5 G1VMR DE	$4\text{-}BrC_6H_4NHCO\text{-}CH_2Cl$	EtOH 50	PY	HMe_3NI 0.05	–	–	–	DME/SCE	0-0	C=1
DC18	C_8H_7ClINO	4'-iodo-2-chloroacetanilide C.A. 2564-00-3 IR DMV1G	$4\text{-}IC_6H_4NHCO\text{-}CH_2Cl$	EtOH 50	PY	HMe_3NI 0.05	–	–	–	DME/SCE	0-0	C=1
DC19	$C_8H_7Cl_2NO$	2,2-dichloroacetanilide C.A. 3289-76-7 GYGVMR	$C_6H_5NHCOCHCl_2$	EtOH 50	PY	HMe_3NI 0.05	–	–	–	DME/SCE	0-0	C=1

TABLE I. Electrochemical Data $C_8H_7Cl_2NO$ DC19

Ref.	C/M	Charact. Potential		Response Const.		n Tech.	n	Electrokinetic Data			Products and Identification	Description and Remarks	Code No.
		Value	vs.		Value			Parameter	Value	From			
JA094 0640	69c	E_p	-1.14F	SCE	i_p	21F	QE	1.01	-	-	-	α-cyano-4-nitro-benzyl anion,red-dish-purple,VA	C,r	DC13
			-1.04F			10F		-					A	
			-0.01F			3F							A	
			-0.04F			66F		-	-	-	-		A,p	
			0.16F			11F							A	
JE030 0289	69d	E_p	-0.97	SCE	i_p	200F	ER	1	-	-	-	1,2-bis(4-nitro-phenyl)ethane,QE, sepn.,4-nitrotolu-ene,QE,sepn.	C,W,decreases on subsequent scans,r	DC14
			-1.24			250F		2					C,W,decreases on subsequent scans	
			-1.18			90F		-					A,W,increases on subsequent scans	
ER003 0170	-	$E_{\frac{1}{2}}$	-0.65F	SCE	i_d	8.0F	-	-	-	-	-	-	C,i_d,Mx,p	DC15 am01 cd27
			-1.0F			5.5F	-	-	-	-	-	-	C,i_d,p	
			-0.65F			5.5F	-	-	-	-	-	-	C,i_d,p	
			-0.95F			5.5F	-	-	-	-	-	-	C,i_d,X,p	
			-0.7F			5.5F	-	-	-	-	-	-	C,i_d,p	
			-0.95F		i_ℓ	3.2F	-	-	-	-	-	-	C,Mx,p	
			-1.20			2.5F							C	
			-0.97F			1.9F	-	-	-	-	-	-	C,p	
			-1.22			3.6F							C	
			-0.97F			0.4F	-	-	-	-	-	-	C,p	
			-1.28			5.0F							C	
RG035 0843	14a	$E_{\frac{1}{2}}$	-1.33	SCE	-	-	-	-	-	-	-	-	C,⇌,p	DC16
RG035 0843	14a	$E_{\frac{1}{2}}$	-1.34	SCE	-	-	-	-	-	-	-	-	C,⇌,p	DC17
RG035 0843	14a	$E_{\frac{1}{2}}$	-1.15	SCE	-	-	-	-	-	-	-	-	C,⇌,p	DC18
			-1.56										C	
RG035 0843	14b	$E_{\frac{1}{2}}$	-1.00	SCE	-	-	-	-	-	-	-	-	C,⇌,p	DC19
			-1.52										C	

DC20 $C_8H_7Cl_2NO$

Code No.	Empirical Formula	Name and WLN Code	Structural Formula	Solvent	Tech.	Medium		μ, M	pH	T, °C	Electrodes	App.	Experimental Parameters
DC20	$C_8H_7Cl_2NO$	2,3'-dichloro-acetanilide C.A. 2564-05-8 G1VMR CG	3-ClC_6H_4NHCO-CH_2Cl	EtOH 50	PY	HMe_3NI	0.05	–	–	–	DME/SCE	O-O	C=1
DC21 am25	C_8H_7N	4-methylbenzo-nitrile C.A. 104-85-8 CNR D	4-$CH_3C_6H_4CN$	DMF	IR	Et_4NClO_4	0.1	–	–	–	PBE/SCE	125A F0	E_{app} = -2.60, t=0.004-8
					VR	Et_4NClO_4	0.1	–	–	–	PBE/SCE	125A F02	v=80.6, Pt Aux
DC22 cd31	$C_8H_7NO_2$	β-nitrostyrene C.A. 102-96-5 WN1U1R	$C_6H_5CH:CHNO_2$	EtOH 10	PY	HCl KCl	?	–	1.6	25±1	DME/SCE	OAO	C=1.5,EC=1.445,t=0.63
						BR			4.1				
				EtOH 50	PY	HCl KCl	?	–	1.6	25±1	DME/SCE	OAO	C=1.5,EC=1.445,t=0.63
DC23	$C_8H_7NO_3S$	methyl 5-(thiocyan-atomethyl)-2-furoate T5OJ BVO1 E1 SCN	TABLE II.	EtOH 10	PQ	BR		0.45	≈5.0	25±0.1	DME/SCE	OCO	C=1,m=4.88, t=1.85, E_{app}=-1.0
				EtOH 86	PY	BR		0.45	≈2.4	25.0±0.1	DME/SCE	OCO	C=0.2,m=1.56,t=0.22
									≈4.0				
									≈6.9				
									≈8.0				
									≈8.9				
									≈9.8				
									≈10.8				
									≈11.9				
DC24 am45 cd37	C_8H_8	cyclooctatetraene C.A. 629-20-9 L8J	TABLE II-2.	MeCN	PY	Me_4NPF_6	0.1	–	–	–	DME/SCE	---	C=0.16
						Me_4NPF_6 H_2O	0.1 1E-4						
						Bu_4NPF_6	0.16						C=0.15
				THF	PY	Bu_4NPF_6	0.19	–	–	–	DME/SCE	---	C=0.24
						Bu_4NClO_4	0.1	–	–	–	DME Ag/Ag^+ 0.1	O1AFO	C=0.504,m=0.63,t=6-12,h=40.0
CONT						Bu_4NClO_4	0.2						C=0.145

TABLE I. Electrochemical Data C_8H_8 (CONT.) DC24

Ref.	C/M	Charact. Potential		Response Const.		n Tech.	n	Electrokinetic Data			Products and Identification	Description and Remarks	Code No.
		Value	vs.		Value			Parameter	Value	From			
RG035 0843	14a	$E_{\frac{1}{2}}$	-1.35	SCE	-	-	-	-	-	-	-	-	$C,\frac{1}{\ast},p$	DC20
JA094 7526	-	-	-	-	$it^{\frac{1}{2}}/C$	42	sttd	1	-	-	-	-	C,R,r	DC21 am25
JA094 7526	-	E_p	-2.36	SCE	-	-	sttd	1	$E_p-E_{p/2}$	60	calc	radical anion	C,R,r	
RR021 0322	-	$E_{\frac{1}{2}}$	-0.22 -1.06	SCE	I	5.78 2.89	-	-	-	-	-	-	C,p C	DC22 cd31
			-0.35F -1.2F		i_ℓ	25 15	-	-	-	-	-	-	C,p C	
RR021 0322	-	$E_{\frac{1}{2}}$	-0.3F -1.1F	SCE	i_ℓ	- 8	-	-	-	-	-	-	C,Mx,p C,X	
RG035 0773	126 b	-	-	-	-	-	-	2.2	-	-	-	methyl (mercapto-methyl)-2-furoate, MPT;CN⁻ rxn with $Cu(C_2H_3O_2)$ and benzidine acetate	C,p	DC23
RG035 0773	126 b	$E_{\frac{1}{2}}$	-0.75	SCE	i_d	0.78	llk	2.2	-	-	-	-	$C,i_d,R,E_{\frac{1}{2}}$ and $i_\ell \neq f(C),p$	
			-0.78			0.78		2.2	-	-	-	-	C,i_d,R,p	
			-0.80			0.71		2.2	-	-	-	-	C,i_d,R,p	
			-0.77			0.65		2.2	-	-	-	-	C,i_d,R,p	
			-0.75			0.65		2.2	-	-	-	-	C,i_d,R,p	
			-0.78 -1.08			0.60 -		2.2 -	-	-	-	-	C,R,p C	
			-0.75 -1.08			0.52 -		2.2 -	-	-	-	-	C,i_k,R,p C	
			-0.75 -1.08			0.17 -		2.2 -	-	-	-	-	C,i_k,R,p C	
JA094 2521	366 a	$E_{\frac{1}{2}}$	-1.87 -2.65	SCE	$I\eta^{\frac{1}{2}}$	2.5 0.3	-	-	-	-	-	-	C,r C	DC24 am45 cd37
			-1.81 -2.62			4.6 1.4	-	-	-	-	-	-	C,r C	
			-1.80 -1.97 -2.60			2.6 2.4 1.5	-	-	-	-	-	-	C,r C C	
JA094 2521	366 a	$E_{\frac{1}{2}}$	-1.96 -2.16 -2.78	SCE	$I\eta^{\frac{1}{2}}$	3.0 1.5 0.6	-	-	-	-	-	-	C,r C C	
JA094 4915	-	$E_{\frac{1}{2}}$	-2.11 -2.27	SCE	I	3.07 2.75	sttd	1 1	-	-	-	-	C,R,r C,R	
			-1.97 -2.15			0.49 0.26	-	-	-	-	-	-	C,i_d,r C,two le waves reported together	
			-2.67			0.34							C	

CONT

DC24 (CONT.) C_8H_8 CRC Handbook Series in Organic Electrochemistry

Code No.	Empirical Formula	Name and WLN Code	Structural Formula	Solvent	Tech.	Medium		μ, M	pH	T, °C	Electrodes	App.	Experimental Parameters
DC24 am45 cd37	C_8H_8	cyclooctatetraene		THF	PY	Bu_4NClO_4 H_2O	0.2 0.1	-	-	-	DME Ag/Ag^+	O1AFO	C=0.145,m= 0.63,t=6-12, h=40.0
DC25 am46	C_8H_8	styrene C.A. 100-42-5 1U1R	$C_6H_5CH:CH_2$	MeOH	VY	$NaClO_4$ Na tosylate	? ?	-	-	25	RCCE SCE RCGC SCE	O6-0	C=1,A=0.8
				MeCN	VY	$NaClO_4$	?	-	-	25	RCCE SCE RPE/SCE	O6-0	C=1,A=0.8, v=20 C=1,v=20
DC26	$C_8H_8Br_2$	(1,2-dibromoethyl)-benzene C.A. 93-52-7 E1YER	$C_6H_5CHBrCH_2Br$	MeOH 80 C_6H_6 20	PY	NH_4NO_3	2%	-	-	-	DME/-	O-O	C≈2,EC= 2.22,t=2.9
DC27	C_8H_8ClNO	chloroacetanilide C.A. 587-65-5 G1VMR	$C_6H_5NHCOCH_2Cl$	EtOH 50	PY	HMe_3NI	0.05	-	-	-	DME/SCE	O-O	C=1
DC28	C_8H_8O	phenylacetaldehyde C.A. 122-78-1 VH1R	$C_6H_5CH_2CHO$	EtOH 40	PY	ACET GEL	0.003	-	≈1.3	-	DME/MP	OAO	C=3.18
				EtOH 60	PY	ACET GEL	0.003	-	≈1.3	-	DME/MP	OAO	C=3.18
				EtOH 80	PY	ACET GEL	0.003	-	≈1.3	-	DME/MP	OAO	C=3.18
				EtOH 90	PY	ACET GEL	0.003	-	≈1.3	-	DME/MP	OAO	C=3.18
				EtOH 96	PY	ACET GEL	0.003	-	≈1.3	-	DME/MP	OAO	C=3.18
DC29 cd47	$C_8H_8O_2S$	phenyl vinyl sulfone C.A. 5535-48-8 1U1SWR	$C_6H_5SO_2CH:CH_2$	MeOH 20	PY	LiCl	0.07	-	-	25	DME/SCE	O-O	C=0.5,m= 1.63,t=2.5
DC30	C_8H_9NO	phenylacetaldoxime C.A. 7028-48-0 QNU2R	$C_6H_5CH_2CH:NOH$	EtOH 10	PY	HCl KCl	?	-	1.6	25±1	DME/SCE	OAO	C=1.5,EC= 1.445,t= 0.63,h=5.5
				EtOH 50	PY	Buffer HCl KCl	?	-	1.6	25±1	DME/SCE	OAO	C=1.5,EC= 1.445,t= 0.63,h=5.5
DC31	$C_8H_9NO_2$	1-phenyl-2-nitro-ethane C.A. 30179-51-2 WN2R	$C_6H_5CH_2CH_2NO_2$	EtOH 10	PY	HCl KCl	?	-	1.6	25±1	DME/SCE	OAO	C=1.5,EC= 1.445,t= 0.63
DC32 CONT	$C_8H_9NO_3$	1-phenyl-2-nitro-ethanol C.A. 15990-45-1 WN1YQR	$C_6H_5CH(OH)CH_2NO_2$	MeOH 9.1	PY	HCl KCl BOR ACET	? ?	0.5	0.7 2 3.5	25± 0.1	DME/SCE	OA-	C=0.35,m= 0.68,t=0.33, EC=0.64

TABLE I. Electrochemical Data $C_8H_9NO_3$ (CONT.) DC32

Ref.	C/M	Charact. Potential		Response Const.		n		Electrokinetic Data			Products and Identification	Description and Remarks	Code No.
		Value	vs.		Value	Tech.		Parameter	Value	From			
JA094 4915	–	$E_{\frac{1}{2}}$	-1.97	SCE	I	0.57	–	±	–	–	–	–	C, i_{cat}, r	DC24 am45 cd37
			-2.14			0.26							C, two 1e waves reported together, i_{cat}	
			-2.66			0.62							C, i_{cat}	
EA017 1595	372 b	$E_{\frac{1}{2}}$	1.425	SCE	–	–	i:i	2.3	–	–	–	–	A,p	DC25 am46
			1.480					2.3				–	A,p	
			1.78				–	–				–	A,p	
EA017 1595	372 c	$E_{\frac{1}{2}}$	1.540	SCE	i_ℓ	1900F	i:i	2.2	–	–	–	–	A,p	
			1.900					1.9	α	0.5	sttd	–	A,p	
RL033 0029	–	$E_{\frac{1}{2}}$	-0.47	SCE	i_d	18F	–	–	–	–	–	–	$C, i_\ell=kC, p$	DC26
RG035 0843	14a	$E_{\frac{1}{2}}$	-1.45	SCE	–	–	–	–	–	–	–	–	$C, \neq, p$	DC27
ER002 0042	–	–	–	–	I	0.13	–	–	–	–	–	–	C,p	DC28
ER002 0042	–	–	–	–	I	0.16	–	–	–	–	–	–	C,p	
ER002 0042	–	–	–	–	I	0.19	–	–	–	–	–	–	C,p	
ER002 0042	–	–	–	–	I	0.20	–	–	–	–	–	–	C,p	
ER002 0042	–	–	–	–	I	0.19	–	–	–	–	–	–	C,p	
DS171 0860	43a	$E_{\frac{1}{2}}$	-1.55	SCE	I	2.98	Ilk	≈2	αn_a	0.5	Elog	ethene	$C, i_d, \neq, r$	DC29 cd47
RR021 0322	–	$E_{\frac{1}{2}}$	-1.14	SCE	I	2.97	QE at -1.3V	2	–	–	–	β-phenylethylamine, 12%, GLC, N-(β-phenyl)ethylhydroxylamine	C, i_d, wave composed of two waves at -0.76 V and -1.2 V due to syn- and anti-isomers, $i_\ell \propto h^{\frac{1}{2}}$, Tc=1.1 for wave 1 and 1.54 for wave 2(T=25-45°C), r	DC30
RR021 0322	–	$E_{\frac{1}{2}}$	-1.3F	SCE	i_d	15	QE	2	–	–	–	–	C,X,p	
RR021 0322	–	$E_{\frac{1}{2}}$	-0.67F	SCE	–	–	–	–	–	–	–	–	C,p	DC31
ER003 0263	–	$E_{\frac{1}{2}}$	-0.55F	SCE	i_ℓ	1.31F	i:i	5.5 ± 0.5	Elog	90	sttd	1-phenyl-2-aminoethanol	C, i_d, Tc=1.66, Mx for C>0.7, r	DC32
			-0.55F			1.19F	QE	5.6		90		–	C, i_d, r	
			-0.55F			1.06F	PQ	3.97 ± 0.18	–	–	–	2-phenyl-2-hydroxyethylhydroxylamine	C, i_d, r	

CONT

DC32 (CONT.) $C_8H_9NO_3$

Code No.	Empirical Formula	Name and WLN Code	Structural Formula	Solvent	Tech.	Medium		μ, M	pH	T, °C	Electrodes	App.	Experimental Parameters
DC32	$C_8H_9NO_3$	1-phenyl-2-nitro-ethanol	$C_6H_5CH(OH)CH_2NO_2$	MeOH 9.1	PY	BOR ACET	?	0.5	5.3	25±0.1	DME/SCE	OAO	C=0.35, m=0.68, t=0.33, EC=0.64
									6.6				
									7.4				
									8.5				
								0.2	9.3				
									9.6				
DC33 cd54	$C_8H_9NO_3$	pyridoxal C.A. 66-72-8 T6NJ B CQ DVH E1Q	TABLE II-3.	H_2O	IL	AM	?	-	9.2	22±0.1	XEl/SCE	2BO	C=0.001-0.2, v=10, d=0.6, Pt Aux
DC34	$C_8H_9NO_4$	3-hydroxy-5-(hydroxymethyl)-2-methyl-4-pyridinecarboxylic acid C.A. 82-82-6 T6N B1 CQ DVQ E1Q	TABLE II.	H_2O	IL	AM	?	-	9.2	22±0.1	XEl/SCE	2BO	C=0.001-0.2, v=10, d=0.6, Pt Aux
DC35 an29	C_8H_{10}	1,3,5-cyclooctatriene C.A. 1871-52-9 L8U CU EUTJ	TABLE II-2.	THF	PY	Bu_4NClO_4	0.2	-	-	-	DME Ag/Ag$^+$ 0.1	256A F2	-
DC36 cd63	$C_8H_{10}NO_6P$	pyridoxal 5-phosphate C.A. 54-47-7 T6NJ B CQ DVH E1OPQ QO	TABLE II-3.	H_2O	IL	AM	?	-	9.2	22±0.1	XEl/SCE	2BO	C=0.001-0.2, v=10, d=0.6, Pt Aux
DC37	$C_8H_{10}O$	benzyl methyl ether C.A. 538-86-3 1O1R	$C_6H_5CH_2OCH_3$	MeCN	QE	$LiClO_4$	0.1	-	-	-	Pt Ag/Ag$^+$ 0.1	25A-	C=41, A=12.9, E_{app}=1.90, stainless steel Aux
DC38 an56	$C_8H_{10}O$	2,6-dimethylphenol C.A. 576-26-1 QR B F	TABLE II-2.	MeCN	IL	$LiClO_4$	0.5	-	-	30	Pt Ag/0.1 $AgNO_3$	245A 1	C=10, Pt Aux
													C=50
													C=100
DC39 an62	$C_8H_{10}O_2$	1,2-dimethoxybenzene C.A. 91-16-7 1OR BO1	2-$CH_3OC_6H_4OCH_3$	CH_2Cl_2 90 CF_3COOH 10	VA	Bu_4NBF_4	?	-	-	-	-	-	0→+1.3→0.3→, C=1, v=150

TABLE I. Electrochemical Data $C_8H_{10}O_2$ DC39

Ref.	C/M	Charact. Potential		Response Const.		n		Electrokinetic Data			Products and Identification	Description and Remarks	Code No.
		Value	vs.		Value	Tech.	n	Parameter	Value	From			
ER003 0263	-	$E_{\frac{1}{2}}$	-0.58F	SCE	i_ℓ	1.22F	-	-	-	-	-	-	$C, i_d, T_c=1.70, r$	DC32
			-0.66F			1.28F	-	-	-	-	-	-	C, i_d, compound decomposes to CH_3NO_2 and C_6H_5CHO, r	
			-0.73F			-	i:i	5.5 ± 0.5	-	50	-	-	$C, i_\ell=f(t), i_\ell$ extrapolated to $t=0, i_d, r$	
			-0.76F			1.42F	-	-	-	-	-	-	$C, i_\ell=f(t), i_\ell$ extrapolated to $t=0, i_d, r$	
			-			1.26F	QE	5-6	-	-	-	-	$C, i_\ell=f(t), i_\ell$ extrapolated to $t=0, i_d, r$	
			-			1.13F	-	-	-	-	-	-	$C, i_\ell=f(t), i_\ell$ extrapolated to $t=0, i_d$, $pK_a=10.19, r$	
AN100 0349	-	$E_{p/2}$	0.591	SCE	-	-	-	-	-	-	-	-	$A, i_p \propto C$, XEl=CPE with wax and silicone oil, r	DC33 cd54
AN100 0349	-	$E_{p/2}$	0.553	SCE	-	-	-	-	-	-	-	-	$A, i_p \propto C$, XEl=CPE with wax and silicone oil, r	DC34
JA095 2198	-	$E_{\frac{1}{2}}$	-2.77	SCE	$i\eta^{\frac{1}{2}}$	0.092	sttd	1	Elog	67	sttd	-	C, r	DC35 an29
AN100 0349	-	$E_{p/2}$	0.747	SCE	-	-	-	-	-	-	-	-	$A, i_p \propto C$, XEl=CPE with wax and silicone oil, r	DC36 cd63
JA094 6812	360 b	-	-	-	-	-	QE	2.8	-	-	-	benzaldehyde, 46%; benzoic acid, 14%; NMR, GLC	A, p	DC37
EA017 1391	314 b	$E_{\frac{1}{2}}$	1.0F	Ag/Ag$^+$	i_ℓ	4F	-	-	-	-	-	1,2-bis(2-hydroxy-3-methylphenyl)ethene, TLC, IRS	A, r?	DC38 an56
			1.2F			16F	-	-	-	-	-	-	A, r?	
			1.55F			50F	-	-	-	-	-	-	A, r?	
JA094 4749	-	E_p	1.02F	SCE	i_p	22F	QE	2.3	-	-	-	hexamethoxytriphenylene, radical cation, VA	A, p	DC39 an62
			0.39F			2.5F	-	-	-	-	-	hexamethoxytriphenylene, NMR, IRS	R	
			0.50F			1.5F							A	

105

DC40 $C_8H_{11}NO_3$

Code No.	Empirical Formula	Name and WLN Code	Structural Formula	Solvent	Tech.	Medium		μ, M	pH	T, °C	Electrodes	App.	Experimental Parameters
DC40 cd78	$C_8H_{11}NO_3$	pyridoxine C.A. 65-23-6 T6NJ B1 CQ D1Q E1Q	TABLE II.	H_2O	IL	AM	?	-	9.2	22.0 ±0.1	XEl/SCE	2B0	C=0.1,v=10, d=0.6,Pt Aux
DC41	$C_8H_{12}N_2$	1,6-diazabicyclo-[4.4.0]deca-3,8-diene T66N FN CU HUTJ	TABLE II.	MeCN	VR	$NaClO_4$	0.1	-	-	-	AuBE SCE	2A0	C=0.2-0.9, d=0.15,v=62, MP Aux,PE
DC42 cd83	$C_8H_{12}N_2O_2$	pyridoxamine C.A. 85-87-0 T6NJ B CQ D1Z E1Q	TABLE II-3.	H_2O	IL	AM	?	-	9.2	22± 0.1	XEl/SCE	2B0	C=0.001-0.2, v=10,d=0.6, Pt Aux
DC43	$C_8H_{14}N_2$	1,3-diazatricyclo-[3.3.1.1^{3,7}]decane C.A. 281-29-8 T666/BI/DJ/HJ A J 2BF J BN DNTJ	TABLE II.	MeCN	VR	$NaClO_4$	0.1	-	-	-	AuBE SCE	2A0	C=0.58,d= 0.15,v=80, MP Aux,PE
DC44	$C_8H_{14}N_2$	2,3-dimethyl-2,3-diazabicyclo[2.2.2]-octa-5-ene C.A. 14287-91-3 T66 A B ANN DUTJ A1 B1	TABLE II.	MeCN	VR	$NaClO_4$	0.1	-	-	-	AuBE SCE	2A0	C=0.2-0.9, d=0.15,v= 61,MP Aux, PE
DC45	$C_8H_{14}O_4S_2$	glycol dimercapto-propionate C.A. 22504-50-3 SH2VO2OV2SH	[HSCH$_2$CH$_2$C(O)O-CH$_2$]$_2$	EtOH 10	PY	ACET	?	-	3.54	30± 0.1	DME/SCE	03A0	C=1,m=1.832, t=2.81,EC= 1.736
									4.30				
									5.40				
						CL			6.70				
									8.86				
									10.00				
									11.80				
DC46	$C_8H_{15}N_3$	1-methyl-3,5,7-tri-azatricyclo[3.3.1.-1^{3,7}]decane C.A. 38705-10-1 T666/BI/DJ/HJ A J 2 BF J BN DN FN HXTJ H1	TABLE II.	MeCN	VR	$NaClO_4$	0.1	-	-	-	AuBE SCE	2A0	C=1.38,d= 0.15,v=110, MP Aux,PE
DC47	$C_8H_{16}N_2$	1,1'-bipyrrolidine C.A. 18389-95-2 T5NTJ A- AT5NTJ	TABLE II.	MeCN	VR	$NaClO_4$	0.1	-	-	-	AuBE SCE	2A0	C=0.2-0.9, d=0.15,v= 52,MP Aux, PE
DC48	$C_8H_{16}N_2$	1,6-diazabicyclo-[4.4.0]decane T66N FNTJ	TABLE II.	MeCN	VR	$NaClO_4$	0.1	-	-	-	AuBE SCE	2A0	C=0.2-0.9, d=0.15,v= 56,MP Aux, PE

TABLE I. Electrochemical Data $C_8H_{16}N_2$ DC48

Ref.	C/M	Charact. Potential		Response Const.		n Tech.	n	Electrokinetic Data			Products and Identification	Description and Remarks	Code No.
		Value	vs.		Value			Parameter	Value	From			
AN100 0349	-	$E_{p/2}$ 0.535	SCE	i_p	5F	-	-	-	-	-	-	A, $i_p \propto C$ for C=0.001 -0.2, XEl=CPE with wax and silicone oil, r	DC40 cd78
JA094 7108	-	XE 0.30 -	SCE	-	-	sttd	1 1	ΔE_p	100 -	sttd	radical cation, ESR	A,R, XE=$\frac{1}{2}(E_{p,A}+E_{p,C})$, r C,R	DC41
AN100 0349	-	$E_{p/2}$ 0.543	SCE	-	-	-	-	-	-	-	-	A, $i_p \propto C$, XEl=CPE with wax and silicone oil, r	DC42 cd83
JA094 7114	-	E_p 0.86	SCE	-	-	-	-	$E_p - E_{p/2}$	80	sttd	-	A, p	DC43
JA094 7108	-	XE 0.10 -	SCE	-	-	sttd	1 1	ΔE_p	80 -	sttd	radical cation, ESR	A,R, XE=$\frac{1}{2}(E_{p,A}+E_{p,C})$, r C,R	DC44
EA017 2009	-	$E_{\frac{1}{2}}$ -0.27F	SCE	-	-	-	-	Elog	34	-	-	A, $i_\ell \propto h^{\frac{1}{2}}$, Tc=1.2, i_ℓ=f[EtOH], $i_\ell \propto C$, r	DC45
		-0.33F		-	-	-	-		30	-	-	A, $i_\ell \propto h^{\frac{1}{2}}$, Tc=1.2, i_ℓ=f[EtOH], $i_\ell \propto C$, r	
		-0.38F		-	-	-	-		30	-	-	A, $i_\ell \propto h^{\frac{1}{2}}$, Tc=1.2, i_ℓ=f[EtOH], $i_\ell \propto C$, r	
		-0.45F		-	-	-	-		30	-	-	A, $i_\ell \propto h^{\frac{1}{2}}$, Tc=1.2, i_ℓ=f[EtOH], $i_\ell \propto C$, r	
		-0.56F		-	-	-	-		50	-	-	A, r	
		-0.56F		-	-	-	-		72	-	-	A, r	
		-0.56F		-	-	-	-		76	-	-	A, r	
JA094 7114	-	E_p 1.02	SCE	-	-	-	-	$E_p - E_{p/2}$	75	sttd	-	A, p	DC46
JA094 7108	-	XE -0.04 -	SCE	-	-	sttd	1 1	ΔE_p	85 -	sttd	radical cation, ESR	A,R, XE=$\frac{1}{2}(E_{p,A}+E_{p,C})$, r C,R	DC47
JA094 7108	-	XE 0.24 -	SCE	-	-	sttd	1 1	ΔE_p	65 -	sttd	radical cation, ESR	A,R, XE=$\frac{1}{2}(E_{p,A}+E_{p,C})$, r C,R	DC48

Code No.	Empirical Formula	Name and WLN Code	Structural Formula	Solvent	Tech.	Medium		μ, M	pH	T, °C	Electrodes	App.	Experimental Parameters
DC49	$C_8H_{16}N_2$	2,3-dimethyl-2,3-diazabicyclo[2.2.2]-octane C.A. 14287-92-4 T66 A B ANNTJ A1 B1	TABLE II.	MeCN	VR	$NaClO_4$	0.1	–	–	–	AuBE SCE	2AO	C=0.2-0.9, d=0.15, v=62, MP Aux, PE
DC50	$C_8H_{16}N_2O$	tetrahydro-1,3-dimethyl-1H,3H-[1,3,-4]oxadiazolo[3,4-a]-pyridazine C.A. 38704-98-2 T56N CO ENTJ B1 D1 A&D B&B	TABLE II.	MeCN	VR	$NaClO_4$	0.1	–	–	–	AuBE SCE	2AO	C=3.05, d=0.15, v=54, MP Aux, PE
DC51	$C_8H_{16}N_4$	1,3,6,8-tetraazatricyclo[4.4.1.1³,⁸]-dodecane C.A. 51-46-7 T777/BK/EL/JL A L 2BF L BN EN GN JNTJ	TABLE II.	MeCN	VR	$NaClO_4$	0.1	–	–	–	AuBE SCE	2AO	C=0.88, d=0.15, v=60, MP Aux, PE v=1210
DC52	$C_8H_{18}N_2$	1,2-<u>cis</u>-3,6-tetramethylperhydro-pyridazine C.A. 26171-64-2 T6NNTJ A1 B1 C1 F1 B&CF	TABLE II.	MeCN	VR	$NaClO_4$	0.1	–	–	–	AuBE SCE	2AO	C=0.2-0.9, d=0.15, v=50, MP Aux, PE
DC53	$C_8H_{18}N_2$	1,2-<u>trans</u>-3,6-tetramethylperhydro-pyridazine C.A. 38704-91-5 T6NNTJ A1 B1 C1 F1 A&C B&F	TABLE II.	MeCN	VR	$NaClO_4$	0.1	–	–	–	AuBE SCE	2AO	C=2.06, d=0.15, v=52, MP Aux, PE
DC54	$C_8H_{18}N_2O$	3,4-diisopropyl-1,3,4-oxadiazolidine C.A. 6400-42-6 T5NN DOTJ AY1&1 BY1⊃&1	TABLE II.	MeCN	VR	$NaClO_4$	0.1	–	–	–	AuBE SCE	2AO	C=0.2-0.9, d=0.15, v=8, MP Aux, PE
DC55	$C_8H_{18}N_4$	2,4,6,8-tetramethyl-2,4,6,8-tetraazabicyclo[3.3.0]octane C.A. 38705-09-8 T55 BN DN FN HNTJ B1 D1 F1 H1	TABLE II.	MeCN	VR	$NaClO_4$	0.1	–	–	–	AuBE SCE	2AO	C=2.94, d=0.15, v=92, MP Aux, PE
DC56	$C_9H_5F_3MoO_3$	η-cyclopentadienyl-trifluoromethyltricarbonylmolybdenum-(0) C.A. 12152-60-2 L5ØJ Ø-MO-XFFF/CO 3	$(\pi-C_5H_5)Mo(CO)_3$-CF_3	MG	PY	Bu_4NClO_4	0.1	–	–	22	DME Ag/Ag⁺	25-O	C=2
					PA	Bu_4NClO_4	0.1	–	–	22	HMDE Ag/Ag⁺	25-2	C=2, v=1E3
DC57	$C_9H_5F_3N_2O$	3-trifluoromethyl-ω-diazoacetophenone C.A. 17263-70-6 NUNU1VR CXFFF	$3\text{-}F_3CC_6H_4COCHN_2$	EtOH	1 PY	BR TX100	0.002	0.2	6.87	25.0 ±0.1	DME/SCE	2AO	t=2.66
DC58	$C_9H_5F_6FeS_2$	η-cyclopentadienyl-perfluorodimethyl-ethylene-1,2-disulfidoiron(III) C.A. 12261-16-4 T5S-FE-SJ DXFFF EXF⊃FF B- ØL5ØJ	TABLE II.	MG	PY	Bu_4NClO_4	0.1	–	–	22	DME Ag/Ag⁺	25AO	Pt Aux
					PA	Bu_4NClO_4	0.1	–	–	–	HMDE Ag/Ag⁺	25AO	v=1E3, Pt Aux
DC59	$C_9H_5F_6IrS_2$	η-cyclopentadienyl-perfluorodimethyl-ethylene-1,2-disulfidoiridium(III) C.A. 12169-93-6 T5S-IR-SJ DXFFF EXF⊃FF B- ØL5ØJ	TABLE II.	MG	PY	Bu_4NClO_4	0.1	–	–	22	DME Ag/Ag⁺	25AO	C=2
					PA	Bu_4NClO_4	0.1	–	–	22	HMDE Ag/Ag⁺	25-2	C=2

TABLE I. Electrochemical Data $C_9H_5F_6IrS_2$ DC59

Ref.	C/M	Charact. Potential		Response Const.		n Tech.		Electrokinetic Data			Products and Identification	Description and Remarks	Code No.
		Value	vs.		Value			Parameter	Value	From			
JA094 7108	–	XE	0.00	SCE	–	–	sttd	1	ΔE_p	70	sttd	radical cation, ESR	A,R,XE=$\frac{1}{2}$($E_{p,A}$+$E_{p,C}$), r	DC49
		–						1		–			C,R	
JA094 7108	–	XE	0.42	SCE	–	–	sttd	1	ΔE_p	90	sttd	radical cation, ESR	A,R,XE=$\frac{1}{2}$($E_{p,A}$+$E_{p,C}$), r	DC50
		–						1		–			C,R	
JA094 7114	–	E_p	0.58	SCE	–	–	–	–	E_p-$E_{p/2}$	55	sttd	–	A,p	DC51
			–							–			C,o	
			0.56		–	–	–	–	ΔE_p	70	–	–	A,C,p	
JA094 7108	–	XE	0.12	SCE	–	–	sttd	1	ΔE_p	100	sttd	radical cation, ESR	A,R,XE=$\frac{1}{2}$($E_{p,A}$+$E_{p,C}$), r	DC52
		–						1		–			C,R	
JA094 7108	–	XE	0.13	SCE	–	–	sttd	1	ΔE_p	130	sttd	radical cation, ESR	A,R,XE=$\frac{1}{2}$($E_{p,A}$+$E_{p,C}$), r	DC53
		–						1		–			C,R	
JA094 7108	–	XE	0.56	SCE	–	–	sttd	1	ΔE_p	130	sttd	radical cation, ESR	A,R,XE=$\frac{1}{2}$($E_{p,A}$+$E_{p,C}$), r	DC54
		–						1		–			C,R	
JA094 7114	–	E_p	1.03	SCE	–	–	–	–	E_p-$E_{p/2}$	140	sttd	–	A,p	DC55
JA088 0471	–	$E_{\frac{1}{2}}$	-2.1	Ag/Ag$^+$	–	–	QE	1.6	–	–	–	–	C,p	DC56
JA088 0471	–	E_p	-2.5 -1	Ag/Ag$^+$	–	–	–	–	–	–	–	–	C,p A,broad multiple peak	
JL70B 0034	21b	$E_{\frac{1}{2}}$	-0.8F -1.1F -1.3F -1.6$_5$F	SCE	i_ℓ	40F 30F 10F 10F	–	–	Elog	non-linear	–	–	C,$\frac{1}{4}$(PK),r C C C	DC57
JA090 2001	–	$E_{\frac{1}{2}}$	-2.1	Ag/Ag$^+$	–	–	QE	1.6	–	–	–	–	C,p	DC58
JA090 2001	–	$E_{p/2}$	-2.2 -1.8	Ag/Ag$^+$	–	–	–	–	–	–	–	–	C,p A	
JA088 5112	–	$E_{\frac{1}{2}}$	-1.7	Ag/Ag$^+$	–	–	QE	0.8	–	–	–	–	C,R,p	DC59
JA088 5112	–	E_p	-2.1 -1.7	Ag/Ag$^+$	–	–	QE	0.8	–	–	–	–	C,R,p A,R	

DC60 $C_9H_5F_6NiS_2$ CRC Handbook Series in Organic Electrochemistry

Code No.	Empirical Formula	Name and WLN Code	Structural Formula	Solvent	Tech.	Medium		μ, M	pH	T, °C	Electrodes	App.	Experimental Parameters
DC60	$C_9H_5F_6NiS_2$	η-cyclopentadienyl-perfluorodimethyl-ethylene-1,2-di-sulfidonickel(III) C.A. 12169-95-8 T5S-NI-SJ DXFFF EXFC FF B- ØL5ØJ	TABLE II.	MG	PY	Bu_4NClO_4	0.1	-	-	22	DME Ag/Ag^+	25-0	C=2
					PA	Bu_4NClO_4	0.1	-	-	22	HMDE Ag/Ag^+	25-2	C=2, v=1E3
DC61	$C_9H_5F_6RhS_2$	η-cyclopentadienyl-perfluorodimethyl-ethylene-1,2-di-sulfidorhodium(III) C.A. 12169-96-9 T5S-RH-SJ DXFFF EXFC FF B- ØL5ØJ	TABLE II.	MG	PY	Bu_4NClO_4	0.1	-	-	22	DME Ag/Ag^+	25A0	Pt Aux
					PA	Bu_4NClO_4	0.1	-	-	22	HMDE Ag/Ag^+	25A2	v=1E3, Pt Aux
DC62	$C_9H_7Br_3O_2$	2,2,2-tribromoethyl benzoate C.A. 4998-93-0 EXEE1OVR	$C_6H_5COOCH_2CBr_3$	MeOH	PY	$LiClO_4$	0.1	-	-	-	DME/SCE	---	ns
DC63	$C_9H_7Cl_3O_2$	2,2,2-trichloro-ethyl benzoate C.A. 37934-99-9 GXGG1OVR	$C_6H_5COOCH_2CCl_3$	MeOH	PY	$LiClO_4$	0.1	-	-	-	DME/SCE	---	ns
				DMF	CP	$LiClO_4$	0.1	-	-	-	Pt gauze SCE	35A-	E_{app} = -1.65, Pt Aux
DC64	$C_9H_8BrClO_2$	3-bromo-2-chloro-3-phenylpropanoic acid QVYGYER	$C_6H_5CHBr-CHClCOOH$	H_2O	PY	HCl	2	-	-	-	DME/SCE	4A0	C=0.5
DC65	$C_9H_8Cl_2O_2$	2,2-dichloroethyl benzoate C.A. 37934-98-8 GYG1OVR	$C_6H_5COOCH_2CHCl_2$	MeOH	PY	$LiClO_4$	0.1	-	-	-	DME/SCE	---	ns
DC66	$C_9H_8Cl_3NO$	3'-methyl-2,2,2-tri-chloroacetanilide C.A. 2563-96-4 GXGGVMR C1	$3-CH_3C_6H_4NHCO-CCl_3$	EtOH 50	PY	HMe_3NI	0.05	-	-	-	DME/SCE	0-0	C=1
DC67	$C_9H_8Cl_3NO$	4'-methyl-2,2,2-tri-chloroacetanilide C.A. 2564-09-2 GXGGVMR D1	$4-CH_3C_6H_4NHCO-CCl_3$	EtOH 50	PY	HMe_3NI	0.05	-	-	-	DME/SCE	0-0	C=1
DC68	$C_9H_8N_2O_3$	1,3-dimethylbenz-imidazolone-4,7-quinone T56 BNVN FV IVJ B1 D1	TABLE II.	MeOH	PY	Et_4NClO_4	0.05	-	-	-	DME/MP	0-0	EC=1.23
				DMF	PY	Et_4NClO_4	0.1	-	-	-	DME/MP	0-0	EC=1.23
DC69	$C_9H_8N_2O_3$	1,3-dimethylbenz-imidazolone-5,6-quinone T56 BNVN GVVJ B1 D1	TABLE II.	MeOH	PY	Et_4NClO_4	0.05	-	-	-	DME/MP	0-0	EC=1.23
				DMF	PY	Et_4NClO_4	0.1	-	-	-	DME/MP	0-0	EC=1.23

TABLE I. Electrochemical Data $C_9H_8N_2O_3$ DC69

Ref.	C/M	Charact. Potential		Response Const.		n Tech.		Electrokinetic Data			Products and Identification	Description and Remarks	Code No.
		Value	vs.		Value			Parameter	Value	From			
JA088 0471	-	$E_{\frac{1}{2}}$	-0.96 -2.4	Ag/Ag$^+$	-	-	QE	1.0 -	-	-	-	-	C,R,p C	DC60
JA088 0471	-	E_p	-1.1 -0.8	Ag/Ag$^+$	-	-	-	-	-	-	-	-	C,R,p A,R	
JA090 2001	-	$E_{\frac{1}{2}}$	-1.4	Ag/Ag$^+$	-	-	QE	1.2	-	-	-	-	C,p	DC61
JA090 2001	-	$E_{p/2}$	-1.5 -1.2	Ag/Ag$^+$	-	-	-	-	-	-	-	-	C,p A	
JA094 5139	355 a	$E_{\frac{1}{2}}$	-0.60	SCE	-	-	-	-	-	-	-	CP at -0.70 V → benzoic acid after addn. of H$^+$, 85%, MPT	C,Mx,p	DC62
JA094 5139	355 b	$E_{\frac{1}{2}}$	-1.28	SCE	-	-	-	-	-	-	-	CP at -1.65 V → benzoic acid after addn. of H$^+$, 87%, MPT	C,p	DC63
JA094 5139	355 b	-	-	-	-	-	-	-	-	-	-	benzoic acid after addn. of H$^+$, 91%, MPT	C,p	
AN095 0387	-	$E_{\frac{1}{2}}$	-0.31F	SCE	i_ℓ	3F	i:i	2	-	-	-	-	C,W,r	DC64
JA094 5139	355 a	$E_{\frac{1}{2}}$	-1.91	SCE	-	-	-	-	-	-	-	-	C,p	DC65
RG035 0843	14c	$E_{\frac{1}{2}}$	-0.48 -0.94 -1.54	SCE	-	-	-	-	-	-	-	-	C,≠,p C C	DC66
RG035 0843	14c	$E_{\frac{1}{2}}$	-0.47 -0.97 -1.56	SCE	-	-	-	-	-	-	-	-	C,≠,p C C	DC67
RY002 1096	-	$E_{\frac{1}{2}}$	-0.29 -	SCE	-	-	-	-	-	-	-	-	C,p C,X	DC68
RY002 1096	-	$E_{\frac{1}{2}}$	-0.59 -1.22	SCE	-	-	-	-	-	-	-	-	C,p C,X	
RY002 1096	-	$E_{\frac{1}{2}}$	-0.31 -	SCE	-	-	-	-	-	-	-	-	C,p C,X	DC69
RY002 1096	-	$E_{\frac{1}{2}}$	-0.63 -1.09	SCE	-	-	-	-	-	-	-	-	C,p C,X	

Code No.	Empirical Formula	Name and WLN Code	Structural Formula	Solvent	Tech.	Medium		μ, M	pH	T, °C	Electrodes	App.	Experimental Parameters
DC70	$C_9H_8N_6$	1-methyl-5-phenyl-azonia-1,2,3-triazolo[4,5-d]-1,2,3-triazole C.A. 30597-87-6 T55 BNNN FNNNJ B1 GR	TABLE II.	DMF	PY	Bu_4NI	0.155	-	-	22	DME/MP?	OA?O	C=?, t=4.36, h=60
DC71	$C_9H_8N_6$	2-methyl-5-phenyl-azonia-1,2,3-triazolo[4,5-d]-1,2,3-triazole anion C.A. 30597-88-7 T55 BNNN FNNNJ C1 GR &11/1∅	TABLE II.	DMF	PY	Bu_4NI	0.155	-	-	22	DME/MP?	OA?O	C=?, t=4.36, h=60
DC72 ap00	C_9H_8O	cinnamaldehyde C.A. 14371-10-9 VH1U1R -T	$C_6H_5CH:CHCHO$	EtOH 25	PY	BR		-	2.3	-	DME/SCE	4AO	C=0.2, v=33
									2.7				
									3.55				
									6.3				
									7.7				
									8.85				
									9.5				
									11.8				
				DMF	PY	Bu_4NBF_4	0.50	-	-	25	DME/SCE	23-0	C=2.0, Pt Aux
DC73 ap06	$C_9H_9BrN_2O_4$	5-bromo-3,6-dinitro-1,2,4-trimethylbenzene WNR BE C1 E1 F1 DNW	TABLE II-2.	MeOH 60	PY	BR		-	1.9	-	DME/SCE	0-4	EC=2.71
									3.5				
									4.6				
									6.2				
									8.2				
									9.7				
									10.8				
									11.5				
DC74	$C_9H_9Cl_2NO$	3'-methyl-2,2-dichloroacetanilide C.A. 2563-98-6 GYGVMR C1	$3-CH_3C_6H_4NHCO-CHCl_2$	EtOH 50	PY	HMe_3NI	0.05	-	-	-	DME/SCE	0-0	C=1
DC75	$C_9H_9Cl_2NO$	4'-methyl-2,2-dichloroacetanilide C.A. 2842-11-7 GYGVMR D1	$4-CH_3C_6H_4NHCO-CHCl_2$	EtOH 50	PY	HMe_3NI	0.05	-	-	-	DME/SCE	0-0	C=1

TABLE I. Electrochemical Data $C_9H_9Cl_2NO$ DC75

Ref.	C/M	Charact. Potential		Response Const.		n Tech.	n	Electrokinetic Data			Products and Identification	Description and Remarks	Code No.
		Value	vs.		Value			Parameter	Value	From			
BJ047 1490	-	$E_{\frac{1}{2}}$	-1.06 -1.88	MP	i_ℓ/C	4.3 6.7	-	-	Elog	53 -	-	-	$C, E_{\frac{1}{2}} \propto$ LUMO, R?, r C	DC70
BJ047 1490	-	$E_{\frac{1}{2}}$	-0.90 -1.5	MP	i_ℓ/C	3.4 5	-	-	Elog	58 -	sttd	-	$C, E_{\frac{1}{2}} \propto$ LUMO, R?, r C	DC71
TF065 1668	156 g	$E_{\frac{1}{2}}$	-0.84F	SCE	i	0.3F	-	-	-	-	-	-	C,W,r	DC72 ap00
			-0.88F			0.3F	-	-	-	-	-	-	C,W,r	
			-0.76F -1.3F			0.3F 0.3F	-	-	-	-	-	-	C,W,r C,M	
			-0.94F -1.28F			0.3F 0.3F	-	-	-	-	-	-	C,W,r C,W	
	156 h		-1.04F -1.24F			0.3F 0.3F	-	-	-	-	-	-	C,W,r C,W	
			-1.00F - -1.6F			0.6F - 0.1F	-	-	-	-	-	-	C,W,r O C	
			-1.20F -1.8F			0.6F 0.2F	-	-	-	-	-	-	C,W,r C,W	
			-1.20F -1.7F			0.5F 0.3F	-	-	-	-	-	-	C,W,r C,dr	
JA094 8471	-	$E_{\frac{1}{2}}$	-1.53 -2.08	SCE	-	-	-	-	αn_a	1.2 1.0	Elog	-	C,r C	
RR020 1055	-	$E_{\frac{1}{2}}$	-0.51	SCE	i_ℓ	1.49	-	-	-	-	-	-	$C, i_d, D=8.1, r$	DC73 ap06
			-0.57			1.38	PQ(in ACET) IIk	2.7 2.5	$dE_{\frac{1}{2}}/dpH \neq 0$	-	-	-	$C, i_d, i_\ell \propto (h^{\frac{1}{2}}), i_\ell = f(T), r$	
			-0.60			1.21	-	-	-	-	-	-	C, i_d, r	
			-0.72			1.26	-	-	-	-	-	-	C, i_d, r	
			-0.71 -0.91			0.52 0.57	-	-	-	-	-	-	C, i_d, r C	
			-0.73 -0.92			0.57 0.46	-	-	-	-	-	-	C, i_d, r C	
			-0.72 -0.90			0.34 0.29	-	-	-	-	-	-	C, i_d, r C	
			-0.72 -0.90			0.29 0.17	-	-	-	-	-	-	C, i_d, r C	
RG035 0843	14b	$E_{\frac{1}{2}}$	-0.94 -1.54	SCE	-	-	-	-	-	-	-	-	$C, \neq, p$ C	DC74
RG035 0843	14b	$E_{\frac{1}{2}}$	-0.94 -1.54	SCE	-	-	-	-	-	-	-	-	$C, \neq, p$ C	DC75

Code No.	Empirical Formula	Name and WLN Code	Structural Formula	Solvent	Tech.	Medium		μ, M	pH	T, °C	Electrodes	App.	Experimental Parameters
DC76	$C_9H_9IO_2$	2-iodoethyl benzoate C.A. 1829-28-3 I2OVR	$C_6H_5COOCH_2CH_2I$	MeOH	PY	$LiClO_4$	0.1	-	-	-	DME/SCE	---	ns
DC77	$C_9H_9NO_3S$	methyl 2-methyl-5-(thiocyanatomethyl)-3-furoate T5OJ B1 CVO1 E1SCN	TABLE II.	EtOH 86	PY	BR		0.45	≈ 2.4 ≈ 4.0 ≈ 5.0 ≈ 5.9 ≈ 8.0 ≈ 9.8 ≈ 10.8 ≈ 11.9	25.0 ±0.1	DME/SCE	OCO	C=0.2,m=1.56, t=0.22
	$C_9H_9NO_7$ see DB82												
DC78	C_9H_{10}	cis-bicyclo[6.1.0]-nona-2,4,6-triene C.A. 26132-66-1 L38T&J	TABLE II.	MeCN	PY	Me_4NPF_6	0.1	-	-	-	DME Ag/0.1 Ag^+	256A F2	ns
				THF	PY	Bu_4NClO_4	0.2	-	-	-	DME Ag/0.1 Ag^+	256A F2	ns
DC79	C_9H_{10}	trans-bicyclo-[6.1.0]nona-2,4,6-triene C.A. 18012-46-9 L38T&J &1 &1	TABLE II.	THF	PY	Bu_4NClO_4	0.2	-	-	-	DME Ag/0.1 Ag^+	256A F2	ns
					VR	Bu_4NClO_4	0.2	-	-	-	PBE Ag/0.1 Ag^+	256A F2	v=40 v=100 v=600
DC80	C_9H_{10}	α-methylstyrene C.A. 25013-15-4 1UY1&R	$C_6H_5C(CH_3):CH_2$	MeOH	VY	$NaClO_4$ Na tosylate	? ?	-	-	25	RCCE SCE	06-0	C=1,A=0.8
				MeCN	VY	$NaClO_4$	?	-	-	25	RCCE SCE RPE/SCE	06-0	C=1,A=0.8, v=20 A=?
DC81	$C_9H_{10}ClNO$	N-benzylchloro-acetamide C.A. 2564-06-9 G1VM1R	$C_6H_5CH_2NHCO-CH_2Cl$	EtOH 50	PY	HMe_3NI	0.05	-	-	-	DME/SCE	0-0	C=1
DC82	$C_9H_{10}O_2$	benzyl acetate C.A. 1333-46-6 1VO1R	$C_6H_5CH_2OCOCH_3$	MeCN	QE	$LiClO_4$	0.1	-	-	-	Pt Ag/0.1 Ag^+	25A-	C=3.3,A=12.9,E_{app}=1.90,stainless steel/Aux

TABLE I. Electrochemical Data $C_9H_{10}O_2$ DC82

Ref.	C/M	Charact. Potential		Response Const.		Tech.	n	Electrokinetic Data			Products and Identification	Description and Remarks	Code No.
		Value	vs.		Value			Parameter	Value	From			
JA094 5139	355 a	$E_{\frac{1}{2}}$	>-2.20	SCE	-	-	-	-	-	-	-	-	C,M,p	DC76
RG035 0773	126 b	$E_{\frac{1}{2}}$	-1.16	SCE	i_ℓ	0.45	Ilk	2.2	-	-	-	-	$C,i_d,R,E_{\frac{1}{2}}$ and $i_\ell \neq f(C),r$	DC77
			-1.15			0.55		2.2	-	-	-	-	C,i_d,R,r	
			-1.16			0.57		2.2	-	-	-	-	C,i_d,R,r	
			-1.18			0.60		2.2	-	-	-	-	$C,i_d,R,Mc(pH),r$	
			-1.18			0.58		2.2	-	-	-	-	C,i_d,R,r	
			-1.20			0.58		2.2	-	-	-	-	C,i_d,R,r	
			-1.22			0.57		2.2	-	-	-	-	C,i_d,R,r	
			-1.24			0.55		2.2	-	-	-	-	C,i_d,R,r	
JA095 2198	-	$E_{\frac{1}{2}}$	-2.60	SCE	$I\eta^{\frac{1}{2}}$	0.158	sttd	2	Elog	100	sttd	-	$C,i_\ell \propto C,r$	DC78
JA095 2198	-	$E_{\frac{1}{2}}$	-2.55	SCE	$I\eta^{\frac{1}{2}}$	0.095	sttd	1	Elog	80	sttd	-	$C,\neq$ from $VR,i_\ell \propto C,r$	
			-2.79			0.095		1		80			$C,\neq$ from $VR,i_\ell \propto C$	
JA095 2198	-	$E_{\frac{1}{2}}$	-2.50	SCE	$I\eta^{\frac{1}{2}}$	0.139	sttd	1-2	Elog	110	sttd	-	$C,\neq,I\eta^{\frac{1}{2}}\downarrow$ as $C\uparrow,r$	DC79
JA095 2198	-	E_p	-2.35F	SCE	i_p	2.6	-	-	-	-	-	-	$C,\neq,r$	
			-2.36F -2.22F			4F 0.2F	-	-	-	-	-	-	C,r A	
			-2.40F 2.21F			6.7F 0.7F	-	-	-	-	-	-	C,r A	
EA017 1595	372 b	$E_{\frac{1}{2}}$	1.350	SCE	-	-	i:i	2.4	-	-	-	-	A,p	DC80
			1.410		-	-		2.4	-	-	-	-	A,p	
EA017 1595	372 a	$E_{\frac{1}{2}}$	1.480	SCE	-	-	i:i	2.3	-	-	-	-	A,p	
			1.760		-	-		1.8	-	-	-	-	A,p	
RG035 0843	14a	$E_{\frac{1}{2}}$	-1.49	SCE	-	-	-	-	-	-	-	-	$C,\neq,p$	DC81
JA094 6812	360 b	-	-	-	-	-	QE	3.0	-	-	-	benzaldehyde,30%, NMR,GLC,IRS	A,p	DC82

DC83 C$_9$H$_{10}$O$_2$S

Code No.	Empirical Formula	Name and WLN Code	Structural Formula	Solvent	Tech.	Medium	μ, M	pH	T, °C	Electrodes	App.	Experimental Parameters
DC83	C$_9$H$_{10}$O$_2$S	benzyl vinyl sulfone C.A. 15753-89-6 1U1SW1R	C$_6$H$_5$CH$_2$SO$_2$CH:CHCH$_2$	MeOH 20	PY	LiCl 0.07	-	-	25	DME/SCE	0-0	C=0.5,m=1.63, t=2.5
DC84 ap46	C$_9$H$_{11}$Cl$_2$N$_3$O$_4$S$_2$	6-chloro-3-chloromethyl-3,4-dihydro-2-methyl-7-sulfamoyl-1,2,4-benzothiadiazine 1,1-dioxide C.A. 135-07-9 T66 BM DNSW CHJ C1G D HSWZ IG	TABLE II-2.	H$_2$O	PY	BOR ? Me$_4$NBr 0.1	-	10.3 12.0	21± 0.2	DME/SCE	OAO	C=0.56,m= 1.58,t=3.3 C=ns
DC85	C$_9$H$_{11}$NO$_2$	ethyl 2-cyano-2,4-hexadienoate 2U2UYCN&VO2	TABLE II.	DMF	PY	Et$_4$NClO$_4$ 0.1 Et$_4$NClO$_4$ 0.1 H$_2$O 1	-	-	-	DME/SCE	--0	C=1,m=1.58, t=4.20
DC86 ap87	C$_9$H$_{13}$N	N,N-dimethyl-4-toluidine C.A. 99-97-8 1N1&R D	4-H$_3$CC$_6$H$_4$N(CH$_3$)$_2$	H$_2$O	IL	BR p-dimethylaminobenzaldehyde 5.0E-4	-	6.8	-	Pt/SCE	0-0	C=0.4,v=3.3
DC87	C$_9$H$_{13}$NO$_2$	3,3-diethyl-2,4-(1H,3H)pyridinedione C.A. 77-04-3 T6NVXVJ C2 C2	TABLE II.	H$_2$O	DI	BR	-	8.0	-	DME/SCE	O4AO	C=0.05,m=1.73,t=0.5, v=10,ΔE=100
					PT	BOR ?	-	9.2	-	DME/SCE	O4AO	C=0.1,m=1.73, t=1,h=80
DC88 ap97	C$_9$H$_{13}$N$_3$O$_5$	cytidine C.A. 65-46-3 T6NVNJ DZ A- ET5OTJ B1Q EQ DQ	TABLE II-2	H$_2$O	PY	MB	0.13	5.0	0.0	DME/SCE	235A 03	C=0.1,m=1.0, t=6-7,Pt Aux
					PV	ACET ?	-	4.2	25	DME/SCE	235A 03	C=1.0,m=1.0, t=6-7,f=50, Pt Aux
					PVI	MB	0.13	5.0	0	DME/SCE	235A 03	C=0.1,m=1.0, t=6-7,f=50, Δe=3.5rms, Pt Aux C=1.0
DC89	C$_9$H$_{14}$N$_3$O$_8$P	cytidine monophosphate C.A. 27214-06-8 T6NVNJ DZ A- BT5OTJ CQ DQ E1OPQQO	TABLE II.	H$_2$O	PY	ACET ?	0.5	4.1	0 25	DME/SCE	235A 03	C=0.49,m= 1.0,t=6-7, Pt Aux C=0.99
						MB	0.13	5.0	0			C=0.1
					PV	ACET ?	-	4.2	25	DME/SCE	235A 03	C=1.0,m=1.0, t=6-7,f=50, Pt Aux
					PVI	ACET ?	-	4.2	25	DME/SCE	235A 03	C=1.0,m=1.0, t=6-7,f=50, Pt Aux
						MB	0.13	5.0	0			C=0.1,m=1.0, t=6-7,f=50, Δe=3.5rms, Pt Aux
					VR	ACET ?	0.5	4.1	25	HMDE SCE	235A 03	C=0.99,A= 0.022,v=20, Pt Aux v=200

CONT

TABLE I. Electrochemical Data $C_8H_{14}N_3O_8P$ (CONT.) DC89

Ref.	C/M	Charact. Potential		Response	Const.	n		Electrokinetic Data			Products and	Description and	Code
		Value	vs.		Value	Tech.		Parameter	Value	From	Identification	Remarks	No.
DS171 0860	43a	$E_{\frac{1}{2}}$ -1.67	SCE	I	2.98	Ilk	≈ 2	αn_a	0.50	Elog	-	$C, i_d, \neq, r$	DC83
AA080 0017	-	$E_{\frac{1}{2}}$ -1.64	SCE	-	-	-	-	-	-	-	-	C,r?	DC84 ap46
		-1.78F		-	-	-	-	-	-	-	-	C,r?	
JS116 0743	71a	$E_{\frac{1}{2}}$ -1.37	SCE	I	1.00	-	-	-	-	-	-	C,r	DC85
		-1.37			0.84	-	-	-	-	-	-	C,r	
JP067 0862	-	E_p $\approx$0.5F	SCE	i_p	1	-	-	-	-	-	-	$A, \neq, E_{P/2}=f(v),r$	DC86 ap87
		$\approx$0.9F			2	-	-	-	-	-	-	$A, \neq, E_{P/2}=f(v)$	
AA080 0233	-	E_{su} -1.20	SCE	-	-	-	-	-	-	-	-	C,r	DC87
AA080 0233	-	$E_{\frac{1}{2}}$ -0.63F	SCE	i_ℓ	0.06F	-	-	-	-	-	-	C,r	
JA095 0991	89c	$E_{\frac{1}{2}}$ -1.38	-	I	4.1	-	-	-	-	-	-	C,D=5.1,Tc=1.7; in ACET(pH=4.1) D=12.0 and I=6.3,r	DC88 ap97
JA095 0991	89c	E_{su} -1.42F	SCE	i_{su}	0.48F	-	-	-	-	-	-	C,p	
JA095 0991	89c	E_{su} -1.40	SCE	i_{su}	0.05	-	-	-	-	-	-	C,p	
		-1.43			0.52	-	-	-	-	-	-	C,p	
JA095 0991	89c	$E_{\frac{1}{2}}$ -1.37	SCE	I	3.0	QE at pH=3	2.6 to 4.6	-	-	-	-	C,value of n depends on base line used,r	DC89
		-1.37			4.6	-	-	-	-	-	-	C,D=6.4,Tc=1.7,r	
		-1.46			3.2	-	-	-	-	-	-	C,D=3.1,r	
JA095 0991	89c	E_{su} -1.42F	SCE	i_{su}	0.38F	-	-	-	-	-	-	C,r	
JA095 0991	89c	E_{su} -1.43	SCE	i_{su}	0.37	-	-	-	-	-	-	C,r	
		-1.50			0.02	-	-	-	-	-	-	C,r	
JA095 0991	89c	-	-	-	$i_p/ACv^{\frac{1}{2}}$	26.0F	-	-	-	-	-	-	C,D=4-6,r	
		E_p -1.42F			22.6F	-	-	-	-	-	-	C,r	CONT

Code No.	Empirical Formula	Name and WLN Code	Structural Formula	Solvent	Tech.	Medium		μ, M	pH	T, °C	Electrodes	App.	Experimental Parameters
DC89	$C_9H_{14}N_3O_8P$	cytidine monophosphate	TABLE II.	H_2O	VR	ACET	?	0.5	4.1	25	HMDE SCE	235A 03	C=0.99,v=500, Pt Aux
													v=4.4E3
													v=1E4
DC90	$C_9H_{15}NO_2$	3,3-diethyl-2,4-dioxopiperidine C.A. 77-03-2 T6MVXVTJ C2 C2	TABLE II.	H_2O	PT	BOR	?	-	9.2	-	DME/SCE	O4AO	C=0.1,m=1.73, t=1,h=80
DC91	$C_9H_{16}N_2$	6,6-cyclopentamethylene-1,5-diazabicyclo[3.1.0]hexane T35 ANXNTJ B-& AL6XOTJ	TABLE II.	MeCN	VR	$NaClO_4$		0.1	-	-	-	AuBE SCE	2AO	C=1.32,d=0.15,v=111, MP Aux,PE
DC92	$C_9H_{17}N_2$	2-tert-butyl-2-azonia-3-aza-2-norbornene ion C.A. 41322-56-9 T55 A CNNTJ CX1&1&1 &9	TABLE II.	H_2O	VR	ns		-	10-13	-	ns	---	ns
				MeCN	VR	ns		-	-	-	ns	---	ns
DC93	$C_9H_{18}HgO_4$	2-butoxy-2-methoxyethylmercuric acetate C.A. 42414-57-3 4OYO1&1-HG-OV1	TABLE II.	MeOH 50	PY	ACET TX100	? 0.002	-	5.56	-	DME/SCE	-AO	C=0.2,m=1.65,t=3.34, h=55
						BOR TX100	? 0.002		9.55				
						NaOH TX100	? 0.002		12.9				
DC94	$C_9H_{18}HgO_4$	2-(2-methylpropoxy)-2-methoxyethylmercuric acetate 1Y1&1OYO1&1-HG-OV1	TABLE II.	MeOH 50	PY	ACET TX100	? 0.002	-	5.56	-	DME/SCE	-AO	C=0.2,m=1.65,t=3.34, h=55
						BOR TX100	? 0.002		9.55				
						NaOH TX100	? 0.002		12.9				
DC95	$C_9H_{18}NNaS_2$	sodium dibutyldithiocarbamate C.A. 136-30-1 SUCS&N4&4 &-NA-	$(CH_3CH_2CH_2CH_2)_2$-NCSSNa	H_2O	PY	KOH KCl TX100	0.01 0.3 0.008	-	-	25± 0.2	DME/SCE	O1AO	C=1.0,m=2.52,t=3.00, h=40
					IL	KOH KCl	0.01 0.3	-	-	25± 0.2	HMDE SCE	O1AO	C=1.0,v=20
DC96	$C_9H_{18}N_2$	2-tert-butyl-2,3-diazanorbornane C.A. 40953-63-7 T55 A CMNTJ DNX1&1&1	TABLE II.	MeCN	VR	ns		-	-	-	ns	---	ns
DC97	$C_9H_{20}N_2$	1,2-diisopropylpyrazolidine C.A. 38704-87-9 T5NNTJ AY1&1 BY1&1	TABLE II.	MeCN	VR	$NaClO_4$		0.1	-	-	-	AuBE SCE	2AO	C=0.2-0.9, d=0.15,v=50,MP Aux, PE

TABLE I. Electrochemical Data $C_9H_{20}N_2$ DC97

Ref.	C/M	Charact. Potential		Response Const.		n Tech.	Electrokinetic Data			Products and Identification	Description and Remarks	Code No.
		Value	vs.		Value		Parameter	Value	From			
JA095 0991	89c	E_p	-1.45F	SCE	$i_p/ACv^{1/2}$	20F	-	-	-	-	-	C,r	DC89
			-1.51F			17.5F	-	-	-	-	-	C,r	
			-			17.4F	-	-	-	-	-	C,r	
AA080 0233	-	$E_{1/2}$	-0.80F	SCE	i_ℓ	0.07F	-	-	-	-	-	A,r	DC90
JA094 7108	-	E_p	1.15	SCE	-	-	-	-	-	-	-	A,⇌,p	DC91
JA095 5422	363 a	$E_{1/2}$	-0.75 ±0.01	SCE	-	-	-	-	-	-	radical	C,p	DC92
JA095 5422	363 a	$E_{1/2}$	-0.72	SCE	-	-	-	-	-	-	radical	C,p	
JE025 0397	135 a	$E_{1/2}$	-0.26 -0.35	SCE	-	-	-	-	-	-	-	C,i_d,p C,i_d	DC93
	135 b		-0.38 -0.91		i	0.91	-	-	-	-	-	C,i_d,p C,i_d	
			-0.58 -0.88		i	0.97	-	-	-	-	-	C,i_d,p C,i_d	
JE025 0397	135 a	$E_{1/2}$	-0.26 -0.36	SCE	-	-	-	-	-	-	-	C,i_d,p C,i_d	DC94
	135 b		-0.39 -0.92		i	0.96	-	-	-	-	-	C,i_d,p C,i_d	
			-0.58 -0.89		i	0.94	-	-	-	-	-	C,i_d,p C,i_d	
AJ028 0021	-	$E_{1/2}$	-0.692 -0.567	SCE	-	3.0F -	-	-	-	-	-	A,i_a,Pr,r A	DC95
AJ028 0021	-	E_p	-0.562	SCE	-	-	-	-	$E_p-E_{p/2}$	26	-	-	A,r	DC96
JA095 5422	363 a	E_p	-0.08	SCE	-	-	sttd	2	-	-	-	monocation	A,⇌,ece,p	DC96
JA094 7108	-	XE	0.00	SCE	-	-	sttd	1	ΔE_p	120	sttd	radical cation, ESR	A, $XE=\frac{1}{2}(E_{p,A}+E_{p,C})$, R, r	DC97

Code No.	Empirical Formula	Name and WLN Code	Structural Formula	Solvent	Tech.	Medium		μ, M	pH	T, °C	Electrodes	App.	Experimental Parameters
DC98	$C_9H_{20}N_2$	1,2-dimethyl-3-$\underline{n}$-hexyldiaziridine C.A. 17043-14-0 T3NN CHJ A1 B1 C6	TABLE 11.	MeCN	VR	$NaClO_4$	0.1	-	-	-	AuBE SCE	2A0	C=0.5, d=0.15, v=80, MP Aux, PE
DC99	$C_{10}ClCo_3O_9$	tris-[tricarbonyl-cobalt(I)]methyl chloride C.A. 13682-02-5 OC 3-CO- 31G	$[Co(CO)_3]_3CCl$	MG	PY	Bu_4NClO_4	0.1	-	-	22	DME Ag/Ag$^+$	25-0	C=2
					PA	Bu_4NClO_4	0.1	-	-	22	HMDE Ag/Ag$^+$	25-2	C=2
DD00 aq50	$C_{10}H_6O_2$	1,2-naphthoquinone C.A. 524-42-5 L66 BVVJ	TABLE 11-2.	MeCN	PY	Et_4NClO_4	0.1	-	-	-	DME/MP	0-0	EC=1.23
DD01 aq51	$C_{10}H_6O_2$	1,4-naphthoquinone C.A. 130-15-4 L66 BV EVJ	TABLE 11-2.	DMF	PY	Et_4NClO_4	0.1				DME/MP		
DD02	$C_{10}H_7MnO_4$	acetyl-η-cyclopenta-dienyltricarbonyl-manganese(11) C.A. 12116-28-8 L50J AV1 Ø-MN-/CO 3	TABLE 11.	EtOH 40	PY	PW		-	2 4.21	25	DME/SCE	0-0	C=1
DD03 aq80 cf20	$C_{10}H_8$	naphthalene C.A. 91-20-3 L66J	TABLE 11-2.	NH_3	VR	KI or $MeBu_3NI$	0.1 0.1	-	-	-	AuDE Ag/0.01 $AgNO_3$, 0.1 KI	125F2	C=1, A=0.5, Pt Aux
DD04	$C_{10}H_8FeN_4O_2$	dinitrosyl-2,2'-bi-pyridineiron C.A. 36454-18-9 D B656 GN-FE-NJ HNO HNO	TABLE 11.	MG	PY	Bu_4NClO_4	?	-	-	-	DME Ag/0.001 Ag$^+$	5A0	t=3
DD05 aq89	$C_{10}H_8N_2$	2,2'-bipyridine C.A. 366-18-7 T6NJ B- 2	TABLE 11-2.	H_2O	PY	BOR	?	-	9.6	27	DME/SCE	---	ns
				dioxane 20	PY	HCl NaOAc	?	-	0.76	25±0.1	DME/SCE	05A0	C=0.057, m=1.51, t=3.38, h=75
						PHTH HCl	?		2.21				
						ACET	?		5.65				
						BOR?	?		10.30				
				MG	PY	Bu_4NClO_4	?	-	-	-	DME Ag/0.001 Ag$^+$	5A0	t=3
DD06	$C_{10}H_9ClO_4S_2$	3-methyl-5-phenyl-1,2-dithiolylium perchlorate C.A. 41467-62-3 T5SSTJ C1 ER & G-04	TABLE 11.	CH_2Cl_2	VR	Bu_4NBF_4	0.2	-	-	-	PBE/SCE	26A2	v=150
DD07 cf26	$C_{10}H_9N$	2-naphthylamine C.A. 91-59-8	TABLE 11-3.	DMF	PY	Et_4NClO_4	?				DME/SCE		

TABLE I. Electrochemical Data $C_{10}H_9N$ DD07

Ref.	C/M	Charact. Potential Value	vs.	Response Const.	Value	Tech.	n	Electrokinetic Data Parameter	Value	From	Products and Identification	Description and Remarks	Code No.
JA094 7108	-	E_p >2	SCE	-	-	-	-	-	-	-	-	A,≠,p	DC98
JA088 5112	-	$E_{\frac{1}{2}}$ -1.1 -2.0	Ag/Ag$^+$	-	-	QE	1.0 -	-	-	-	-	C,p C	DC99
JA088 5112	-	E_p -1.2 -0.9	Ag/Ag$^+$	-	-	QE	1.0 -	-	-	-	-	C,p A	
RY002 1096	-	$E_{\frac{1}{2}}$ -0.56 -	SCE	-	-	-	-	-	-	-	-	C,p C,X	DD00 aq50
RY002 1096	-	-	-	-	-	-	-	-	-	-	-	see AQ51	DD01 aq51
BF630 1655	65d	$E_{\frac{1}{2}}$ -1.17	SCE	-	-	-	-	-	-	-	-	C,p	DD02
		-1.26		-	-	-	-	-	-	-	-	C,p	
JA095 3495	-	E_p -2.20	Ag/0.01 Ag$^+$	-	-	sttd	1	-	-	-	-	C,R,p	DD03 aq80 cf20
JA094 0738	-	$E_{\frac{1}{2}}$ -0.56	Ag/0.001 Ag$^+$	-	-	QE	1	-	-	-	radical cation, orange-pink,ESR	A,R,parent cmpd. brown,p	DD04
		-2.14					1				radical anion, orange-brown,ESR	C,R	
		-2.78					1					C,R	
JA095 3411	-	$E_{\frac{1}{2}}$ -0.76F	NHE	-	-	-	-	-	-	-	-	C,p	DD05 aq89
BJ045 0685	364 b	$E_{\frac{1}{2}}$ -0.88	SCE	I	7.33	-	-	-	-	-	-	C,i_d,$E_{\frac{1}{2}}$=f(pH),r	
		-0.95			6.98	-	-	-	-	-	-	C,i_d,$E_{\frac{1}{2}}$=f(pH),r	
		-1.17			0.52	-	-	-	-	-	-	C,i_d,$E_{\frac{1}{2}}$=f(pH),r	
		-1.3			0.7							C,dr	
		-1.45			6.20	-	-	-	-	-	-	C,i_d,r	
		-1.8			0.7							C,dr	
JA094 0738	-	$E_{\frac{1}{2}}$ -2.80 -3.11	Ag/0.001 Ag$^+$	-	-	QE	1 1	-	-	-	-	C,R,p C,R	
JA095 4373		E_p -0.26	SCE	-	-	sttd	1	-	-	-	-	C,p	DD06
BJ046 0147	-	-	-	-	-	-	-	-	-	-	-	see CF26	DD07 cf26

Code No.	Empirical Formula	Name and WLN Code	Structural Formula	Solvent	Tech.	Medium		μ, M	pH	T, °C	Electrodes	App.	Experimental Parameters
DD08	$C_{10}H_{10}BrClHgO_2$	ethyl α-bromomercury-(4-chlorophenyl)-acetate C.A. 10338-77-9 GR DY-HG-EVO2	4-ClC_6H_4CH(Hg-Br)$COOC_2H_5$	MeOH 50	PY	KNO_3 BR	0.1 0.05	-	7.0	2.5±0.5	DME/SCE	OA1	C=1,m=0.67,t=3
DD09	$C_{10}H_{10}BrFHgO_2$	ethyl α-bromomercury-(4-fluorophenyl)-acetate C.A. 716-54-1 FR DY-HG-EVO2	4-FC_6H_4CH(Hg-Br)$COOC_2H_5$	MeOH 50	PY	KNO_3 BR	0.1 0.05	-	7.0	2.5±0.5	DME/SCE	OA1	C=1,m=0.67,t=3
DD10	$C_{10}H_{10}BrHgIO_2$	ethyl α-bromomercury-(4-iodophenyl)acetate C.A. 10338-79-1 IR DY-HG-EVO2	4-IC_6H_4CH(Hg-Br)$COOC_2H_5$	MeOH 50	PY	KNO_3 BR	0.1 0.05	-	7.0	2.5±0.5	DME/SCE	OA1	C=1,m=0.67,t=3
DD11	$C_{10}H_{10}Br_2HgO_2$	ethyl α-bromomercury-(2-bromophenyl)acetate C.A. 947-69-3 ER BY-HG-EVO2	2-BrC_6H_4CH(Hg-Br)$COOC_2H_5$	MeOH 50	PY	KNO_3 BR	0.1 0.05	-	7.0	2.5±0.5	DME/SCE	OA1	C=1,m=0.67,t=3
DD12	$C_{10}H_{10}Br_2HgO_2$	ethyl α-bromomercury-(3-bromophenyl)acetate C.A. 833-29-4 ER CY-HG-EVO2	3-BrC_6H_4CH(Hg-Br)$COOC_2H_5$	MeOH 50	PY	KNO_3 BR	0.1 0.05	-	7.0	2.5±0.5	DME/SCE	OA1	C=1,m=0.67,t=3
DD13	$C_{10}H_{10}Br_2HgO_2$	ethyl α-bromomercury-(4-bromophenyl)acetate C.A. 10338-78-0 ER DY-HG-EVO2	4-BrC_6H_4CH(Hg-Br)$COOC_2H_5$	MeOH 50	PY	KNO_3 BR	0.1 0.05	-	7.0	2.5±0.5	DME/SCE	OA1	C=1,m=0.67,t=3
DD14 ar36 cf30	$C_{10}H_{10}Fe$	ferrocene C.A. 102-54-5 L50J Ø- 2-FE-	$(\pi-C_5H_5)_2Fe$	MeOH	VY	$NaClO_4$ Na tosylate	? ?	-	-	25	RCCE SCE RPE/SCE	06-0	C=1,A=0.8 C=1,v=20,A=0.2 A=ns
DD15	$C_{10}H_{10}N_2O_2$	isatin 3-(O-ethyl-oxime) C.A. 38469-75-9 T56 BMVYJ DUNO2	TABLE II.	H_2O	PY	BR		-	2.12 4.2 5.68 7.89 8.66 11.07	22±0.1	DME/SCE	-AO	C=0.2,EC=1.63,h=80

TABLE I. Electrochemical Data $C_{10}H_{10}N_2O_2$ DD15

Ref.	C/M	Charact. Potential		Response Const.		n Tech.	n	Electrokinetic Data			Products and Identification	Description and Remarks	Code No.
		Value	vs.		Value			Parameter	Value	From			
RY003 0218	8f	$E_{\frac{1}{2}}$	- -0.025 -1.79	SCE	-	-	sttd	1 1 1	-	-	-	-	C,Pr C,$E_{\frac{1}{2}}\neq f(pH)$ C,$\Delta E_{\frac{1}{2}}=2(0.095)\sigma$	DD08
RY003 0218	8f	$E_{\frac{1}{2}}$	-0.040 -0.197	SCE	-	-	sttd	1 1	-	-	-	-	C,p C,$\Delta E_{\frac{1}{2}}=2(0.095)\sigma$	DD09
RY003 0218	8f	$E_{\frac{1}{2}}$	-0.040 -0.162	SCE	-	-	sttd	1 1	-	-	-	-	C,p C,$\Delta E_{\frac{1}{2}}=2(0.095)\sigma$	DD10
RY003 0218	8f	$E_{\frac{1}{2}}$	-0.030 -0.128	SCE	-	-	sttd	1 1	-	-	-	-	C,p C,$\Delta E_{\frac{1}{2}}=2(0.095)\sigma$	DD11
RY003 0218	8f	$E_{\frac{1}{2}}$	-0.025 -0.145	SCE	-	-	sttd	1 1	-	-	-	-	C,p C,$\Delta E_{\frac{1}{2}}=2(0.095)\sigma$	DD12
RY003 0218	-	$E_{\frac{1}{2}}$	-0.08 -0.177	SCE	-	-	sttd	1 1	-	-	-	-	C,p C,$\Delta E_{\frac{1}{2}}=2(0.095)\sigma$	DD13
EA017 1595	-	$E_{\frac{1}{2}}$	0.370	SCE	-	-	-	-	-	-	-	-	A,p	DD14 ar36 cf30
			0.960		-	-	-	-	-	-	-	-	A,p	
			0.325		-	-	-	-	-	-	-	-	A,p	
			0.325		-	-	-	-	-	-	-	-	A,p	
EA017 1524	-	$E_{\frac{1}{2}}$	-0.2F	SCE	i_ℓ	2.2F	QE	3.9 ± 0.1	-	-	-	3-aminooxindole,UVS	C,$i\propto C$,i_d,$i\neq f(pH)$,r	DD15
			-0.42F			2.2F		3.9 ± 0.1	-	-	-	3-aminooxindole,UVS	C,$i\propto C$,i_d,r	
			-0.58F			2.2		3.9 ± 0.1	-	-	-	3-aminooxindole,UVS	C,$i\propto C$,i_d,r	
			-0.77F			2.2F		3.9 ± 0.1	-	-	-	3-aminooxindole,UVS	C,$i\propto C$,i_d,r	
			-0.83F			2.2F		3.9 ± 0.1	-	-	-	3-aminooxindole,UVS	C,$i\propto C$,i_d,r	
			-0.98F			2.2F		3.9 ± 0.1	-	-	-	3-aminooxindole,UVS	C,$i\propto C$,i_d,r	

DD16 $C_{10}H_{10}O_2$

Code No.	Empirical Formula	Name and WLN Code	Structural Formula	Solvent	Tech.	Medium		μ, M	pH	T, °C	Electrodes	App.	Experimental Parameters
DD16 ar55	$C_{10}H_{10}O_2$	methyl cinnamate C.A. 103-26-4 1OV1U1R	$C_6H_5CH:CHCOOCH_3$	MeCN	PY	Bu_4NClO_4	0.1	–	–	23	DME Ag/0.01 Ag^+	2AO	C=2, Pt Aux
DD17	$C_{10}H_{10}O_2$	methyl trans-cinnamate C.A. 1754-62-7 1OV1U1R -T	trans-C_6H_5CH: $CHCOOCH_3$	DMF	PY	Bu_4NBF_4	0.50	–	–	25	DME/SCE	23-0	C=1.6-2.7, Pt Aux
DD18	$C_{10}H_{11}BrHgO_2$	ethyl α-bromomercuryphenylacetate C.A. 831-08-3 E-HG-YR&VO2	TABLE II.	MeOH 50	PY	$LiClO_4$	0.1	–	–	25± 0.5	DME/SCE	OA1	C=0.080
													C=0.210
													C=0.500
						$NaBF_4$	0.1						C=0.051, m= 0.67, t=3
													C=0.255
													C=0.640
						$NaBF_4$ NaBr	0.1 0.01						C=0.051
													C=0.385
													C=0.640
						BR	0.05		4.0				C=ns
									6.2				
									9.5				
						KNO_3 BR	0.1 0.05		7.0				C=1
DD16 ar55		methyl cinnamate	$C_6H_5CH:CHCOOCH_3$	MeCN	PY	Bu_4NClO_4	0.1			23	DME Ag/0.01 Ag^+	2AO	C=2, Pt Aux

TABLE I. Electrochemical Data $C_{10}H_{11}BrHgO_2$ DD18

Ref.	C/M	Charact. Potential		Response Const.		n Tech.	n	Electrokinetic Data			Products and Identification	Description and Remarks	Code No.
		Value	vs.		Value			Parameter	Value	From			
JA095 1703	8f	$E_{\frac{1}{2}}$	-2.14	Ag/0.01 Ag$^+$	I	3.18	sttd	1.29	αn_a	1.05	Tomeš	-	C,r	DD16
JA094 8471	-	$E_{\frac{1}{2}}$	-1.81 -2.27	SCE	-	-	-	-	αn_a	1.2 1.2	Elog	-	C,p C	DD17
RY003 0218		$E_{\frac{1}{2}}$	-0.080	SCE	i_ℓ	0.140	sttd	1	-	-	-	-	C,$E_{\frac{1}{2}}\neq f(pH)$,r	DD18
			-0.320			0.190		1					C,$E_{\frac{1}{2}}=f(pH)$	
			0.093			0.380		1					C,$E_{\frac{1}{2}}=f(pH)$,r	
			0.345			0.370		1					C,$E_{\frac{1}{2}}=f(pH)$	
			-0.115			0.850		1	-	-	-	-	C,$E_{\frac{1}{2}}\neq f(pH)$,r	
			-0.370			0.850		1					C,$E_{\frac{1}{2}}=f(pH)$	
			-0.070			0.065		1	-	-	-	-	C,i_d ($i_\ell \propto h^{\frac{1}{2}}, \propto C, Tc=1.3$),$E_{\frac{1}{2}}\neq f(pH)$, Mc(medium),r	
			-0.285			0.082		1					C,i_d ($i_\ell \propto h^{\frac{1}{2}}, \propto C, Tc=1.3$),$E_{\frac{1}{2}}=f(pH)$	
			-0.075			0.280		1	-	-	-	-	C,$E_{\frac{1}{2}}=f(pH)$	
			-0.315			0.320		1					C,$E_{\frac{1}{2}}=f(pH)$	
			-0.087			0.910		1	-	-	-	-	C,$E_{\frac{1}{2}}\neq f(pH)$,r	
			-0.350			0.975		1					C,$E_{\frac{1}{2}}=f(pH)$	
			-0.153			0.070		1	-	-	-	-	C,$E_{\frac{1}{2}}\neq f(pH)$,r	
			-0.255			0.080		1					C,$E_{\frac{1}{2}}=f(pH)$	
			-0.147			0.650		1	Elog $dE_{\frac{1}{2}}/dlogC$	60 -36 +12	(in absence of NaBr) (presence NaBr)	-	C,$E_{\frac{1}{2}}\neq f(pH)$,r	
			-0.267			0.680		1	Elogi^2	30 60	in NaBr in LiCl or NaBF$_4$		C,$E_{\frac{1}{2}}=f(pH)$	
			-0.140			0.990		1	-	-	-	-	C,$E_{\frac{1}{2}}\neq f(pH)$,r	
			-0.280			1.100		1					C,$E_{\frac{1}{2}}=f(pH)$	
			-0.05F			-		1	-	-	-	-	C,$E_{\frac{1}{2}}\neq f(pH)$,r	
			-0.09F					1					C,$E_{\frac{1}{2}}=f(pH)$	
			-0.055					1	-	-	-	-	C,$E_{\frac{1}{2}}\neq f(pH)$,r	
			-0.18F					1					C,$E_{\frac{1}{2}}=f(pH)$	
			-0.05F					1	-	-	-	-	C,$E_{\frac{1}{2}}\neq f(pH)$,r	
			-0.27F					1					C,$E_{\frac{1}{2}}=f(pH)$	
			0.050					1	-	-	-	-	C,r	
			-0.225					1					C	

Code No.	Empirical Formula	Name and WLN Code	Structural Formula	Solvent	Tech.	Medium		μ, M	pH	T, °C	Electrodes	App.	Experimental Parameters
DD19	$C_{10}H_{11}N_3O_2$	1-methyl-2-dimethyl-aminobenzimidazole-4,7-quinone T56 BN DN FV IVJ B1 CN1&1	TABLE II.	MeOH	PY	Et_4NClO_4	0.05	-	-	-	DME/MP	0-0	EC=1.23
				DMF	PY	Et_4NClO_4	0.1	-	-	-	DME/MP	0-0	EC=1.23
DD20	$C_{10}H_{11}N_3O_2$	1-methyl-2-dimethyl-aminobenzimidazole-5,6-quinone T56 BN DN GVVJ B1 CN1&1	TABLE II.	MeOH	PY	Et_4NClO_4	0.05	-	-	-	DME/MP	0-0	EC=1.23
				DMF	PY	Et_4NClO_4	0.1	-	-	-	DME/MP	0-0	EC=1.23
DD21	$C_{10}H_{11}N_3O_2$	1-methyl-2-dimethyl-aminobenzimidazole-6,7-quinone T56 BN DN FVVJ CN1&1 D1	TABLE II.	DMF	PY	Et_4NClO_4	0.1	-	-	-	DME/MP	0-0	EC=1.23
DD22	$C_{10}H_{12}$	cis-bicyclo[6.2.0]-deca-2,4,6-triene C.A. 34784-43-5 L48T&J	TABLE II.	MeCN	PY	Me_4NPF_6	0.1	-	-	-	DME Ag/0.1 Ag^+	256A F2	ns
				THF	PY	Bu_4NClO_4	0.2	-	-	-	DME Ag/0.1 Ag^+	256A F2	ns
					VR	Bu_4NClO_4	0.2	-	-	-	PBE Ag/0.1 Ag^+	256A F2	v = 500
DD23	$C_{10}H_{12}ClNO$	2',3'-dimethyl-2-chloroacetanilide G1VMR B1 C1	TABLE II.	EtOH 50	PY	HMe_3NI	0.05	-	-	-	DME/SCE	0-0	C=1
DD24	$C_{10}H_{12}ClNO$	3',4'-dimethyl-2-chloroacetanilide G1VMR C1 D1	TABLE II.	EtOH 50	PY	HMe_3NI	0.05	-	-	-	DME/SCE	0-0	C=1
DD25	$C_{10}H_{13}NO$	3,8-dimethyl-2-methoxyazocine C.A. 20205-53-2 T8NJ BO1 C1 H1	TABLE II.	THF	PY	Bu_4NPF_6	0.2	-	-	-	DME Ag/0.1 Ag^+	OAF-	t=6-12
					VA	Bu_4NPF_6	0.2	-	-	-	PBE Ag/0.1 Ag^+	OAF-	ns
DD26	$C_{10}H_{14}MoO_2Sn$	dicarbonyl-(η-cyclopentadienyl)-trimethylstannio-(III)molybdenum(I) C.A. 12193-46-3 L5ØJ Ø-MO-CO&CO&-SN-1&1&1	(π-C_5H_5)Mo(CO)$_2$-Sn(CH$_3$)$_3$	MG	PY	Bu_4NClO_4	0.1	-	-	22	DME Ag/Ag^+	2-0	C=2
DD27	$C_{10}H_{14}O$	ethyl 1-phenylethyl ether C.A. 3299-05-6 2OY1&R	$CH_3CH(C_6H_5)O$-C_2H_5	MeCN	QE	$LiClO_4$	0.1	-	-	-	Pt Ag/0.1 Ag^+	25A-	C=20, A=12.9, E_{app}=1.90, stainless steel Aux

TABLE I. Electrochemical Data $C_{10}H_{14}O$ DD27

Ref.	C/M	Charact. Potential		Response Const.		n Tech.	n	Electrokinetic Data			Products and Identification	Description and Remarks	Code No.
		Value	vs.		Value			Parameter	Value	From			
RY002 1096	-	$E_{\frac{1}{2}}$	-0.40 -	SCE	-	-	-	-	-	-	-	-	C,p C,X	DD19
RY002 1096	-	$E_{\frac{1}{2}}$	-0.67 -1.29	SCE	-	-	-	-	-	-	-	-	C,p C,X	
RY002 1096	-	$E_{\frac{1}{2}}$	-0.31 -	SCE	-	-	-	-	-	-	-	-	C,p C,X	DD20
RY002 1096	-	$E_{\frac{1}{2}}$	-0.71 -0.91	SCE	-	-	-	-	-	-	-	-	C,p C,X	
RY002 1096	-	$E_{\frac{1}{2}}$	-0.64 -	SCE	-	-	-	-	-	-	-	-	C,p C,X	DD21
JA095 2198	-	$E_{\frac{1}{2}}$	-2.59	SCE	$I\eta^{\frac{1}{2}}$	0.120	sttd	1	Elog	50	sttd	-	C,r	DD22
JA095 2198	-	$E_{\frac{1}{2}}$	-2.83	SCE	$I\eta^{\frac{1}{2}}$	0.114	sttd	1	Elog	60	sttd	-	C,r	
JA095 2198	-	E_p	-2.95F	SCE	i_p	1.5	-	-	-	-	-	-	C,r	
RG035 0843	14a	$E_{\frac{1}{2}}$	-1.46	SCE	-	-	-	-	-	-	-	-	C,$\frac{1}{2}$,r	DD23
RG035 0843	14a	$E_{\frac{1}{2}}$	-1.45	SCE	-	-	-	-	-	-	-	-	C,$\frac{1}{2}$,r	DD24
JA094 4907	-	$E_{\frac{1}{2}}$	-2.28	SCE	I	5.2	I:I	2	Elog	99	plot	-	C,r	DD25
JA094 4907	-	$E_{p/2}$	-2.43 -1.59	SCE	-	-	-	-	-	-	-	-	C,r A	
JA088 5117	-	$E_{\frac{1}{2}}$	-1.9	Ag/Ag$^+$	-	-	QE	1	-	-	-	Mo$^-$ and Sn$^\cdot$,HCl	C,p	DD26
JA094 6812	360 c	-	-	-	-	-	QE	7.2	-	-	-	methyl phenyl ketone, 41%,NMR,GLC,IRS	A,p	DD27

127

DD28 $C_{10}H_{14}O$

Code No.	Empirical Formula	Name and WLN Code	Structural Formula	Solvent	Tech.	Medium		μ, M	pH	T, °C	Electrodes	App.	Experimental Parameters
DD28	$C_{10}H_{14}O$	1,2,3,4,4a,5,6,7-octahydronaphthalen-1-one C.A. 24037-79-4 L66 BV JUTTJ	TABLE II.	DMF	PY	Bu_4NBF_4	0.50	-	-	25	DME/SCE	23-0	C=1.8-3.6, Pt Aux
DD29	$C_{10}H_{14}O$	1,2,3,4,5,6,7,8-octahydronaphthalen-1-one C.A. 18631-96-4 L66 BV AU-FTTJ	TABLE II.	DMF	PY	Bu_4NBF_4	0.50	-	-	25	DME/SCE	23-0	C=3.2-4.9, Pt Aux
DD30	$C_{10}H_{14}O_2$	2,3,5,6-tetramethyl-hydroquinone C.A. 527-18-4 QR DQ B1 C1 E1 F1	TABLE II.	MeCN	VR	Et_4NClO_4	0.1	-	-	-	Pt/SCE	12A2	C=2.12,v=1.80E3, Pt Aux
													v=1.80E4
													v=1.8E5
													v=6E5
DD31	$C_{10}H_{14}O_4$	3-(2-methoxyphen-oxy)-1,2-propanediol C.A. 93-14-1 Q1YQ1OR BO1	TABLE II.	H_2O	PO	KOH KOH	1 10	-	-	-	ns	---	C=1
DD32 as58 cf66	$C_{10}H_{16}N_2$	Wurster's Blue C.A. 100-22-1 1N1&R DN1&1	TABLE II-2.	DMF	VR	Bu_4NClO_4	0.1	-	-	-	PDE/SCE	3F-	C=0.1-1, v=220
DD33	$C_{10}H_{16}N_2O_2S$	5-ethyl-5(1-methyl-propyl)-2-thio-barbituric acid C.A. 2095-57-0 T6MVXVMYJ C2 CY2&1 FUS	TABLE II.	H_2O	DI	Et_4NClO_4 BR	0.01	-	8.0	-	DME/SCE	04A0	C=0.05,m=1.73,t=0.5, v=10,ΔE=100
DD34	$C_{10}H_{18}O_2$	2-methoxy-3,5,5-trimethyl-2-cyclo-hexen-1-one C.A. 5682-76-8 L6V CX EUTJ C1 C1 E1 FO1	TABLE II.	DMF	PY	Bu_4NBF_4	0.50	-	-	25	DME/SCE	23-0	C=1.6-7.2, Pt Aux
DD35	$C_{10}H_{16}O_4$	diethyl isopropyl-idenemalonate C.A. 6802-75-1 2OVYVO2&UY1&1	$(CH_3)_2C:C(COO-C_2H_5)_2$	DMF	PY	Bu_4NBF_4	0.50	-	-	25	DME/SCE	23-0	C=1.4-6.1, Pt Aux
DD36	$C_{10}H_{17}Br$	4-bromobornane L55 ATJ BE C D D	TABLE II.	DMF	PY	Et_4NBr	?	-	-	25±0.1	DME/SCE	01A0	ns
DD37	$C_{10}H_{17}NO_2$	3,3-diethyl-5-methyl-2,4-dioxopiper-idine C.A. 125-64-4 T6MVXVTJ C2 C2 E1	TABLE II.	H_2O	PT	BOR	?	-	9.2	-	DME/SCE	04A0	C=0.1,m=1.73, t=1,h=80

TABLE I. Electrochemical Data $\quad C_{10}H_{17}NO_2 \quad DD37$

Ref.	C/M	Charact. Potential		Response Const.		n		Electrokinetic Data			Products and Identification	Description and Remarks	Code No.
		Value	vs.		Value		Tech.	Parameter	Value	From			
JA094 8471	-	$E_{\frac{1}{2}}$	-2.11	SCE	-	-	-	-	αn_a	0.7	Elog	-	C,p	DD28
JA094 8471	-	$E_{\frac{1}{2}}$	-2.34	SCE	-	-	-	-	αn_a	1.3	Elog	-	C,p	DD29
JS117 0186	119 a	E_p	0.7F 0.0F -0.83	SCE	i_p	70F 30F 20F	-	-	-	-	-	-	A,W,⅓,p C,W,⅓ C,X	DD30
			1.0F 0.0F -0.83			190F 60F 60F	-	-	-	-	-	-	A,W,⅓,p C,W,⅓ C,W	
			1.0F 0.0F -0.83			600F 200F 300F	-	-	-	-	-	-	A,W,⅓,p C,W,X C	
			1.0 - -0.83			1300F - 600F	-	-	-	-	-	-	A,W,⅓,p 0 C,W	
CZ018 0427	-	Q	0.61	-	-	-	-	-	-	-	-	-	C,p	DD31
			0.78	-	-	-	-	-	-	-	-	-	C,p	
JA094 0691	-	E_p	0.24	SCE	-	-	-	-	-	-	-	-	A,R,p	DD32 as58 cf66
AA080 0233	-	E_{su}	-1.32	SCE	-	-	-	-	-	-	-	-	C,r	DD33
			-1.29	-	-	-	-	-	-	-	-	-	C,r	
JA094 8471	-	$E_{\frac{1}{2}}$	-2.20	SCE	-	-	-	-	αn_a	1.3	Elog	-	C,p	DD34
JA094 8471	-	$E_{\frac{1}{2}}$	-2.13	SCE	-	-	-	-	αn_a	1.2	Elog	-	C,p	DD35
JA086 3155	-	-	-	-	-	-	-	-	-	-	-	-	0,p	DD36
AA080 0233	-	$E_{\frac{1}{2}}$	-0.73F	SCE	i_ℓ	0.1F	-	-	-	-	-	-	A,r	DD37

Code No.	Empirical Formula	Name and WLN Code	Structural Formula	Solvent	Tech.	Medium		μ, M	pH	T, °C	Electrodes	App.	Experimental Parameters
DD38	$C_{10}H_{19}NO_2$	diethylaminoethyl methacrylate C.A. 105-16-8 2N2&2OVY1&U1	$CH_2:C(CH_3)COO-CH_2CH_2N(C_2H_5)_2$	MeOH 92	PY	Et_4NI	0.05	-	-	-	DME/SCE	0-0	C=0.24,EC= 1.204,t=10.4
				DMF	PY	Et_4NI	0.05	-	-	-	DME/SCE	0-0	C=0.26,EC= 0.897,t= 1.86,E=-2.0
DD39	$C_{10}H_{20}N_2$	2-tert-butyl-3-methyl-2,3-diazanorbornane C.A. 42842-99-9 T55 A CNNTJ C1 DX1& 1&1	TABLE II.	MeCN	VR	ns		-	-	-	-	---	ns
DD40	$C_{10}H_{20}N_2$	1,4-dimethyl-1,4-diazaspiro[4.5]-decane C.A. 38704-82-4 T5NXNTJ A1 C1 B-AL6XTJ	TABLE II.	MeCN	VR	$NaClO_4$	0.1	-	-	-	AuBE SCE	2AO	C=1.64,d= 0.15,v=52, MP Aux,PE
DD41	$C_{10}H_{22}N_2O$	3,4-di-tert-butyl-1,3,4-oxadiazolidine C.A. 38786-33-3 T5NN DOTJ AX1&1&1 BX1&1&1	TABLE II.	MeCN	VR	$NaClO_4$	0.1	-	-	-	AuBE SCE	2AO	C=18.9,d= 0.15,v=1340, MP Aux,PE
DD42	$C_{10}H_{24}N_2$	1,2,4,5-tetraethyl-perhydro-s-tetrazine C.A. 37882-95-4 T6NN DNNTJ A2 B2 D2 E2	TABLE II.	MeCN	VR	$NaClO_4$	0.1	-	-	-	AuBE SCE	2AO	C=0.2-0.9, d=0.15,v=13, MP Aux,PE
DD43	$C_{10}H_{30}B_{18}Ni$	bis[π-μ-1,2-trimethylene-1,2-dicarbaundecaborane-(9)]nickel(IV) (orange form) C.A. 38904-48-2	$(B_9C_5H_{15})_2Ni^{IV}$	MeCN	VR	Et_4NClO_4	0.1	-	-	0	PBE/SCE	--03	v=6E3
DD44	$C_{10}H_{30}B_{18}Ni$	bis[π-μ-1,3-trimethylene-1,2-dicarbaundecaborane-(9)]nickel(IV) (yellow form) C.A. 38903-46-7	$(B_9C_5H_{15})_2Ni^{IV}$	MeCN	VR	Et_4NClO_4	0.1	-	-	-	PBE/SCE	--03	v=6E3
DD45	$C_{11}H_8CoN_3O_2$	carbonyl(nitrosyl)-2,2'-bipyridine-cobalt C.A. 36454-21-4 D B656 GN-CO-NJ HNO HCO	TABLE II.	MG	PY	Bu_4NClO_4	?	-	-	-	DME Ag/0.001 Ag^+	5AO	t=3
DD46	$C_{11}H_8FeN_4O_3$	dinitrosyl[di-(2-pyridyl) ketone]iron C.A. 36454-20-3 D C666 AN-FE-N IVJ BNO BNO	TABLE II.	MG	PY	Bu_4NClO_4	?	-	-	-	DME Ag/0.001 Ag^+	5AO	t=3
DD47	$C_{11}H_8N_2O$	di-2-pyridyl ketone C.A. 19437-26-4 T6NJ BV- BT6NJ	TABLE II.	MG	PY	Bu_4NClO_4	?	-	-	-	DME Ag/0.001 Ag^+	5AO	t=3

TABLE I. Electrochemical Data $C_{11}H_8N_2O$ DD47

Ref.	C/M	Charact. Potential		Response Const.		n Tech.	n	Electrokinetic Data			Products and Identification	Description and Remarks	Code No.
		Value	vs.		Value	Tech.		Parameter	Value	From			
RR020 0469	71d	$E_{1/2}$	-2.03	SCE	i_ℓ	5.60	-	2.12	-	-	-	-	C,r	DD38
RR020 0469	71d	$E_{1/2}$	-2.26	SCE	i_ℓ	4.07	-	4.07	-	-	-	-	C,r	
JA095 5422	-	$E_{1/2}$	0.17	SCE	-	-	sttd	1	-	-	-	radical cation	A,p	DD39
JA094 7114	-	E_p	0.85	SCE	-	-	-	-	$E_p-E_{p/2}$	210	sttd	-	A,p	DD40
JA094 7108	-	XE	0.96	SCE	-	-	sttd	1	ΔE_p	100	sttd	radical anion, unstable	A,XE=½$(E_{p,A}+E_{p,c})$,R,r	DD41
JA094 7108	-	XE	0.25	SCE	-	-	sttd	1	ΔE_p	65	sttd	radical cation, ESR	A,XE=½$(E_{p,A}+E_{p,c})$,R,r	DD42
JA094 4882	368 a	E_p	0.19F	SCE	i_p	20F	-	-	-	-	-	-	C,R,p	DD43
			-0.88F			20F							C,R	
			-0.83F			26F							A,R	
			-0.22F			24F							A,R	
		$E_{p/2}$	0.30		-	-	-	-	-	-	-	-	C,R,p	
			-1.20										C,R	
JA094 4882	368 a	$E_{p/2}$	0.02	SCE	-	-	-	-	-	-	-	-	C,R,p	DD44
			-1.20										C,R	
JA094 0738	-	$E_{1/2}$	-0.50	Ag/0.001 Ag$^+$	-	-	QE	1	-	-	-	radical cation	A,R,parent compound violet,p	DD45
			-2.26					1				radical anion, red-violet, ESR	C,R	
			-2.81					1					C,R	
JA094 0738	-	$E_{1/2}$	-0.30	Ag/0.001 Ag$^+$	-	-	QE	1	-	-	-	radical cation, light brown, ESR	A,R,parent compound green,p	DD46
			-1.68					1				radical anion, purple, ESR	C,R	
			-2.40					1				dianion, deep purple	C,R	
JA094 0738	-	$E_{1/2}$	-2.31	Ag/0.001 Ag$^+$	-	-	QE	1	-	-	-	-	C,R,p	DD47
			-2.92					1					C,R	

DD48 $C_{11}H_8O_2$

Code No.	Empirical Formula	Name and WLN Code	Structural Formula	Solvent	Tech.	Medium		μ, M	pH	T, °C	Electrodes	App.	Experimental Parameters
DD48 as95	$C_{11}H_8O_2$	2-methyl-1,4-naphthoquinone C.A. 58-27-5 L66 BV EVJ C	TABLE II-2.	MeOH 25	DI	ACET	0.1	-	5.0	25±0.1	DME/SCE	2A0	C=0.001, t=2, ΔE=25, Pt Aux
	$C_{11}H_{10}MoO_4$ (see DG32)												
DD49	$C_{11}H_{10}N_2O_2$	4-methoxy-2,2'-bipyridyl 1-oxide C.A. 14163-05-4 T6NJ AO D01 B- BT6NJ	TABLE II.	dioxane 20	PY	HCl NaOAc	?	-	0.76	25±0.1	DME/SCE	05A0	C=0.08, m=1.51, t=3.38, h=75
						PHTH HCl	?		2.21				
						ACET	?		5.65				
						BOR	?		9.50				
						NaOH	0.1		-				C=0.044
DD50	$C_{11}H_{13}BrHgO_2$	ethyl α-bromomercury(2-methylphenyl)acetate C.A. 883-17-0 E-HG-YVO2&R B1	2-$CH_3C_6H_4$CH(HgBr)$COOC_2H_5$	MeOH 50	PY	BR KNO_3	0.05 0.1	-	7.0	25±0.5	DME/SCE	0A1	C=ns, m=0.67, t=3
DD51	$C_{11}H_{13}BrHgO_2$	ethyl α-bromomercury(3-methylphenyl)acetate C.A. 779-01-1 E-HG-YVO2&R C1	3-$CH_3C_6H_4$CH(HgBr)$COOC_2H_5$	MeOH 50	PY	BR KNO_3	0.1	-	7.0	25±0.5	DME/SCE	0A1	C=1, m=0.67, t=3
DD52	$C_{11}H_{13}NO_7$	N-(2-hydroxymethyl-3,4-dihydroxy-5-tetrahydrofuryl)-3-carboxypyridine-(1H)2-one C.A. 42576-34-1 T6NVJ CVQ A- ET5OTJ B1Q CQ DQ	TABLE II.	EtOH ≈40	DI	PHOS	≈0.1	-	5.0	-	DME/SCE	25A0	-1.10 → -1.80 V, C=0.07, t=0.5, EC=1.561, v=5, Δe=50, Pt Aux
DD53	$C_{11}H_{16}O$	benzyl tert-butyl ether C.A. 3459-80-1 1X1&1&O1R	$C_6H_5CH_2OC(CH_3)_3$	MeCN	QE	$LiClO_4$	0.1	-	-	-	Pt Ag/0.1 Ag^+	25A-	C=30.4, A=12.9, E_{app}=1.95, stainless steel Aux
DD54	$C_{11}H_{17}N$	4-(tert-butyl)-1-cyanocyclohexane C.A. 7370-14-1 L6UTJ ACN DX1&1&1	TABLE II.	DMF	PY	Bu_4NBF_4	0.50	-	-	25	DME/SCE	23-0	C=1.1-2.0, Pt Aux
DD55 au08	$C_{11}H_{19}ClHgO$	cis-2-chloromercury-3-ethoxycyclononene C.A. 33217-24-2 L9UTJ B-HG-G COZ	TABLE II-2.	dioxane 50	PY	ACET HCl TX100	? 0.001	-	1.35	30±0.1	DME/SCE	03A	C=ns, m=2.12, t=3.2, h=65
CONT						PHOS	?		7.00				

TABLE I. Electrochemical Data $C_{11}H_{19}ClHgO$ (CONT.) DD55

Ref.	C/M	Charact. Potential Value	vs.	Response Value	Const.	Tech.	n	Electrokinetic Data Parameter	Value	From	Products and Identification	Description and Remarks	Code No.
AN100 0377	-	E_{su} -0.05F	SCE	i_{su}	0.032F	-	-	-	-	-	-	C,r	DD48 as95
BJ045 0685	364 a	$E_{\frac{1}{2}}$ -0.58	SCE	I	3.79	1:1	2	-	-	-	4-methoxy-2,2'-bi-pyridyl,QE	$C,i_d,E_{\frac{1}{2}}=f(pH),Tc=$ 3.2,$i \propto C$,r	DD49
		-0.95			7.18		2					$C,i_d,E_{\frac{1}{2}}=f(pH),Tc=$ 2.9,$i \propto C$	
		-0.64			3.61		2				4-methoxy-2,2'-bi-pyridyl,QE	$C,i_d,i \propto C,E_{\frac{1}{2}}=f(pH)$, Tc=3.2,r	
		-1.00			5.06		2					$C,i_d,i \propto C,E_{\frac{1}{2}}=f(pH)$, Tc=2.9	
		-0.94			3.73		2				4-methoxy-2,2'-bi-pyridyl,QE	$C,i_d,i \propto C,E_{\frac{1}{2}}=f(pH)$, Tc=3.2,r	
		-1.17			6.07		2					$C,i_d,i \propto C,E_{\frac{1}{2}}=f(pH)$, Tc=2.9	
		-1.18			0.22		2				4-methoxy-2,2'-bi-pyridyl,QE	$C,i_d,i \propto C,E_{\frac{1}{2}}=f(pH)$, Tc=3.2,r	
		-1.42			1.31		2					$C,i_d,i \propto C,E_{\frac{1}{2}}=f(pH)$, Tc=2.9	
		-1.62F		i_ℓ	0.6F	-	-	-	-	-	-	C,r	
RY003 0218	8f	$E_{\frac{1}{2}}$ - -0.040 -0.180	SCE	-	-	sttd	- 1 1	-	-	-	-	C,Pr,p C $C,\Delta E_{\frac{1}{2}} \neq 2(0.095)\sigma$	DD50
RY003 0218	8f	$E_{\frac{1}{2}}$ - 0.075 -0.225	SCE	-	-	sttd	- 1 1	-	-	-	-	C,Pr,p C $C,\Delta E_{\frac{1}{2}}=2(0.095)\sigma$	DD51
AA066 0023	-	E_{su} -1.45	SCE	i_{su}	1.1	-	-	-	-	-	N-(2-hydroxymethyl-3,4-dihydroxy-5-tetrahydrofuryl)-3-carboxy-1,2-dihydro-2-hydroxypyridine	C,p	DD52
JA094 6812	360 b	-	-	-	-	QE	2.1	-	-	-	benzaldehyde,46%, NMR,GLC	A,p	DD53
JA094 8471	-	$E_{\frac{1}{2}}$ -2.55	SCE	-	-	-	-	αn_a	0.9	Elog	-	C,p	DD54
EA017 2077	-	$E_{\frac{1}{2}}$ -1.20 -0.912	SCE	i_ℓ	6.30 0.26	-	-	-	-	-	-	C,r C	DD55 au08
		-0.097 -1.25			0.26 0.32	-	-	-	-	-	-	C,r C	CONT

DD55 (CONT.) $C_{11}H_{19}ClHgO$

Code No.	Empirical Formula	Name and WLN Code	Structural Formula	Solvent	Tech.	Medium		μ, M	pH	T, °C	Electrodes	App.	Experimental Parameters
DD55 au08	$C_{11}H_{19}ClHgO$	cis-2-chloromercury-3-ethoxycyclononene		dioxane 50	PY	NaOH TX100	? 0.001	-	11.00	30± 0.1	DME/SCE	03A0	C=ns,m=2.12, t=3.2,h=65
DD56	$C_{11}H_{19}ClHgO$	cis-2-chloromercury-3-methoxycyclodecene C.A. 32393-41-2 L-10-UTJ B-HG-G C01	TABLE II.	dioxane 50	PY	ACET HCl TX100	? 0.001	-	1.35	30± 0.1	DME/SCE	03A0	C=ns,m=2.12, t=3.2,h=65
						PHOS TX100	? 0.001		7.00				
						NaOH TX100	? 0.001		11.00				
DD57	$C_{11}H_{19}ClO_4S_2$	3,5-di(tert-butyl)-1,2-dithiolylium perchlorate C.A. 35610-83-4 T5SSTJ CX1&1&1 EX1& 1&1 & G-O4	TABLE II.	MeCN	VR	Bu_4NBF_4	0.1	-	-	-	PBE/SCE	26A2	v=150
				CH_2Cl_2	VR	Bu_4NBF_4	0.2	-	-	-	PBE/SCE	26A2	v=150
DD58	$C_{11}H_{20}N_2$	8,8-cyclopenta-methylene-1,5-diaza-bicyclo[3.2.1]-octane C.A. 38705-05-4 T56 A AX BN ENTJ A-& L6XTJ	TABLE II.	MeCN	VR	$NaClO_4$	0.1	-	-	-	AuBE SCE	2A0	C=3.40,d= 0.15,v=51, MP Aux,PE
DD59	$C_{12}H_5FeMnO_7$	pentacarbonylmanganese(0)(dicarbonyl-η-cyclopentadienyl)-iron(I) C.A. 12088-73-2 L5ØJ Ø-FE-CO&CO&-MN-CO&CO&CO&CO&CO	$(OC)_5MnFe(CO)_2$ (C_5H_5)	MG	PY	Bu_4NClO_4	0.1	-	-	22	DME Ag/ $AgClO_4$	2-0	C=2
DD60 ba15	$C_{12}H_8$	acenaphthylene C.A. 208-96-8 L566 1A LJ	TABLE II-2.	MeCN or PrCN	VR	Et_4NClO_4	0.1	-	-	-	HMDE SCE PBE/SCE	6A02	C=1
DD61	$C_{12}H_8Br_2$	1,2-dibromoace-naphthene C.A. 25226-58-8 L566 1A LT&&J CE DE	TABLE II.	MeCN or PrCN	VR	Et_4NClO_4	0.1	-	-	-	HMDE SCE	6A02	C=1
				MeCN or PrCN	VR	Et_4NClO_4	0.1	-	-	-	PBE/SCE	6A02	C=1
DD62	$C_{12}H_8ClN_3OS$	7-chloro-1,3-di-hydro-5-(2-thiazol-yl)-2H-1,4-benzo-diazepin-2-one C.A. 39254-69-8 T67 GMV JN IHJ CG K-BT5N CSJ	TABLE II.	MeOH 33	DI	PHOS	1	-	3.0	-	DME/SCE	25A0	0.00→ -1.100 V,C= 0.12,EC= 2.295,m= 2.75,t=0.5, ΔE=50,v=5, Pt Aux
									7.0				0.00→ -1.750 V
DD63	$C_{12}H_8FeN_4O_2$	dinitrosyl-1,10-phenanthrolineiron C.A. 36454-19-0 D566 B6 2AB CN-FE-N OJ DNO DNO	TABLE II.	MG	PY	Bu_4NClO_4	?	-	-	-	DME Ag/0.001 Ag^+	5A0	t=3
DD64 ba21	$C_{12}H_8N_2$	1,10-phenanthroline C.A. 66-71-7 T B666 CN NNJ	TABLE II-2.	MG	PY	Bu_4NClO_4	?	-	-	-	DME Ag/0.001 Ag^+	5A0	t=3

TABLE I. Electrochemical Data $C_{12}H_8N_2$ DD64

Ref.	C/M	Charact. Potential		Response Const.		Tech.	n	Electrokinetic Data			Products and Identification	Description and Remarks	Code No.
		Value	vs.		Value			Parameter	Value	From			
EA017 2077	–	$E_{\frac{1}{2}}$	-0.187 -1.311	SCE	i_ℓ	0.28 0.30	–	–	–	–	–	–	C,r C	DD55 au08
EA017 2077	–	$E_{\frac{1}{2}}$	-0.087 -0.812	SCE	i_ℓ	0.30 0.26	–	–	–	–	–	–	C,r C	DD56
			-0.087 -1.250			0.30 0.30	–	–	–	–	–	–	C,r C	
			-0.187 -1.250			0.28 0.30	–	–	–	–	–	–	C,r C	
JA095 4373	370 a	E_p	-0.58 -0.64	SCE	–	–	sttd	1 1	–	–	–	–	A,p C	DD57
JA095 4373	370 a	E_p	-0.51 -0.64	SCE	–	–	sttd	1 1	–	–	–	–	A,p C	
JA094 7114	–	E_p	0.84	SCE	–	–	–	–	$E_p-E_{p/2}$	200	sttd	–	A,p	DD58
JA088 5117 (JA088 5124)	–	$E_{\frac{1}{2}}$	-1.6 -2.2	Ag/AgClO$_4$	–	–	QE	1 –	–	–	–	Mn⁻ and Fe·,sttd $[(C_5H_5)Fe(CO)_2]_2$,PY	C C,W	DD59
JA095 2646	–	E_p	-1.74 -2.33	SCE	–	–	–	–	–	–	–	–	C,R,p C,⅟,broad peak	DD60 ba15
			-1.74		–	–	–	–	–	–	–	–	C,R,p	
JA095 2646	–	E_p	-2.33	SCE	–	–	QE at -1.0	2	–	–	–	acenaphthylene	C,⅟,p	DD61
JA095 2646	–	E_p	-0.73 -1.74	SCE	–	–	–	–	–	–	–	–	C,⅟,p C,R	
AA074 0367	–	E_{su}	-0.400 -1.040	SCE	i_{su}	8.12 –	–	–	–	–	–	–	C,r C	DD62
			-0.650			3.31	–	–	–	–	–	–	C,r	
JA094 0738	–	$E_{\frac{1}{2}}$	-0.60	Ag/0.001 Ag$^+$	–	–	QE	1	–	–	–	radical cation, orange,ESR	A,R,parent compound orange-brown,p	DD63
			-2.16					1				radical anion,gray-green,ESR	C,R	
			-2.80					1					C,R	
JA094 0738	–	$E_{\frac{1}{2}}$	-2.72 -3.24	Ag/0.001 Ag$^+$	–	–	QE	1 1	–	–	–	–	C,R,r C,R	DD64 ba21

DD65 $C_{12}H_8N_2O_2$

Code No.	Empirical Formula	Name and WLN Code	Structural Formula	Solvent	Tech.	Medium		μ, M	pH	T, °C	Electrodes	App.	Experimental Parameters
DD65	$C_{12}H_8N_2O_2$	1-methyl(2,3-d)-naphthoimidazole-4,9-quinone C.A. 14882-62-3 T C566 BV DN FN HVJ D1	TABLE II.	DMF	PY	Et_4NClO_4	0.1	-	-	-	DME/MP	O-O	EC=1.23
DD66	$C_{12}H_8S$	dibenzothiophene C.A. 132-65-0 T B656 HSJ	TABLE II.	HOAc	IL	H_2SO_4	0.18	-	-	25±0.1	Pt/SCE (0)	2AO	C=11, A=0.25, v=24, Pt Aux
					VA	H_2SO_4	0.18	-	-	25±0.1	Pt/SCE (0)	2AO	-0.15→1.80 V, C=11, A=0.25, v=80, Pt Aux
													0.90→1.80 V
DD67 ba42 cg75	$C_{12}H_8S_2$	thianthrene C.A. 92-85-3 T C666 BS ISJ	TABLE II-2.	MeCN	VR	Bu_4NClO_4	0.1	-	-	-	Pt/SCE	2--	C=1, A=0.17, v=100, Pt Aux
DD68	$C_{12}H_9NO$	phenoxazine T C666 BM IOJ	TABLE II.	MeCN	IL	Et_4NClO_4	0.1	-	-	-	Pt/SCE	2A-	C=0.5-1, E_{app} = 0.65 at Pt
DD69 ba62	$C_{12}H_9N_3O_2$	4-nitroazobenzene C.A. 2491-52-3 WNR DNUNR	$4-NO_2C_6H_4N:NC_6H_5$	MeCN	PY	Et_4NClO_4	0.1	-	-	-	DME/SCE	25AO	C=ns, m=1.556, t=3.20, h=50, Pt Aux
						Et_4NClO_4 H_2SO_4	0.1 0.0025						C=1.06
						Et_4NClO_4 H_2SO_4	0.1 0.002						
						Et_4NClO_4 H_2SO_4	0.1 0.0005						
						Et_4NClO_4 C_6H_5OH	0.1 0.001						C=1.0
						Et_4NClO_4 C_6H_5OH	0.1 0.002						
DD70	$C_{12}H_{10}$	benzocyclooctatetraene C.A. 265-49-6 L68J	TABLE II.	THF	PY	Bu_4NClO_4	0.2	-	-	-	DME Ag/0.1 Ag^+	O1AFO	C=0.38, m=0.63, t=6-12, h=40.0
						Bu_4NClO_4 H_2O	0.2 5E-5						
						Bu_4NClO_4 H_2O	0.2 8.5E-2						
						Bu_4NPF_6	0.2						C=0.31
						Bu_4NPF_6 LiCl	0.2 6E-3						
					VA	Bu_4NClO_4	0.2	-	-	-	PBE Ag/0.1 Ag^+	O1AFO	C=0.34, A=0.051
DD71	$C_{12}H_{10}HgS_2$	mercury phenylmercaptide RS-HG-SR	$(C_6H_5S)_2Hg$	EtOH	PY	LiCl	0.1	-	-	-	DME/SCE	O--	ns

TABLE I. Electrochemical Data $C_{12}H_{10}HgS_2$ DD71

Ref.	C/M	Charact. Potential		Response Const.		n Tech.	n	Electrokinetic Data			Products and Identification	Description and Remarks	Code No.
		Value	vs.		Value			Parameter	Value	From			
RY002 1096	–	$E_{\frac{1}{2}}$	-0.87 –	SCE	–	–	–	–	–	–	–	–	C,Mc(name),p C,X	DD65
EA017 2145	371 a	E_p	1.44	SCE	–	–	–	–	–	–	–	dibenzothiophen 5,5-dioxide,CP,IRS,CHA	A,r	DD66
EA017 2145	371 a	E_p	1.45F	SCE	j_p	0.73F	–	–	–	–	–	dibenzothiophene 5,5-dioxide,CP,IRS,CHA	A,r	
			1.50F			0.58F	–	–	–	–	–	dibenzothiopen-5,5-dioxide,CP,IRS,CHA	A,r	
JA094 1522	–	E_p	1.25	SCE	$i_p/v^{\frac{1}{2}}C$ D	0.25 29	–	–	–	–	–	–	A,R,p	DD67 ba42 cg75
JA094 5538	–	$E_{\frac{1}{2}}$	0.59	SCE	–	–	–	–	–	–	–	–	A,R,p	DD68
AJ026 1251 (AJ026 1669)	198 d	$E_{\frac{1}{2}}$	-0.86 -1.195	SCE	I	2.63 2.64	QE	1 1	–	–	–	azobenzene dianion, PY	C,R,r C,R	DD69
	198 f		0.30F		i_p	11F	–	–	–	–	–	–	C,r	
			0.318 -0.68F			9F 0.75F	–	–	–	–	–	–	C,r C	
			0.350 -0.76F -1.20F			2.5F 7.0F 0.5F	–	–	–	–	–	–	C,r C C	
	198 e		-0.85F -1.18F			7.0F 2.0F	–	–	–	–	–	–	C,r C	
			-0.78F -1.20F			8.6F 0.4F	–	–	–	–	–	–	C,r C	
JA094 4915	366 b	$E_{\frac{1}{2}}$	-2.13	SCE	I i_ℓ	3.46 1.09	I:I	1	–	–	–	–	C,R,r	DD70
			-2.95			0.35		<1					C,ece	
	366 c		-2.12 -2.95			1.54 0.54	–	–	–	–	–	–	C,i_{cat},r C,i_{cat}	
			-2.09 -2.87			1.5 1.2	–	–	–	–	–	–	C,i_{cat},r C,i_{cat}	
	366 b		-2.09			1.35	–	–	–	–	–	–	C,r	
			-0.34 -2.09 -2.44			5.4 1.2 6.0	–	–	–	–	–	Hg_2Cl_2	A,r C C,Li^+ redn.	
JA094 4915	366 b	E_p	-2.32F -2.11F	SCE	i_p	6F <2F	–	–	–	–	–	–	C,r A	
ER003 0623	–	$E_{\frac{1}{2}}$	– -0.35	SCE	–	–	PQ	2	–	–	–	–	C,i_a,p C	DD71

DD72 $C_{12}H_{10}N_2$

Code No.	Empirical Formula	Name and WLN Code	Structural Formula	Solvent	Tech.	Medium	μ, M	pH	T, °C	Electrodes	App.	Experimental Parameters
DD72 ba88 cg92	$C_{12}H_{10}N_2$	azobenzene C.A. 103-33-3 RNUNR	$C_6H_5N:NC_6H_5$	MeCN	PY	Et_4NClO_4 0.1	-	-	25±0.1	DME/SCE	25A0	C=1.9, m=1.56, t=3.20, h=50, Pt Aux
						Et_4NClO_4 0.1 H_2O 0.020						C=1.0
						Et_4NClO_4 0.1 H_2O 0.150						
						Et_4NClO_4 0.1 H_2O 0.350						
						Et_4NClO_4 0.1 H_2SO_4 0.0005						C=1.025
						Et_4NClO_4 0.1 H_2SO_4 0.0015						
						Et_4NClO_4 0.1 H_2SO_4 0.003						
						Et_4NClO_4 0.1 HOAc 0.0004						C=1.0
						Et_4NClO_4 0.1 HOAc 0.001						
						Et_4NClO_4 0.1 HOAc 0.002						
						Et_4NClO_4 0.1 HOAc 0.004						
						Et_4NClO_4 0.1 HOAc 0.002 Et_4NAc 0.002						
						Et_4NClO_4 0.1 HOAc 0.002 Et_4NAc 0.004						
						Et_4NClO_4 0.1 HOAc 0.002 Et_4NAc 0.010						
						Et_4NClO_4 0.1 HOAc 0.002 Et_4NAc 0.015						
						Et_4NClO_4 0.1 C_6H_5OH 0.001						
						Et_4NClO_4 0.1 C_6H_5OH 0.002						
DD73 ba93	$C_{12}H_{10}N_2O$	4-hydroxyazobenzene C.A. 1689-82-3 QR DNUNR	$4\text{-}HOC_6H_4N:NC_6H_5$	MeCN	PY	Et_4NClO_4 0.1	-	-	25±0.01	DME/SCE	125A0	C=1.2, m=1.56, t=3.20, h=50, Pt Aux
DD74 CONT	$C_{12}H_{10}N_2O$	N-nitrosodiphenyl-amine C.A. 86-30-6 ONNR&R	$(C_6H_5)_2NNO$	H_2O	DI	BR	-	2.0	-	DME/SCE	04A0	C=0.05, m=1.73, t=1, h=80, v=10, ΔE=100
								10.0				

TABLE I. Electrochemical Data $C_{12}H_{10}N_2O$ (CONT.) DD74

Ref.	C/M	Charact. Potential		Response Const.		Tech.	n	Electrokinetic Data			Products and Identification	Description and Remarks	Code No.
		Value	vs.		Value			Parameter	Value	From			
AJ026 1251 (AJ026 1669)	198 d	$E_{\frac{1}{2}}$ -1.405	SCE	I	3.10	QE	1.06 ± 0.02	-	-	-	azobenzene radical anion,UVS,VIS	C,R,r	DD72 ba88 cg92
		-1.755			2.97		0.96 ± 0.02				monoprotonated azo- benzene dianion, UVS,VIS,PY	C	
AJ026 1251	198 e	$E_{\frac{1}{2}}$ -1.37F -1.68F	SCE	I	5.2F 3.8F	-	-	-	-	-	-	C,r C	
		-1.37F -1.65F			8.5F 0.5F	-	-	-	-	-	-	C,r C	
		-1.34F			10F	-	-	-	-	-	-	C,r	
	198 f	0.26F			2F	-	-	-	-	-	-	C,i_d,r	
		-1.07F			1.5F							$C,i_d,E_{\frac{1}{2}}=f[H_2SO_4]$	
		-1.4F -1.75F			3.3F 2F							C C	
		0.26F			5F	-	-	-	-	-	-	C,i_d,r	
		-1.09F			4.5F							$C,i_d,E_{\frac{1}{2}}=f[H_2SO_4]$	
		0.26F			11F	-	-	-	-	-	-	C,i_d,r	
		-1.20F			2F							C,H_2SO_4 redn.	
	198 e	-0.9F			1.2F	-	-	-	-	-	-	$C,\neq,i_d,r$	
		-1.23F			0.4F							$C,\neq,i_d$	
		-1.40 -1.76			3.6F 4.0F							C C	
		-0.9F -1.23F -1.40F -1.76F			2.8F 0.8F 2.0F 2.4F	-	-	-	-	-	-	C,r C C C	
		-0.9F -1.26F			5.0F 4.0F	-	-	-	-	-	-	C,r C	
		-0.9F			8.8F	-	-	-	-	-	-	C,i_d,r	
		-1.0 V -1.26			3.0F 6.0F	-	-	-	-	-	-	C,r C	
		-1.06F -1.26F			1.8F 7.2F	-	-	-	-	-	-	C,r C	
		-1.08F -1.26F			0.8F 8.2F	-	-	-	-	-	-	C,r C	
		-1.26F			9.0F	-	-	-	-	-	-	C,r	
		-1.15F -1.32F -1.40F			2.8F 1.6F 1.6F	-	-	-	-	-	-	C,r C C	
		-1.15F -1.30F			4.0F 4.0F	-	-	-	-	-	-	C,r C	
AJ027 1215	-	$E_{\frac{1}{2}}$ -0.80 -1.10 -1.67 -1.98	SCE	i_ℓ	1.4F 4.0F 2.7F 10.3F	-	-	-	-	-	-	$C,\neq,X,r$ $C,=,X$ $C,\neq,X$ $C,\neq$	DD73 ba93
AA078 0081	82a	E_{su} -0.6	SCE	i_{su}	2.5	QE	4	-	-	-	-	C,r	DD74
	82e	-0.95			0.50		2	-	-	-	-	C,r	CONT

Code No.	Empirical Formula	Name and WLN Code	Structural Formula	Solvent	Tech.	Medium	μ, M	pH	T, °C	Electrodes	App.	Experimental Parameters
DD74	$C_{12}H_{10}N_2O$	N-nitrosodiphenyl-amine	$(C_6H_5)_2NNO$	H_2O	PT	BR	-	2	-	DME/SCE	04A0	C=0.05, m=1.73, t=1, h=80, v=10
DD75 bb02	$C_{12}H_{10}N_2O_3S$	azobenzene-4-sulfonic acid C.A. 2484-88-0	$4\text{-}HOSO_2C_6H_4N\!:\!N\text{-}C_6H_5$	MeCN	PY	Et_4NClO_4 0.1	-	-	25±0.01	DME/SCE	125A0	C=0.73, m=1.56, t=3.20, h=50, Pt Aux
DD76 ch03	$C_{12}H_{10}O_4$	quinhydrone C.A. 106-34-3 T -T6Y DYJ DUOHOR-DOHO4- A-14-J	$C_6H_4O_2 \cdot C_6H_4(OH)_2$	MeOH	PY	LiCl 0.1 C_6H_5COOH 0→6E-3 $(C_6H_5)_3PS$ 1%	0.1	-	-	DME/SCE	0A-	C=2
				MeCN	VR	Et_4NClO_4 0.1	-	-	-	Pt/SCE	12A0	[HQ]=500, [benzoquinone]=20, v=191, Pt Aux
												v=1.12E3
												v=1.12E4
				DMSO	PY	Et_4NClO_4 0.1	-	-	25	DME/SCE	13A0	t=6.02, m=1.03
				PYR	VA	$LiClO_4$ 0.5	-	-	25.0±0.1	PGE Ag/AgNO_3	123A0	0.6→-0.4→0.6, C=2, d=0.40, v=20, Pt Aux
DD77 bb14	$C_{12}H_{10}S$	diphenyl sulfide C.A. 139-66-2 RSR	$(C_6H_5)_2S$	HOAc 80	IL	H_2SO_4 0.18	-	-	25±0.1	Pt/SCE (0)	2A0	C=0.82, A=0.25, v=5, Pt Aux
						H_2SO_4 0.18 NaBr 0.0022						C=3.2
						H_2SO_4 0.18 NaCl 0.0043						
					VA	H_2SO_4 0.18	-	-	25±0.1	Pt/SCE (0)	2A0	0.90→1.75 V, C=16, A=0.25, v=70, Pt Aux
												-0.05→1.75 V
						H_2SO_4 0.18 NaCl 0.013						v=75

TABLE I. Electrochemical Data $\quad C_{12}H_{10}S \quad DD77$

Ref.	C/M	Charact. Potential		Response Const.		n Tech.	n	Electrokinetic Data			Products and Identification	Description and Remarks	Code No.
		Value	vs.		Value			Parameter	Value	From			
AA078 0081	82a	$E_{\frac{1}{2}}$	-0.6	SCE	-	-	QE	4	-	-	-	-	C,r	DD74
AJ027 1215	198 h	$E_{\frac{1}{2}}$	0.30 0.01 -1.340 -1.67	SCE	i_ℓ	2.0F 1.0F 2.0F 2.3F	-	-	-	-	-	-	C,r C C C	DD75 bb02
JE024 0230	119 c	$E_{\frac{1}{2}}$	-	SCE	-	-	change of wave shape in presence of acid or base	2	-	-	-	p-benzoquinone	A,O in absence of H^+ acceptor,r	DD76 ch03
			-0.18F		-	-		2	-	-	-	hydroquinone	C,i_{cat} in presence of H^+	
JS117 0186	119 i	E_p	0.17F -0.23F	SCE	i_p	30F 180F	-	-	-	-	-	radical anion,ESR	C,X,r $A,W,i_a+i_k,i_p \neq f[C_6H_4O_2]$	
			0.26F -0.40F			130F 480F	-	-	-	-	-	benzoquinone radical anion,ESR	C,X,r $A,W,i_a+i_k,i_p \neq f[C_6H_4O_2]$, $4-HOC_6H_4O^-$ oxidn.	
			0.55F -0.63F			210F 650F	-	-	-	-	-	benzoquinone radical anion,ESR	C,X,r $A,X,i_a+i_k,i_p \neq f[4-HOC_6H_4OH]$, $4-HOC_6H_4O^-$ oxidn.	
JS108 0980	119 h	$E_{\frac{1}{2}}$	-0.40 -1.24 -	SCE	-	-	sttd	1 -	-	-	-	-	C,W,R,p C,W,X A,O	
JS112 1215	-	E_p	-0.25F -0.08F 0.60F	SCE	i_p	24F 16F 44F	-	-	-	-	-	-	C,W,r A,W A,dr	
EA017 1421	371 a	E_p	1.3F	SCE	j_p	8.0F	QE	2	-	-	-	diphenylsulfoxide, CP,IRS,MAS	$A,j_p \propto C,r$	DD77 bb14
			0.98F			35F	-	-	-	-	-	-	A,r	
	371 b		0.98F			34F	-	-	-	-	-	-	A,r	
EA017 1421	371 a	E_p	1.34F	SCE	j_p	1100F	-	-	-	-	-	-	A,r	
			1.25			1380F	-	2	-	-	-	diphenylsulfoxide, IRS,MAS	A,r	
	371 b		1.03			2500F	-	-	-	-	-	-	$A,E_p=f[Cl^-]$ for $[Cl^-]<3$, r	

Code No.	Empirical Formula	Name and WLN Code	Structural Formula	Solvent	Tech.	Medium		μ, M	pH	T, °C	Electrodes	App.	Experimental Parameters
DD78	$C_{12}H_{11}NO$	2-methoxybenz[b]-azocine C.A. 37908-51-3 T68 GNJ H01	TABLE II.	THF	PY	Bu_4NPF_6	0.2	-	-	-	DME Ag/0.1 Ag^+	0AF-	C=0.19, t=6-12
					VA	Bu_4NPF_6	0.2	-	-	-	PBE Ag/0.1 Ag^+	0AF-	C=0.19, v=50
DD79	$C_{12}H_{11}NO$	4-methoxybenz[d]-azocine C.A. 37908-50-2 T68 INJ J01	TABLE II.	THF	PY	Bu_4NPF_6	0.2	-	-	-	DME Ag/0.1 Ag^+	0AF-	C=0.16, t=6-12
					VA	Bu_4NPF_6	0.2	-	-	-	PBE Ag/0.1 Ag^+	0AF-	C=0.16, v=40
DD80	$C_{12}H_{11}N_3O_5$	5-ethyl-5-(3-nitrophenyl)barbituric acid C.A. 509-85-3 T6MVMVXVJ E2 ER CNW	TABLE II.	H_2O	DI	PHOS	1	-	7.0	-	DME/SCE	2A0	-0.200 → -0.600 V, C=1.29E-2, EC=1.56, m=2.32, t=0.5, ΔE=50, v=5, Pt Aux
DD81 bb58	$C_{12}H_{12}Ni_2$	bis-[π-(η-cyclopentadienylnickel-(I)]acetylene C.A. 52445-55-3 L5ØJ Ø-NI- Y1UU1 Y-NI-- ØL5ØJ	$[(C_6H_5)Ni]_2HC\equiv CH$	MG	PY	Bu_4NClO_4	0.1	-	-	22	DME Ag/Ag ClO_4	2-0	C=2
DD82	$C_{12}H_{12}O$	2-phenyl-2-cyclohexenone C.A. 4556-09-6 L6V BUTJ BR	TABLE II.	DMF	PY	Bu_4NBF_4	0.50	-	-	25	DME/SCE	23-0	C=2.0-2.2, Pt Aux
DD83	$C_{12}H_{14}Br_2N_2$	1,1'-dimethyl-4,4'-bipyridylium bromide C.A. 3240-78-6 T6NJ A1 D- DT6NJ A1 &E &E	TABLE II.	H_2O	PY	BOR		-	9.6	27	DME/NHE	---	ns
DD84 bb97	$C_{12}H_{15}BrHgO_2$	ethyl α-bromomercury(4-ethylphenyl)-acetate C.A. 781-99-7 E-HG-YVO2&R D2	$4-C_2H_5C_6H_4CH-$ $(HgBr)COOC_2H_5$	MeOH 50	PY	KNO_3 BR	0.1 0.05	-	7.0	25±0.5	DME/SCE	0A1	C=1, m=0.67, t=3
DD85	$C_{12}H_{16}O$	2,4a,5,6,7,8-hexahydro-4a,8-dimethyl-naphthalen-2-one C.A. 3451-65-8 L66 AX DV&TJ C1 B&A	TABLE II.	DMF	PY	Bu_4NBF_4	0.50	-	-	25	DME/SCE	23-0	C=2.1-2.6, Pt Aux

TABLE I. Electrochemical Data

Ref.	C/M	Charact. Potential		Response Const.		n		Electrokinetic Data			Products and Identification	Description and Remarks	Code No.
		Value	vs.		Value	Tech.		Parameter	Value	From			
JA094 4907	-	$E_{1/2}$	-2.11	SCE	I	2.8-4.4	I:I	1-2	Elog	60	plot	-	C,p	DD78
			-2.31			1.2		<1		60			C,Po, $i \propto C$, $i_d/C = 1.2 \to 0.9$ as $C = 0.05 \to 0.2$, $i_d/h^{\frac{1}{2}} = (10.7 \to 9.5)E-3$ as $h = 22.0 \to 47.8$, i_k	
JA094 4907	-	$E_{p/2}$	-2.14 -2.3 -2.3 -1.25	SCE	i_p	34F - - 14F	-	-	-	-	-	-	C,p C,Po A,Po A	
JA094 4907	-	$E_{1/2}$	-2.13	SCE	I	3.8-6.0	I:I	1-2	Elog	90	plot	radical anion, $t_{\frac{1}{2}} \sim$ 10 s	C,$\frac{1}{4}$, I=f(C),r	DD79
JA094 4907	-	$E_{p/2}$	-2.14 -2.1	SCE	i_p	47F $\approx$3F	-	-	-	-	-	-	C,r A	
AA064 0165	-	E_{su}	-0.380	SCE	i_{su}	0.75F	-	-	-	-	-	-	C, $i \propto C$, $E_p=f(pH)$, r?	DD80
JA088 5117	-	$E_{1/2}$	-2.2	Ag/AgClO$_4$	-	-	QE	2	-	-	-	dianion, stable	C,p	DD81
JA094 8471	-	$E_{1/2}$	-1.96	SCE	-	-	-	-	αn_a	1.3	Elog	-	C,p	DD82
JA095 3411	-	$E_{1/2}$	0.79F	NHE	-	-	-	-	-	-	-	-	C,p	DD83
RY003 0218	8f	$E_{1/2}$	- -0.090 -0.244	SCE	-	-	sttd	- 1 1	-	-	-	-	C,Pr,p C C, $\Delta E_{\frac{1}{2}}=2(0.095)_\sigma$	DD84 bb97
JA094 8471	-	$E_{1/2}$	-2.01	SCE	-	-	-	-	αn_a	0.8	Elog	-	C,p	DD85

Code No.	Empirical Formula	Name and WLN Code	Structural Formula	Solvent	Tech.	Medium	μ, M	pH	T, °C	Electrodes	App.	Experimental Parameters
DD86	$C_{12}H_{18}CdO_2S_4$	bis(O-ethyl thio-acetothioacetato)-cadmium(II) C.A. 34867-02-2 SY1&UYUS&O2 2-CD-	$Cd[SC(CH_3):CH-C(S)OC_2H_5]_2$	Me_2CO	PY	Et_4NClO_4 0.1	-	-	20	DME Ag/AgCl, 0.1 LiCl	12A-	m=1.48, t=4.70, h=40, EC=1.68, W Aux
					PV	Et_4NClO_4 0.1	-	-	20	DME Ag/AgCl, 0.1 LiCl	12A-	EC=1.68, t=0.32, f=250, W Aux
DD87 ch28	$C_{12}H_{18}HgO_2S_4$	bis(O-ethyl thio-acetothioacetato)-mercury(II) C.A. 34867-03-3 SY1&UYUS&O2 2-HG-	$Hg[SC(CH_3):CHC-(S)OC_2H_5]_2$	Me_2CO	PY	Et_4NClO_4 0.1	-	-	20	DME Ag/AgCl, 0.1 LiCl	12A-	m=1.48, t=4.70, h=40, EC=1.68, W Aux
												t=0.32
					PV	Et_4NClO_4 0.1	-	-	20	DME Ag/AgCl, 0.1 LiCl	12A-	EC=1.68, t=0.16, f=300, W Aux
DD88	$C_{12}H_{18}O_2S_4Zn$	bis(O-ethyl thio-acetothioacetato)-zinc(II) C.A. 34867-01-1 SY1&UYUS&O2 2-ZN-	$Zn[SC(CH_3):CH-C(S)OC_2H_5]_2$	Me_2CO	PY	Et_4NClO_4 0.1	-	-	20	DME Ag/AgCl, 0.1 LiCl	12A-	m=1.48, t=4.70, h=40, EC=1.68, W Aux
					PV	Et_4NClO_4 0.1	-	-	20	DME Ag/AgCl,	12A-	EC=1.68, t=1.25, f=900, W Aux
												f=100, t=0.91
DD89	$C_{12}H_{19}LiO_2$	lithium 4-tert-butylcyclohexyl-ideneacetate C.A. 37881-12-2 L6YTJ A1U1VO DX1&1&1 &-LI-	TABLE II.	DMF	PY	Bu_4NBF_4 0.50	-	-	25	DME/SCE	23-0	C=2.6-4.4, Pt Aux
DD90	$C_{12}H_{20}O$	1-acetyl-4-tert-butyl-1-cyclohexene L6UTJ AV1 DX1&1&1	TABLE II.	DMF	PY	Bu_4NBF_4 0.50	-	-	25	DME/SCE	23-0	C=1.3-4.6, Pt Aux
DD91 ch37	$C_{12}H_{20}O_2$	3(2-methylpropoxy)-5,5-dimethyl-2-cyclo-hexen-1-one C.A. 15466-96-3 L6V CX EUTJ C C EO1Y	TABLE II-3.	DMF	PY	Bu_4NBF_4 0.50	-	-	25	DME/SCE	23-0	C=ns, Pt Aux
DD92	$C_{12}H_{20}O_2$	methyl 4-tert-butyl-1-cyclohex-1-enoate C.A. 22173-19-9 L6UTJ AVO1 DX1&1&1	TABLE II.	DMF	PY	Bu_4NBF_4 0.50	-	-	25	DME/SCE	23-0	C=1.5-3.1, Pt Aux

TABLE I. Electrochemical Data $C_{12}H_{20}O_2$ DD92

Ref.	C/M	Charact. Potential		Response Const.		n Tech.	n	Electrokinetic Data			Products and Identification	Description and Remarks	Code No.
		Value	vs.		Value			Parameter	Value	From			
JA095 1449	335 d	$E_{\frac{1}{2}}$	0.166	Ag/AgCl, 0.1LiCl	-	-	sttd	4	Tomeš Elog	47 48	sttd	-	$A, i_\ell \propto h^{\frac{1}{2}}, i_d, \neq, E_{\frac{1}{2}}$ at t=1.85,r	DD86
			-0.70					-	Tomeš	130	at t= 0.16		$C, E_{\frac{1}{2}}$ at t=1.72	
			-1.22							85	at t= 0.16		$C, E_{\frac{1}{2}}$ at t=1.32	
JA095 1449	335 d	E_{su}	0.23	Ag/AgCl, 0.1LiCl	-	-	sttd	4	$\Delta E_{su/2}$	≈160	sttd	-	A,X,r	
JA095 1449	335 c	$E_{\frac{1}{2}}$	0.145	Ag/AgCl, 0.1LiCl	-	-	QE	1.91 to 2.02	Tomeš	50	sttd	-	$A, i_\ell \propto h^{\frac{1}{2}}, i_d, E_{\frac{1}{2}}$ at t=6.00,r	DD87 ch28
			-0.20					-		≈68			$C, Pr, E_{\frac{1}{2}}$ at t=1.85	
			-							62			$C, E_{\frac{1}{2}}$ at t=1.95	
			0.118F		$i_\ell(u)$	5F	-	-	-	-	-	-	A, i_k, r	
			0.287F			2.5F							A, i_k	
JA095 1449	335 c	E_{su}	0.28	Ag/AgCl, 0.1LiCl	-	-	sttd	2	$\Delta E_{su/2}$	210	sttd	-	A,r	
			0.11					2		?			A	
			-0.54					2		155			C	
JA095 1449	335 e	$E_{\frac{1}{2}}$	0.244	Ag/AgCl, 0.1LiCl	-	-	QE	4	Tomeš Elog $Elog^2$ $Elog^3$	50 53 31 23	sttd	-	$A, i_\ell \propto h^{\frac{1}{2}}, i_d$, Elog, $Elog^2$ and $Elog^3$ are nonlinear, $\neq, E_{\frac{1}{2}}$ at t=1.80,r	DD88
			-1.143				i:i	1	Tomeš	59	sttd		$C, E_{\frac{1}{2}}$ at t=1.56, R at t=0.16, i_k	
			-1.45					1		≈63			$C, E_{\frac{1}{2}}$ at t=1.71	
JA095 1449	335 e	E_{su}	-1.14	Ag/AgCl, 0.1LiCl	-	-	sttd	1	$\Delta E_{su/2}$	100	sttd	-	C,r	
			-1.47					1		130			C	
			0.22				QE	4		175			A,r	
JA094 8471	-	$E_{\frac{1}{2}}$	-2.37	SCE	-	-	-	-	αn_a	0.8	Elog	-	C,p	DD89
JA094 8471	-	$E_{\frac{1}{2}}$	-2.24	SCE	-	-	-	-	αn_a	1.2	Elog	-	C,p	DD90
JA094 8471	-	$E_{\frac{1}{2}}$	-2.43	SCE	-	-	-	-	αn_a	1.0	Elog	-	C,p	DD91 ch37
JA094 8471	-	$E_{\frac{1}{2}}$	-2.50	SCE	-	-	-	-	αn_a	1.1	Elog	-	C,p	DD92

Code No.	Empirical Formula	Name and WLN Code	Structural Formula	Solvent	Tech.	Medium		μ, M	pH	T, °C	Electrodes	App.	Experimental Parameters
DD93	$C_{12}H_{21}ClHgO$	trans-2-chloromercury-3-methoxycyclo-undecane C.A. 32393-42-3 L-11-UTJ B-HG-G C O1	Table II.	dioxane 50	PY	ACET HCl TX100	? 0.001	-	1.35	30±0.1	DME/SCE	O3AO	C=ns,m=2.12, t=3.2,h=65
						PHOS TX100	? 0.001		7.00				
						NaOH TX100	? 0.001		11.00				
DD94	$C_{12}H_{22}N_2$	9,9-cyclopentamethylene-1,5-diazabicyclo[3.3.1]-nonane C.A. 38705-06-5 T66 A AXN FNTJ A-& AL6XTJ	TABLE II.	MeCN	VR	$NaClO_4$	0.1	-	-	-	AuBE SCE	2AO	C=1.35,d= 0.15,v=52, MP Aux,PE
DD95	$C_{12}H_{22}N_2O$	dicyclohexyl-N-nitrosoamine C.A. 947-92-2 L6TJ ANNO&- AL6TJ	TABLE II.	H_2O	PT	BR		-	2.0	-	DME/SCE	O4AO	C=0.05,m= 1.73,t=1, h=80,v=10
DD96	$C_{12}H_{28}FeN_4$	2,3-dimethyl-1,4,8,-11-tetraazacyclotetradecaneiron(II)dication D5656 1A O BM EM IM LMTJ C1 D1 &∅ &∅	TABLE II.	MeCN	VR	Et_4NClO_4	0.1?	-	-	-	Ag/ $0.1Ag^+$	---	ns
DD97	$C_{13}H_6Cl_6O_2$	bis(3,4,6-trichloro-2-hydroxyphenyl)-methane C.A. 70-30-4 QR CG DG FG B1R BQ CG EG FG	TABLE II.	H_2O	PY	PHOS	?	-	8.0	25±0.1	DME/SCE	2AO	C=0.1,Pt Aux
					PV	PHOS	?	-	8.0	25±0.1	DME/SCE	2AO	C=0.001-0.1,Δe=10, f=50,Pt Aux
					VA	PHOS	?	-	8.0	25±0.6	HMDE SCE	2AO	C=0.1,v=10, Pt Aux
													v=50
													v=200
DD98	$C_{13}H_8CoN_3O_2$	carbonyl(nitrosyl)-1,10-phenanthroline-cobalt C.A. 36454-22-5 D B6566 2AB CN-CO-N OJ DCO DNO	TABLE II.	MG	PY	Bu_4NClO_4	?	-	-	-	DME Ag/ $0.001Ag^+$	5AO	t=3
DD99 bc67	$C_{13}H_8O$	9-fluorenone C.A. 486-25-9 L B565 HVJ	TABLE II-2.	DMF	PY	Bu_4NBF_4	0.50	-	-	25	DME/SCE	23-0	C=1.4-2.7, Pt Aux

TABLE I. Electrochemical Data $C_{13}H_8O$ DD99

Ref.	C/M	Charact. Potential		Response Const.		n Tech.	n	Electrokinetic Data			Products and Identification	Description and Remarks	Code No.
		Value	vs.		Value			Parameter	Value	From			
EA017 2077	174 b	$E_{\frac{1}{2}}$	-0.080 -0.787	i_ℓ	0.28 0.28	-	-	-	-	-	-	C,r C	DD93
			-0.090 -1.245		0.32 0.28	-	-	-	-	-	-	C,r C	
			-0.112 -1.32		0.32 0.32	-	-	-	-	-	-	C,r C	
JA094 7114	-	E_p	0.59	SCE	-	-	-	-	$E_p-E_{p/2}$	100	sttd	-	A,p	DD94
AA078 0081	82 bd	$E_{\frac{1}{2}}$	-0.75	SCE	-	-	QE	4	-	-	-	-	C,i_d,r	DD95
JA094 5502	-	$E_{\frac{1}{2}}$	0.27	Ag/0.1Ag$^+$	-	-	-	-	-	-	-	-	A,R,p	DD96
AA061 0320	-	$E_{\frac{1}{2}}$	-1.020	SCE	i_ℓ	0.39F	-	-	-	-	-	-	C,i_a,$i_\ell \propto h$, $E_{\frac{1}{2}} \neq f(pH)$ for pH=6-11,r	DD97
AA061 0320	-	E_{su}	-1.100	SCE	i_{su}/C	5.24	-	-	-	-	-	-	C,$i_{su} \propto h$,i_a,$E_{su} \neq f(pH)$ for pH=6-11,r	
AA061 0320	-	-	-	-	i_p	0.093	-	-	ΔE_p	75	sttd	-	C,$i_p/Cv^{\frac{1}{2}} \uparrow$ as $v\uparrow$, i_p/Cv=9.36,Av(5), $i_{p,A}/i_{p,C} \downarrow$ as $v\uparrow$, i_a,r	
						0.055				-			A	
			-			0.513 0.163				75			C,r A	
		E_p	-1.23F -1.09F	SCE		1.288 0.338	-	-		135 -			C,r A	
JA094 0738	-	$E_{\frac{1}{2}}$	-0.36	Ag/ 0.001Ag$^+$	-	-	QE	1	-	-	-	-	A,R,parent compound dark maroon,p	DD98
			-2.19					1				radical anion,gray-green,ESR	C,R	
			-2.82					1					C,R	
JA094 8471	-	$E_{\frac{1}{2}}$	-1.29 -1.95	SCE	-	-	-	-	αn_a	1.1 0.6	Elog	-	C,p C	DD99 bc67

DE00 $C_{13}H_{10}N_2O_4$

Code No.	Empirical Formula	Name and WLN Code	Structural Formula	Solvent	Tech.	Medium		μ, M	pH	T, °C	Electrodes	App.	Experimental Parameters
DE00	$C_{13}H_{10}N_2O_4$	6-methoxy-1-hydroxy-phenazine-5,10-dioxide C.A. 13925-12-7 T C666 BN INJ BO DQ IO KO1	TABLE II.	H_2O	PY	BR		–	3.17	–	DME Ag/AgCl	OAO	C=0.125?,m=3.60,t=2.68, h=56
									5.08				
									6.87				
									8.80				
									10.75				
DE01	$C_{13}H_{10}N_2O_4$	Thalidomide C.A. 50-35-1 T56 BVNVJ C- DT6VMVTJ	TABLE II.	MeOH 40	PO	BARB LiCl	0.1 0.1	–	4.7	–	DME/MP	0-2	ns
				MeOH 50 THFA 50	PY	PHOS GEL	0.01	–	1	20	DME/MP	OAO	C=7.75,m=1.97,t=3.0
									4				
									7				
DE02 bd15 ch75	$C_{13}H_{10}O$	benzophenone C.A. 119-61-9 RVR	$C_6H_5COC_6H_5$	NH_3	VR	KI	0.1	–	–	-50	AuDE Ag/0.01 Ag$^+$,0.1 KI	125F 2	C=1,A=0.5, v=200, Pt Aux
						KI MeOH	0.1 1.6E-2						
DE00 CONT						KI MeOH	0.1 3.9E-2						

TABLE I. Electrochemical Data $\quad C_{13}H_{10}O$ (CONT.) DE02

Ref.	C/M	Charact. Potential		Response Const.		Tech.	n	Electrokinetic Data			Products and Identification	Description and Remarks	Code No.
		Value	vs.		Value			Parameter	Value	From			
AA063 0415	365 a	$E_{\frac{1}{2}}$ 0.15$_8$	Ag/AgCl	i_ℓ	-	QE	2	αn_a $p=$ }	1.36± 0.03 0.97	Elog $dE_{\frac{1}{2}}/$ dpH	6-methoxy-1-phena- zinol-10-oxide,TLC, UVS,VIS,PY	$C,\frac{1}{2},i_d,r$	DE00
		0.06$_8$			2.12		2	}	2.03± 0.10 1.8		6-methoxy-1-phena- zinol,UVS,VIS	$C,\frac{1}{2},i_d$	
		-0.18$_2$			0.70		-	-	-	-		$C,\frac{1}{2},i_d,\alpha n_a=f(pH)$	
	365 b	0.06$_8$			1.04		2	αn_a $p=$ }	1.36± 0.03 0.97	Elog $dE_{\frac{1}{2}}/$ dpH	6-methoxy-5-phena- zinol-10-oxide,TLC, UVS,VIS,PY	$C,\frac{1}{2},i_d,r$	
		0.04$_5$			1.08		2	}	2.03± 0.10 1.8		6-methoxy-1-phena- zinol,UVS,VIS	$C,\frac{1}{2},i_d$	
		-0.29$_2$			0.84		-	-	-	-		$C,\frac{1}{2},i_d,\alpha n_a=f(pH)$	
		-			1.06		2	-	-	-	6-methoxy-5-phena- zinol-10-oxide,TLC, UVS,VIS,PY	$C,\frac{1}{2},i_d,\alpha n_a=f(pH),r$	
					1.10		2	αn_a $p=$ }	2.03± 0.10 1.8	Elog $dE_{\frac{1}{2}}/$ dpH	6-methoxy-1-phena- zinol,UVS,VIS	$C,\frac{1}{2},i_d$	
					0.74		-	-	-	-		$C,\frac{1}{2},i_d,\alpha n_a=f(pH)$	
		-0.14$_0$			1.10		2	-	-	-	6-methoxy-1-phena- zinol-10-oxide,TLC, UVS,VIS,PY	$C,\frac{1}{2},i_d,\alpha n_a=f(pH),r$	
		-0.23$_4$			1.10		2	αn_a p }	2.03± 0.10 1.8	Elog $dE_{\frac{1}{2}}/$ dpH	6-methoxy-1-phena- zinol,UVS,VIS	$C,\frac{1}{2},i_d$	
		-0.51$_9$			0.72		-	-	-	-		$C,\frac{1}{2},i_d,\alpha n_a=f(pH)$	
		- -0.30$_4$ -0.65$_0$			- 2.10 1.02		-	-	-	-		$C,0,r$ $C,\frac{1}{2},i_d$ $C,\frac{1}{2},i_d,\alpha n_a=f(pH)$	
CZ018 0422	-	Q 0.66	-	-	-	-	-	-	-	-	-	C,p	DE01
AA030 0313	-	$E_{\frac{1}{2}}$ -0.84	MP	i_ℓ	1.40	-	-	-	-	-	phthalic acid mono- amide,2,6-piper- idinedione	C,W,p	
		-1.02			1.78	-	-	-	-	-	-	C,W,p	
		-1.2			2.17	-	-	-	-	-	-	$C,W,$slow decomposi- tion,p	
JA095 3495	65 r	E_p -1.20	Ag/0.1 Ag$^+$	i_p	2.8	sttd	1	-	-	-	-	$A,R,E_p\neq f(v)$ for $v<$ 5E3,$E_p\neq f$(supp elect),r	DE02 bd15 ch75
		-1.25			2.9		1				radical anion	$C,R,E_p\neq f(v)$ for $v<$ 5E3	
		-1.73			2.9		1					$A,R,E_p\neq f(v)$ for $v<$ 5E3	
		-1.78			2.9		1				dianion	$C,R,E_p\neq f(v)$ for $v<$ 5E3	
		-1.20 -1.25 -1.65F			-	-	-	-	-	-	-	A,R,r C,R $C,\frac{1}{2}$	
		-1.20 -1.25 -1.63F			-	QE	-	αn_a	-	-	-	A,R,r C,r $C,\frac{1}{2}$	CONT

149

DE02 (CONT.) $C_{13}H_{10}O$

Code No.	Empirical Formula	Name and WLN Code	Structural Formula	Solvent	Tech.	Medium	μ, M	pH	T, °C	Electrodes	App.	Experimental Parameters
DE02 bd15 ch75	$C_{13}H_{10}O$	benzophenone	$C_6H_5COC_6H_5$	NH_3	VR	KI 0.1 MeOH 0.16	–	–	–50	AuDE Ag/0.01 Ag^+,0.1 KI	125F 2	C=1, A=0.5, Pt Aux
						KI 0.1 EtOH 2.8E-3						
						KI 0.1 EtOH 2.8E-2						
						KI 0.1 EtOH 0.1						v=50-500
						KI 0.1 $HOCH_2CH_2OH$ 3E-3						
						KI 0.1 $HOCH_2CH_2OH$ 6.1E-2						
						KI 0.1 $HOCH_2CH_2OH$ 2						
						KI 0.1 i-PrOH 4.5E-3						
						KI 0.1 i-PrOH 4.2E-2						
						KI 0.1 i-PrOH 0.76						
						KI 0.1 t-BuOH 3.5E-3						
						KI 0.1 t-BuOH 3.3E-2						
						KI 0.1 t-BuOH 0.32						
DE03	$C_{13}H_{11}ClN_4O$	7-chloro-1,3-di-hydro-5-(1-methyl-1H-pyrazol-5-yl)-2H-1,4-benzodiazepin-2-one C.A. 39264-08-9 T67 GMV JN IHJ CG K-DT5NNJ A1	TABLE II.	MeOH 3.3	DI	PHOS 1	–	3.0	–	DME/SCE	25A0	0.00 → –1.100, C=0.12, EC=2.28, m=2.75, t=0.5, $\Delta E \approx 50$, v=5, Pt Aux
								7.0				0.00 → –1.750 V
DE04 ch91	$C_{13}H_{11}NS$	10-methylphenothiazine C.A. 1207-72-3 T C666 BN IOJ B	TABLE II-3.	DMF	VR	Bu_4NClO_4 0.1	–	–	–	PDE/SCE	3F-	C=0.1-1, v=100
						Bu_4NClO_4 0.1	–	–	–	PDE/SCE	3F-	v=200
				CH_2Cl_2	VR	Bu_4NClO_4 0.2	–	–	–	Pt/SCE	25F-	C=1, v=100

TABLE I. Electrochemical Data　　　$C_{13}H_{11}NS$　DE04

Ref.	C/M	Charact.	Potential Value	vs.	Response	Const. Value	Tech.	n	Electrokinetic Data Parameter	Value	From	Products and Identification	Description and Remarks	Code No.
JA095 3495	65 r	E_p	-1.20 -1.25 -1.59F	Ag/0.1 Ag$^+$	-	-	-	-	-	-	-	-	A,R,r C,R C,⟊	DE02 bd15 ch75
			-1.20 -1.25 -1.670		i_p	2.8F 2.8F 2.2F	-	-	$dE_p/dlogv$	35±5	sttd	-	A,R,r C,R C,⟊	
			-1.20 -1.25 -1.615				-	-	$dE_p/dlog[EtOH]$ $dE_p/dlogv$	55±10 38±5	sttd	-	A,R,r C,R C,⟊	
			-1.20 -1.25 -1.525		-	-	-	-	-	-	-	-	A,R,r C,R C,⟊	
			-1.20 -1.25 -1.71F		-	-	-	-	-	-	-	-	A,R,r C,R C,⟊	
			-1.20 -1.25 -1.55F		-	-	-	-	-	-	-	-	A,R,r C,R C,⟊	
			-1.20 -1.25 -1.45F		-	-	-	-	-	-	-	-	A,R,r C,R C,⟊	
			-1.20 -1.25 -1.77F		-	-	-	-	-	-	-	-	A,R,r C,R C,⟊	
			-1.20 -1.25 -1.65F		-	-	-	-	-	-	-	-	A,R,r C,R C,⟊	
			-1.20 -1.25 -1.57F		-	-	-	-	-	-	-	-	A,R,r C,R C,⟊	
			-1.20 -1.25 -1.79F		-	-	-	-	-	-	-	-	A,R,r C,R C,⟊	
			-1.20 -1.25 -1.76F		-	-	-	-	-	-	-	-	A,R,r C,R C,⟊	
			-1.20 -1.25 -1.70F		-	-	-	-	-	-	-	-	A,R,r C,R C,⟊	
AA074 0367	-	E_{su}	-0.680	SCE	i_{su}	7.97	-	-	-	-	-	-	C,r	DE03
			-0.875			11.09	-	-	-	-	-	-	C,r	
JA094 0691	-	E_p	0.82	SCE	-	-	-	-	-	-	-	-	A,R,r	DE04 ch91
JA094 4790			0.90 -	Ag/AgCl	-	-	-	-	-	-	-	-	A,R,p C,0 to -2.6 V	
JA094 4872	-	E_p	0.84 0.92	SCE	-	-	-	-	ΔE_p	57 -	sttd	-	C,R,p A,R	

151

Code No.	Empirical Formula	Name and WLN Code	Structural Formula	Solvent	Tech.	Medium		μ, M	pH	T, °C	Electrodes	App.	Experimental Parameters
DE05	$C_{13}H_{12}$	benzo[b]bicyclo-[6.1.0]nona-2,4,6-triene C.A. 35335-01-4 L B386T&&J	TABLE II.	MeCN	PY	Me_4NPF_6	0.1	-	-	-	DME Ag/0.1 Ag^+	256A F2	ns
					VR	Me_4NPF_6	0.1	-	-	-	PBE Ag/0.1 Ag^+	256A F2	ns
DE06	$C_{13}H_{12}Cl_2O_4$	[2,3-dichloro-4-(2-methylene-1-oxo-butyl)phenoxy]acetic acid C.A. 58-54-8 QV1OR BG CG DVY2&U1	TABLE II.	EtOH 2	PY	BR		-	1.89	-	DME/SCE	OAO	C=0.2, t=3.4, h=80
									2.93				
									3.87				
									5.11				
									7.69				
									9.66				
						PHOS	?	0.1	1.09				
								0.2	1.92				
								0.1	3.06				
								0.50	5.51				
									6.09				
									6.67				
									7.65				
								0.80	10.37				
									11.37				
						BOR	?	0.02	7.72				
									8.80				
									9.92				
								0.45	8.10				
									8.99				
									9.95				
DE05 CONT		benzo[b]bicyclo-[6.1.0]nona-2,4,6-triene				AM	?	0.45	8.59				

TABLE I. Electrochemical Data $C_{13}H_{12}Cl_2O_4$ (CONT.) DE06

Ref.	C/M	Charact. Potential		Response Const.		Tech.	n	Electrokinetic Data			Products and Identification	Description and Remarks	Code No.
		Value	vs.		Value			Parameter	Value	From			
JA095 2198	-	$E_{\frac{1}{2}}$	-2.33	SCE	$I\eta^{\frac{1}{2}}$ 0.095	sttd	1	Elog	60	sttd	-	C,$\not\equiv$(VR),r	DE05
JA095 2198	-	$E_{p/2}$	-2.36	SCE	-	-	-	-	-	-	-	C,$\not\equiv$,r	
AA073 0337	-	$E_{\frac{1}{2}}$	-0.835	SCE	I	1.54	sttd	1	-	-	-	-	C,i_a,r	DE06
			-1.015			1.41		1				saturated ketone,PY	C,redn. of anion	
			-0.910			1.80		1				-	C,i_a,r	
			-1.055			1.49		1				saturated ketone,PY	C	
			-1.035			3.31	QE	2				saturated ketone,PY	C,i_a,r	
			-1.10			3.54		2				saturated ketone,PY	C,i_a,r	
			-1.230			3.88		2				saturated ketone,PY	C,i_a,r	
			-1.280			3.67		2				saturated ketone,PY	C,i_a,r	
			-0.64		i_ℓ	-	-	-	-	-	-	-	C,i_a,Pr,$i \neq f(C)$ for C>0.07,r	
			-0.77			0.44							C,i_a	
			-0.95			0.40						saturated ketone,PY	C	
			-0.69			-	-	-	-	-	-	-	C,i_a,Pr,$i \neq f(C)$ for C>0.07,r	
			-0.83			0.48							C,i_a	
			-0.97			0.41						saturated ketone,PY	C	
			-0.76			-	-	-	-	-	-	-	C,i_a,Pr,r	
			-0.89			0.46							C,i_a	
			-1.015			0.52						saturated ketone,PY	C	
			-1.09			1.06	QE	2			-	saturated ketone,PY	C,i_a,r	
			-1.30			0.24		-					C,i_a,X	
			-1.16			1.14		2	-	-	-	saturated ketone,PY	C,i_a,r	
			-1.30			0.27		-					C,i_a,X	
			-1.18			1.14		2				saturated ketone,PY	C,i_a,r	
			-1.31			0.38		-					C,i_a,X	
			-1.20			1.06		2	-	-	-	saturated ketone,PY	C,i_a,r	
			-1.31			0.22		-					C,i_a,X	
			-1.26			1.14		2				saturated ketone,PY	C,i_a,r	
			-1.58			0.34		-					C	
	373 cd		-1.30			1.15		2				saturated ketone,PY	C,i_a,r	
			-1.55			0.56		-					C	
	-		-1.24			1.21		2	-	-	-	saturated ketone,PY	C,i_a,r	
			-1.28			1.13		2				saturated ketone,PY	C,i_a,r	
			-1.33			1.10		2				saturated ketone,PY	C,i_a,r	
			-1.17			0.95		2	-	-	-	saturated ketone,PY	C,i_a,r	
			-1.21			1.18		2	-	-	-	saturated ketone,PY	C,i_a,r	
			-1.25			1.16		2	-	-	-	saturated ketone,PY	C,i_a,r	
JA095			-1.16		$I\eta^{\frac{1}{2}}$	1.11		2	-	-	-	saturated ketone,PY	C,i_a,r	
			-1.52			0.42		-					C	CONT

Code No.	Empirical Formula	Name and WLN Code	Structural Formula	Solvent	Tech.	Medium	μ, M	pH	T, °C	Electrodes	App.	Experimental Parameters
DE06	$C_{13}H_{12}Cl_2O_4$	[2,3-dichloro-4-(2-methylene-1-oxo-butyl)phenoxy]acetic acid	TABLE II.	EtOH 2	PY	AM ?	0.45	9.12	-	DME/SCE	OAO	C=0.2,t=3.4, h=80
								10.09				
						CARB ?	1.40	9.16				
								9.94				
								11.09				
				EtOH 60	PY	BR ?	-	2.70	-	DME/SCE	OAO	C=0.2,t=3.4, h=80
								3.92				
								5.60				
								6.40				
								8.20				
								11.00				
						PHOS ?	0.1	1.95				
								3.20				
								4.30				
							0.5	6.30				
								7.22				
								7.70				
						BOR ?	0.02	9.70				
								11.20				
								12.30				
							0.45	9.80				
								10.78				
								11.82				
						AM ?		8.38				
								9.15				
								10.05				
						CARB ?	1.40	10.26				
								11.37				
DE07 bd54 CONT	$C_{13}H_{12}N_2O$	4-methoxyazobenzene C.A. 2396-60-3 1OR DNUNR	$4-CH_3OC_6H_4N=NC_6H_5$	MeCN	PY	Et_4NClO_4 0.1	-	-	25±0.1	DME/SCE	125AO	C=1.04,m= 1.556,t= 3.20,h=50, Pt Aux

TABLE I. Electrochemical Data $C_{13}H_{12}N_2O$ (CONT.) DE07

Ref.	C/M	Charact. Potential		Response Const.		Tech.	n	Electrokinetic Data			Products and Identification	Description and Remarks	Code No.
		Value	vs.		Value			Parameter	Value	From			
AA073 0337	-	$E_{\frac{1}{2}}$	-1.21	SCE	i_ℓ	1.24	QE	2	-	-	-	saturated ketone, PY	c, i_a, r	DE06
			-1.59			0.21		-	-	-	-		c	
			-1.27			1.24		2	-	-	-	saturated ketone, PY	c, i_a, r	
			-1.63			0.18		-	-	-	-		c	
			-1.19			1.14		2	-	-	-	saturated ketone, PY	c, i_a, r	
			-1.54			0.26		-	-	-	-		c	
			-1.21			1.18		2	-	-	-	saturated ketone, PY	c, i_a, r	
			-1.56			0.36		-	-	-	-		c	
	373 cd		-1.26			1.18		2	-	-	-	saturated ketone, PY	c, i_a, r	
			-1.58			0.73		-	-	-	-		c	
AA073 0337	373 a	$E_{\frac{1}{2}}$	0.905	SCE	I	0.89	sttd	1	-	-	-	-	c, r	
			-1.015			0.85		1	-	-	-	-	c, r	
	373 b		-1.160			0.71		1	-	-	-	-	c, i_a, r	
			-			0.33		-	-	-	-	-	c, i_a, r	
			-1.41			1.66	QE	2	-	-	-	saturated ketone, PY	c, i_a	
			-			0.13		-	-	-	-	-	c, i_a, r	
			-1.41			1.79		2	-	-	-	saturated ketone, PY	c, i_a	
			-1.42			1.87		2	-	-	-	saturated ketone, PY	c, i_a, r	
	373 a		-0.86			0.36	sttd	1	-	-	-	-	c, i_a, r	
			-1.67			0.53		-	-	-	-		c	
			-0.95			0.38		1	-	-	-	-	c, i_a, r	
	373 b		-1.02			0.38		1	-	-	-	-	c, i_a, r	
			-1.28			0.85	QE	2	-	-	-	saturated ketone, PY	c, i_a, r	
			-1.30			0.78		2	-	-	-	saturated ketone, PY	c, i_a, r	
			-1.35			0.74		2	-	-	-	saturated ketone, PY	c, i_a, r	
			-1.45			0.71		2	-	-	-	saturated ketone, PY	c, i_a, r	
			-1.47			0.76		2	-	-	-	saturated ketone, PY	c, i_a, r	
			-1.48			0.72		2	-	-	-	saturated ketone, PY	c, i_a, r	
			-1.34			0.80		2	-	-	-	saturated ketone, PY	c, i_a, r	
			-1.36			0.70		2	-	-	-	saturated ketone, PY	c, i_a, r	
			-1.40			0.76		2	-	-	-	saturated ketone, PY	c, i_a, r	
			-1.23			0.82		2	-	-	-	saturated ketone, PY	c, i_a, r	
			-1.29			0.80		2	-	-	-	saturated ketone, PY	c, i_a, r	
			-1.33			0.82		2	-	-	-	saturated ketone, PY	c, i_a, r	
			-1.32			0.88		2	-	-	-	saturated ketone, PY	c, i_a, r	
			-1.34			0.76		2	-	-	-	saturated ketone, PY	c, i_a, r	
AJ026 1669	198 g	$E_{\frac{1}{2}}$	-1.5F	SCE	i_ℓ	5F	-	-	-	-	-	-	c, r	DE07 bd54
			-1.8F			4.5F			-	-	-		c	

CONT

DE07 (CONT.) $C_{13}H_{12}N_2O$

Code No.	Empirical Formula	Name and WLN Code	Structural Formula	Solvent	Tech.	Medium	μ, M	pH	T, °C	Electrodes	App.	Experimental Parameters
DE07 bd54	$C_{13}H_{12}N_2O$	4-methoxyazobenzene	4-$CH_2OC_6H_4$N:N-C_6H_5	MeCN	PY	Et_4NClO_4 0.1 H_2SO_4 0.0005	-	-	25± 0.1	DME/SCE	125A0	c=1.04,m= 1.56,t=3.20, h=50,Pt Aux
						Et_4NClO_4 0.1 H_2SO_4 0.001						
						Et_4NClO_4 0.1 H_2SO_4 0.002						
						Et_4NClO_4 0.1 H_2SO_4 0.0026						
DE08	$C_{13}H_{12}N_4O_2S$	2,3-dihydro-5-(5-isothiazolyl)-1-methyl-7-nitro-1H-1,4-benzodiazepine C.A. 39254-76-7 T67 GN JN JU&TJ CNW G1 K- CT5NSJ	TABLE II.	MeOH 3.3	DI	PHOS	1	3.0	-	DME/SCE	25A0	0.00 → -1.100, C=0.12,EC= 2.28,m= 2.75,t=0.5, ΔE=50,v=5, Pt Aux
								7.0				0.00 → -1.750 V
DE09	$C_{13}H_{12}N_4O_2S$	2,3-dihydro-1-methyl-7-nitro-5-(2-thiazolyl)-1H-1,4-benzodiazepine C.A. 39264-20-5 T67 GN JN JU&TJ CNW G1 K- BT5N CSJ	TABLE II.	MeOH 3.3	DI	PHOS	1	3.0	-	DME/SCE	25A0	0.00 → -1.100 V, C= 0.12,EC= 2.28,m=2.75, t=0.5,ΔE=50, v=5,Pt Aux
								7.0				0.00 → -1.750 V
DE10	$C_{13}H_{12}N_4S$	diphenylthiocarbazone C.A. 60-10-6 RNUNCUS&MMR	C_6H_5N:NCSNHNH-C_6H_5	EtOH 20	PY	buffer	-	7.20	25	DME/SCE	01-0	C=0.5
								10.5				
								13.3				
					VA	buffer	-	13.30	-	HMDE SCE	01-0	C=0.5
DE11 ch96	$C_{13}H_{13}N_3$	1,3-diphenylguanidine C.A. 102-06-7 MUYMR&MR	C_6H_5NHC(:NH)NH-C_6H_5	m-cresol	PY	Et_4NClO_4 0.1	-	-	-	DME Ag/AgCl, 0.1 Me_4NCl	216A0	C=0.5-1.5, m=2.87,t= 2.70,h=56, MP Aux
DE12	$C_{13}H_{14}Br_2N_2$	5,9-diazonia-5,6,7-trihydrodibenzo-[a,c]azepine bromide T B676 GN KN&T&J &E &E &9 &12	TABLE II.	H_2O	PY	BOR	-	9.6	27	DME/NHE	---	ns
DE13 bd91	$C_{13}H_{15}NO_2$	3-ethyl-3-phenyl-2,6-dioxopiperidine C.A. 77-21-4	TABLE II-2.	H_2O	PT	BOR	-	9.2	-	DME/SCE	04A0	C=0.1,m=1.73, t=1,h=80

TABLE I. Electrochemical Data $C_{13}H_{15}NO_2$ DE13

Ref.	C/M	Charact. Potential		Response Const.		n Tech.	Electrokinetic Data			Products and Identification	Description and Remarks	Code No.
		Value	vs.		Value		Parameter	Value	From			
AJ026 1699	198 g	$E_{\frac{1}{2}}$	0.30 -0.2 -1.105 -1.5F -1.8F	SCE	i_ℓ	1F 0.25F 2.5F 3.5F 2.0F	-	-	-	-	-	-	C,r C C C C	DE07 bd54
			0.3F -0.4F -1.180 -1.5F			3F 1F 3F 2.5F	-	-	-	-	-	-	C,r C C C	
			0.3F -0.45 -1.20F			6F 2F 3F	-	-	-	-	-	-	C,r C C	
			0.3F -0.76			8F 1.5F	-	-	-	-	-	-	C,r C	
AA074 0367	-	E_{su}	-0.210 -0.590	SCE	i_{su}	17.62 6.67	-	-	-	-	-	-	C,r C	DE08
			-0.520 -0.975			0.70 0.39	-	-	-	-	-	-	C,r C	
AA074 0367	-	E_{su}	-0.260 -0.320 -1.130	SCE	i_{su}	15.06 5.27 -	-	-	-	-	-	-	C,r C C	DE09
			-0.535 -0.650			3.39 3.01	-	-	-	-	-	-	C,r C	
AA070 0411	-	$E_{\frac{1}{2}}$	-0.43F -1.18F	SCE	i_ℓ	2.0F 0.7F	Elog QE	2 -	$dE_{\frac{1}{2}}/dpH$ -	59.6 -	sttd -	$C_6H_5NHN:C(S^-)NHNH-C_6H_5$,sttd	C,R,r C,i_{cat}?	DE10
			-0.39F -0.59F			0.5F 2.0F		- 2	-	- 59.6	-	- $C_6H_5NHN:C(S^-)NHNH-C_6H_5$,sttd	A,r C,R,i_d,Tc=1.7	
			-1.25F			0.7F		-		-			C,i_{cat}?,r	
			-0.50F -0.75F			0.5F 2.0F		- 2		- 59.6		-	A,r C,R,i_d	
			-1.43F			0.5F		-		-			C,i_{cat}?	
AA070 0411	-	E_p	-0.2F -0.58F -0.75F	SCE	i_p	0.6F 0.7F 6.7F	-	-	- $E_p-E_{p/2}$	- 30	- sttd	-	A,r A,R C,R	
AA072 0169	-	$E_{\frac{1}{2}}$	0.38	Ag/AgCl	I	0.376	-	-	-	-	-	-	A,$\frac{1}{2}$,m-cresolate oxidn,r	DE11 ch96
JA095 3411	-	$E_{\frac{1}{2}}$	-0.92F	NHE	-	-	-	-	-	-	-	-	C,p	DE12
AA080 0233	-	$E_{\frac{1}{2}}$	-0.76F	SCE	i_ℓ	0.06F	-	-	-	-	-	-	A,r	DE13 bd91

157

DE14 $C_{13}H_{16}O$

Code No.	Empirical Formula	Name and WLN Code	Structural Formula	Solvent	Tech.	Medium		μ, M	pH	T, °C	Electrodes	App.	Experimental Parameters
DE14 be01 ci01	$C_{13}H_{16}O$	trans-4,4-dimethyl-1-phenyl-1-penten-3-one C.A. 29569-91-3 1X&&V1U1R -T	$C_6H_5CH:CHCOC-(CH_3)_3$	DMF	PY	Bu_4NBF_4	0.50	-	-	25	DME/SCE	23-0	C=1.0-2.0, Pt Aux
DE15	$C_{13}H_{16}O$	3-phenyl-3-hepten-2-one C.A. 40006-62-0 4UYR&V1	TABLE II.	DMF	PY	Bu_4NBF_4	0.50	-	-	25	DME/SCE	23-0	C=1.1, Pt Aux
DE16 be04	$C_{13}H_{17}BrHgO_2$	ethyl 2-bromomercurio-2-(4-isopropylphenyl)acetate	TABLE II-2.	MeOH 50	PY	KNO_3 BR	0.1 0.05				DME/SCE		
DE17	$C_{13}H_{18}O$	benzyl cyclohexyl ether C.A. 16224-09-2 L6TJ A01R	TABLE II.	MeCN	QE	$LiClO_4$	0.1	-	-	-	Pt Ag/0.1 Ag^+	25A-	C=26, A=12.9, E_{app}=1.8, stainless steel Aux
DE18	$C_{13}H_{22}N_2$	9,9-pentamethylene-1,5-diazatricyclo-[3.3.1.1^{3,7}]decane C.A. 19531-64-7 T666/BI/DJ/HJ A J 2BF J AXN FNTJ A-& AL6XTJ	TABLE II.	MeCN	VR	$NaClO_4$	0.1	-	-	-	AuBE SCE	2A0	C=0.39, d=0.15, v=50, MP Aux, PE v=5050
DE19	$C_{13}H_{24}N_2$	2,2-diethyl-5,8-ethanoperhydropyrazolo[1,2-a]pyridazine C.A. 23211-28-11 T566/FK 2AE AN CX EN KTJ C2 C2	TABLE II.	MeCN	VR	$NaClO_4$	0.1	-	-	-	AuBE SCE	2A0	C=0.2-0.9, d=0.15, v=60, MP Aux, PE
DE20	$C_{13}H_{26}O_3$	tert-butyl peroxy-nonanoate C.A. 22913-02-6 8VOOX1&1&1	$CH_3(CH_2)_7C(O)O-OC(CH_3)_3$	MeOH 50 C_6H_6 50	PY	LiCl		-	-	-	DME/SCE	---	ns
DE21 be46	$C_{14}H_7NaO_5S$ (see also DE26)	sodium 1-anthraquinonesulfonate C.A. 27600-99-3 L C666 BV IVJ DSWO &-NA- &22/18	TABLE II-2.	H_2O	PY	BOR		-	9.6	27	DME/NHE	---	ns
DE22 be48	$C_{14}H_7NaO_7S$	Alizarine Red S C.A. 130-22-3 L C666 BV IVJ DQ EQ FSWO &-NA&	TABLE II.	H_2O	PY	buffer		-	1.83	-	DME/SCE	0-0	EC=3.06
						ACET			4.60				
						buffer			6.75				
									11.20				
					PD	ACET		-	4.6	-	DME/SCE	0-0	EC=3.06
CONT					IL	BR	0.02	0.1	1.81	-	HMDE SCE	05A0	0→0.6, C=4E-4, v=33.3

TABLE I. Electrochemical Data $C_{14}H_7NaO_7S$ (CONT.) DE22

Ref.	C/M	Charact. Potential		Response Const.		n Tech.	n	Electrokinetic Data			Products and Identification	Description and Remarks	Code No.
		Value	vs.		Value			Parameter	Value	From			
JA094 8471	-	$E_{\frac{1}{2}}$	-1.65 -2.12	SCE	-	-	-	-	αn_a	1.5 1.2	Elog	-	C,p C	DE14 be01 ci01
JA094 8471	-	$E_{\frac{1}{2}}$	-2.07	SCE	-	-	-	-	αn_a	0.9	Elog	-	C,p	DE15
RY003 0218	-	-	-	-	-	-	-	-	-	-	-	-	see BE04	DE16 be04
JA094 6812	360 b	-	-	-	-	QE	3.5	-	-	-	benzaldehyde,25%; cyclohexanol,53%; benzoic acid,31%; NMR,GLC	A,p	DE17
JA094 7114	-	E_p	0.70	SCE	-	-	-	-	$E_p-E_{p/2}$	95	sttd	-	A,redn. product observed at higher scan rate,p	DE18
		XE	0.70	-	-	-	-	-	ΔE_p	70	-	-	A,R,XE=$\frac{1}{2}(E_{p,A}+E_{p,C})$,r	
JA094 7108	-	E_p	-0.28 -	SCE	-	-	sttd	1 1	ΔE_p	90 -	sttd	radical cation,ESR	A,XE=$\frac{1}{2}(E_{p,A}+E_{p,C})$,r C	DE19
AC035 0880	-	$E_{\frac{1}{2}}$	-0.96	SCE	i_l/C	10.7	-	-	-	-	-	-	C,i_d,p	DE20
JA095 3411	-	$E_{\frac{1}{2}}$	-1.14F	NHE	-	-	-	-	-	-	-	-	C,p	DE21 be46
IV008 0069	-	$E_{\frac{1}{2}}$	0.17 -0.33	SCE	-	-	-	-	-	-	-	1,2,9,10-tetrahydroxyanthracene-3-sulfonate	A,p C	DE22 be48
			0.17 -0.57		i_l	- 4.5	-	-	-	-	-	-	A,p C	
			0.17 -0.86			-	-	-	-	-	-	-	A,p C	
			0.17 -0.87			-	-	-	-	-	-	-	A,p C	
RI650 0069	-	E_p	0.17 -0.6	SCE	i_l	- 6.0F	-	-	-	-	-	-	A,p C	
AA063 0175	-	E_p	-0.03 -0.29	SCE	i_p	0.006 0.120	-	-	-	-	-	-	C,$i \propto A$,r C	

CONT

DE22 (CONT.) $C_{14}H_7NaO_7S$ CRC Handbook Series in Organic Electrochemistry

Code No.	Empirical Formula	Name and WLN Code	Structural Formula	Solvent	Tech.	Medium		μ, M	pH	T, °C	Electrodes	App.	Experimental Parameters
DE22 be48	$C_{14}H_7NaO_7S$	Alizarine Red S	TABLE II-2.	H_2O	IL	BR	0.02	0.1	4.10	-	HMDE SCE	05A0	$0 \rightarrow 0.6$, C=4E-4, v=33.3
									6.80				
									9.62				
					VV	BR	0.02	0.1	1.81	-	HMDE SCE	05A0	C=4E-4, v=6.6, Δe=20
									4.10				
									6.80				
									9.62	25±0.2			
DE23	$C_{14}H_8N_2$	4,4'-dicyanobiphenyl C.A. 1591-30-6 NCR DR DCN	TABLE II.	DMF	VR	Et_4NClO_4		0.1	-	-	-	PBE/SCE	125A F02	v=80.6, Pt Aux
DE24	$C_{14}H_8N_2NiO_2$	dicarbonyl-1,10-phenanthrolinenickel(0) C.A. 36454-23-6 D566 B6 2AB O CN-NI -NJ DCO DCO	TABLE II.	MG	PY	Bu_4NClO_4		?	-	-	-	DME Ag/0.001 Ag^+	5A0	t=3
DE25	$C_{14}H_8N_2O_4$	9,10-dinitroanthracene C.A. 33685-60-8 L C666J BNW INW	TABLE II.	DMF	PY	Et_4NClO_4		0.1	-	-	25±0.1	DME/SCE	05A0	C=0.4-0.8, m=1.52, t=5.06, h=45
					IR	Et_4NClO_4		0.1	-	-	25±0.1	HMDE? SCE	05A0	ns
					PV	Et_4NClO_4		0.1	-	-	25±0.1	DME/SCE	05A0	m=1.52, t=5.06, h=45
					VA	Et_4NClO_4		0.1	-	-	25±0.1	HMDE SCE	05A0	C=0.4, v=30.3
						$NaClO_4$		0.1						
						$Mg(ClO_4)_2$		0.05						
DE26 be65	$C_{14}H_8O_5S$ (see also DE21)	anthraquinone-1-sulfonic acid C.A. 30637-95-7 L C666 BV IVJ DSWQ	TABLE II-2.	i-PrOH 40	PY	PHOS		?	-	7.6	25	DME/NCE	---	C=0.4
DE27 be66	$C_{14}H_8O_5S$	anthraquinone-2-sulfonic acid C.A. 84-48-0 L C666 BV IVJ ESWQ	TABLE II-2.	i-PrOH 40	PY	PHOS		?	-	7.6	25	DME/NCE	---	C=0.4
DE28 be71	$C_{14}H_8O_8S_2$	anthraquinone-1,5-disulfonic acid C.A. 117-14-6 L C666 BV IVJ DSQ KSQ	TABLE II-2.	i-PrOH 40	PY	PHOS		?	-	7.6	25	DME/NCE	---	C=0.4

TABLE I. Electrochemical Data $C_{14}H_8O_8S_2$ DE28

Ref.	C/M	Charact. Value	Potential vs.	Response Const.	Value	Tech.	n	Electrokinetic Data Parameter	Value	From	Products and Identification	Description and Remarks	Code No.
AA063 0175	-	E_p	-0.16 -0.40	SCE	i_p	0.008 0.134	-	-	-	-	-	-	$C, i \propto A, r$ C	DE22 be48
			-0.23 -0.58 -1.04			0.025 0.118 0.015	-	-	-	-	-	-	$C, i \propto A, r$ C C	
			-0.12 -0.67 -0.75			0.012 0.031 0.042	-	-	-	-	-	-	$C, i \propto A$ C C	
AA063 0175	-	E_{su}	-0.03 -0.27 -0.88	SCE	i_{su}	0.008 0.360 0.016	-	-	-	-	-	-	C C C	
			-0.20 -0.40 -0.86			0.060 0.608 0.040	-	-	-	-	-	-	C C C	
			- -0.42 -0.58			- 0.024 0.260	-	-	-	-	-	-	O, r C C	
			-0.12 -0.58 -0.72 -0.83			0.180 0.012 0.060 0.016	-	-	-	-	-	-	C C C C	
JA094 7526	-	E_p	-1.63 -2.06	SCE	-	-	sttd	1 1	$E_p-E_{p/2}$	60 -	calc.	radical anion	C, R, r C	DE23
JA094 0738	-	$E_{\frac{1}{2}}$	-0.74 -2.42 -2.76	Ag/ 0.001 Ag$^+$	-	-	QE	1 1 1	-	-	-	-	A, R, r C, R C, R	DE24
BJ046 3792	-	$E_{\frac{1}{2}}$	-0.545	SCE	I	2.65	I:I	2	Elog	33	sttd	-	C, r	DE25
BJ046 3792	-	-	-	-	$it^{\frac{1}{2}}/C$	18.3	-	-	-	-	-	-	$C, r?$	
BJ046 3792	-	E_{su}	-0.562	SCE	$(i_{su}/i_d)t^{\frac{1}{2}}n$	130	-	-	$\Delta E_{su/2}$	67	-	-	$C, r?$	
BJ046 3792	-	E_p	-0.560	SCE	i_p	1.16F	-	-	$E_p-E_{p/2}$	34	sttd	-	$C, i_p \propto v^{\frac{1}{2}}, i_{p,C}/i_{p,A}=1, r$	
			-0.50F			1.16F	-	-	-	-	-	-	C, p	
			-0.45F			0.96F	-	-	-	-	-	-	C, p	
BP068 0003	-	$E_{\frac{1}{2}}$	-0.620	NCE	-	-	-	-	-	-	-	-	$C, \Delta E_{\frac{1}{2}}=\Delta\beta, p$	DE26 be65
BP068 0003	-	$E_{\frac{1}{2}}$	-0.540	NCE	-	-	-	-	-	-	-	-	$C, \Delta E_{\frac{1}{2}}=\Delta\beta, p$	DE27 be66
BP068 0003	-	$E_{\frac{1}{2}}$	-0.770	NCE	-	-	-	-	-	-	-	-	$C, \Delta E_{\frac{1}{2}}=\Delta\beta, p$	DE28 be71

DE29 $C_{14}H_8O_8S_2$

Code No.	Empirical Formula	Name and WLN Code	Structural Formula	Solvent	Tech.	Medium		μ, M	pH	T, °C	Electrodes	App.	Experimental Parameters
DE29	$C_{14}H_8O_8S_2$	anthraquinone-2,6-disulfonic acid C.A. 84-50-4 L C666 BV IVJ ESWQ LSWQ	TABLE II.	i-PrOH 40	PY	PHOS	?	-	7.6	25	DME/NCE	---	C=0.4
DE30	$C_{14}H_9ClN_6$	5-(4-chlorophenyl)-2-phenyl-1,2,3-triazolo[4,5-d]triazolium anion C.A. 30597-85-4 T55 BNNN FNNNJ CR& GR DG &7/6	TABLE II.	DMF	PY	Bu_4NI	0.155	-	-	22	DME/MP?	0A?0	t=4.36,h=60
DE31 ci44	$C_{14}H_9NO_2$	9-nitroanthracene C.A. 602-60-8 L C666J BNW	TABLE II-3.	DMF	PY	Et_4NClO_4	0.1	-	-	25±0.1	DME/SCE	05A0	C=0.4-0.8, m=1.52,t=5.06,h=45
						$KClO_4$	0.1						
						$NaClO_4$	0.1						
						$LiClO_4$	0.1						
						$Mg(ClO_4)_2$	0.05						
					IR	Et_4NClO_4	0.1	-	-	25±0.1	HMDE? SCE	05A0	ns
					PV	Et_4NClO_4	0.1	-	-	25±0.1	DME/SCE	05A0	m=1.52,t=5.06,h=45
					VA	Et_4NClO_4	0.1	-	-	-	PDE/SCE	05A0	C=0.67,v=30.3,A=0.0665
DE32 be82 ci50	$C_{14}H_{10}$	anthracene C.A. 120-12-7 L C666J	TABLE II-2.	DMF	VR	Bu_4NClO_4	0.1	-	-	-	PDE/SCE	3F-	C=0.1-1, v=224
DE33	$C_{14}H_{10}ClN_3O$	7-chloro-1,3-dihydro-5-(2-pyridinyl)-2H-1,4-benzodiazepin-2-one C.A. 1812-32-4 T67 GMV JN IHJ CG K-BT6NJ	TABLE II.	MeOH 3.3	DI	PHOS	1	-	3.0	-	DME/SCE	25A0	0.00 → -1.100 V, C=0.12, EC=2.285, m=2.75, t=0.5, ΔE=50, v=5, Pt Aux
									7.0				0.00 → -1.750 V
DE34	$C_{14}H_{10}Fe_2O_4$	bis[dicarbonyl-(η-cyclopentadienyl)-iron(I)] C.A. 12154-95-9 L5ØJ Ø-FE-CO&CO&-FE-CO&CO&- ØL5ØJ	$[(C_5H_5)Fe(CO)_2]_2$	MG	PY	Bu_4NClO_4	0.1	-	-	22	DME Ag/AgClO_4	2-0	C=2
DE35 be90 ci55	$C_{14}H_{10}N_2O$	2,5-diphenyl-1,3,4-oxadiazole C.A. 725-12-2 T5	TABLE II-2.	MeCN	VR	Bu_4NClO_4	0.1	-	-	-	Pt/SCE	2--	C=1, A=0.17, v=100, Pt Aux

TABLE I. Electrochemical Data $C_{14}H_{10}N_2O$ DE35

Ref.	C/M	Charact. Potential		Response Const.		n		Electrokinetic Data			Products and Identification	Description and Remarks	Code No.
		Value	vs.		Value	Tech.		Parameter	Value	From			
BP068 0003	-	$E_{\frac{1}{2}}$	-0.525	NCE	-	-	-	-	-	-	-	-	$C, \Delta E_{\frac{1}{2}} = \Delta\beta, p$	DE29
BJ047 1490	-	$E_{\frac{1}{2}}$	-0.57 -1.27	MP	i_ℓ/C	2.4 2.4	-	-	Elog	62 -	sttd	-	$C, E_{\frac{1}{2}} \propto LUMO, r$ C	DE30
BJ046 3792	-	$E_{\frac{1}{2}}$	-0.969 -1.49	SCE	I	1.40 1.15	I:I	1 1	Tomeš	44 70	-	-	$C, i_\ell \propto C, i_d, p$ C, i_d	DE31 ci44
			-0.98F -1.45F			1.44F 1.44F	-	-	-	-	-	-	C,p C	
			-0.93F -1.38F			1.44F 1.55F	-	-	-	-	-	-	C,p C	
			-0.90F -1.28F -1.65F			1.44F 2.22F 0.89F	-	-	-	-	-	-	C,p C C	
			-0.83F -1.0F -1.45F			1.44F 2.80F 2.00F	-	-	-	-	-	-	C,p C C	
BJ045 3792	-	-	-	-	$it^{\frac{1}{2}}/C$	11.3 19.7	-	-	-	-	-	-	C,r? C	
BJ045 3792	-	E_{su}	-1.019 -1.53	SCE	$(i_{su}/i_d)t^{\frac{1}{2n}}$	38 17	-	-	$\Delta E_{su/2}$	120 110	-	-	C,r? C	
BJ046 3792	-	E_p	-1.0F -1.56F	SCE	i_p	5.8F 4.6F	-	-	$E_p - E_{p/2}$	60 -	sttd	-	C,p C	
JA094 0691	-	E_p	1.36 -2.00	SCE	-	-	-	-	-	-	-	-	$A, \frac{1}{\pm}, r$ C,R	DE32 be82 ci50
AA074 0367	-	E_{su}	-0.400 -1.120	SCE	i_{su}	8.81 -	-	-	-	-	-	-	C,r C	DE33
			-0.760			11.21	-	-	-	-	-	-	C,r	
JA088 5117	-	$E_{\frac{1}{2}}$	-2.2	Ag/AgClO$_4$	-	-	QE	2	-	-	-	2Fe$^-$	$C, \frac{1}{\pm}, p$	DE34
JA094 1522	-	E_p	-2.17 -	SCE	D	28	-	-	-	-	-	-	C,R,p $A, \frac{1}{\pm}$, filming of electrode	DE35 be90 ci55

Code No.	Empirical Formula	Name and WLN Code	Structural Formula	Solvent	Tech.	Medium	μ, M	pH	T, °C	Electrodes	App.	Experimental Parameters
DE36	$C_{14}H_{10}N_6$	2,5-diphenyl-1,2,3,4,5,6-hexaazapentalene C.A. 28705-30-8 T55 BNNN FNNNJ CR& GR &7/6	TABLE 11.	DMF	PY	Bu_4NI 0.155	-	-	-	DME/MP?	0A?0	t=4.36,h=60
DE37	$C_{14}H_{10}O_4Ru_2$	bis[dicarbonyl-(η-cyclopentadienyl)-ruthenium(I)] C.A. 12132-87-5 L50J Ø-RU-CO&CO&-RU-CO&CO&- ØL50J	$[(C_5H_5)Ru(CO)_2]_2$	MG	PY	Bu_4NClO_4 0.1	-	-	22	DME Ag/AgClO$_4$	2-0	C=2
DE38	$C_{14}H_{12}Br_2$	1,2-dibromo-1,2-diphenylethane C.A. 5789-30-0 EYR&YER	TABLE 11.	DMF	PY	Bu_4NClO_4 0.1	-	-	-	DME Ag/satd AgNO$_3$	5A0	ns
					VR	Bu_4NClO_4 0.1	-	-	-	Pt Ag/satd AgNO$_3$	5A0	C=0.5,v=500
DE39 bf46	$C_{14}H_{12}N_2O_4$	1,2-bis(4-nitrophenyl)ethane C.A. 736-30-1 WNR D2R DNW	TABLE 11-2.	MeCN	IL	Et_4NClO_4 0.1	-	-	-	Pt/SCE	12F-	ns
DE40	$C_{14}H_{12}O$	trans-2,3-diphenyl-oxirane C.A. 1439-07-2 T30TJ BR& CR	TABLE 11.	MeCN	QE	$LiClO_4$ 0.1	-	-	-	Pt Ag/0.1 Ag$^+$	25A-	C=25.5,A=12.9,E_{app}=1.70,stainless steel Aux
DE41	$C_{14}H_{13}N_5O_2$	2,3-dihydro-1-methyl-7-nitro-5-(2-pyrimidinyl)-1H-1,4-benzodiazepine C.A. 26440-32-4 T67 GN JN JU&TJ CNW G1 K- BT6N CNJ	TABLE 11.	MeOH 3.3	DI	PHOS 1	-	3.0	-	DME/SCE	25A0	0.00→-1.100 V,C=0.12,EC=2.28,m=2.75,t=0.5,ΔE=50,v=5, Pt Aux
								7.0				0.00→-1.750 V
DE42	$C_{14}H_{14}N_2O$	dibenzyl-N-nitroso-amine C.A. 5336-53-8 ONN1R&1R	$(C_6H_5CH_2)_2NNO$	H_2O	DI	BR	-	10.0	-	DME/SCE	04A0	C=0.05,m=1.73,t=1,h=80,v=10,ΔE=100
					PT	BR	-	2.0	-	DME/SCE	04A0	C=0.05,m=1.73,t=1,v=10
DE43	$C_{14}H_{14}O$	benzhydryl methyl ether C.A. 1016-09-7 1OYR&R	$(C_6H_5)_2CHOCH_3$	MeCN	QE	$LiClO_4$ 0.1	-	-	-	Pt Ag/0.1 Ag$^+$	25A-	C=50,A=12.9,E_{app}=1.70,stainless steel Aux
DE44	$C_{14}H_{14}O$	benzyl ether C.A. 103-50-4 R1O1R	$(C_6H_5CH_2)_2O$	MeCN	QE	$LiClO_4$ 0.1	-	-	-	Pt Ag/0.1 Ag$^+$	25A-	C=31,A=12.9,E_{app}=1.90,stainless steel Aux
DE45	$C_{14}H_{15}N_3$	x-dimethylaminoazo-benzene C.A. 29387-92-6 1N1&R YNUNR	x-$(CH_3)_2NC_6H_4$-N:NC$_6H_5$	m-cresol	PY	Et_4NClO_4 0.1	-	-	-	DME Ag/AgCl, 0.1 Me$_4$NCl	216A0	C=0.5-1.5,m=2.87,t=2.70,h=56,MP Aux

TABLE I. Electrochemical Data $C_{14}H_{15}N_3$ DE45

Ref.	C/M	Charact. Potential		Response Const.		n Tech.	n	Electrokinetic Data			Products and Identification	Description and Remarks	Code No.
		Value	vs.		Value			Parameter	Value	From			
BJ047 1490	-	$E_{\frac{1}{2}}$	-0.62 -1.37	MP	i_ℓ/C	3.2 3.0	-	-	Elog	65 -	sttd	-	$C,\underline{E_{\frac{1}{2}}} \propto LUMO, R?, r$ C	DE36
JA088 5117	-	$E_{\frac{1}{2}}$	-2.6	$Ag/AgClO_4$	-	-	QE	2	-	-	-	$2Ru^-$	$C, \neq, p$	DE37
JA094 9020	-	$E_{\frac{1}{2}}$	-0.2 -2.1	SCE	-	-	sttd	2 -	-	-	-	-	C,p C	DE38
JA094 9020	-	E_p	1.10F 2.15F 2.15F 0.60F	SCE	i_p	27.5F 27.5F 27F 12F	-	-	-	-	-	-	C,p C A A	
JA094 0640	69b	$E_{\frac{1}{2}}$	-1.20	SCE	-	-	-	-	-	-	-	-	C,p	DE39 bf46
JA094 6812	-	-	-	-	-	-	QE	1.3	-	-	-	benzophenone,27%; diphenylacetic acid,17%;NMR,GLC, IRS	A,p	DE40
AA074 0367	-	E_{su}	-0.215 -0.300	SCE	i_{su}	17.38 12.65	-	-	-	-	-	-	C,r C	DE41
			-0.520 -0.595			19.40 12.18	-	-	-	-	-	-	C,r C	
AA078 0081	82c	E_{su}	-1.35	SCE	i_{su}	0.65	QE	2	-	-	-	-	C,r	DE42
AA078 0081	82a	$E_{\frac{1}{2}}$	-0.73	SCE	-	-	QE	4	-	-	-	-	C,r	
JA094 6812	360 a	-	-	-	-	-	QE	2.8	-	-	-	benzophenone,77%, NMR,GLC	A,p	DE43
JA094 6812	360 b	-	-	-	-	-	QE	4.2	-	-	-	benzaldehyde,46%; benzoic acid,33%; NMR,GLC	A,p	DE44
AA072 0169	-	$E_{\frac{1}{2}}$	0.52	Ag/AgCl	I	0.398	-	-	-	-	-	-	$A,\neq,\underline{m}$-cresolate oxidn.,$pK_a=11$,r	DE45

DE46 $C_{14}H_{15}N_3$

Code No.	Empirical Formula	Name and WLN Code	Structural Formula	Solvent	Tech.	Medium	μ, M	pH	T, °C	Electrodes	App.	Experimental Parameters
DE46 bg06	$C_{14}H_{15}N_3$	4-(dimethylamino)-azobenzene C.A. 60-11-7 1N1&R DNUNR	4-$(CH_3)_2NC_6H_4$-N:NC_6H_5	MeCN	PY	Et_4NClO_4 0.1 H_2SO_4 0.0005	-	-	25± 0.1	DME/SCE	125A0	C=1.05,m= 1.56,t=3.20, h=50,Pt Aux
						Et_4NClO_4 0.1 H_2SO_4 0.001						
						Et_4NClO_4 0.1 H_2SO_4 0.0017						
						Et_4NClO_4 0.1 H_2SO_4 0.0026						
DE47	$C_{14}H_{15}N_5O_2$	2,3-dihydro-1-methyl-5-(1-methyl-1H-imidazol-2-yl)-7-nitro-1H-1,4-benzodiazepine C.A. 39264-38-5 T67 GN JN JU&TJ CNW G1 K- BT5N CNJ A1	TABLE II.	MeOH 3.3	DI	PHOS 1	-	3.0	-	DME/SCE	25A0	0.00→ -1.100 V,C= 0.12,EC= 2.28,m=2.75, t=0.5,ΔE= 50,v=5, Pt Aux
								7.0				0.00→ -1.750 V
DE48	$C_{14}H_{15}N_5O_2$	2,3-dihydro-1-methyl-5-(1-methyl-1H-pyrazol-5-yl)-7-nitro-1H-1,4-benzodiazepine C.A. 39264-24-9 T67 GN JN JU&TJ CNW G1 K- CT5NNJ B1	TABLE II.	MeOH 3.3	DI	PHOS 1	-	3.0	-	DME/SCE	25A0	0.00→ -1.100 V,C= 0.12,EC= 2.28,m=2.75, t=0.5,ΔE= 50,v=5, Pt Aux
								7.0				0.00→ -1.750 V
DE49	$C_{14}H_{16}Br_2N_2$	2,7-dimethyl-9,10-dihydro-8a,10a-diazonia phenanthrene bromide T B666 GN JN&T&J E1 L1 &E &E &9 &12	TABLE II.	H_2O	PY	BOR	-	9.6	27	DME/NHE	---	ns
DE50	$C_{14}H_{16}Br_2N_2$	3,6-dimethyl-9,10-dihydro-8a,10a-diazonia phenanthrene bromide T B666 GN JN&T&J D1 M1 &E &E &9 &12	TABLE II.	H_2O	PY	BOR	-	9.6	27	DME/NHE	---	ns
DE51	$C_{14}H_{16}N_6$	2-cyclohexyl-5-phenyl-1,2,3-triazolo[4,5-d]-1,2,3-triazolium anion C.A. 30597-89-8 T55 BNNN FNNNJ CR& G- L6TJ &7/6	TABLE II.	DMF	PY	Bu_4NI 0.155	-	-	22	DME/MP?	0A?0	t=4.36,h=60
DE52	$C_{14}H_{16}O$	5,5-dimethyl-3-phenyl-2-cyclohexenone C.A. 36047-17-3 T6V CX EUTJ C1 C1 ER	TABLE II.	DMF	PY	Bu_4NBF_4 0.50	-	-	25	DME/SCE	23-0	C=1.5-2.0, Pt Aux

TABLE I. Electrochemical Data $C_{14}H_{16}O$ DE52

Ref.	C/M	Charact. Potential		Response Const.		n Tech.	n	Electrokinetic Data			Products and Identification	Description and Remarks	Code No.
		Value	vs.		Value			Parameter	Value	From			
AJ026 1669	198 g	$E_{\frac{1}{2}}$	0.07F -0.45F -1.16F -1.62F -1.9F	SCE	i_ℓ	1.5F 0.4F 2.0F 6.0F 3.0F	-	-	-	-	-	-	$c, i_d + i_{cat}, r$ c, i_d c, i_d c, i_d c, i_d	DE46 bg06
			0.12F -0.42F -1.15F -1.6F			3F 0.5F 5.5F 3.5F	-	-	-	-	-	-	$c, i_d + i_{cat}, r$ c, i_d c, i_d c, i_d	
			0.14F -0.40F -1.15F			5.0F 0.75F 7.5F	-	-	-	-	-	-	$c, i_d + i_{cat}, r$ c, i_d c, i_d	
			0.2F -0.48F -1.18F			8.4F 3.0F 3.0F	-	-	-	-	-	-	$c, i_d + i_{cat}, r$ c, i_d c, i_d	
AA074 0367	-	E_{su}	-0.240 -0.450	SCE	i_{su}	18.08 6.58	-	-	-	-	-	-	c, r c	DE47
			-0.540 -0.750			17.45 12.53	-	-	-	-	-	-	c, r c	
AA074 0367	-	E_{su}	-0.220 -0.650	SCE	i_{su}	15.26 8.18	-	-	-	-	-	-	c, r c	DE48
			-0.520 -0.920			17.61 13.71	-	-	-	-	-	-	c, r c	
JA095 3411	-	$E_{\frac{1}{2}}$	-0.86F	NHE	-	-	-	-	-	-	-	-	c, p	DE49
JA095 3411	-	$E_{\frac{1}{2}}$	-0.69F	NHE	-	-	-	-	-	-	-	-	c, p	DE50
BJ047 1490	-	$E_{\frac{1}{2}}$	-0.94 -1.47	MP	i_ℓ/C	2.7 3.5	-	-	Elog	64 -	-	-	$c, E_{\frac{1}{2}} \propto LUMO, R?, r$ c	DE51
JA094 8471	-	$E_{\frac{1}{2}}$	-1.71 -2.06	SCE	-	-	-	-	αn_a	1.2 1.3	Elog	-	c, p c	DE52

167

DE53 $C_{14}H_{18}BC_4F_4N_4$ CRC Handbook Series in Organic Electrochemistry

Code No.	Empirical Formula	Name and WLN Code	Structural Formula	Solvent	Tech.	Medium		μ, M	pH	T, °C	Electrodes	App.	Experimental Parameters
DE53	$C_{14}H_{18}BCuF_4N_4$	5,7,12,14-tetramethyl-1,4,8,11-tetraazacyclotetradeca-2,5,7,9,12,14-hexaenecopper(I) tetrafluoroborate C.A. 39011-19-3 D5656 1A O A-CU-N EN IN LNJ F1 H1 M1 O1 &FBFFF	TABLE II.	MeCN	VY	Bu_4NBF_4	0.05	-	-	25	RPE/SCE	O--	$C \approx 1$
DE54	$C_{14}H_{18}BF_4N_4Ni$	5,7,12,14-tetramethyl-1,4,8,11-tetraazacyclotetradeca-2,5,7,9,12,14-hexaenenickel(I) tetrafluoroborate C.A. 39018-12-7 D5656 1A O A-NI-N EN IN LNJ F1 H1 M1 O1 &FBFFF	TABLE II.	MeCN	VY	Bu_4NBF_4	0.05	-	-	25	RPE/SCE	O--	$C \approx 1$
DE55	$C_{14}H_{18}CuN_4$	5,7,12,14-tetramethyl-1,4,8,11-tetraazacyclotetradeca-2,5,7,9,12,14-hexaenecopper(II) C.A. 39018-17-2 D5656 1A O A-CU-N EN IN LNJ F1 H1 M1 O1	TABLE II.	MeCN	VR	Bu_4NBF_4	0.05	-	-	25	Pt/SCE	O--	$C \approx 1, v=500$
					VY	Bu_4NBF_4	0.05	-	-	25	RPE/SCE	O--	$C \approx 1$
DE56	$C_{14}H_{18}FeO$	1-hydroxybutylferrocene L5ØJ Ø-FE-- ØL5ØJ AYQ3	TABLE II.	MeCN	EE	$LiClO_4$	?	-	-	25	Pt/SCE	---	ns
DE57	$C_{14}H_{18}N_2$	7-(1,3-dihydro-2-isoindolyl)-7-azabicyclo[4.1.0]heptane C.A. 31863-29-3 T56 CNT&J C- BT36 BNTJ	TABLE II.	MeCN	VR	$NaClO_4$	0.1	-	-	-	AuBE SCE	2AO	C=0.301, d=0.15, v=58, MP Aux, PE
DE58	$C_{14}H_{18}N_4Ni$	5,7,12,14-tetramethyl-1,4,8,11-tetraazacyclotetradeca-2,5,7,9,12,14-hexaenenickel(II) C.A. 39018-13-8 D5656 1A O A-NI-N EN IN LNJ F1 H1 M1 O1	TABLE II.	MeCN	VR	Bu_4NBF_4	0.05	-	-	25	Pt/SCE	O--	$C \approx 1, v=500$
					VY	Bu_4NBF_4	0.05	-	-	25	RPE/SCE	O--	$C \approx 1$
DE59 bg36	$C_{14}H_{19}BrHgO_2$	ethyl 2-bromomercury-2-(3-tert-butylphenyl)acetate XR CY-HG-EVO2	TABLE II-2.	MeOH 50	PY	KNO_3 BR	0.1 0.05	-	-	-	DME/SCE	---	ns
DE60	$C_{14}H_{20}O_8$	tetraethyl ethenetetracarboxylic acid C.A. 6174-95-4 20VYVO2&UYVO2&VO2	TABLE II.	DMF	PY	Bu_4NBF_4	0.50	-	-	25	DME/SCE	23-0	C=1.1-2.0, Pt Aux

TABLE I. Electrochemical Data $C_{14}H_{20}O_8$ DE60

Ref.	C/M	Charact. Potential		Response Const.		Tech.	n	Electrokinetic Data			Products and Identification	Description and Remarks	Code No.
		Value	vs.		Value			Parameter	Value	From			
JA094 4529	–	$E_{\frac{1}{2}}$	-0.07 0.50	SCE	i_d/C	26 25	i:i	1 1	Tomeš	60 57	sttd	–	C,R,p A,R	DE53
JA094 4529	–	$E_{\frac{1}{2}}$	0.10 0.70	SCE	i_d/C	33 26	i:i	1.3 1.0	Tomeš	59 52	sttd	–	C,R,p A,R	DE54
JA094 4529	–	E_p	0.04F -0.53F -0.44 0.11F	SCE	i_p	8F 8.5F 9F 8F	sttd	1 1 1 1	–	–	–	–	C,R or Q,p C,R or Q A,R or Q A,R or Q	DE55
JA094 4529	–	$E_{\frac{1}{2}}$	-0.04 0.50	SCE	i_d/C	25 26	i:i	1 1	Tomeš	58 –	sttd	–	A,R,p A,R	
JA086 1382	–	$E_{\tau/4}$	0.324	SCE	–	–	–	–	–	–	–	–	$A, \Delta E_{\tau/4} = \rho\sigma, p$	DE56
JA094 7108	–	E_p	0.63	SCE	–	–	–	–	–	–	–	–	A,≠,p	DE57
JA094 4529	–	E_p	-0.14F -0.75F -0.69F -0.06F	SCE	i_p	6F 7F 6F 6F	sttd	1 1 1 1	–	–	–	–	C,R or Q,p C,R or Q A,R or Q A,R or Q	DE58
JA094 4529	–	$E_{\frac{1}{2}}$	0.11 0.70	SCE	i_d/C	29 24	i:i	1.2 1	Tomeš	64 54	sttd	–	A,R,p A,R	
RY003 0218	–	–	–	–	–	–	–	–	–	–	–	–	see BG36	DE59 bg36
JA094 8471	–	$E_{\frac{1}{2}}$	-0.97 -1.10	SCE	–	–	–	–	αn_a	1.1 1.0	Elog	–	C,p C	DE60

Code No.	Empirical Formula	Name and WLN Code	Structural Formula	Solvent	Tech.	Medium	μ, M	pH	T, °C	Electrodes	App.	Experimental Parameters
DE61	$C_{14}H_{22}CuN_4$	5,7,12,14-tetramethyl-1,4,8,11-tetra-azacyclotetradeca-5,7,12,14-tetraene-copper(II) C.A. 39060-36-1 D5656 1A O A-CU-N EN IN LNT&T&J F1 H1 M1 O1	TABLE II.	DMF	VY	Bu_4NBF_4 0.05	-	-	25	RPE/SCE	O--	$C \approx 1$
DE62	$C_{14}H_{22}N_4Ni$	5,7,12,14-tetramethyl-1,4,8,11-tetra-azacyclotetradeca-5,7,12,14-tetraene-nickel(II) C.A. 39060-35-0 D5656 1A O A-NI-N EN IN LNT&T&J F1 H1 M1 O1	TABLE II.	DMF	VY	Bu_4NBF_4 0.05	-	-	25	RPE/SCE	O--	$C \approx 1$
DE63	$C_{14}H_{22}N_4Zn$	5,7,12,14-tetramethyl-1,4,8,11-tetra-azacyclotetradeca-5,7,12,14-tetraene-zinc(II) C.A. 31075-97-5 D5656 1A O A-ZN-N EN IN LNT&T&J F1 H1 M1 O1	TABLE II.	MeCN	VY	Bu_4NBF_4 0.05	-	-	25	RPE/SCE	O--	$C \approx 1$
DE64	$C_{14}H_{22}N_6$	2,5-dicyclohexyl-1,2,3-triazolo-[4,5d]-1,2,3-triazolium anion C.A. 36597-90-1 T55 BNNN FNNNJ C-AL6TJ& G- AL6TJ &7/6	TABLE II.	DMF	PY	Bu_4NI 0.155	-	-	22	DME/MP?	OA?O	t=4.36, h=60
DE65	$C_{14}H_{24}O_2$	ethyl 4-tert-butyl-cyclohexylidene acetate C.A. 13733-50-1 L6YTJ AU1VO2 DX1&1&1	TABLE II.	DMF	PY	Bu_4NBF_4 0.50	-	-	25	DME/SCE	23-O	C=1.9-3.0, Pt Aux
DE66	$C_{14}H_{25}ClHgO$	trans-2-chloromercury-3-methoxycyclo-tridecene C.A. 32462-45-6 L-13-UTJ B-HG-G CO1	TABLE II.	dioxane 50	PY	ACET HCl TX100 0.001	-	1.4	30± 0.1	DME/SCE	O3AO	C=0.216, m=2.12, t=3.2, h=65
						$C_6H_4(COOH)_2$ HCl TX100 0.001		2.6				
						ACET TX100 0.001		5.4				
						BOR TX100 0.001		9.8				
						NaOH TX100 0.001		11.3				

TABLE I. Electrochemical Data $\quad C_{14}H_{25}ClHgO \quad DE66$

Ref.	C/M	Charact. Potential		Response Const.		n Tech.	n	Electrokinetic Data			Products and Identification	Description and Remarks	Code No.
		Value	vs.		Value			Parameter	Value	From			
JA094 4529	-	$E_{\frac{1}{2}}$	0.17	SCE	i_d/C	15	i:i	1.25	Tomeš	61	sttd	-	A,R,p	DE61
JA094 4529	-	$E_{\frac{1}{2}}$	0.11	SCE	i_d/C	14	i:i	1.2	Tomeš	64	sttd	-	A,R,p	DE62
JA094 4529	-	$E_{\frac{1}{2}}$	0.23	SCE	i_d/C	22	i:i	0.9	Tomeš	86	sttd	-	A,Q,p	DE63
BJ047 1490	-	$E_{\frac{1}{2}}$	-1.32 -1.7	MP	i_ℓ/C	2.7 6	-	-	Elog	61 -	sttd	-	$C, E_{\frac{1}{2}} \propto LUMO, R?, r$ C	DE64
JA094 8471	-	$E_{\frac{1}{2}}$	-2.45	SCE	-	-	-	-	αn_a	1.2	Elog	-	C,p	DE65
EA017 2077	174 b	$E_{\frac{1}{2}}$	-0.075	SCE	I	1.23	-	-	-	-	-	-	$C, i \propto h^{\frac{1}{2}}, Tc<3.5, R, E_{\frac{1}{2}} \neq f(C), i_\ell \propto C, r$	DE66
			-0.750			1.33							$C, i_\ell \propto (h^{\frac{1}{2}}, C), Tc < 3.5, \neq, E_{\frac{1}{2}}=f(C)$	
			-0.050			1.03	-	-	-	-	-	-	$C, i \propto h^{\frac{1}{2}}, i_\ell \propto C, Tc < 3.5, R, E_{\frac{1}{2}} \neq f(C), r$	
			-0.937			1.36							$C, i_\ell \propto (h^{\frac{1}{2}}, C), Tc < 3.5, \neq, E_{\frac{1}{2}} \neq f(C)$	
			-0.050			1.01	-	-	-	-	-	-	$C, i_\ell \propto h^{\frac{1}{2}}, C, Tc < 3.5, R, E_{\frac{1}{2}} \neq f(C), r$	
			-1.200			1.37							$C, i \propto (h^{\frac{1}{2}}, C), Tc < 3.5, \neq, E_{\frac{1}{2}}=f(C)$	
			-0.125			1.07							$C, i_\ell \propto (h^{\frac{1}{2}}, C), Tc < 3.5, \neq, E_{\frac{1}{2}}=f(C)$	
			-1.380			1.36							$C, i_\ell \propto (h^{\frac{1}{2}}, C), Tc < 3.5, \neq, E_{\frac{1}{2}}=f(C)$	
			-0.212			1.03							$C, i_\ell \propto (h^{\frac{1}{2}}, C), Tc < 3.5, R, E_{\frac{1}{2}}=f(C), r$	
			-1.380			1.35							$C, i_\ell \propto (h^{\frac{1}{2}}, C), Tc < 3.5, \neq, E_{\frac{1}{2}}=f(C)$	

Code No.	Empirical Formula	Name and WLN Code	Structural Formula	Solvent	Tech.	Medium		μ, M	pH	T, °C	Electrodes	App.	Experimental Parameters
DE67	$C_{14}H_{25}NO_2$	2-(2'-ethyl-6'-methylpiperidyl)ethyl methacrylate T6NTJ A2OVY1&U1 B2 F1	TABLE II.	MeOH 92	PY	Et_4NI	0.05	-	-	-	DME/SCE	O-O	C=0.24,EC=1.204,t=10.4
DE68	$C_{14}H_{32}N_4$	1,2,4,5-tetraisopropylperhydro-s-tetrazine C.A. 38704-93-7 T6NN DNNTJ AY1&1 BY1&1 DY1&1 EY1&1	TABLE II.	MeCN	VR	$NaClO_4$	0.1	-	-	-	AuBE SCE	2AO	C=0.2-0.9, d=0.15,v=52, MP Aux,PE
DE69	$C_{14}H_{42}B_{18}CoN$	tetramethylammonium bis[π-μ-trimethylene-(3)-1,2-dicarbaundecaboranyl-(9)]-cobaltate(III) C.A. 38904-47-1	$[(CH_3)_4N]^+[(B_9\text{-}C_5H_{15})_2Co(III)]^-$	MeCN	VR	Et_4NClO_4	0.1	-	-	-	PBE/SCE	--O3	v=6E3
DE70	$C_{14}H_{42}B_{18}NNi$	tetramethylammonium bis[π-μ-1,2-trimethylene-(3)-1,2-dicarbaundecaboranyl-(9)nickelate(III) C.A. 36683-55-3	$[(CH_3)_4N]^+[(B_9\text{-}C_5H_{15})_2Ni(III)]^-$	MeCN	VA	Et_4NClO_4	0.1	-	-	-8	PBE/SCE	--O3	v=6E3
					VR	Et_4NClO_4	0.1	-	-	-8	PBE/SCE	--O3	v=6E3
DE71	$C_{15}H_{10}BrClO_4S_2$	3-(4-bromophenyl)-5-phenyl-1,2-dithiolylium perchlorate C.A. 42877-41-8 T5SSTJ CR& ER DE & G-O4	TABLE II.	MeCN	VR	Bu_4NBF_4	0.1	-	-	-	PBE/SCE	26A2	v=150
				CH_2Cl_2	VR	Bu_4NBF_4	0.2	-	-	-	PBE/SCE	26A2	v=150
DE72 cj40	$C_{15}H_{10}ClFN_2O$	7-chloro-1,3-dihydro-5-(2-fluorophenyl)-2H-1,4-benzodiazepin-2-one C.A. 2886-65-9 T67 GMV JN IHJ CG KR BF	TABLE II-3.	MeOH 3.3	DI	PHOS	1	-	3.0	-	DME/SCE	25AO	0.00→-1.100 V,C=0.11,EC=2.285,m=2.75,t=0.5,ΔE=50,v=5, Pt Aux
									7.0				0.00→-1.750 V
DE73	$C_{15}H_{10}Cl_2N_2O$	7-chloro-5-(2-chlorophenyl)-2,3-dihydro-1H-1,4-diazepin-2-one C.A. 2894-67-9 T67 GMV JN IHJ CG KR BG	TABLE II.	MeOH 3.3	DI	PHOS	1	-	3.0	-	DME/SCE	25AO	0.00→-1.100 V,C=0.11,EC=2.285,m=2.75,t=0.5,ΔE=50,v=5, Pt Aux
									7.0				0.00→-1.750 V
DE74 CONT	$C_{15}H_{10}Cl_2N_2O_2$	7-chloro-5-(2-chlorophenyl)-2,3-dihydro-3-hydroxy-1H-1,4-benzodiazepin-2-one T67 GMV JN IHJ CG IQ KR BG		H_2O	PY	H_2SO_4		-	0.6 4.1	25	DME/SCE	OAO	C=0.4

TABLE I. Electrochemical Data $C_{15}H_{10}Cl_2N_2O_2$ (CONT.) DE74

| Ref. | C/M | Charact. Potential | | | Response Const. | | n | | Electrokinetic Data | | | Products and | Description and | Code |
		Value	vs.		Value	Tech.		Parameter	Value	From	Identification	Remarks	No.
RR020 0469	71d	$E_{1/2}$	-2.00	SCE	i_ℓ	5.56	-	2.21	-	-	-	-	C,Mc(name),p	DE67
JA094 7108	-	XE	0.18	SCE	-	-	sttd	1	ΔE_p	110	sttd	radical cation,ESR	A,R,XE=½($E_{p,A}$+$E_{p,C}$), r	DE68
			-					1		-			C	
JA094 4882	-	$E_{p/2}$	-1.00	SCE	-	-	-	-	-	-	-	-	C,R,p	DE69
JA094 4882	368 a	E_p	0.55F -0.50F -0.36F 0.66F	SCE	i_p	15F 19F 20F 22F	-	-	-	-	-	-	C,R,p C,R A,R A,R	DE70
JA094 4882	368 a	$E_{p/2}$	0.60 -0.45	SCE	-	-	-	-	-	-	-	-	A,R C,R	
JA095 4373	370 a	E_p	-0.13 -0.19 -0.30 -0.94 -1.47 -1.55	SCE	-	-	sttd	1 1 1 1 1 1	-	-	-	-	A,p C A C A C	DE71
JA095 4373	370 a	E_p	-0.07 -0.13 -1.10	SCE	-	-	sttd	1 1 -	-	-	-	-	A,p C C	
AA074 0367	-	E_{su}	-0.705	SCE	i_{su}	4.14	-	-	-	-	-	-	C,r	DE72 cj40
			-0.440			7.12	-	-	-	-	-	-	C,r	
AA074 0367	-	E_{su}	-0.675	SCE	i_{su}	8.31	-	-	-	-	-	-	C,r	DE73
			-0.950			3.73	-	-	-	-	-	-	C,r	
AA066 0427	219 f	$E_{1/2}$	0.60F	SCE	i_ℓ	3.65F	sttd	4	$dE_{1/2}/dpH$	40	sttd	-	C,i_d,r	DE74
			-0.87F		-	-	-	-		90		-	C,i_a,r	
			-1.12F							0			C,i_{cat},r	

CONT

Code No.	Empirical Formula	Name and WLN Code	Structural Formula	Solvent	Tech.	Medium	μ, M	pH	T, °C	Electrodes	App.	Experimental Parameters
DE74	$C_{15}H_{10}Cl_2N_2O_2$	7-chloro-5-(2-chlorophenyl)-2,3-dihydro-3-hydroxy-1H-1,4-benzodiazepin-2-one	TABLE II.	H_2O	PY	BR	—	8.0	25	DME/SCE	OAO	C=0.4
								11				C=0.2
DE75 bg83	$C_{15}H_{10}FeMoO_5$	tricarbonyl-η-cyclopentadienylmolybdenum(I) dicarbonyl-η-cyclopentadienyliron(II) C.A. 12130-13-1 L5ØJ Ø-MO-CO&CO&CO&-FE-CO&CO&- OL5ØJ	$(C_5H_5)Mo(CO)_3$-$Fe(CO)_2(C_5H_5)$	MG	PY	Bu_4NClO_4 0.1	—	—	22	DME Ag/Ag+	2-0	C=2
DE76	$C_{15}H_{10}I_2N_2O$	7,9-diiodo-1,3-dihydro-5-phenyl-2H-1,4-benzodiazepin-2-one C.A. 55098-62-9 T67 GMV JN IHJ CI EI KR	TABLE II.	MeOH 3.3	DI	PHOS 1	—	3.0	—	DME/SCE	25AO	0.00 → -1.100 V, C= 0.07, EC= 2.285, m=2.75, t=0.5, ΔE=50, v=5, Pt Aux
								7.0				0.00 → -1.750 V
DE77	$C_{15}H_{10}N_4O_6$	5,5-bis(3-nitrophenyl)-2,4-imidazolidinedione T5MVMXVJ DR CNW& DR CNW	TABLE II.	H_2O	DI	NaOH 0.1	—	—	—	DME/SCE	2AO	-0.400 to -0.900 V, C= 1.19E-2, EC= 1.561, m= 2.32, t=0.5, ΔE=50, v=5 Pt Aux
					IL	NaOH 0.1	—	—	—	HMDE SCE	2AO	-0.540 to -0.700 V, C= 1.19E-2, v= 100, Pt Aux
DE78 bg88	$C_{15}H_{10}O_2$	2-phenyl-1,3-indandione C.A. 83-12-5 L56 BV DVJ CR	TABLE II-2.	EtOH 5	PY	CITR 0.15	—	1.2	25± 0.1	DME Ag/AgCl, satd KCl	2AO	C=0.25, m= 2.10, t=3.43, h=59, W Aux
								3.0				
								4.5				
								5.5				
					IL	CITR ?	—	4.9	25± 0.1	HMDE Ag/AgCl, satd KCl	2AO	C=0.25, v= 100, W Aux
DE79	$C_{15}H_{11}BrN_2O$	7-bromo-1,3-dihydro-5-phenyl-2H-1,4-benzodiazepin-2-one C.A. 2894-61-3 T67 GMV JN IHJ CE KR	TABLE II.	MeOH 3.3	DI	PHOS 1	—	3.0	—	DME/SCE	25AO	0.00 → -1.100 V, C= 0.11, EC= 2.285, m=2.75, t=0.5, ΔE=50, v=5, Pt Aux
								7.0				0.00 → -1.750 V

TABLE I. Electrochemical Data $C_{15}H_{11}BrN_2O$ DE79

Ref.	C/M	Charact. Potential Value	vs.	Response Const.	Value	n Tech.	n	Electrokinetic Data Parameter	Value	From	Products and Identification	Description and Remarks	Code No.
AA066 0427	219 g	$E_{\frac{1}{2}}$	-1.05F	SCE	-	3.0F	-	-	$dE_{\frac{1}{2}}/dpH$	35	-	-	C, i_a, r	DE74
	219 gh		-1.16F -1.43F		-	-	-	-		35 90	-	-	C, i_a, r C, i_{cat}, r	
JA088 5117	-	$E_{\frac{1}{2}}$	-1.4	Ag/Ag$^+$	-	-	QE	1	-	-	-	Mo$^-$+?,no radical anion,ESR	C, p	DE75 bg83
AA074 0367	-	E_{su}	-0.775	SCE	i_{su}	3.39	-	-	-	-	-	-	$C, \Delta E_{su}=-0.172\sigma_m, r$	DE76
			-0.970			0.39	-	-	-	-	-	-	$C, \Delta E_{su}=-0.134\sigma_m, r$	
AA064 0165	-	E_{su}	-0.625	SCE	i_{su}	1.25F	-	-	-	-	-	-	$C, i_{su} \propto C, E_{su}=f(pH), r?$	DE77
AA064 0165	-	E_p	-0.610 -0.635	SCE	-	-	-	-	-	-	-	-	$C, r?$ C	
AA062 0405	358 a	$E_{\frac{1}{2}}$	-0.700	Ag/AgCl	i_ℓ	1.42	QE	1.9	Tomeš	70	sttd	-	$C, i_d, i_\ell \propto h^{\frac{1}{2}}, E_{\frac{1}{2}}=f(pH), r$	DE78 bg88
			-0.820			1.40		1.9		70	-	-	$C, i_d, i_\ell \propto h^{\frac{1}{2}}, E_{\frac{1}{2}}=f(pH), r$	
			-1.22			1.4		1.8		-			$C, i_d + i_a, E_{\frac{1}{2}}=f(pH)$	
			-0.925			1.42		1.9		60			$C, i_d, i_\ell \propto h^{\frac{1}{2}}, E_{\frac{1}{2}}=f(pH), r$	
			-1.255			1.34		1.8		60			$C, i_d + i_a, E_{\frac{1}{2}}=f(pH)$	
			-0.990			1.43		1.9		60			$C, i_d, i_\ell \propto h^{\frac{1}{2}}, E_{\frac{1}{2}}=f(pH), r$	
			-1.290			1.17		1.8		45			$C, i_d + i_a, E_{\frac{1}{2}}=f(pH), r$	
AA062 0405	358 a	E_p	-0.915	Ag/AgCl	$i_p/Cv^{\frac{1}{2}}$	1.16	QE	1.9	-	-	-	-	$C, i_a, i_p/Cv^{\frac{1}{2}} = f(v), r$	
			-0.985			0.24		1.8					C, i_d	
			-1.265			1.64		-					$C, i_p/Cv^{\frac{1}{2}} \neq f(v)$	
AA074 0367	-	E_{su}	-0.715	SCE	i_{su}	5.34	-	-	-	-	-	-	$C, \Delta E_{su}=-0.172\sigma_m, r$	DE79
			-0.975			5.10	-	-	-	-	-	-	$C, \Delta E_{su}=-0.134\sigma_m, r$	

175

Code No.	Empirical Formula	Name and WLN Code	Structural Formula	Solvent	Tech.	Medium	μ, M	pH	T, °C	Electrodes	App.	Experimental Parameters
DE80 bg95	$C_{15}H_{11}ClN_2O$	7-chloro-1,3-di-hydro-5-phenyl-2H-1,4-benzodiazepin-2-one C.A. 1088-11-5 T67 GMV JN IHJ CG KR	TABLE II-2.	MeOH 3.3	DI	PHOS 1	-	3.0	-	DME/SCE	25A0	0.00 → -1.100 V, C= 0.12, EC= 2.285, m=2.75, t=0.5, ΔE=50, v=5, Pt Aux
								7.0				0.00 → -1.750 V
DE81	$C_{15}H_{11}ClN_2O_2$	Oxazepam C.A. 604-75-1 T67 GMV JN IHJ CG IQ KR	TABLE II.	H_2O	PY	H_2SO_4	-	1	25	DME/SCE	OA0	C=0.4
								3.2				
								4.7				
								8.4				
								11				
DE82	$C_{15}H_{11}ClO_4S_2$	3,5-diphenyl-1,2-dithiolylium perchlorate C.A. 1270-66-2 T5SSTJ CR& ER & G-O4	TABLE II.	MeCN	VR	Bu_4NBF_4 0.2	-	-	-	PBE/SCE	26A2	v=300
									-31			v=15
												v=56
												v=240
												v=600
				CH_2Cl_2	QE	Bu_4NBF_4 0.2	-	-	-80	Pt gauze/SCE	26A2	ns
					VR	Bu_4NBF_4 0.2	-	-	-	PBE/SCE	26A2	v=150
									-70			C=1
DE83	$C_{15}H_{11}FN_2O$	1,3-dihydro-7-fluoro-5-phenyl-2H-1,4-benzodiazepin-2-one C.A. 2648-00-2 T67 GMV JN IHJ CF KR	TABLE II.	MeOH 3.3	DI	PHOS 1	-	3.0	-	DME/SCE	25A0	0.00 → -1.100 V, C= 0.11, EC= 2.285, m=2.75, t=0.5, ΔE=50, v=5, Pt Aux
								7.0				0.00 → -1.750 V
DE84 bh09 CONT	$C_{15}H_{11}N_3O_3$	Nitrazepam C.A. 146-22-5 T67 GMV JN IHJ CNW KR	TABLE II-2.	H_2O	PY	PHOS ? decylamine (trace)	-	6.7	25± 0.1	DME Ag/AgCl, satd KCl	2A0	C=0.1, m= 2.23, t=2.95, h=3.4, W Aux

TABLE I. Electrochemical Data $C_{15}H_{11}N_3O_3$ (CONT.) DE84

| Ref. | C/M | Charact. Potential | | Response Const. | | n | Electrokinetic Data | | | Products and | Description and | Code |
		Value	vs.		Value	Tech.	Parameter	Value	From	Identification	Remarks	No.
AA074 0367	-	E_{su}	-0.725	SCE	i_{su}	9.79	-	-	-	-	-	-	$C, \Delta E_{su}=-0.172\sigma_m, r$	DE80 bg95
			-0.970			6.79	-	-	-	-	-	$C, \Delta E_{su}=-0.134\sigma_m, r$	
AA066 0427	219 f	$E_{\frac{1}{2}}$	-0.65F	SCE	i_ℓ	3.7F	sttd	4	$dE_{\frac{1}{2}}/dpH$	50	sttd	-	C, i_d, r	DE81
			-0.83F			3.6F		4		90		-	C, i_d, r	
			-0.90F			3.3F		4		90		-	C, i_d, r	
			-1.16F			0.09F		4		0			C, i_{cat}	
	219 g		-1.08F			3.0F	-	-		40			C, i_a, r	
	219 gh		-1.19F -1.48F			0.16F 0.075F		4 2		40 60		-	C, r C	
JA095 4373	370 a	E_p	-0.20F	SCE	i_p	8F	QE	1.0	-	-	-	-	$A, Q, i_{p,A}/i_{p,C}=1$ for $v<3E4$	DE82
			-0.24F			20F		1.0				radical, deep green, VIS(400,650 nm), ESR, stable(VR)	C, Q	
			-0.37F -1.08F			22F 19F		1.0 1.0				anion, deep red, VIS (483 nm), stable(VR)	$A, \neq$ $C, \neq$	
			-1.63F -1.68F			22F 20F	sttd	1 1				dianion unstable, QE	A, R C, R	
			-		$i_p/Cv^{\frac{1}{2}}$	13F 11.5F	-	-		-		-	C, p C	
			-			12.2F 8F	-	-		-		-	C, p C	
			-			11.3F 5F	-	-		-		-	C, p C	
			-			10.3F 2.4F	-	-		-		-	C, p C	
JA095 4373	370 a	-	-	-	-	-	QE	1	-	-	-	at -80°C, ESR signal absent, appears on warming	C, p	
JA095 4373	370 a	E_p	-0.10 -0.17 -1.14	SCE	-	-	sttd	1 1 -	-	-	-	-	A, R, p C, R C	
			-0.10 -0.17 -1.14		-	-		1 1 -				dimer, ESR	$A, \neq, p$ $C, \neq$ C	
AA074 0367	-	E_{su}	-0.720	SCE	i_{su}	9.22	-	-	-	-	-	-	$C, \Delta E_{su}=-0.172\sigma_m, r$	DE83
			-0.985			11.12	-	-	-	-	-	-	$C, \Delta E_{su}=-0.134\sigma_m, r$	
AA059 0127	-	$E_{\frac{1}{2}}$	-0.38	Ag/AgCl	i_ℓ	1.20F	QE	3.98	αn_a p=	2.1 1.8	Tomeš $dE_{\frac{1}{2}}/dpH$	-	$C, i_d, r?$	DE84 bh09
			-1.02			0.90F		1.95		1.3 0.81		-	C	

CONT

177

DE84 (CONT.) $C_{15}H_{11}N_3O_3$

Code No.	Empirical Formula	Name and WLN Code	Structural Formula	Solvent	Tech.	Medium		μ, M	pH	T, °C	Electrodes	App.	Experimental Parameters
DE84	$C_{15}H_{11}N_3O_3$	Nitrazepam	TABLE II.	H_2O	VR	PHOS decylamine (trace)	?	-	6.7	25± 0.1	HMDE Ag/AgCl, satd KCl	2AO	C=0.1, v=100, W Aux
				MeOH 3.3	DI	PHOS	1	-	3.0	-	DME/SCE	25AO	0.00 → -1.100 V, C= 0.11, EC= 2.285, m=2.75, t=0.5, ΔE=50, v=5, Pt Aux
									7.0				0.00 → -1.750 V
DE85 cj61	$C_{15}H_{12}ClN_3O$	2-amino-7-chloro-5-phenyl-3H-1,4-benzodiazepine 4-N-oxide C.A. 7722-15-8 T67 GN JN IHJ CG HZ JO KR	TABLE II-3.	MeOH 3.3	DI	PHOS	1	-	3.0	-	DME/SCE	25AO	0.00 → -1.100 V, C= 0.12, EC= 2.285, m=2.75, t=0.5, ΔE=50, v=5, Pt Aux
									7.0				0.00 → -1.750 V
DE86	$C_{15}H_{12}MoO_3$	α-tolylcyclopentadienyltricarbonyl-molybdenum(II) C.A. 12194-07-9 L5ØJ Ø-MO-CO&CO&CO& 1R	$[(C_5H_5)Mo(CO)_3$- $CH_2C_6H_5]$	MG	PY	Bu_4NClO_4	0.1	-	-	22	DME Ag/Ag+	2-O	C=2
DE87 bh21	$C_{15}H_{12}N_2O$	1,3-dihydro-5-phenyl-2H-1,4-benzodiazepin-2-one C.A. 2898-08-0 T67 GMV JN IHJ KR	TABLE II-2.	MeOH 3.3	DI	PHOS	1	-	3.0	-	DME/SCE	25AO	0.00 → -1.100 V, C= 0.11, EC= 2.285, m=2.75, t=0.5, ΔE=50, v=5, Pt Aux
									7.0				0.00 → -1.750 V
DE88 bh22 cj65	$C_{15}H_{12}O$	chalcone C.A. 614-47-1 RV1U1R	trans-C_6H_5CH: CHCOC_6H_5	DMF	PY	Bu_4NBF_4	0.50	-	-	25	DME/SCE	23-O	C=1.9-3.2, Pt Aux
DE89	$C_{15}H_{12}O$	3,3-diphenylprop-2-enal C.A. 1210-39-5 VH1UYR&R	$(C_6H_5)_2$C:CHCHO	DMF	PY	Bu_4NBF_4	0.50	-	-	25	DME/SCE	23-O	C=2.0, Pt Aux
DE90	$C_{15}H_{13}ClN_2$	7-chloro-2,3-dihydro-5-phenyl-1H-1,4-benzodiazepine C.A. 1694-78-6 T67 GM JN JU&TJ CG KR	TABLE II.	MeOH 3.3	DI	PHOS	1	-	3.0	-	DME/SCE	25AO	0.00 → -1.100 V, C= 0.13, EC= 2.285, m=2.75, t=0.5, ΔE=50, v=5, Pt Aux
									7.0				0.00 → -1.750 V
DE91	$C_{15}H_{13}ClN_4O$	7-chloro-1,3-dihydro-2-hydrazino-5-phenyl-2H-1,4-benzodiazepine 4-N-oxide C.A. 18091-88-8 T67 GN JN IHJ CG HMZ JO KR	TABLE II.	MeOH 3.3	DI	PHOS	1	-	3.0	-	DME/SCE	25AO	0.00 → -1.100 V, C= 0.11, EC= 2.285, m=2.75, t=0.5, ΔE=50, v=5, Pt Aux
									7.0				0.00 → -1.750 V

TABLE I. Electrochemical Data — $C_{15}H_{13}ClN_4O$ — DE91

Ref.	C/M	Charact. Potential Value	vs.	Response Const. Value		n Tech.		Electrokinetic Data Parameter	Value	From	Products and Identification	Description and Remarks	Code No.
AA059 0127	-	E_p -0.45F -1.05F	Ag/AgCl	i_p	15F 5F	-	-	-	-	-	-	C,r?	DE84
AA074 0367	-	E_{su} -0.765	SCE	i_{su}	0.39	-	-	-	-	-	-	$C, \Delta E_{su} \neq -0.172\sigma_m, r$	
		-1.015			2.52	-	-	-	-	-	-	$C, \Delta E_{su} \neq -0.134\sigma_m, r$	
AA074 0367	-	E_{su} -0.380 -0.705 -1.080	SCE	i_{su}	5.99 8.62 5.51	-	-	-	-	-	-	C,r C C	DE85 cj61
		-0.725 -0.910 -1.300			1.12 1.68 0.7	-	-	-	-	-	-	C,r C C	
JA088 5117	-	$E_{\frac{1}{2}}$ -2.1	Ag/Ag⁺	-	-	QE	1	-	-	-	Mo⁻ + benzyl radical	C,p	DE86
AA074 0367	-	E_{su} -0.780	SCE	i_{su}	10.34	-	-	-	-	-	-	$C, \Delta E_{su} = -0.172\sigma_m, r$	DE87 bh21
		-1.030			11.81	-	-	-	-	-	-	$C, \Delta E_{su} = -0.134\sigma_m, r$	
JA094 8471	-	$E_{\frac{1}{2}}$ -1.41 -2.1	SCE	-	-	-	-	αn_a	1.5 -	Elog	-	C,p C	DE88 bh22 cj65
JA094 8471	-	$E_{\frac{1}{2}}$ -1.51 -1.93	SCE	-	-	-	-	αn_a	1.0 1.2	Elog	-	C,p C	DE89
AA074 0367	-	E_{su} -0.835	SCE	i_{su}	9.63	-	-	-	-	-	-	C,r	DE90
		-0.945			8.94	-	-	-	-	-	-	C,r	
AA074 0367	-	E_{su} -0.385 -0.690 -1.075	SCE	i_{su}	5.50 8.63 10.0	-	-	-	-	-	-	C,r C C	DE91
		-0.730 -0.925 -1.300			8.02 11.9 14.5	-	-	-	-	-	-	C,r C C	

Code No.	Empirical Formula	Name and WLN Code	Structural Formula	Solvent	Tech.	Medium		μ, M	pH	T, °C	Electrodes	App.	Experimental Parameters
DE92	$C_{15}H_{14}ClN_3$	1-amino-7-chloro-2,3-dihydro-5-phenyl-1H-1,4-benzodiazepine C.A. 55098-43-6 T67 GN JN JU&TJ CG GZ KR	TABLE II.	MeOH 3.3	DI	PHOS	1	-	3.0	-	DME/SCE	25A0	0.00→ -1.100 V,C= 0.12,EC= 2.285,m=2.75, t=0.5,ΔE=50, v=5,Pt Aux
									7.0				0.00→ -1.750 V
DE93	$C_{15}H_{14}N_4O_2$	2,3-dihydro-1-methyl-7-nitro-5-(4-pyridyl)-1H-1,4-benzodiazepine C.A. 55098-64-1 T67 GN JN JU&TJ CNW G1 K- DT6NJ	TABLE II.	MeOH 3.3	DI	PHOS	1	-	3.0	-	DME/SCE	25A0	0.00→ -1.100 V,C= 0.12,EC= 2.285,m=2.75, t=0.5,ΔE=50, v=5,Pt Aux
									7.0				0.00→ -1.750 V
DE94	$C_{15}H_{14}O_2$	1-phenylethyl benzoate C.A. 13358-49-1 1YR&OVR	$C_6H_5C(O)OCH-(CH_3)C_6H_5$	MeCN	QE	$LiClO_4$	0.1	-	-	-	Pt Ag/0.1 Ag^+	25A-	C=10,A=12.9, E_{app}=1.95, stainless steel Aux
DE95	$C_{15}H_{18}ClN_3O_4$	4-trimethylammonio-azobenzene perchlorate 1K1&1&R DNUNR & G-O4	$4-(CH_3)_3N^+$-$C_6H_4N:NC_6H_5$ ClO_4^-	MeCN	PY	Et_4NClO_4 0.1 H_2SO_4 0.0003 Et_4NClO_4 0.1 H_2SO_4 0.0015		-	-	25± 0.1	DME/SCE	125A0	C=1.06,m= 1.56,t=3.26, h=50,Pt Aux
DE96	$C_{15}H_{24}O$	2-octyl benzyl ether C.A. 38523-68-1 6Y1&O1R	$C_6H_5CH_2OC(CH_3)$-$HCH_2(CH_2)_4CH_3$	MeCN	QE	$LiClO_4$	0.1	-	-	-	Pt Ag/0.1 Ag^+	25A-	C=45,A=12.9, E_{app}=1.90, stainless steel Aux
DE97	$C_{15}H_{25}ClN_2O_2$	benzyloxycarbonyl-methyl(dimethyl)(2-dimethylaminoethyl)-ammonium chloride 1N1&2K1&1&1VO1R &G	$(CH_3)_2NCH_2CH_2\overset{+}{N}$-$(CH_3)_2CH_2COO$-$CH_2C_6H_5$ Cl^-	H_2O	PY	Et_4NI	0.5	-	5.0-6.0	20	DME/SCE	0-0	C=1,m=0.86, t=3.1 C=5 C=10
DE98	$C_{16}H_{10}$	diphenylbutadiyne C.A. 886-66-8 R1UU2UU1R	$C_6H_5C\equiv CC\equiv CC_6H_5$	MeCN	VY	$NaClO_4$		-	-	25	RPE/SCE	06-0	C=1,v=20
DE99 bi06	$C_{16}H_{10}$	fluoranthene C.A. 206-44-0 L C6566 1A PJ	TABLE II-2.	DMF	VR	Bu_4NClO_4 0.1 Bu_4NClO_4 0.1		-	-	-	PDE/SCE PDE Ag/AgCl, satd KCl	3F-	C=0.1-1, v=1.00 C=ns,v=200
DF00 bi09	$C_{16}H_{10}Cr_2HgO_6$	bis[(tricarbonyl)-(η-cyclopentadienyl)-chromium(I)]mercury-(0) C.A. 12194-11-5 L50J Ø-CR-CO&CO&CO& 2-HG-	$[(C_5H_5)Cr-(CO)_3]_2Hg$	MG	PY	Bu_4NClO_4	0.1	-	-	22	DME Ag/Ag^+	2-0	C=2

TABLE I. Electrochemical Data $C_{16}H_{10}Cr_2HgO_6$ DF00

Ref.	C/M	Charact. Potential Value	vs.	Response Const. Value		n Tech.		Electrokinetic Data Parameter	Value	From	Products and Identification	Description and Remarks	Code No.
AA074 0367	-	-	-	-	-	-	-	-	-	-	-	O,p	DE92
		-	-	-	-	-	-	-	-	-	-	O,p	
AA074 0367	-	E_{su} -0.215 -0.270 -1.030	SCE	i_{su}	13.56 8.74 -	-	-	-	-	-	-	C,r C C	DE93
		-0.510 -0.600 -1.190			15.16 13.12 -	-	-	-	-	-	-	C,r C C	
JA094 6812	360 c	-	-	-	-	QE	6.2	-	-	-	acetophenone,50%; benzoic acid,50%; NMR,GLC,IRS	A,p	DE94
AJ026 1669	198 f	$E_{\frac{1}{2}}$ 0.320 -0.818	SCE	-	-	-	-	-	-	-	-	C,r C	DE95
		0.365		-	-	-	-	-	-	-	-	C,r	
JA094 6812	360 b	-	-	-	-	QE	1.9	-	-	-	benzaldehyde,50%;2-octanol,60%;NMR, GLC,IRS	A,p	DE96
RG036 0622	-	$E_{\frac{1}{2}}$ -1.92	SCE	i_ℓ	1.05	IIk	1	Elog	122 F	plot	-	C,p	DE97
		-1.97			5.05		1		122 F		-	C,p	
		-2.11			9.71		1		122 F		-	C,p	
EA017 1595	-	$E_{\frac{1}{2}}$ 1.850	SCE	-	-	i:i	2	-	-	-	-	A,p	DE98
JA094 0691	-	E_p -1.76	SCE	-	-	-	-	-	-	-	-	C,R,r	DE99 bi06
JA094 4790		E_p -1.70	Ag/AgCl	-	-	-	-	-	-	-	-	C,p	
JA088 5117	-	$E_{\frac{1}{2}}$ -1.3	Ag/Ag$^+$	-	-	QE	2	-	-	-	2Cr$^-$ + Hg,sttd	C, Mc(formula),p	DF00 bi09

Code No.	Empirical Formula	Name and WLN Code	Structural Formula	Solvent	Tech.	Medium		μ, M	pH	T, °C	Electrodes	App.	Experimental Parameters
DF01	$C_{16}H_{10}Cr_2O_6$	bis[(tricarbonyl)(η-cyclopentadienyl)-chromium(I)] C.A. 12194-12-6 L5ØJ Ø-CR-CO&CO&CO& -CR-CO&CO&CO&- ØL5ØJ	$[(C_5H_5)Cr(CO)_3]_2$	MG	PY	Bu_4NClO_4	0.1	–	–	22	DME Ag/Ag$^+$	2-O	C=2
DF02 bi12	$C_{16}H_{10}HgMo_2O_6$	bis[(tricarbonyl)(η-cyclopentadienyl)-molybdo(I)]mercury-(O) C.A. 12194-13-7 L5ØJ Ø-MO-CO&CO&CO& 2-HG-	$[\pi-(C_5H_5)Mo-(CO)_3]_2Hg$	MG	PY	Bu_4NClO_4	0.1	–	–	22	DME Ag/Ag$^+$	2-O	C=2
DF03 bi13	$C_{16}H_{10}Mo_2O_6$	bis[(tricarbonyl)(η-cyclopentadienyl)-molybdo(I)] C.A. 12091-64-4 L5ØJ Ø-MO-CO&CO&CO& 2	$[\pi-(C_5H_5)Mo-(CO)_3]_2$	MG	PY	Bu_4NClO_4	0.1	–	–	22	DME Ag/Ag$^+$	2-O	C=2
DF04 bi16	$C_{16}H_{11}F_3N_2O$	1,3-dihydro-5-phenyl-7-trifluoromethyl-2H-1,4-benzodiazepin-2-one C.A. 2285-16-7 T67 GMV JN IHJ CXFFF KR	TABLE II-2.	MeOH 3.3	DI	PHOS	1	–	3.0	–	DME/SCE	25AO	0.00 → -1.100 V,C= 0.11,EC= 2.285,m=2.75, t=0.5,ΔE=50, v=5,Pt Aux
									7.0				0.00 → -1.750 V
DF05	$C_{16}H_{12}$	dibenzo[a,e]cyclooctatetraene C.A. 262-89-5 L D686J	TABLE II.	THF	PY	Bu_4NClO_4	0.1	–	–	–	DME Ag/0.1 Ag$^+$	O1AF O	C=0.615, m=0.63,t= 6-12,h=40.0
						Bu_4NClO_4 H_2O	0.2 1.0						C=0.60
DF06	$C_{16}H_{13}ClN_2O$	7-chloro-2,3-dihydro-1-methanoyl-5-phenyl-1H-1,4-benzodiazepine C.A. 1694-79-7 T67 GN JN JU&TJ CG GVH KR	TABLE II.	MeOH 3.3	DI	PHOS	1	–	3.0	–	DME/SCE	25AO	0.00 → -1.100 V,C= 0.12,EC= 2.285,m=2.75, t=0.5,ΔE=50, v=5,Pt Aux
									7.0				0.00 → -1.750 V
DF07	$C_{16}H_{13}ClN_2O$	7-chloro-1,3-dihydro-5-(2-methylphenyl)-2H-1,4-benzodiazepin-2-one C.A. 5358-35-0 T67 GMV JN IHJ CG KR B1	TABLE II.	MeOH 3.3	DI	PHOS	1	–	3.0	–	DME/SCE	25AO	0.00 → -1.100 V,C= 0.11,EC= 2.285,m=2.75, t=0.5,ΔE=50, v=5,Pt Aux
									7.0				0.00 → -1.750 V
DF08	$C_{16}H_{13}ClN_2O$	7-chloro-2-methyl-5-phenyl-3H-1,4-benzodiazepine 4-N-oxide C.A. 7713-21-5 T67 GN JN IHJ CG H1 JO KR	TABLE II.	MeOH 3.3	DI	PHOS	1	–	3.0	–	DME/SCE	25AO	0.00 → -1.100 V,C= 0.11,EC= 2.285,m=2.75, t=0.5,ΔE=50, v=5,Pt Aux
									7.0				0.00 → -1.750 V

TABLE I. Electrochemical Data $C_{16}H_{13}ClN_2O$ DF08

Ref.	C/M	Charact. Potential Value	vs.	Response Const.	Value	Tech.	n	Electrokinetic Data Parameter	Value	From	Products and Identification	Description and Remarks	Code No.
JA088 5117	-	$E_{\frac{1}{2}}$	-1.3	Ag/Ag^+	-	-	QE	2	-	-	-	$2Cr^-$, sttd	C,p	DF01
JA088 5117	-	$E_{\frac{1}{2}}$	-1.3	Ag/Ag^+	-	-	QE	2	-	-	-	$2Mo^-$ + Hg, sttd	C,p	DF02
JA088 5117	-	$E_{\frac{1}{2}}$	-1.4	Ag/Ag^+	-	-	QE	2	-	-	-	$2Mo^-$, sttd	C,p	DF03 bi13
AA074 0367	-	E_{su}	-0.680	SCE	i_{su}	7.37	-	-	-	-	-	-	C, $\Delta E_{su}=-0.172\sigma_m$, r	DF04 bi16
			-0.950			1.47	-	-	-	-	-	-	C, $\Delta E_{su}=-0.134\sigma_m$, r	
JA094 4915	366 b	$E_{\frac{1}{2}}$	-2.29	SCE	i_ℓ I D	1.99 3.70 16	I:I	1	Elog	56	plot	radical anion	C,R,r	DF05
	366 c		-2.24 -2.56		i_ℓ	2.28 1.97	-	-	-	-	-	-	C,i_{cat},r C,i_{cat}	
AA074 0367	-	E_{su}	-0.61 -0.835	SCE	i_{su}	2.48 0.93	-	-	-	-	-	-	C, $\Delta E_{su}=-0.417\sigma_m$, r C	DF06
			-0.890			6.56	-	-	-	-	-	-	C, $\Delta E_{su}=-0.133\sigma_m$, r	
AA074 0367	-	E_{su}	-0.790	SCE	i_{su}	8.54	-	-	-	-	-	-	C,r	DF07
			-1.030			9.56	-	-	-	-	-	-	C,r	
AA074 0367	-	E_{su}	-0.480 -0.615 -0.855	SCE	i_{su}	5.54 2.32 1.27	-	-	-	-	-	-	C,r C C	DF08
			-0.750 -0.870 -1.000			2.99 6.79 2.24	-	-	-	-	-	-	C,r C C	

DF09 $C_{16}H_{13}ClN_2O$

Code No.	Empirical Formula	Name and WLN Code	Structural Formula	Solvent	Tech.	Medium		μ, M	pH	T, °C	Electrodes	App.	Experimental Parameters
DF09	$C_{16}H_{13}ClN_2O$	Diazepam C.A. 439-14-5 T67 GNV JN IHJ CG G1 KR	TABLE II.	H_2O	PY	H_2SO_4	0.1	-	1	25±0.1	DME Ag/AgCl, satd KCl	OAO	C=0.35
					DI	H_2SO_4	0.1	-	-	25±0.1	DME Ag/AgCl, satd KCl	OAO	C=4E-4, t=1, v=5, ΔE=100
DF10	$C_{16}H_{13}ClN_2O_2$	7-chloro-1,3-dihydro-5-(2-methoxyphenyl)-2H-1,4-benzodiazepin-2-one C.A. 3023-44-7 T67 GMV JN IHJ CG KR B01	TABLE II.	MeOH 3.3	DI	PHOS	1	-	3.0	-	DME/SCE	25AO	0.00 → -1.100 V, C=0.11, EC=2.285, m=2.75, t=0.5, ΔE=50, v=5, Pt Aux
									7.0				0.00 → -1.750 V
DF11	$C_{16}H_{13}ClN_2O_2$	7-chloro-2-methoxy-5-phenyl-3H-1,4-benzodiazepine 4-N-oxide C.A. 3897-18-5 T67 GN JN IHJ CG H01 JO KR	TABLE II.	MeOH 3.3	DI	PHOS	1	-	3.0	-	DME/SCE	25AO	0.00 → -1.100 V, C=0.11, EC=2.285, m=2.75, t=0.5, ΔE=50, v=5, Pt Aux
									7.0				0.00 → -1.750 V
DF12	$C_{16}H_{13}ClN_2O_4$	Chlorazepam T67 GMX JN IHJ CG HQ HQ IVQ KR	TABLE II.	H_2O	PT	BR		-	2.0	-	DME/SCE	O4AO	C=0.1, m=1.73, t=1, h=80, v=10
									4.0				
									6.0				
									9.0				
									12.2				
DF13	$C_{16}H_{13}ClO_4S_2$	3-(4-methylphenyl)-5-phenyl-1,2-dithiol-1-ium perchlorate C.A. 42877-42-9 T5SSTJ CR& ER D1 & G-O4	TABLE II.	MeCN	VR	Bu_4NBF_4	0.1	-	-	-	PBE/SCE	26A2	v=150
				CH_2Cl_2	VR	Bu_4NBF_4	0.2	-	-	-	PBE/SCE	26A2	v=150
DF14	$C_{16}H_{13}ClO_5S_2$	3-(4-methoxyphenyl)-5-phenyl-1,2-dithiol-1-ium perchlorate C.A. 29776-28-1 T5SSTJ CR& ER DO1 & G-O4	TABLE II.	MeCN	VR	Bu_4NBF_4	0.1	-	-	-	PBE/SCE	26A2	v=150
				CH_2Cl_2	VR	Bu_4NBF_4	0.2	-	-	-	PBE/SCE	26A2	v=150

TABLE I. Electrochemical Data $C_{16}H_{13}ClO_5S_2$ DF14

Ref.	C/M	Charact. Potential		Response Const.		n Tech.	n	Electrokinetic Data			Products and Identification	Description and Remarks	Code No.
		Value	vs.		Value			Parameter	Value	From			
AA060 0472	-	$E_{\frac{1}{2}}$	-0.59F	Ag/AgCl	i_ℓ/C	34.2F	sttd	2	-	-	-	-	C,r	DF09
AA064 0473	-	E_{su}	-0.61	Ag/AgCl	i_{su}	0.080	sttd	2	-	-	-	-	$C, i_{su} \propto C$	
AA074 0367	-	E_{su}	-0.670	SCE	i_{su}	7.79	-	-	-	-	-	-	C,r	DF10
			-0.990			6.85	-	-	-	-	-	-	C,r	
AA074 0367	-	E_{su}	-0.410 -0.720	SCE	i_{su}	6.02 8.31	-	-	-	-	-	-	C,r C	DF11
			-0.730 -0.925 -1.300			8.02 11.9 14.5	-	-	-	-	-	-	C,r C C	
AA076 0289	219 b	$E_{\frac{1}{2}}$	-0.67F	SCE	i_ℓ	3.0F	-	-	$dE_{\frac{1}{2}}/dpH$ αn_a $p=$	27 1.7 0.85	pH= 0-3 Elog	-	C, i_d, r	DF12
	219 c		-0.75F			3.0F	-	-	$dE_{\frac{1}{2}}/dpH$ αn_a $p=$	75 1.5 1.7	pH= 3-5.5 Elog	-	C, i_d, r	
	219 d		-0.80F			3.0F	-	-	$dE_{\frac{1}{2}}/dpH$ αn_a $p=$	90 1.9 2.5	pH= 5.5-9 Elog	-	C, i_d, r	
			-1.05F			3.0F	-	-	$dE_{\frac{1}{2}}/dpH$ αn_a $p=$	90 1.9 2.5	pH= 3.5-9 Elog	-	C, i_d, r	
	219 e		-1.20F			3.0F	-	-	$dE_{\frac{1}{2}}/dpH$ αn_a $p=$	47 1.07 0.80	pH=9- 12.2 Elog	-	C, i_d, r	
JA095 4373	370 a	E_p	-0.22 -0.30 -0.40 -1.19 -1.63 -1.73	SCE	-	-	sttd	1 1 1 1 1 1	-	-	-	-	A,p C A C A C	DF13
JA095 4373	370 a	E_p	-0.13 -0.20 -1.25	SCE	-	-	sttd	1 1 -	-	-	-	-	A,p C C	
JA095 4373	370 a	E_p	-0.26 -0.32 -0.42 -1.15 -1.64 -1.72	SCE	-	-	sttd	1 1 1 1 1 1	-	-	-	-	A,p C A C A C	DF14
JA095 4373	370 a	E_p	-0.17 -0.24 -0.96	SCE	-	-	sttd	1 1 -	-	-	-	-	A,p C C	

Code No.	Empirical Formula	Name and WLN Code	Structural Formula	Solvent	Tech.	Medium		μ, M	pH	T, °C	Electrodes	App.	Experimental Parameters
DF15	$C_{16}H_{13}Cl_3N_2$	7-chloro-5-(2,3-dichlorophenyl)-2,3-dihydro-1-methyl-1H-1,4-benzodiazepine C.A. 55098-48-1 T67 GN JN JU&TJ CG G1 KR BG CG	TABLE II.	MeOH 3.3	DI	PHOS	1	-	3.0	-	DME/SCE	25A0	0.00 → -1.100 V,C=0.1, EC=2.285,m= 2.75,t=0.5, ΔE=50,v=5, Pt Aux
									7.0				0.00 → -1.750 V
DF16	$C_{16}H_{13}Cl_3N_2$	7-chloro-5-(2,6-dichlorophenyl)-2,3-dihydro-1-methyl-1H-1,4-benzodiazepine C.A. 51102-49-9 T67 GN JN JUTJ CG G1 KR BG FG	TABLE II.	MeOH 3.3	DI	PHOS	1	-	3.0	-	DME/SCE	25A0	0.00 → -1.100 V,C= 0.1,EC= 2.285,m=2.75, t=0.5,ΔE=50, v=5,Pt Aux
									7.0				0.00 → -1.750 V
DF17	$C_{16}H_{13}FIN_3O$	5-(2-fluorophenyl)-7-iodo-2-methyl-amino-3H-1,4-benzodiazepine 4-N-oxide C.A. 55098-59-4 T67 GN JN IHJ CI HM1 JO KR BF	TABLE II.	MeOH 3.3	DI	PHOS	1	-	3.0	-	DME/SCE	25A0	0.00 → -1.100 V,C= 0.08,EC= 2.285,m=2.75, t=0.5,ΔE=50, v=5,Pt Aux
									7.0				0.00 → -1.750 V
DF18	$C_{16}H_{13}NO$	6-methoxydibenz-[b,f]azocine C.A. 37908-48-8 T D686 BNJ CO1	TABLE II.	THF	PY	Bu_4NClO_4	0.2	-	-	23±1	DME Ag/0.1 Ag^+	0AF-	t=6-12
					PA	Bu_4NClO_4	0.2	-	-	-	DME Ag/0.1 Ag^+	0AF-	t=6-12, v=2E3
DF19 bi81	$C_{16}H_{14}$	1,4-diphenyl-1,3-butadiene C.A. 886-65-7 R1U2U1R	$C_6H_5CH:CHCH:CH$-C_6H_5	MeCN	VY	$NaClO_4$	?	-	-	25	RPE/SCE	06-0	C=1,v=20
											RCCE SCE		A=0.8
DF20	$C_{16}H_{14}BrN_3O$	7-bromo-2-methyl-amino-5-phenyl-3H-1,4-benzodiazepine 4-N-oxide C.A. 26868-58-6 T67 GN JN IHJ CE HM1 JO KR	TABLE II.	MeOH 3.3	DI	PHOS	1	-	3.0	-	DME/SCE	25A0	0.00 → -1.100 V,C= 0.10,EC= 2.285,m=2.75, t=0.5,ΔE=50, v=5,Pt Aux
										7.0			0.00 → -1.750 V
DF21	$C_{16}H_{14}ClIN_2$	5-(2-chlorophenyl)-2,3-dihydro-7-iodo-1-methyl-1H-1,4-benzodiazepine C.A. 55153-72-5 T67 GN JN JU&TJ CI G1 KR BG	TABLE II.	MeOH 3.3	DI	PHOS	1	-	3.0	-	DME/SCE	25A0	0.00 → -1.100 V,C= 0.08,EC= 2.285,m=2.75, t=0.5,ΔE=50, v=5,Pt Aux
DF22 bi86 ck18	$C_{16}H_{14}ClN_3O$	7-chloro-2-methyl-amino-5-phenyl-3H-1,4-benzodiazepine 4-N-oxide C.A. 58-25-3 T67 GN JN IHJ CG HM1 JO KR	TABLE II-2.	MeOH 3.3	DI	PHOS	1	-	3.0	-	DME/SCE	25A0	0.00 → -1.100 V,C= 0.11,EC= 2.285,m=2.75, t=0.5,ΔE=50, v=5,Pt Aux
										7.0			0.00 → -1.750 V

TABLE I. Electrochemical Data $C_{16}H_{14}ClN_3O$ DF22

Ref.	C/M	Charact. Potential Value	vs.	Response Const. Value		n Tech.	n	Electrokinetic Data Parameter	Value	From	Products and Identification	Description and Remarks	Code No.
AA074 0367	-	E_{su}	-0.730	SCE	i_{su}	7.63	-	-	-	-	-	-	-	$C, \Delta E_{su}=-0.132\sigma_m, r$	DF15
			-0.960		-	0.23	-	-	-	-	-	-	$C, \Delta E_{su}=-0.027\sigma_m, r$	
AA074 0367	-	E_{su}	-0.750	SCE	i_{su}	8.18	-	-	-	-	-	-	$C, \Delta E_{su}=-0.132\sigma_m, r$	DF16
			-		-	-	-	-	-	-	-	-	O,r	
AA074 0367	-	E_{su}	-0.400 -0.670 -1.170	SCE	i_{su}	4.67 5.69 11.85	-	-	-	-	-	-	C,r $C, \Delta E_{su}=-0.118\sigma_m$ C	DF17
			-0.740 -0.860 -1.280			1.25 2.65 0.78	-	-	-	-	-	-	C,r $C, \Delta E_{su}=-0.171\sigma_m$ C	
JA094 4907	-	$E_{\frac{1}{2}}$	-2.08 -2.32	SCE	I	5.20 -	1:1	2 <1	Elog	57	plot	-	C,≠,ece,r C,Po	DF18
JA094 4907	-	$E_{p/2}$	-2.42 -2.3 -2.3 -1.19	SCE	i_p	20F 6F 16F	-	-	-	-	-	-	C,r C,Po? A,Po? A	
EA017 1595	-	$E_{\frac{1}{2}}$	1.140	SCE	-	-	i:i	1	-	-	-	-	A,p	DF19 bi81
			1.140		-	-		1	-	-	-	-	A,p	
AA074 0367	-	E_{su}	-0.390 -0.705 -1.155	SCE	i_{su}	5.96 8.55 9.17	-	-	-	-	-	-	C,r $C, \Delta E_{su}=-0.118\sigma_m$ C	DF20
			-0.715 -0.915 -1.350			1.49 2.67 2.20	-	-	-	-	-	-	C,r $C, \Delta E_{su}=-0.171\sigma_m$ C	
AA074 0367	-	E_{su}	-0.765	SCE	i_{su}	5.64	-	-	-	-	-	-	$C, \Delta E_{su}=-0.056\sigma_m, r$	DF21
AA074 0367	-	E_{su}	-0.375 -0.705 -1.165	SCE	i_{su}	6.54 10.40 8.82	-	-	-	-	-	-	C,r $C, \Delta E_{su}=-0.118\sigma_m$ C	DF22 bi86 ck18
			-0.760 -0.940 -1.370			3.31 7.01 4.72	-	-	-	-	-	-	C,r $C, \Delta E_{su}=-0.171\sigma_m$ C	

DF23 $C_{16}H_{14}ClN_3O_2$

Code No.	Empirical Formula	Name and WLN Code	Structural Formula	Solvent	Tech.	Medium	μ, M	pH	T, $^\circ$C	Electrodes	App.	Experimental Parameters
DF23	$C_{16}H_{14}ClN_3O_2$	7-chloro-2,3-dihydro-1-methyl-5-(2-nitrophenyl)-1H-1,4-benzodiazepine C.A. 55098-49-2 T67 GN JN JU&TJ CG G1 KR BNW	TABLE II.	MeOH 3.3	DI	PHOS 1	–	3.0	–	DME/SCE	25A0	0.00→ -1.100 V,C= 0.1,EC=2.285, m=2.75,t=0.5, ΔE=50,v=5, Pt Aux
								7.0				0.00→ -1.750 V
DF24	$C_{16}H_{14}ClN_3O_2$	7-chloro-5-(4-hydroxyphenyl)-2-methylamino-3H-1,4-benzodiazepine 4-N-oxide T67 GN JN IHJ CG HM1 JO KR DQ	TABLE II.	MeOH 3.3	DI	PHOS 1	–	3.0	–	DME/SCE	25A0	0.00→ -1.100 V,C= 0.11,EC= 2.285,m=2.75, t=0.5,ΔE=50, v=5,Pt Aux
								7.0				0.00→ -1.750 V
DF25	$C_{16}H_{14}N_2O$	3,4-dihydro-2-methyl-4-oxo-3-<u>o</u>-tolyl-quinazoline C.A. 72-44-6 T66 BVN ENJ CR B1& D1	TABLE II.	H_2O	DI	Et_4NClO_4 0.01	–	–	–	DME/SCE	04A0	C=0.05,m= 1.73,t=0.5, v=10,ΔE=100
DF26	$C_{16}H_{14}N_2O$	1,3-dihydro-7-methyl-5-phenyl-2H-1,4-benzodiazepin-2-one C.A. 5571-63-1 T67 GMV JN IHJ C1 KR	TABLE II.	MeOH 3.3	DI	PHOS 1	–	3.0	–	DME/SCE	25A0	0.00→ -1.100 V,C= 0.13,EC= 2.285,m=2.75, t=0.5,ΔE=50, v=5,Pt Aux
								7.0				0.00→ -1.750 V
DF27	$C_{16}H_{14}N_2OS$	1,3-dihydro-7-methylthio-5-phenyl-2H-1,4-benzodiazepin-2-one C.A. 2891-12-5 T67 GMV JN IHJ CS1 KR	TABLE II.	MeOH 3.3	DI	PHOS 1	–	3.0	–	DME/SCE	25A0	0.00→ -1.100 V,C= 0.11,EC= 2.285,m=2.75, t=0.5,ΔE=50, v=5,Pt Aux
								7.0				0.00→ -1.750 V
DF28	$C_{16}H_{14}N_2O_2$	1,3-dihydro-7-methoxy-5-phenyl-2H-1,4-benzodiazepin-2-one C.A. 5358-96-3 T67 GMV JN IHJ CO1 KR	TABLE II.	MeOH 3.3	DI	PHOS 1	–	3.0	–	DME/SCE	25A0	0.00→ -1.100 V,C= 0.11,EC= 2.285,m=2.75, t=0.5,ΔE=50, v=5,Pt Aux
								7.0				0.00→ -1.750 V
DF29	$C_{16}H_{15}ClN_2$	7-chloro-2,3-dihydro-1-methyl-5-phenyl-1H-1,4-benzodiazepine C.A. 2898-12-6 T67 GN JN JU&TJ CG G1 KR	TABLE II.	MeOH 3.3	DI	PHOS 1	–	3.0	–	DME/SCE	25A0	0.00→ -1.100 V,C= 0.12,EC= 2.285,m=2.75, t=0.5,ΔE=50, v=5,Pt Aux
								7.0				0.00→ -1.750 V

TABLE I. Electrochemical Data $C_{16}H_{15}ClN_2$ DF29

Ref.	C/M	Charact. Potential		Response Const.		n Tech.		Electrokinetic Data			Products and Identification	Description and Remarks	Code No.
		Value	vs.		Value			Parameter	Value	From			
AA074 0367	-	E_{su}	-0.140 -0.780	SCE	i_{su}	1.14 -	-	-	-	-	-	-	C,r C,X,$\Delta E_{su} \neq -0.132\sigma_m$	DF23
			-0.375 -0.975			2.36 0.87	-	-	-	-	-	-	C,r C,$\Delta E_{su} \neq -0.027\sigma_m$	
AA074 0367	-	E_{su}	-0.375 -0.750 -1.165	SCE	i_{su}	5.77 8.35 7.72	-	-	-	-	-	-	C,r C,$\Delta E_{su}=-0.118\sigma_m$ C	DF24
			-0.740 -0.970 -1.345			4.37 8.43 6.40	-	-	-	-	-	-	C,r C,$\Delta E_{su}=-0.171\sigma_m$ C	
AA080 0233	-	E_{su}	-1.36	SCE	-	-	-	-	-	-	-	-	C,p	DF25
AA074 0367	-	E_{su}	-0.790	SCE	i_{su}	8.54	-	-	-	-	-	-	C,$\Delta E_{su}=-0.172\sigma_m$,r	DF26
			-1.030			9.56	-	-	-	-	-	-	C,$\Delta E_{su}=-0.134\sigma_m$,r	
AA074 0367	-	E_{su}	-0.740	SCE	i_{su}	7.66	-	-	-	-	-	-	C,$\Delta E_{su}=-0.172\sigma_m$,r	DF27
			-0.980			0.38	-	-	-	-	-	-	C,$\Delta E_{su}=-0.134\sigma_m$,r	
AA074 0367	-	E_{su}	-0.750	SCE	i_{su}	8.80	-	-	-	-	-	-	C,$\Delta E_{su}=-0.172\sigma_m$,r	DF28
			-1.005			6.18	-	-	-	-	-	-	C,$\Delta E_{su}=-0.134\sigma_m$,r	
AA074 0367	-	E_{su}	-0.790 -0.835	SCE	i_{su}	7.50 8.06	-	-	-	-	-	-	C,$\Delta E_{su}=-0.132\sigma_m$,r C	DF29
			-0.975			4.66	-	-	-	-	-	-	C,$\Delta E_{su}=-0.027\sigma_m$,r	

Code No.	Empirical Formula	Name and WLN Code	Structural Formula	Solvent	Tech.	Medium		μ, M	pH	T, °C	Electrodes	App.	Experimental Parameters
DF30	$C_{16}H_{15}ClN_2O$	7-chloro-2,3-dihydro-2-hydroxy-1-methyl-5-phenyl-1H-1,4-benzodiazepine C.A. 28739-21-1 T67 GN JN JU&TJ CG G1 HQ KR	TABLE II.	MeOH 3.3	DI	PHOS	1	-	3.0	-	DME/SCE	25A0	0.00→ -1.100 V, C= 0.12, EC= 2.285, m=2.75, t=0.5, ΔE=50, v=5, Pt Aux
									7.0				0.00→ -1.750 V
DF31	$C_{16}H_{15}ClN_2O_2S$	7-chloro-2,3-dihydro-1-methylsulfonyl-5-phenyl-1H-1,4-benzodiazepine C.A. 16290-29-2 T67 GN JN JU&TJ CG GSW1 KR	TABLE II.	MeOH 3.3	DI	PHOS	1	-	3.0	-	DME/SCE	25A0	0.00→ -1.100 V, C= 0.11, EC= 2.285, m=2.75, t=0.5, ΔE=50, v=5, Pt Aux
									7.0				0.00→ -1.750 V
DF32	$C_{16}H_{15}NO$	6-methoxy-11,12-dihydrodibenz[b,f]azocine C.A. 37908-52-4 T D686 BN BU&T&J CO1	TABLE II.	THF	PY	Bu_4NClO_4	0.2	-	-	-	DME Ag/0.1 Ag$^+$	02F-	t=6-12
DF33	$C_{16}H_{15}N_3O$	2-methylamino-5-phenyl-3H-1,4-benzodiazepine 4-N-oxide C.A. 55098-57-2 T67 GN JN IHJ HM1 JO KR	TABLE II.	MeOH 3.3	DI	PHOS	1	-	3.0	-	DME/SCE	25A0	0.00→ -1.100 V, C= 0.13, EC= 2.285, m=2.75, t=0.5, ΔE=50, v=5, Pt Aux
									7.0				0.00→ -1.750 V
DF34	$C_{16}H_{15}N_3O_2$	2,3-dihydro-1-methyl-7-nitro-5-phenyl-1H-1,4-benzodiazepine C.A. 2898-19-3 T67 GN JN JU&TJ CNW G1 KR	TABLE II.	MeOH 3.3	DI	PHOS	1	-	3.0	-	DME/SCE	25A0	0.00→ -1.100 V, C= 0.12, EC= 2.285, m=2.75, t=0.5, ΔE=50, v=5, Pt Aux
									7.0				0.00→ -1.750 V
DF35	$C_{16}H_{16}ClN_3$	5-(2-aminophenyl)-7-chloro-2,3-dihydro-1-methyl-1H-1,4-benzodiazepine T67 GN JN JU&TJ CG G1 KR BZ	TABLE II.	MeOH 3.3	DI	PHOS	1	-	3.0	-	DME/SCE	25A0	0.00→ -1.100 V, C= 0.12, EC= 2.285, m=2.75, t=0.5, ΔE=50, v=5, Pt Aux
									7.0				0.00→ -1.750 V
DF36	$C_{16}H_{16}Fe_2O_4Sn$	bis[dicarbonyl-η-cyclopentadienyliron-(I)]dimethyltin(II) C.A. 12091-98-4 L50J Ø-FE-CO&CO&-SN- -1&1&-FE-CO&CO&- ØL 50J	$[(C_5H_5)Fe-(CO)_2]_2Sn(CH_3)_2$	MG	PY	Bu_4NClO_4	0.1	-	-	22	DME Ag/Ag$^+$	2-0	C=2
DF37	$C_{16}H_{16}N_2$	2,3-dihydro-1-methyl-5-phenyl-1H-1,4-benzodiazepine C.A. 2898-21-7 T67 GN JN JU&TJ G1 KR	TABLE II.	MeOH 3.3	DI	PHOS	1	-	3.0	-	DME/SCE	25A0	0.00→ -1.100 V, C= 0.14, EC= 2.285, m=2.75, t=0.5, ΔE=50, v=5, Pt Aux
									7.0				0.00→ -1.750 V

TABLE I. Electrochemical Data $C_{16}H_{16}N_2$ DF37

Ref.	C/M	Charact. Potential		Response Const.		n Tech.	n	Electrokinetic Data			Products and Identification	Description and Remarks	Code No.
		E_{su} Value	vs.		Value			Parameter	Value	From			
AA074 0367	-	E_{su}	-0.795	SCE	i_{su}	8.91	-	-	-	-	-	-	C,r	DF30
			-0.980			0.47	-	-	-	-	-	-	C,r	
AA074 0367	-	E_{su}	-0.640	SCE	i_{su}	1.18	-	-	-	-	-	-	C,$\Delta E_{su} \neq -0.417\sigma_m$,r	DF31
			-0.820			6.86	-	-	-	-	-	-	C	
			-0.920			0.63	-	-	-	-	-	-	C,$\Delta E_{su} \neq -0.133\sigma_m$,r	
JA094 4907	-	$E_{\frac{1}{2}}$	-2.87	-	I	3.0	l:l	1	Elog	50	plot	-	C,R,r	DF32
AA074 0367	-	E_{su}	-0.390	SCE	i_{su}	6.02	-	-	-	-	-	-	C,r	DF33
			-0.745			9.98	-	-	-	-	-	-	C,$\Delta E_{su}=-0.118\sigma_m$	
			-1.275			7.47	-	-	-	-	-	-	C	
			-0.770			4.42	-	-	-	-	-	-	C,r	
			-0.985			9.60	-	-	-	-	-	-	C,$\Delta E_{su}=-0.171\sigma_m$	
			-1.485			4.42	-	-	-	-	-	-	C	
AA074 0367	-	E_{su}	-0.240	SCE	i_{su}	13.39	-	-	-	-	-	-	C,r	DF34
			-0.770			6.89	-	-	-	-	-	-	C,$\Delta E_{su}=-0.056\sigma_m$	
			-0.525			0.71	-	-	-	-	-	-	C,r	
			-1.010			0.71	-	-	-	-	-	-	C	
AA074 0367	-	E_{su}	-0.770	SCE	i_{su}	7.63	-	-	-	-	-	-	C,r	DF35
			-0.965			1.10	-	-	-	-	-	-	C,r	
JA088 5117	-	$E_{\frac{1}{2}}$	-2.7	Ag/Ag$^+$	-	-	QE	2	-	-	-	2Fe$^-$ + dimethyltin, sttd	C,p	DF36
AA074 0367	-	E_{su}	-0.810	SCE	i_{su}	9.94	-	-	-	-	-	-	C,$\Delta E_{su}=-0.056\sigma_m$,r	DF37
			-1.010			9.40	-	-	-	-	-	-	C,r	

191

DF38 $C_{16}H_{18}ClN_3S$

Code No.	Empirical Formula	Name and WLN Code	Structural Formula	Solvent	Tech.	Medium	μ, M	pH	T, °C	Electrodes	App.	Experimental Parameters
DF38 bj40	$C_{16}H_{18}ClN_3S$	Methylene Blue	TABLE II-2.	H_2O	PY	BOR ?				DME/NHE		
DF39	$C_{16}H_{22}Fe$	1,2,2-trimethyl-1-propyl)ferrocene L5ØJ Ø-FE-- ØL5ØJ AY1&X1&1&1	TABLE II.	MeCN	ER	$LiClO_4$ 0.2	–	–	–	Pt/SCE	---	C=1.074
DF40	$C_{16}H_{22}N_8Ni$	7,14-dimethyl-bis-(1,2-cyclohexylenyl-[c,j])-1,2,5,6,8,9,-12,13-octaaza-2,4,-7,9,11,14-cyclo-tetradecahexaene nickel(II) C.A. 36869-44-0 D5656 F6 Q6 1A W A-NI-N DNN LNN ONN WN&&&&TTJ C1 N1	TABLE II.	CH_2Cl_2	VY	Bu_4NClO_4 0.1	–	–	25	RPE/SCE	O--	C≈1
DF41	$C_{16}H_{28}FeN_4$	5,5,7,12,12,14-hexamethyl-1,4,8,11-tetraaza-1,3,7,10-cyclotetradecatetra-eneiron(II) dication D5656 1A O A-FE-N EN HXN LN GH JH&&&TJ F1 F1 H1 M1 M1 O1	TABLE II.	MeCN	VR	Et_4NClO_4 0.1?	–	–	–	ns Ag/0.1 Ag^+	---	ns
DF42	$C_{16}H_{28}FeN_4$	5,5,7,12,12,14-hexamethyl-1,4,8,11-tetraaza-1,3,8,10-cyclotetradecatetra-eneiron(II) dication D5656 1A O A-FE-N E NX IN LNX&T&TJ F1 F1 H1 M1 M1 O1	TABLE II.	MeCN	VR	Et_4NClO_4 0.1?	–	–	–	ns Ag/0.1 Ag^+	---	ns
DF43	$C_{16}H_{28}FeN_4$	5,5,7,12,14,14-hexa-methyl-1,4,8,11-tetraaza-1,3,7,11-cyclotetradecatetra-eneiron(II) dication D5656 1A O A-FE-N E NX IN LN OX GH NH&& T&J F1 F1 H1 M1 O1 O1	TABLE II.	MeCN	VR	Et_4NClO_4 0.1?	–	–	–	ns Ag/0.1 Ag^+	---	ns
DF44	$C_{16}H_{28}FeN_4$	5,7,7,12,14,14-hexa-methyl-1,4,8,11-tetraaza-1,4,8,11-cyclotetradecatetra-eneiron(II) dication D5656 1A O A-FE-N E NX IN LNX CH GH JH NHJ F1 F1 H1 M1 M1 O1	TABLE II.	MeCN	VR	Et_4NClO_4 0.1?	–	–	–	ns Ag/0.1 Ag^+	---	ns
DF45	$C_{16}H_{30}FeN_4$	5,7,7,12,14,14-hexa-methyl-1,4,8,11-tetraaza-1,3,8-cy-clotetradecatriene-iron(II) dication D5656 1A O A-FE-M E NX IN LNX CH GH&&TJ F1 F1 H1 M1 M1 O1	TABLE II.	MeCN	VR	Et_4NClO_4 0.1?	–	–	–	ns Ag/0.1 Ag^+	---	ns

TABLE I. Electrochemical Data $C_{16}H_{30}FeN_4$ DF45

| Ref. | C/M | Charact. Potential | | Response Const. | | n | | Electrokinetic Data | | | Products and Identification | Description and Remarks | Code No. |
		Value	vs.		Value	Tech.		Parameter	Value	From			
JA095 3411	-	-	-	-	-	-	-	-	-	-	-	see BJ40	DF38 bj40
JA083 3949	-	$E_{\tau/4}$ $E_{0.22}$	0.258 0.252	SCE	$i_\tau^{\frac{1}{2}}/C$	612	-	-	-	-	-	-	$A, \Delta E_{\tau/4} = \rho \sigma_p, \tau_A/\tau_C = 3.34, r$	DF39
JA094 4529	-	$E_{\frac{1}{2}}$	-1.20 1.26	SCE	i_ℓ/C	25 43	i:i	1.4 2.4	Tomeš	57 71	sttd	-	C,R,r A,≠	DF40
JA094 5502	-	$E_{\frac{1}{2}}$	0.82	Ag/0.1 Ag$^+$	-	-	-	-	-	-	-	-	A,R,p	DF41
JA094 5502	-	$E_{\frac{1}{2}}$	0.89 -0.80 -1.41 -1.83	Ag/0.1 Ag$^+$	-	-	- sttd	- 1 1 1	-	-	-	-	A,R,p C C C	DF42
JA094 5502	-	$E_{\frac{1}{2}}$	0.72	Ag/0.1 Ag$^+$	-	-	-	-	-	-	-	-	A,R,p	DF43
JA094 5502	-	$E_{\frac{1}{2}}$	0.59	Ag/0.1 Ag$^+$	-	-	-	-	-	-	-	-	A,R,p	DF44
JA094 5502	-	$E_{\frac{1}{2}}$	0.76	Ag/0.1 Ag$^+$	-	-	-	-	-	-	-	-	A,R,p	DF45

Code No.	Empirical Formula	Name and WLN Code	Structural Formula	Solvent	Tech.	Medium	μ, M	pH	T, °C	Electrodes	App.	Experimental Parameters
DF46	$C_{16}H_{30}FeN_4$	5,7,7,12,14,14-hexamethyl-1,4,8,11-tetraaza-1,4,11-cyclotetradecatrieneiron(II) dication D5656 1A O A-FE-N E MX IN LNX GH JH NHT &&&J F1 F1 H1 M1 M1 O1	TABLE II.	MeCN	VR	Et_4NClO_4 0.1?	-	-	-	ns Ag/0.1 Ag^+	---	ns
DF47	$C_{16}H_{32}FeN_4$	5,7,7,12,14,14-hexamethyl-1,4,8,11-tetraaza-4,11-cyclotetradecadieneiron-(II) dication D5656 1A O A-FE-N E MX IN LMX GH NHT&T&J	TABLE II.	MeCN	VR	Et_4NClO_4 0.1?	-	-	-	ns Ag/0.1 Ag^+	---	ns
DF48	$C_{16}H_{32}O_3$	tert-butyl peroxydodecanoate C.A. 2123-88-8 11VOOX1&1&1	$CH_3(CH_2)_{10}C(O)-OOC(CH_3)_3$	MeOH 50 C_6H_6 50	PY	LiCl	?	-	-	DME/SCE	---	ns
DF49	$C_{16}H_{36}FeN_4$	5,5,7,12,12,14-hexamethyl-1,4,8,11-tetraazacyclotetradecaneiron(II) dication bis[tetrafluoroborate] D5656 1A O A-FE-M E MX IM LMXTJ F1 F1 H1 M1 M1 O1 &Ø &Ø	TABLE II.	MeCN	VR	Et_4NClO_4 0.1?	-	-	-	ns Ag/0.1 Ag^+	---	ns
DF50	$C_{17}H_{10}FeO_3$	(anthracene)tricarbonyliron(0) C.A. 12094-57-4 L C6660J Ø-FE-CO&CO &CO	TABLE II.	MG	PY	Bu_4NClO_4 0.1	-	-	22	DME Ag/Ag^+	2AO	C=1, Pt Aux
DF51	$C_{17}H_{12}ClN_3O$	7-chloro-1,3-dihydro-5-(1H-indol-1-yl)-2H-1,4-benzodiazepin-2-one C.A. 55098-63-0 T67 GMV JN IHJ CG K-BT56 BNJ	TABLE II.	MeOH 3.3	DI	PHOS 1	-	3.0 7.0	-	DME/SCE	25AO	0.00 → -1.100 V, C=0.11, EC=2.285, m=2.75, t=0.5, ΔE=50, v=5, Pt Aux 0.00 → -1.750 V
DF52 bj74	$C_{17}H_{12}N_2$	2-methyl-7,8-benzophenazine	TABLE II-2.	DMF	PY	Et_4NBr 0.1				DME/SCE		
DF53	$C_{17}H_{14}ClN_3$	7-chloro-2-cyano-2,3-dihydro-1-methyl-5-phenyl-1H-1,4-benzodiazepine T67 GN JN JU&TJ CG G1 HCN KR	TABLE II.	MeOH 3.3	DI	PHOS 1	-	3.0 7.0	-	DME/SCE	25AO	0.00 → -1.100 V, C=0.12, EC=2.285, m=2.75, t=0.5, ΔE=50, v=5, Pt Aux 0.00 → -1.750 V

TABLE I. Electrochemical Data $C_{17}H_{14}ClN_3$ DF53

| Ref. | C/M | Charact. Potential | | Response Const. | | n | | Electrokinetic Data | | | Products and | Description and | Code |
		Value	vs.		Value	Tech.		Parameter	Value	From	Identification	Remarks	No.
JA094 5502	-	$E_{\frac{1}{2}}$	0.51	Ag/0.1 Ag$^+$	-	-	QE	1.03	-	-	-	-	A,R,p	DF46
JA094 5502	-	$E_{\frac{1}{2}}$	0.44	Ag/0.1 Ag$^+$	-	-	QE	0.96	-	-	-	-	A,R,p	DF47
AC035 0880	-	$E_{\frac{1}{2}}$	-0.87	SCE	i_l/C	10.0	-	-	-	-	-	-	C,i_d,P	DF48
JA094 5502	-	$E_{\frac{1}{2}}$	0.38	Ag/0.1 Ag$^+$	-	-	-	-	-	-	-	-	A,R,p	DF49
JA090 1995	-	$E_{\frac{1}{2}}$	-1.8 -2.4	Ag/Ag$^+$	-	-	-	-	-	-	-	-	C,≠,p C,≠	DF50
AA074 0367	-	E_{su}	-0.915	SCE	i_{su}	8.38	-	-	-	-	-	-	C,r	DF51
			-1.085			1.22	-	-	-	-	-	-	C,r	
VZ011 0144	-	-	-	-	-	-	-	-	-	-	-	-	see BJ74	DF52 bj74
AA074 0367	-	E_{su}	-0.725	SCE	i_{su}	9.02	-	-	-	-	-	-	C,r	DF53
			-0.980			0.47	-	-	-	-	-	-	C,r	

195

DF54 $C_{17}H_{14}N_2O$

Code No.	Empirical Formula	Name and WLN Code	Structural Formula	Solvent	Tech.	Medium	μ, M	pH	T, °C	Electrodes	App.	Experimental Parameters
DF54	$C_{17}H_{14}N_2O$	1-o-tolylazo-2-naphthol C.A. 2646-17-5 L66J BNUNR B1& CQ	TABLE II.	H_2O	PY	MB polyoxyethylene nonylphenylether 0.04	-	3.0	25±0.1	DME/SCE	OAO	C=0.1, m=1.74, h=87.15
								5.88				
						MB Na dodecylsulfate 0.02		2.0				C=0.05
								4.0				
								6.8				
						MB dodecyltrimethylammonium chloride 0.05		2.0				
								3.8				
								5.6				
								8.0				
						MB Na dodecylbenzenesulfonate 0.02		4.0	45±0.1			C=0.02
								5.0				
								6.9				
				EtOH 40	PY	MB	-	2.7	25±0.1	DME/SCE	OAO	C=0.05, m=1.74, h=87.15
								4.3				
								5.3				
DF55	$C_{17}H_{14}N_2O_2$	7-acetyl-1,3-dihydro-5-phenyl-2H-1,4-benzodiazepin-2-one C.A. 36093-53-5 T67 GMV JN IHJ CV1 KR	TABLE II.	MeOH 3.3	DI	PHOS 1	-	3.0	-	DME/SCE	25AO	0.00→ -1.100 V, C=0.11, EC=2.285, m=2.75, t=0.5, ΔE=50, v=5, Pt Aux
								7.0				0.00→ -1.750 V
DF56	$C_{17}H_{15}ClN_2O$	1-acetyl-7-chloro-2,3-dihydro-5-phenyl-1H-1,4-benzodiazepine C.A. 1803-95-8 T67 GN JN JU&TJ CG GV1 KR	TABLE II.	MeOH 3.3	DI	PHOS 1	-	3.0	-	DME/SCE	25AO	0.00→ -1.100 V, C=0.12, EC=2.285, m=2.75, t=0.5, ΔE=50, v=5, Pt Aux
								7.0				0.00→ -1.750 V
DF57	$C_{17}H_{15}ClN_2O_2$	7-chloro-2,3-dihydro-1-methyl-5-phenyl-1H-1,4-benzodiazepin-2-carboxylic acid C.A. 55098-47-0 T67 GN JN JU&TJ CG G1 HVQ KR	TABLE II.	MeOH 3.3	DI	PHOS 1	-	3.0	-	DME/SCE	25AO	0.00→ -1.100 V, C=0.12, EC=2.285, m=2.75, t=0.5, ΔE=50, v=5, Pt Aux
								7.0				0.00→ -1.750 V

TABLE I. Electrochemical Data $C_{17}H_{15}ClN_2O_2$ DF57

Ref.	C/M	Charact. Potential		Response Const.		n Tech.	Electrokinetic Data			Products and Identification	Description and Remarks	Code No.
		Value	vs.		Value		Parameter	Value	From			
BJ047 1093	-	$E_{\frac{1}{2}}$	-0.38F -0.85	SCE	i_ℓ	0.10F 0.06F	-	-	-	-	-	-	$c,i_d?,r?$ c	DF54
			-0.7F			0.1F	-	-	-	-	-	-	$c,i_d?,r?$
			-0.06F			-	-	-	-	-	-	-	$c,i_d?,r?$
			-0.28F			-	-	-	-	-	-	-	$c,i_d?,r?$
			-0.60F			-	-	-	-	-	-	-	$c,i_d?,r?$
			-0.17F			-	-	-	-	-	-	-	$c,i_d?,r?$
			-0.29F			-	-	-	-	-	-	-	$c,i_d?,r?$
			-0.43F			-	-	-	-	-	-	-	$c,i_d?,r?$
			-0.56F			-	-	-	-	-	-	-	$c,i_d?,r?$
			-0.38			-	-	-	-	-	-	-	$c,i_d?,r?$
			-0.55F			-	-	-	-	-	-	-	$c,i_d?,r?$
			-0.74F			-	-	-	-	-	-	-	$c,i_d?,r?$
BJ047 1093	-	$E_{\frac{1}{2}}$	-0.22F	SCE		-	-	-	-	-	-	-	$c,i_d?,r?$
			-0.36F			-	-	-	-	-	-	-	$c,i_d?,r?$
			-0.46F			-	-	-	-	-	-	-	$c,i_d?,r?$
AA074 0367	-	E_{su}	-0.710	SCE	i_{su}	8.18	-	-	-	-	-	-	$c,\Delta E_{su}=-0.172\sigma_m,r$	DF55
			-0.970			4.54	-	-	-	-	-	-	c,r
AA074 0367	-	E_{su}	-0.625 -0.800	SCE	i_{su}	1.91 3.03	-	-	-	-	-	-	$c,\Delta E_{su}=-0.417\sigma_m,r$ c	DF56
			-0.915			4.61	-	-	-	-	-	-	$c,\Delta E_{su}=-0.133\sigma_m,r$
AA074 0367	-	E_{su}	-0.850	SCE	i_{su}	6.60	-	-	-	-	-	-	c,r	DF57
			-1.040			6.76	-	-	-	-	-	-	c,r

DF58 $C_{17}H_{15}ClO_4S_2$

Code No.	Empirical Formula	Name and WLN Code	Structural Formula	Solvent	Tech.	Medium		μ, M	pH	T, °C	Electrodes	App.	Experimental Parameters
DF58	$C_{17}H_{15}ClO_4S_2$	3,5-bis(4-methyl-phenyl)-1,2-dithiol-1-ium perchlorate C.A. 42877-43-0 T5SSTJ CR D1& ER D1 & G-O4	TABLE II.	CH_2Cl_2	VR	Bu_4NBF_4	0.2	-	-	-	PBE/SCE	26A2	v=150
DF59	$C_{17}H_{15}ClO_6S_2$	3,5-bis(4-methoxy-phenyl)-1,2-dithiol-1-ium perchlorate C.A. 29776-30-5 T5SSTJ CR DO1& ER DO1 & G-O4	TABLE II.	MeCN	VR	Bu_4NBF_4	0.1	-	-	-	PBE/SCE	26A2	v=150
				CH_2Cl_2	VR	Bu_4NBF_4	0.2	-	-	-	PBE/SCE	26A2	v=150
DF60	$C_{17}H_{15}NO_3$	6-acetamido-2-phenylchromanone T66 BO EVT&J CR& HM V1	TABLE II.	EtOH 50	PO	$HClO_4$	0.5	-	-	-	SME/ -	--2	C=1
DF61	$C_{17}H_{15}N_3O_2$	7-acetamido-2,3-dihydro-5-phenyl-1H-1,4-benzodiazepin-2-one C.A. 4928-03-4 T67 GMV JN IHJ CMV1 KR	TABLE II.	MeOH 3.3	DI	PHOS	1	-	3.0	-	DME/SCE	25A0	0.00→ -1.100 V, C=0.11, EC= 2.285, m=2.75, t=0.5, ΔE=50, v=5, Pt Aux
									7.0				0.00→ -1.750 V
DF62 bj94	$C_{17}H_{15}Ni_3O_2$	tris(η-cyclopentadienyl)dicarbonyl-trinickel C.A. 12194-69-3	TABLE II-2.	MG	PY	Bu_4NClO_4	0.1				DME Ag/Ag+		
DF63	$C_{17}H_{16}ClN_3O$	2-carbamoyl-7-chloro-2,3-dihydro-1-methyl-5-phenyl-1H-1,4-benzodiazepine C.A. 55098-46-9 T67 GN JN JU&TJ CG G1 HVZ KR	TABLE II.	MeOH 3.3	DI	PHOS	1	-	3.0	-	DME/SCE	25A0	0.00→ -1.100 V, C= 0.11, EC= 2.285, m=2.75, t=0.5, ΔE=50, v=5, Pt Aux
									7.0				0.00→ -1.750 V
DF64	$C_{17}H_{16}ClN_3O$	7-chloro-1-acetamido-2,3-dihydro-5-phenyl-1H-benzodiazepine C.A. 55098-44-7 T67 GN JN JU&TJ CG GMV1 KR	TABLE II.	MeOH 3.3	DI	PHOS	1	-	3.0	-	DME/SCE	25A0	0.00→ -1.100 V, C= 0.11, EC= 2.285, m=2.75, t=0.5, ΔE=50, v=5, Pt Aux
									7.0				0.00→ -1.750 V
DF65	$C_{17}H_{16}ClN_3O$	7-chloro-2,3-dihydro-1-(N-methyl-carbamoyl)-5-phenyl-1H-1,4-benzodiazepine C.A. 22960-08-3 T67 GN JN JU&TJ CG GVM1 KR	TABLE II.	MeOH 3.3	DI	PHOS	1	-	3.0	-	DME/SCE	25A0	0.00→ -1.100 V, C= 0.11, EC= 2.285, m=2.75, t=0.5, ΔE=50, v=5, Pt Aux
									7.0				0.00→ -1.750 V

TABLE I. Electrochemical Data $C_{17}H_{16}ClN_3O$ DF65

Ref.	C/M	Charact. Potential		Response Const.		n Tech.	n	Electrokinetic Data			Products and Identification	Description and Remarks	Code No.
		E_p	vs.		Value			Parameter	Value	From			
JA095 4373	370 a	E_p	-0.17 -0.23 -1.28	SCE	-	-	sttd	1	-	-	-	-	A,p C C	DF58
JA095 4373	370 a	E_p	-0.31 -0.38 -0.45 -1.27 -1.65 -1.73	SCE	-	-	sttd	1 1 1 1 1 1	-	-	-	-	A,p C A C A C	DF59
JA095 4373	370 a	E_p	-0.23 -0.30 -1.01	SCE	-	-	sttd	1 1 -	-	-	-	-	A,p C C	
BF630 2252	-	Q	0.75	-	-	-	-	-	-	-	-	-	C,p	DF60
AA074 0367	-	E_{su}	-0.745	SCE	i_{su}	7.10	-	-	-	-	-	-	C,ΔE_{su}=-0.172σ_m,r	DF61
			-0.995			8.80	-	-	-	-	-	-	C,ΔE_{su}=-0.134σ_m,r	
JA088 5117	-	-	-	-	-	-	-	-	-	-	-	-	see BJ94	DF62 bj94
AA074 0367	-	E_{su}	-0.820	SCE	i_{su}	7.84	-	-	-	-	-	-	C,r	DF63
			-1.040			7.07	-	-	-	-	-	-	C,r	
AA074 0367	-	E_{su}	-0.660	SCE	i_{su}	0.14	-	-	-	-	-	-	C,ΔE_{su}=-0.417σ_m,r	DF64
			-0.940			0.27	-	-	-	-	-	-	C,ΔE_{su}=-0.133σ_m,r	
AA074 0367	-	E_{su}	-0.67	SCE	i_{su}	1.22	-	-	-	-	-	-	C,ΔE_{su}=-0.417σ_m,r	DF65
			-			8.94	-	-	-	-	-	-	C,ΔE_{su}=-0.133σ_m,r	

DF66 $C_{17}H_{16}ClN_3O$

Code No.	Empirical Formula	Name and WLN Code	Structural Formula	Solvent	Tech.	Medium		μ, M	pH	T, °C	Electrodes	App.	Experimental Parameters
DF66	$C_{17}H_{16}ClN_3O$	7-chloro-2-dimethyl-amino-5-phenyl-3H-1,4-benzodiazepine 4-N-oxide C.A. 3693-14-9 T67 GN JN IHJ CG HN 1&1 JO KR	TABLE II.	MeOH 3.3	DI	PHOS	1	−	3.0	−	DME/SCE	25AO	$0.00 \rightarrow$ -1.100 V, C= 0.11, EC= 2.285, m=2.75, t=0.5, ΔE=50, v=5, Pt Aux
									7.0				$0.00 \rightarrow$ -1.750 V
DF67	$C_{17}H_{16}ClN_3O$	7-chloro-2-ethyl-amino-5-phenyl-2H-1,4-benzodiazepine 4-N-oxide C.A. 55098-54-9 T67 GN JN IHJ CG HMZ JO KR	TABLE II.	MeOH 3.3	DI	PHOS	1	−	3.0	−	DME/SCE	25AO	$0.00 \rightarrow$ -1.100 V, C= 0.11, EC= 2.285, m=2.75, t=0.5, ΔE=50, v=5, Pt Aux
									7.0				$0.00 \rightarrow$ -1.750 V
DF68	$C_{17}H_{16}Mo_2O_5Sn$	[dicarbonyl-η-cyclopentadienylmolybdenum(I)][tricarbonyl-η-cyclopentadienylmolybdenum(I)]-dimethyltin(II) L50J Ø-MO-CO&CO&CO&-SN-1&1&-MO-CO&CO&-ØL50J	$(C_5H_5)Mo(CO)_2$-$[Sn(CH_3)_2]Mo$-$(CO)_3(C_5H_5)$	MG	PY	Bu_4NClO_4	0.1	−	−	22	DME Ag/Ag$^+$	2-0	C=2
DF69	$C_{17}H_{17}ClN_2OS$	7-chloro-2,3-dihydro-1-methyl-5-(2-methylsulfinylphenyl)-1H-1,4-benzodiazepine C.A. 55098-52-7 T67 GN JN JU&TJ CG G1 KR BSO&1	TABLE II.	MeOH 3.3	DI	PHOS	1	−	3.0	−	DME/SCE	25AO	$0.00 \rightarrow$ -1.100 V, C= 0.1, EC=2.285, m=2.75, t=0.5, ΔE=50, v=5, Pt Aux
									7.0				$0.00 \rightarrow$ -1.750 V
DF70	$C_{17}H_{17}ClN_2S$	7-chloro-2,3-dihydro-1-methyl-5-(2-methylmercaptophenyl)-1H-1,4-benzodiazepine C.A. 55098-51-6 T67 GN JN JU&TJ CG G1 KR BS1	TABLE II.	MeOH 3.3	DI	PHOS	1	−	3.0	−	DME/SCE	25AO	$0.00 \rightarrow$ -1.100 V, C= 0.11, EC= 2.285, m=2.75, t=0.5, ΔE=50, v=5, Pt Aux
									7.0				$0.00 \rightarrow$ -1.750 V
DF71	$C_{17}H_{17}N_3O$	7-methyl-2-methyl-amino-5-phenyl-3H-1,4-benzodiazepine 4-N-oxide C.A. 55098-61-8 T67 GN JN IHJ C1 HM1 JO KR	TABLE II.	MeOH 3.3	DI	PHOS	1	−	3.0	−	DME/SCE	25AO	$0.00 \rightarrow$ -1.100 V, C= 0.11, EC= 2.285, m=2.75, t=0.5, ΔE=50, v=5, Pt Aux
									7.0				$0.00 \rightarrow$ -1.750 V
DF72	$C_{17}H_{17}N_3OS$	1,2-methylamino-7-methylmercapto-5-phenyl-3H-1,4-benzodiazepine 4-N-oxide C.A. 23193-88-6 T67 GN JN IHJ CS1 HM1 JO KR	TABLE II.	MeOH 3.3	DI	PHOS	1	−	3.0	−	DME/SCE	25AO	$0.00 \rightarrow$ -1.100 V, C= 0.11, EC= 2.285, m=2.75, t=0.5, ΔE=50, v=5, Pt Aux
									7.0				$0.00 \rightarrow$ -1.750 V

TABLE I. Electrochemical Data $\quad C_{17}H_{17}N_3OS \quad DF72$

Ref.	C/M	Charact. Potential Value	vs.	Response Const. Value		Tech.	n	Electrokinetic Data Parameter	Value	From	Products and Identification	Description and Remarks	Code No.
AA074 0367	-	E_{su}	-0.300 -0.630 -1.030	SCE	i_{su}	7.47 11.3 7.85	-	-	-	-	-	-	C,r C C	DF66
			-0.74 -0.940 -1.300			4.34 9.07 6.94	-	-	-	-	-	-	C,r C C	
AA074 0367	-	E_{su}	-0.380 -0.700 -1.160	SCE	i_{su}	5.68 8.20 8.12	-	-	-	-	-	-	C,r C C	DF67
			-0.750 -0.930 -1.360			1.95 3.65 2.43	-	-	-	-	-	-	C,r C C	
JA088 5117	-	$E_{\frac{1}{2}}$	-1.8	Ag/Ag$^+$	-	-	QE	2	-	-	-	2Mo + dimethyltin, sttd	C,p	DF68
AA074 0367	-	E_{su}	0.735	SCE	i_{su}	6.99	-	-	-	-	-	-	C,$\Delta E_{su}=-0.132\sigma_m$,r	DF69
			-0.955			5.40	-	-	-	-	-	-	C,$\Delta E_{su}=-0.027\sigma_m$,r	
AA074 0367	-	E_{su}	-0.790	SCE	i_{su}	8.30	-	-	-	-	-	-	C,$\Delta E_{su}=-0.132\sigma_m$,r	DF70
			-1.020			0.47	-	-	-	-	-	-	C,$\Delta E_{su}=-0.027\sigma_m$,r	
AA074 0367	-	E_{su}	-0.390 -0.730 -1.200	SCE	i_{su}	6.03 11.74 7.40	-	-	-	-	-	-	C,r C,$\Delta E_{su}=-0.118\sigma_m$, C	DF71
			-0.775 -0.990 -1.390			4.65 10.30 7.29	-	-	-	-	-	-	C,r C,$\Delta E_{su}=-0.171\sigma_m$ C	
AA074 0367	-	E_{su}	-0.380 -0.720 -1.175	SCE	i_{su}	6.40 9.20 11.13	-	-	-	-	-	-	C,r C,$\Delta E_{su}=-0.118\sigma_m$ C	DF72
			-0.755 -0.995 -1.375			4.08 8.16 8.14	-	-	-	-	-	-	C,r C,$\Delta E_{su}=-0.171\sigma_m$ C	

DF73 $C_{17}H_{20}N_4NaO_9P$

Code No.	Empirical Formula	Name and WLN Code	Structural Formula	Solvent	Tech.	Medium		μ, M	pH	T, °C	Electrodes	App.	Experimental Parameters
DF73 bk36	$C_{17}H_{20}N_4NaO_9P$	sodium flavin mononucleotide C.A. 130-40-5 T C666 BN DNVMV INJ B1YQYQYQ1OPWQ L M &-NA- &41/32	TABLE II-2.	H_2O	PY	BOR	?	-	9.6	27	DME/NHE	---	ns
DF74 bk37	$C_{17}H_{20}N_4O_6$	riboflavin C.A. 83-88-5 T C666 BN DNVMV INJ B1YQYQYQ1Q L M	TABLE II-2.	H_2O	DI	PHOS	0.1	-	7.2	25±0.1	DME/SCE	2AO	$C=5.3E-4, v=2, t=2, \Delta E=25$, Pt Aux
DF75	$C_{18}H_{10}F_6Mo_2O_6$	bis[(tricarbonyl-η-cyclopentadienyl-trifluoromethyl)-molybdenum] L5ØJ Ø-MO-XFFFCO&CO &CO&-MO-XFFFCO&CO&CO &- ØL5ØJ	$[(C_5H_5)Mo(CO)_3-CF_3]_2$	MG	PY	Bu_4NClO_4	0.1	-	-	22	DME Ag/Ag$^+$	2-0	C=2
DF76	$C_{18}H_{12}N_2O_2S_2$	bis(2-quinolyl)disulfide di-N-oxide C.A. 18470-19-4 T66 BNJ BO CSS- CL66 BNJ BO	TABLE II.	EtOH 50	PT	BR		-	2.0 3.8 5.9 8.5 10.0	21	DME/SCE	OAO	m=2.72, t=3.24
DF77	$C_{18}H_{13}NO_2S$	12-acetylbenzo[a]-phenothiazine 7-oxide T D6 C666 BN MSJ BV1 MO	TABLE II.	EtOH	PY	Me_4NCl CH_3COOH	0.1 4%	-	-	-	DME/MP DME/SCE	0-0	C=1, m=5.4, t=4.66 C=ns
DF78	$C_{18}H_{13}NO_3S$	7-acetylbenzo[c]-phenothiazine 12-dioxide T D6 C666 BSW MNJ MV1	TABLE II.	EtOH	PY	Me_4NCl	0.1	-	-	-	DME/MP	0-0	C=1, m=5.4, t=4.66
DF79	$C_{18}H_{13}NO_3S$	12-acetylbenzo[a]-phenothiazine 7-dioxide T D6 C666 BN MSWJ BV1	TABLE II.	EtOH	PY	Me_4NCl	0.1	-	-	-	DME/MP	0-0	m=5.4, t=4.6

TABLE I. Electrochemical Data $\quad C_{15}H_{13}NO_3S \quad DF79$

Ref.	C/M	Charact. Potential Value	Charact. Potential vs.	Response Const.	Response Const. Value	n Tech.	n	Electrokinetic Data Parameter	Electrokinetic Data Value	Electrokinetic Data From	Products and Identification	Description and Remarks	Code No.
JA095 3411	−	$E_{\frac{1}{2}}$	−0.24F	NHE	−	−	−	−	−	−	−	−	C,p	DF73 bk36
AN100 0377	−	E_{su}	−0.45F	SCE	i_{su}	0.0125	−	−	−	−	−	−	C, $i_p \propto C$ for C=0.003−0.13,r	DF74 bk37
JA088 5117	−	$E_{\frac{1}{2}}$	−2.1	Ag/Ag$^+$	−	−	QE	1	−	−	−	Mo$^-$ + CF$_3$,sttd	C,p	DF75
TA014 0745	141 f	$E_{\frac{1}{2}}$	−0.18	NHE	D	1.73	sttd	2	−	−	−	−	C,W,i_d/C=k,r	DF76
			−0.28			1.73		2	α / $\log k^o_{f,h}$	0.53 / 3.58	$\log k^o_f$ vs. E	−	C,W,i_d/C=k,r	
			−0.41			1.73		2		0.50 / 4.50	$\log k^o_f$ vs. E	−	C,W,i_d/C=k,r	
			−0.46			1.73		2		0.51 / 4.86	$\log k^o_f$ vs. E	−	C,W,i_d/C=k,r	
			−0.46			1.73		2		0.45 / 4.86	$\log k^o_f$ vs. E	−	C,W,i_d/C=k,r	
RG035 0011	−	$E_{\frac{1}{2}}$	−1.63	MP	i_ℓ	0.67	i:i	1	−	−	−	−	C,i_d,Mc(name),p	DF77
			−1.97			1.70		−					C	
			−0.77			3.00		1					C,i_d; $E_{\frac{1}{2}}$ becomes more pos. as [HNO$_3$] ↑, $E_{\frac{1}{2}}$=0.56 in 40% HNO$_3$;p	
RG035 0011		$E_{\frac{1}{2}}$	−1.79	MP	i_ℓ	0.75	i:i	1	−	−	−	−	C,i_d,p	DF78
			−2.10			2.35		−					C	
RG035 0011	−	$E_{\frac{1}{2}}$	−1.83	MP	i_ℓ	1.80		1	−	−	−	−	C,i_d,p	DF79
			−1.95			1.08		−					C	

Code No.	Empirical Formula	Name and WLN Code	Structural Formula	Solvent	Tech.	Medium		μ, M	pH	T, °C	Electrodes	App.	Experimental Parameters
DF80	$C_{18}H_{13}NS$	10-phenylphenothiazine C.A. 7152-42-3 T C666 BN ISJ BR	TABLE II.	DMF	VR	Bu_4NClO_4	0.1	-	-	-	PDE Ag/AgCl, satd KCl	---	v=200
DF81	$C_{18}H_{14}N_4Ni$	dibenzo[b,i]-1,4,8,-11-tetraaza-2,5,7,-9,12,14-cyclotetradecahexaenenickel(II) C.A. 39251-81-5 D5656 C6 1A W A-NI-N IN MN TNJ	TABLE II.	DMSO	VY	Bu_4NBF_4	0.05	-	-	25	RPE/SCE	O--	C≈1
DF82	$C_{18}H_{15}AlCl_4O$	2,6-diphenyl-4-methylpyrylium tetrachloroaluminate T6OJ BR& D1 FR & -A⊃ L-G4	TABLE II.	H_2O	PY	HCl GEL	1	-	0	25.0 ±0.1	DME/SCE	3A4	C=0.5-5,m=2.68,t=5.5,h=60
DF83	$C_{18}H_{15}AsO$	triphenylarsine oxide C.A. 1153-05-5 O-AS-R&R&R	$(C_6H_5)_3AsO$	H_2O	PY	HCl TX100	0.1 0.005	-	-	-	DME/SCE	O4AO	C=0.2,t=2.4,v=1.7
DF84 b107	$C_{18}H_{15}Co_3O_3$	tris(carbonyl-η-cyclopentadienyl-cobalt) C.A. 12194-20-6	$[(C_5H_5)Co(CO)]_3$	MG	PY	Bu_4NClO_4	0.1	-	-	22	DME Ag/Ag$^+$	2-0	C=0.2
DF85	$C_{18}H_{15}IO$	2-methyl-4,6-diphenylpyrylium iodide C.A. 26105-53-3 L6OJ B1 DR& FR &I	TABLE II.	H_2O	PY	HCl GEL	1	-	0	25.0 ±0.1	DME/SCE	3A-	C=0.5-5,m=2.68,t=5.5,h=60
DF86	$C_{18}H_{16}ClN_3O_2$	7-chloro-2-(N-methylacetamido)-5-phenyl-3H-1,4-benzodiazepine 4-N-oxide C.A. 55098-56-1 T67 GN JN IHJ CG HN⊃ 1&V1 JO KR	TABLE II.	MeOH 3.3	DI	PHOS	1	-	3.0 7.0	-	DME/SCE	25AO	0.00→ -1.100 V,C=0.10,EC=2.285,m=2.75,t=0.5,ΔE=50,v=5,Pt Aux 0.00→ -1.750 V
DF87	$C_{18}H_{16}FN_3O_2$	7-acetyl-5-(2-fluorophenyl)-2-methylamino-3H-1,4-benzodiazepine 4-N-oxide C.A. 55098-60-7 T67 GN JN IHJ CV1 HM1 JO KR BF	TABLE II.	MeOH 3.3	DI	PHOS	1	-	3.0 7.0	-	DME/SCE	25AO	0.00→ -1.100 V,C=0.08,EC=2.285,m=2.75,t=0.5,ΔE=50,v=5,Pt Aux 0.00→ -1.750 V
DF88	$C_{18}H_{16}N_4$	1,4-dimethyl-3,6-diphenylpyrazolo[4,5-d]pyrazole C.A. 52887-25-9 T55 BNN FNNJ B1 DR& F1 HR	TABLE II.	DMF	PY	Bu_4NI	0.155	-	-	22	DME/MP?	OA?O	t=4.36,h=60

TABLE I. Electrochemical Data $C_{18}H_{16}N_4$ DF88

Ref.	C/M	Charact. Potential		Response Const.		n		Electrokinetic Data			Products and Identification	Description and Remarks	Code No.
		Value	vs.		Value		Tech.	Parameter	Value	From			
JA094 4790	-	E_p	0.88	Ag/AgCl	-	-	-	-	-	-	-	-	A,p	DF80
JA094 4529	-	$E_{\frac{1}{2}}$	-1.66 0.65	SCE	i_ℓ/C	8.0 12	i:i	1.2 1.8	Tomeš	64 74	sttd	-	C,Q,r A,Q	DF81
JE004 0048	-	$E_{\frac{1}{2}}$	-0.394	SCE	-	-	-	-	-	-	-	-	C,i_d,W,p	DF82
AN100 0584	-	$E_{\frac{1}{2}}$	-0.80F	SCE	i_ℓ	1.15F	QE	2.03 ± 0.15	p=	1	Elogi^a	triphenylarsine,QE, TLC	C,$i_\ell \propto C$ for C=0.1-1.0,i_ℓ=f(pH) for pH=1-2.2,r	DF83
JA088 5117	-	$E_{\frac{1}{2}}$	-1.6	Ag/Ag$^+$	-	-	QE	1	-	-	-	anion,stable	C,X,R(PA),p	DF84 b107
JE004 0048	-	$E_{\frac{1}{2}}$	-0.408	SCE	-	-	-	-	-	-	-	-	C,i_d,W,Tc=0.012 for T=25-40,r	DF85
AA074 0367	-	E_{su}	-0.410 -0.720	SCE	i_{su}	6.03 7.66	-	-	-	-	-	-	C,r C	DF86
			-0.720 -0.900 -1.010			2.41 3.22 1.93	-	-	-	-	-	-	C,r C C	
AA074 0367	-	E_{su}	-0.440 -0.680 -1.190	SCE	i_{su}	4.50 5.88 3.85	-	-	-	-	-	-	C,r C,ΔE_{su}=-0.118σ_m C	DF87
			-0.790 -0.805 -1.30			1.96 5.45 3.56	-	-	-	-	-	-	C,r C C	
BJ047 1490	-	$E_{\frac{1}{2}}$	-1.98	MP	i_ℓ/C	4.1	-	-	Elog	56	sttd	-	C,$E_{\frac{1}{2}} \propto$ LUMO,R?,r	DF88

Code No.	Empirical Formula	Name and WLN Code	Structural Formula	Solvent	Tech.	Medium		μ, M	pH	T, °C	Electrodes	App.	Experimental Parameters
DF89	$C_{18}H_{16}N_4$	2,4-dimethyl-3,6-diphenylpyrazolo[3,4-d]pyrazole C.A. 52887-21-5 T55 BNN FNNJ B1 DR& G1 HR	TABLE II.	DMF	PY	Bu_4NI	0.155	–	–	22	DME/MP?	0A?0	t=4.36,h=60
DF90 b132	$C_{18}H_{16}OSn$	triphenyltin hydroxide C.A. 76-87-9 R-SN-R&R &Q	$(C_6H_5)_3Sn^+$ OH^-	EtOH 50	PY	buffer TX100	0.002	–	3.05	–	DME/SCE	01A0	C=0.5,m=1.65,t=3.34
									4.75				C=ns
									7.3				
									9.6				
									12.45				
									14				
DF91	$C_{18}H_{17}Mo_2O_6P$	μ-hydrido-μ-dimethylphosphinobis[tricarbonyl-(η-cyclopentadienyl)molybdenum(I)] C.A. 12131-42-9 D4H-MO-P-MO-J BCO&C O&CO&R& C1&1 DCO&CO &CO&R	TABLE II.	MG	PY	Bu_4NClO_4	0.1	–	–	22	DME Ag/Ag$^+$	0-0	C=0.2
					PA	Bu_4NClO_4	0.1	–	–	22	HMDE Ag/Ag$^+$	2-2	C=0.2
DF92	$C_{18}H_{17}N_3O_3$	7-acetoxy-2-methylamino-5-phenyl-3H-1,4-benzodiazepine 4-N-oxide C.A. 55098-58-3 T67 GN JN IHJ CV01 HM1 JO KR	TABLE II.	MeOH 3.3	DI	PHOS	1	–	3.0	–	DME/SCE	25A0	0.00→ -1.100 V,C= 0.11,EC= 2.285,m=2.75, t=0.5,ΔE=50, v=5,Pt Aux
									7.0				0.00→ -1.750 V

TABLE I. Electrochemical Data — $C_{18}H_{17}N_3O_3$ DF92

Ref.	C/M	Charact. Potential Value	vs.	Response Const.	Value	Tech.	n	Electrokinetic Data Parameter	Value	From	Products and Identification	Description and Remarks	Code No.
BJ047 1490	-	$E_{\frac{1}{2}}$	-1.82	MP	i_ℓ/C	4.6	-	-	Elog	52	sttd	-	$C, E_{\frac{1}{2}} \propto$ LUMO, R?, r	DF89
AC042 0825	241 a	$E_{\frac{1}{2}}$	-0.50F	SCE	i_ℓ	0.1F	-	-	-	-	-	triphenyltin radical, QE	$C,W,i \neq f(pH),R,p$	DF90 b132
			-0.88F			0.7F			α	0.66 -0.90	sttd		$C,W,i \neq f(pH),\rightleftharpoons$	
			-0.54F			0.1F	-	-	-	-	-	triphenyltin radical, QE	$C,W,i \neq f(pH),R,p$	
			-0.88F			0.7F			α	0.66 -0.90	sttd		$C,W,i \neq f(pH),\rightleftharpoons$	
			-1.35F			0.6F				-			C,W,Mx(suppressed by TX100, $i=f(pH)$, $\rightleftharpoons$, 0 with Bu_4N^+	
			-0.64F			0.1F	-	-	-	-	-	triphenyltin radical, QE	$C,W,i \neq f(pH),R,i/C=k$ for $0.01<C<0.1,p$	
			-0.91F			0.7F		1	α	0.66 -0.90	sttd		$C,W,i \neq f(pH),\rightleftharpoons,i/C=k$ for $0.1<C<1.0, i=kh^{\frac{1}{2}}$	
			-1.35F			0.8F		1		-			C,W,Mx(suppressed by TX100, $i=f(pH)$, $\rightleftharpoons$, 0 with $Bu_4N^+, i/C=k$ for $C<0.5, i=kh^{\frac{1}{2}}$ C?	
			-			-	-							
			-0.79F			0.1F	-	-	-	-	-	triphenyltin radical, QE	$C,W,i \neq f(pH),R,i/C=k,p$	
			-1.01F			0.7F			α	0.66 -0.90	sttd		$C,W,i \neq f(pH),\rightleftharpoons,i/C=k$	
			-1.36F			0.3F				-			$C,W,i \neq f(pH),\rightleftharpoons,0$ with Bu_4N^+, X	
			-1.0F			0.1F	-	-	-	-	-	triphenyltin radical, QE	$C,W,i \neq f(pH),R,i/C=k$	
			-1.2F			0.7F			α	0.66 -0.90	sttd	-	$C,W,i \neq f(pH),\rightleftharpoons,i/C=k$	
			-1.4F			0.3F				-			$C,W,i=f(pH),\rightleftharpoons,0$ with Bu_4N^+, X	
			-			0.9F	-	-	α	0.66 -0.90	sttd	-	$C,W,i \neq f(pH),\rightleftharpoons,i/C=k, S,p$	
			-			-		-		-			$C,W,\rightleftharpoons,p$	
JA088 5112	-	$E_{\frac{1}{2}}$	-2.4	Ag/Ag^+	-	-	QE	1.0	-	-	-	-	C,p	DF91
JA088 5112	-	E_p	3.2 1.5	Ag/Ag^+	-	-	QE	1.0 -	-	-	-	-	C,p A	
AA074 0367	-	E_{su}	-0.400 -0.710 -1.100	SCE	i_{su}	6.20 10.39 6.04	-	-	-	-	-	-	C,r $C,\Delta E_{su}=-0.118\sigma_m$ C	DF92
			-0.765 -0.950 -1.30			4.19 8.94 8.05	-	-	-	-	-	-	C,r $C,\Delta E_{su}=-0.171\sigma_m$ C	

Code No.	Empirical Formula	Name and WLN Code	Structural Formula	Solvent	Tech.	Medium		μ, M	pH	T, °C	Electrodes	App.	Experimental Parameters
DF93	$C_{18}H_{18}ClNO_5S_2$	3-[4-(dimethylamino)-phenyl]-5-(4-methoxyphenyl)-1,2-dithiol-1-ium perchlorate C.A. 33176-67-9 T5SSTJ CR DO1& DR D- N1&1	TABLE 11.	CH_2Cl_2	VR	Bu_4NBF_4	0.2	-	-	-	PBE/SCE	26A2	v=150
DF94	$C_{18}H_{21}BrO_2$	16α-bromo-3-hydroxyestra-1,3,5-[10]-trien-17-one L E5 B666 EXVTTT&J E1 GE OQ -A&G	TABLE 11.	MeOH 5 EtOH 85	PY	ACET	0.1	-	-	25	DME/NCE	013A0	ns
DF95	$C_{18}H_{21}BrO_2$	16β-bromo-3-hydroxyestra-1,3,5(10)-trien-17-one L E5 B666 EXVTTT&J E1 GE OQ -B&G	TABLE 11.	MeOH 5 EtOH 85	PY	ACET	0.1	-	-	25	DME/NCE	013A0	ns
DF96	$C_{18}H_{22}N_4O_5S_2$	4,4'-oxybis[benzenesulfonyl(isopropylidenehydrazide)] C.A. 13279-35-1 1Y1&UNMSWR DOR DSWM- NUY1&1	TABLE 11.	H_2O	PY	H_2SO_4 H_2SO_4 H_2SO_4	0.100 0.050 0.005	-	-	30±0.5	DME/SCE	03-0	m=1.73, t=4.48
				Me_2CO 75 $C_6H_5CH_3$ 15	PY	H_2SO_4	0.5	-	-	30±0.5	DME/SCE	03-0	C=0.2
DF97	$C_{18}H_{24}N_5O_{13}P$	cytidylyl (3'→5') uridine C.A. 2382-64-1 T6VMVNJ D- BT5OTJ CQ DQ E1OPQO&O- CT5OTJ B1Q DQ E- AT6NVNJ DZ	TABLE 11.	H_2O	PY	MB	0.13	4.5	0	DME/SCE	235A03	C=0.094, m=1.0, t=6-7, Pt Aux
					PV	MB	0.5	2.5	25	DME/SCE	235A03	C=0.05, m=1.0, t=6-7, f=50, Δe=3.5, Pt Aux
								5.0				
						MB and CARB		6.0				
								9.2				
					VR	MB	0.1	4.5	25	HMDE SCE	235A03	C=0.094, v=1.5E3, Pt Aux
												v=4.4E3
												v=1E4
												v=2E4
DF98 c137	$C_{18}H_{25}N_6O_{12}P$	cytidylyl (3'→5') cytidine C.A. 2536-99-4 T6NVNJ DZ A- ET5OTJ B1Q DQ -A&BE -B&CD COPQO&O1- BT5OTJ CQ DQ -A&BE -B&CD E- AT6NVNJ DZ	TABLE 11-3.	H_2O	PY	MB	0.13	5.0	0	DME/SCE	235A03	C=0.018-0.89, m=1.0, t=6-7, Pt Aux
CONT							0.5	2.5	25			C=0.05

TABLE I. Electrochemical Data — $C_{18}H_{25}N_6O_{12}P$ (CONT.) DF98

Ref.	C/M	Charact. Potential		Response Const.		n	Tech.	Electrokinetic Data			Products and Identification	Description and Remarks	Code No.
		Value	vs.	Response	Value			Parameter	Value	From			
JA095 4373	370 a	E_p	-0.38 -0.46 -1.04	SCE	-	-	sttd	1 1 -	-	-	-	-	A,p C C	DF93
AC034 1440	-	$E_{\frac{1}{2}}$	-0.16	NCE	I	2.3	-	-	-	-	-	-	C,Br redn.,p	DF94
AC034 1440	-	$E_{\frac{1}{2}}$	-0.28	NCE	I	2.2	-	-	-	-	-	-	C,Br redn.,p	DF95
AC036 0523	-	$E_{\frac{1}{2}}$	-1.14	SCE	-	-	-	-	Elog	44	sttd	-	C,W,⊭,r	DF96
			-1.15		I	6.92	-	-		44		-	C,W,⊭,r	
			-1.22		-	-	-	-		44		-	C,W,⊭,r	
AC036 0523	-	-	-	-	-	5.72	-	-		44		-	C,W,⊭,I Av(4),r	
JA095 0991	89c	$E_{\frac{1}{2}}$	-1.32	SCE	I	2.8	-	-	-	-	-	-	C,r	DF97
JA095 0991	89c	E_{min}	-0.2 to -0.8	SCE	-	-	-	-	-	-	-	-	Mn,largest at pH= 4.5,Mc(name),r	
		E_{su}	-1.15						dE_{su}/dpH	7	sttd		C	
		E_{min}	-0.2 to -0.8		-	-	-	-	-	-	-	-	Mn,largest at pH= 4.5,r	
		E_{su}	-1.17						dE_{su}/dpH	7	sttd		C	
			-1.53		-	-	-	-			38	-	C,r	
			-1.65		-	-	-	-			38	-	C,r	
JA095 0991	89c	E_p	-1.44F	SCE	$i_p/v^{\frac{1}{2}}AC$	21F	-	-	-	-	-	-	C,r	
			-1.52F			21.7F	-	-	-	-	-	-	C,r	
			-			39.7F	-	-	-	-	-	-	C,r	
			-			54.2F	-	-	-	-	-	-	C,r	
JA095 0991	89c	$E_{\frac{1}{2}}$	-1.30	SCE	I	5.5 - 0.4	-	-	-	-	-	-	C,I=f(C),D=5.3,r	DF98 c137
			-1.39 to -1.49			1.6 - 5.9							C,I=f(C)	
			-1.61			1.3 - 1.7							C,I=f(C)	
			-1.143			10	-	-	$dE_{\frac{1}{2}}/dpH$	41	sttd	-	C,D=16.9?,Tc=2.3,r	
									Tomeš	33				
			-1.225						$dE_{\frac{1}{2}}/dpH$	56			C	CONT

209

DF98 (CONT.) $C_{18}H_{25}N_6O_{12}P$

Code No.	Empirical Formula	Name and WLN Code	Structural Formula	Solvent	Tech.	Medium	μ, M	pH	T, °C	Electrodes	App.	Experimental Parameters
DF98 c137	$C_{18}H_{25}N_6O_{12}P$	cytidylyl (3'→5') cytidine	TABLE II.	H_2O	PY	MB	0.5	5.5	25	DME/SCE	235A 03	C=0.05,m=1.0, t=6.7,Pt Aux
								6.5				
								8.0				
								9.0				
						AM	?	9.1				
						CARB	?	10.6				
					PV	MB	—	5.0	0.5	DME/SCE	235A 03	C=0.09,m=1.0, t=6-7,f=50, Pt Aux
							0.5	2.0	25			Δe=3.5
								3.5				
								4.5				
								5.5				
								6.0				
								8				
						AM	?	9				
								10				
					PVI	MB	0.13	4.5	25	DME/SCE	235A 03	C=0.096,m= 1.0,t=6-7, f=50,Δe= 3.5,Pt Aux
								5.0	0			C=0.018
												C=0.89
					VR	MB	—	2.0	25	HMDE SCE	235A 03	C=0.05,A= 0.03,v=300, Pt Aux
								4.0				
								5.5				
								6.5				
								7.5				
								9				
								10				
DF98 c137	$C_{18}H_{25}N_6O_{12}P$	cytidylyl (3'→5') cytidine	TABLE II.	H_2O	PY	MB						

TABLE I. Electrochemical Data $C_{18}H_{25}N_6O_{12}P$ DF98

Ref.	C/M	Charact. Potential		Response Const.		n		Electrokinetic Data			Products and Identification	Description and Remarks	Code No.
		Value	vs.		Value	Tech.		Parameter	Value	From			
JA095 0991	89c	$E_{\frac{1}{2}}$ -1.266	SCE	-	-	-	-	$dE_{\frac{1}{2}}/dpH$	41	sttd	-	C,D=16.9?,Tc=2.3,r	DF98 c137
								Tomeš	33				
		-1.393						$dE_{\frac{1}{2}}/dpH$	56			C	
		-1.476		i_ℓ	0.88F	-	-		104		-	C,r	
		-1.633			1.41F	-	-		104			C,r	
		-1.684		-	-	-	-		51		-	C,r	
		-1.624		-	-	-	-		-	-	-	C,r	
		-1.766		i_ℓ	0.79F	-	-	$dE_{\frac{1}{2}}/dpH$	51	sttd	-	C,r	
JA095 0991	89c	E_{su} -1.33 -1.42 -	SCE	i_{su}	0.025F 0.02F -	-	-	-	-	-	-	C,r C C	
		E_{min} -0.2 to -0.8		i_{min}	0.21	-	-	-	-	-	-	Mn,largest at pH 4.5,r	
		E_{su} -1.18 -1.29			-			dE_{su}/dpH	19 41	sttd		C C	
		E_{min} -0.2 to -0.8			0.21	-	-		-	-	-	Mn,r	
		E_{su} -1.47 -1.35			-				19 41	sttd		C C	
		E_{min} -0.2 to -0.8		-	-	-	-		-		-	Mn,r	
		E_{su} -1.165 -1.51							108			C C	
		E_{min} -0.2 to -0.8		-	-	-	-		-		-	C,r	
		E_{su} -1.18 -1.61							0 108			C C	
		-1.30 -1.67		-	-	-	-		96 108		-	C,r C	
		-1.49		-	-	-	-		96		-	C,r	
		-1.28		-	-	-	-		19		-	C,r	
		-1.30		-	-	-	-		19		-	C,r	
JA095 0991	89c	E_{su} -1.28 -1.4	SCE	-	-	-	-	-	-	-	-	C,r C,r	
		-1.32 -1.43		-	0.013 0.037	-	-	-	-	-	-	C,r C	
		- -1.54 -1.62			0.060 0.50 0.13	-	-	-	-	-	-	C,r C C,0 for C<0.36	
JA095 0991	89c	E_p -1.22	SCE	-	-	-	-	dE_p/dpH	42	sttd	-	C,r	
		-1.31		-	-	-	-		42		-	C,r	
		-1.38		-	-	-	-		210		-	C,r	
		-1.56		-	-	-	-		101		-	C,Mc(Ep),r	
		-1.65		-	-	-	-		101		-	C,Mc(Ep),r	
		-1.764		-	-	-	-		76		-	C,r	
		-1.840		-	-	-	-		76		-	C,r	

Code No.	Empirical Formula	Name and WLN Code	Structural Formula	Solvent	Tech.	Medium	μ, M	pH	T, °C	Electrodes	App.	Experimental Parameters
DF99	$C_{18}H_{26}N_8Ni$	7,14-diethyl-bis(1,-2-cyclohexylenyl[c,j]-1,2,5,6,8,9,12,-13-octaaza-2,4,7,9,-11,14-cyclotetradecahexaenenickel-(II) C.A. 36869-45-1 D5656 F6 Q6 1A W A-NI-N DNN LNN ONN WN &&&&TTJ C2 N2	TABLE II.	CH_2Cl_2	VY	Bu_4NClO_4 0.1	-	-	25	RPE/SCE	0--	$C \approx 1$
DG00	$C_{18}H_{34}O_4$	dinonanoyl peroxide C.A. 762-13-0 8VOOV8	$[CH_3(CH_2)_7$-$C(O)O]_2$	MeOH 50 C_6H_6 50	PY	LiCl	?	-	-	-	DME/SCE	---	ns
DG01	$C_{18}H_{36}O_3$	peroxytetradecanoate C.A. 59710-71-3 13VOOX1&1&1	$CH_3(CH_2)_{12}C(O)$-$OOC(CH_3)_3$	MeOH 50 C_6H_6 50	PY	LiCl	?	-	-	-	DME/SCE	---	ns
DG02	$C_{19}H_{14}N_2$	4-biphenylyl(phenyl)-diazomethane C.A. 30905-13-6 NUNUYR&R DR	TABLE II.	MeCN	VY	$LiClO_4$ 0.1	-	-	20±1	RPDE SCE	3--	d=0.08
DG03	$C_{19}H_{18}ClN_3O_2$	7-chloro-2[N-methyl-propionamido)-5-phenyl-3H-1,4-benzodiazepine 4-N-oxide C.A. 55098-55-0 T67 GN JN IHJ CG HN 1&VZ JO KR	TABLE II.	MeOH 3.3	DI	PHOS	1	3.0	-	DME/SCE	25A0	0.00 → -1.100 V, C= 0.10, EC= 2.285, m=2.75, t=0.5, ΔE=50, v=5, Pt Aux
								7.0				0.00 → -1.750 V
DG04	$C_{19}H_{20}ClN_3O$	2-butylamino-7-chloro-5-phenyl-3H-1,4-benzodiazepine 4-N-oxide C.A. 55098-53-8 T67 GN JN IHJ CG HM4 JO KR	TABLE II.	MeOH 3.3	DI	PHOS	1	3.0	-	DME/SCE	25A0	0.00 → -1.100 V, C= 0.10, EC= 2.285, m=2.75, t=0.5, ΔE=50, v=5, Pt Aux
								7.0				0.00 → -1.7 -1.750 V
DG05	$C_{19}H_{25}N_8O_{12}P$	cytidylyl (3'→5') guanidine C.A. 2382-65-2 T56 BN DN FMYMVJ GUM D- BT50TJ CQ DQ E1 OPQO&O- DT50TJ CQ E1Q B- AT6NVNJ DZ	TABLE II.	H_2O	PV	MB	0.5	2.0	25	DME/SCE	235A 03	50 Hz, C=0.05, m=1.0, t=6-7, f=50, Δe= 3.5, Pt Aux
								3.5				
						MB CARB		4.0				
								6.0				
								9.6				

TABLE I. Electrochemical Data $\quad C_{18}H_{25}N_8O_{12}P \quad DG05$

Ref.	C/M	Charact. Potential		Response Const.		Tech.	n	Electrokinetic Data			Products and Identification	Description and Remarks	Code No.
			Value	vs.		Value			Parameter	Value	From			
JA094 4529	-	$E_{\frac{1}{2}}$	-1.24 1.25	SCE	i_ℓ/C	23 40	i:i	1.3 2.2	Tomeš	61 65	sttd	-	C,R,r A,⇌	DF99
AC035 0880	-	$E_{\frac{1}{2}}$	-0.10	SCE	i_ℓ/C	9.1	-	-	-	-	-	-	C,p	DG00
AC035 0880	-	$E_{\frac{1}{2}}$	-0.82	SCE	i_ℓ/C	10.5	-	-	-	-	-	-	C,p	DG01
EA015 1543	-	$E_{\frac{1}{2}}$	0.89 1.4	SCE	-	-	-	-	Elog	62 -	-	1,2-di(biphenylyl)-1,2-diphenylethylene,QE	A,R,r A,⇌	DG02
AA074 0367	-	E_{su}	-0.400 -0.705	SCE	i_{su}	2.84 3.95	-	-	-	-	-	-	C,r C	DG03
			-0.735 -0.920 -1.350			0.55 1.34 0.7						-	C,r C C	
AA074 0367	-	E_{su}	-0.385 -0.705 -1.075	SCE	i_{su}	5.50 6.76 10.00	-	-	-	-	-	-	C,r C C	DG04
			-0.730 -0.935 -1.300			8.02 - 14.5	-	-	-	-	-	-	C,r C C	
JA095 0991	89c	E_{su}	-0.2 to -0.8	SCE	i_{su}	0.31	-	-	-	-	-	-	Mn,deepest at pH= 5.5	DG05
			-1.14			-			dE_{su}/dpH	15	sttd		C	
			-0.2 to -0.8			0.31	-	-		-		-	Mn,deepest at pH= 5.5,p	
			-1.16			-				15			C	
			-0.2 to -0.8		-	-	-	-		-		-	Mn,p	
			-1.31							18			C	
			-0.2 to -0.8		-	-	-	-		-		-	Mn,p	
			-1.35							18			C	
			-0.2 to -0.8		-	-	-	-		-		-	Mn,p	
			-1.41							18			C	

213

DG06 $C_{20}H_{10}Br_2O_5$

Code No.	Empirical Formula	Name and WLN Code	Structural Formula	Solvent	Tech.	Medium		μ, M	pH	T, °C	Electrodes	App.	Experimental Parameters
DG06	$C_{20}H_{10}Br_2O_5$	9-(2-carboxyphenyl)-4,5-dibromo-6-hydroxy-3H-xanthen-2-one C.A. 596-03-2 T C666 BO EVJ DE IR BVQ& MQ NE	TABLE II.	H_2O	PY	BR		–	3.0 6.0 9.0	–	DME/SCE	05A0	C=0.46, m=2.06, t=4.24, h=40
DG07	$C_{20}H_{10}Cl_2O_5$	9-(2-carboxyphenyl)-4,5-dichloro-6-hydroxy-3H-xanthen-3-one C.A. 2320-96-9 T C666 BO EVJ DG IR BVQ& MQ NG	TABLE II.	H_2O	PY	BR		–	1.85 5.15 7.08 9.0 11.0	–	DME/SCE	05A0	C=0.46, m=2.06, t=4.24, h=40
DG08 bm76	$C_{20}H_{10}I_2O_5$	9-(2-carboxyphenyl)-4,5-diiodo-6-hydroxy-3H-xanthen-3-one C.A. 38577-97-8 T C666 BO EVJ DI IR BVQ& MQ NI	TABLE II.	H_2O	PY	BR		–	4.0 5.15 7.08 10.0	–	DME/SCE	05A0	C=0.46, m=2.06, t=4.24, h=40
DG09	$C_{20}H_{12}N_2Na_2O_7S_2$	naphthyl-1,1'-azo-2-naphthol-4',6-disulfonic acid disodium salt L66J BSWO ENUN- BL6-6J CQ HSWO &-NA- &-NA-	TABLE II.	H_2O	PY	PHOS GEL CITR AM PHOS	? 0.01 ? ? ?	0.8	2.0 4.3 8.5 11.3	25±0.1	DME/SCE	03-1	C=0.5, m=0.909, t=5.9 m=0.908, t=5.80 m=0.910, t=5.58 m=0.919, t=5.48
DG10	$C_{20}H_{15}NO$	6-methoxytribenz-[b,d,f]azocine C.A. 37908-49-9 T H6 B686 NNJ O01	TABLE II.	MeCN	PY	Bu_4NClO_4	0.2	–	–	–	DME Ag/0.1 Ag^+	OAF-	C=0.5, t=6-12
					PA	Bu_4NClO_4	0.2	–	–	–	DME Ag/0.1 Ag^+	OAF-	C=0.5, t=6-12, v=2E4
				THF	PY	Bu_4NPF_6	0.2	–	–	–	DME Ag/0.1 Ag^+	OAF-	C=3.35E-2, t=6-12
					VA	Bu_4NPF_6	0.2	–	–	–	PBE Ag/0.1 Ag^+	OAF-	-2.0→-3.1→-2.0 V, C=3.35E-2, v=400
CONT										23			-2.0→+0.4→-2.0 V

TABLE I. Electrochemical Data $C_{20}H_{15}NO$ (CONT.) DG10

Ref.	C/M	Charact. Potential Value	vs.	Response Const. Value		Tech.	n	Electrokinetic Data Parameter	Value	From	Products and Identification	Description and Remarks	Code No.
EA017 1195	356 a	$E_{\frac{1}{2}}$	−0.45F −0.90F	SCE	i_ℓ	0.5F 0.2F	−	−	−	−	−	−	C,r C	DG06
			−0.88F −1.28F			2.0F 0.5F	−	−	−	−	−	−	C,r C	
			−0.98F −1.54F			1.7F 0.7F	−	−	−	−	−	−	C,r C	
EA017 1195	356 a	$E_{\frac{1}{2}}$	−0.42F −0.60F	SCE	i_ℓ	0.55F 0.25F	−	−	−	−	−	−	C,i_d,r C,i_d	DG07
			−0.75F −1.26F			2.50F 0.50F	−	−	−	−	−	−	C,i_d,r C,i_d	
			−0.90F −1.32F			2.50F 0.25F	−	−	−	−	−	−	C,i_d,r C,i_d	
			−1.05F −1.6F			2.30F 0.35F	−	−	−	−	−	−	C,i_d,r C,i_d	
	356 b		−1.12F −1.60F			1.55 1.05	−	−	−	−	−	−	C,i_d,r C,i_d	
EA017 1195		$E_{\frac{1}{2}}$	−0.60F −1.13F	SCE	i_ℓ	0.50F 0.67F	−	−	−	−	−	−	C,r C	DG08 bm76
			−0.77F −1.02F −1.2F			2.0F 1.0F 3.2F	−	−	−	−	−	−	C,r C C	
	356 c		−0.91F −1.26F −1.63F			3.0F 3.0F 0.3F	−	−	−	−	−	−	C,r C C	
			−1.12F −1.7F			2.6F 2.6F	−	−	−	−	−	−	C,r C	
AN100 0503	−	$E_{\frac{1}{2}}$	−0.08	SCE	I	3.47	D	4	−	−	−	−	C,i_d,$E_{\frac{1}{2}}$=f(C),r	DG09
			−0.26			3.68		4	−	−	−	−	C,i_d,$E_{\frac{1}{2}}$=f(C),r	
			−0.48			3.71		4	−	−	−	−	C,i_d,$E_{\frac{1}{2}}$=f(C),r	
			−0.61			4.08	−	−	−	−	−	−	C,i_d,$E_{\frac{1}{2}}$=f(C),r	
JA094 4907	−	$E_{\frac{1}{2}}$	−2.30F	SCE	I	5.5	I:I	2	−	−	−	−	C,R,r	DG10
JA094 4907	−	$E_{p/2}$	−2.5 −2.5	SCE	i_p	6.7F 6.7F	−	−	−	−	−	−	C,R,M,r A,R	
JA094 4907	−	$E_{\frac{1}{2}}$	−2.81	SCE	I	6.2	I:I	2	Elog	49	plot	−	C,i_k,$E_{\frac{1}{2}}$=f(C),$E_{\frac{1}{2}}$=f(t),r	
JA094 4907	−	E_p	−2.78F	SCE	i_p	85F	−	−	$E_p-E_{p/2}$	50	sttd	−	C,⇌,r	
			0.20F 0.17F −0.88F			48F 57F 28F	−	−	−	−	−	−	A,R,r C,R C	

CONT

DG10 (CONT.) $C_{20}H_{15}NO$

Code No.	Empirical Formula	Name and WLN Code	Structural Formula	Solvent	Tech.	Medium		μ, M	pH	T, °C	Electrodes	App.	Experimental Parameters
DG10	$C_{20}H_{15}NO$	6-methoxytribenz-[b,d,f]azocine	TABLE II.	THF	VR	Bu_4NPF_6	0.2	-	-	23	PBE Ag/0.1 Ag^+	OAF-	0.4→-3.0, C=3.35E-2, v=400
DG11	$C_{20}H_{18}N_4Ni$	[b,i]dibenz-5,12-dimethyl-1,4,8,11-tetraazacyclotetradeca-2,4,6,9,11,13-hexaenenickel(II) C.A. 39018-20-7 D5656 C6 N6 1A W A-NI-N IN MN TNJ J1 U1	TABLE II.	DMSO	VY	Bu_4NBF_4	0.05	-	-	25	RPE/SCE	O--	C≈1
DG12	$C_{20}H_{20}N_2$	1,6-bis(dimethylamino)pyrene C.A. 10075-93-1 L666 B6 2AB PJ GN1&1 NN1&1	TABLE II.	DMF	PY	Pr_4NClO_4	0.1	-	-	-	DME/SCE	---	ns
				MeCN	VY	Pr_4NClO_4	0.1	-	-	-	RPE/SCE	---	ns
DG13	$C_{20}H_{20}N_2O$	5,7-diphenyl-1,3-diazatricyclo-[3.3.1.1^{3,7}]decan-6-one C.A. 19066-35-4 T666/BI/DJ/HJ A J 2BF J BN DN GVTJ FR& HR	TABLE II.	MeCN	VR	$NaClO_4$	0.1	-	-	-	AuBE SCE	2AO	C=1.26, d=0.15, v=55, MP Aux, PE
DG14	$C_{20}H_{22}HgO_4$	bis(α-carbethoxybenzyl)mercury C.A. 10507-36-5 2OVYR&-HG-YR&VO2	[C_2H_5OCOCH-(C_6H_5)]$_2$Hg	MeOH 50	PY	$NaBF_4$	0.1	-	-	25	DME/SCE	O-O	C=0.51, m=0.67, t=3.00
DG15	$C_{20}H_{22}N_2$	5,7-diphenyl-1,3-diazatricyclo-[3.3.1.1^{3,7}]decane C.A. 38705-08-7 T666/BI/DJ/HJ A J 2BF J BN DNTJ FR& HR	TABLE II.	MeCN	VR	$NaClO_4$	0.1	-	-	-	AuBE SCE	2AO	C=0.54, d=0.15, v=52, MP Aux, PE
DG16	$C_{20}H_{23}ClN_2O_4$	[3-(4-chlorophenyl)-3-(2-pyridyl)-1-propyl]dimethylammonium maleate C.A. 113-92-8 T6NJ BYR DG&2N1&1 &C QV1U1VQ	TABLE II.	H_2O	PY	H_2SO_4	?	-	0.2	25±0.1	DME Ag/AgCl, satd KCl	25AO	C=0.25, m=2.65, t=2.78, h=38.3, Pt Aux
						CITR	?		3.0				
						ACET	?		4.4				
						PHOS	?		6.5				
						AM	?		9.1				
CONT													

TABLE I. Electrochemical Data $C_{20}H_{23}ClN_2O_4$ (CONT.) DG16

Ref.	C/M	Charact. Potential		Response Const.		n Tech.	n	Electrokinetic Data			Products and Identification	Description and Remarks	Code No.
		Value	vs.		Value			Parameter	Value	From			
JA094 4907	-	E_p	0.24F 0.19F -0.85F -2.85F	SCE	i_p	51F 48F 37F 85F	-	-	-	-	-	-	A,R,r C,R C,≠ C,≠	DG10
JA094 4529	-	$E_{\frac{1}{2}}$	-1.63 0.58	SCE	i_ℓ/C	7.0 17	i:i	1 2	Tomeš	61 55	sttd	-	C,R,r A,R	DG11
J0032 1322	-	$E_{\frac{1}{2}}$	-2.16	SCE	-	-	sttd	1	-	-	-	$R^{-\cdot}$,sttd;lifetime ≈3s,VA,blue-green electrochemilumines- cence	C,p	DG12
J0032 1322	-	$E_{\frac{1}{2}}$	0.49	SCE	-	-	sttd	1	-	-	-	$R^{+\cdot}$,sttd;lifetime in DMF≈15s,VA, blue-green electro- chemiluminescence	A,p	
JA094 7114	-	E_p	1.15	SCE	-	-	-	-	$E_p-E_{p/2}$	70	sttd	-	A,p	DG13
ER002 0587	23d	$E_{\frac{1}{2}}$	-0.36	SCE	-	-	sttd	2	α	0.77	sttd	-	C,≠,Mc(structure)	DG14
JA094 7114	-	E_p	0.85	SCE	-	-	-	-	$E_p-E_{p/2}$	55	sttd	-	A,p	DG15
AA071 0157	-	$E_{\frac{1}{2}}$	-0.55	Ag/AgCl	i_ℓ	2.07	QE	1.92	Tomeš αn_a p=	48 1.29 1.38	- Elog $dE_{\frac{1}{2}}/dpH$	-	C,i_d,≠,Tc=1.4,maleic acid redn.,r	DG16
			-0.80			2.12		1.92		64 1.29 1.38	- Elog $dE_{\frac{1}{2}}/dpH$	-	C,i_d,≠,Tc=1.4,maleic acid redn.,r	
			-0.98			1.98		1.92	Tomeš	72	sttd	-	C,i_d,≠,Tc=1.4,maleic acid redn.,r	
			-1.3			-		-		-		-	C,X,$E_{\frac{1}{2}}\neq f(pH)$,Cl redn.	
			-1.22			1.45		1.92		63	-	-	C,i_d,≠,Tc=1.4,maleic acid redn.,r	
			-1.3			-		-		-		-	C,X,$E_{\frac{1}{2}}\neq f(pH)$,Cl redn.	
			-1.57			1.35		1.92		-		-	C,i_d,≠,X,M,Tc=1.4, maleic acid redn.,r	

CONT

DG16 (CONT.) $C_{20}H_{23}ClN_2O_4$

Code No.	Empirical Formula	Name and WLN Code	Structural Formula	Solvent	Tech.	Medium		μ, M	pH	T, °C	Electrodes	App.	Experimental Parameters
DG16	$C_{20}H_{23}ClN_2O_4$	[3-(4-chlorophenyl)-3-(2-pyridyl)-1-propyl]dimethyl-ammonium maleate	TABLE II.	H_2O	PV	H_2SO_4	?	-	0.2	-	DME Ag/AgCl, satd KCl	25A0	C=0.25, m=2.65, t=2.78, h=38.3, Δe=10, Pt Aux
						CITR	?		3.0				
						ACET	?		4.4				
						PHOS	?		6.5				
						AM	?		9.1				
					VR	H_2SO_4	0.2	-	-	-	HMDE Ag/AgCl satd KCl	25A0	C=1, v=100, Pt Aux
DG17 bn65	$C_{20}H_{28}O$	vitamin A aldehyde C.A. 116-31-4 L6UTJ A C C B1U1Y&⊃ U2U1Y&U1VH	TABLE II-2.	EtOH 75	IL	H_2SO_4	0.01	-	-	20±0.1	CPDO SCE	26B0	C=0.005-1.0, v=10, Pt Aux
DG18 bn67	$C_{20}H_{28}O_2$	retinoic acid C.A. 302-79-4	TABLE II-2.	EtOH 75	IL	H_2SO_4	0.01	-	-	20±0.1	CPDO SCE	26B0	C=0.005-1.0, v=10, Pt Aux
DG19 bn69	$C_{20}H_{30}O$	retinol C.A. 68-26-8 L6UTJ A C C B1U1Y&U⊃ 2U1Y&U2Q	TABLE II-2.	EtOH 75	IL	H_2SO_4	0.01	-	-	20±0.1	CPDO SCE	26B0	C=0.005-1.0, v=10, Pt Aux
DG20	$C_{20}H_{38}O_4$	didecanoyl peroxide C.A. 762-12-9 9VOOV9	$[CH_3(CH_2)_8C(O)-O]_2$	MeOH 50 C_6H_6 50	PY	LiCl	?	-	-	-	DME/SCE	---	ns
DG21	$C_{21}H_{16}O$	1,1,3-triphenylpropenone C.A. 849-01-4 RYR&U1VR	$(C_6H_5)_2C:CHCO-C_6H_5$	DMF	PY	Bu_4NBF_4	0.50	-	-	25	DME/SCE	23-0	C=1.9-2.5, Pt Aux
DG22	$C_{21}H_{16}O$	cis-1,2,3-triphenyl-2-propenone C.A. 7512-67-6 RVYR&U1R -C	TABLE II.	DMF	PY	Bu_4NBF_4	0.50	-	-	25	DME/SCE	23-0	C=2.4-2.6, Pt Aux
DG23	$C_{21}H_{16}O$	trans-1,2,3-triphenyl-2-propenone C.A. 7474-65-9 RVYR&U1R -T	TABLE II.	DMF	PY	Bu_4NBF_4	0.50	-	-	25	DME/SCE	23-0	C=2.4-2.5, Pt Aux
DG24	$C_{21}H_{22}N_2O_2$	strychnine C.A. 57-24-9 T656 C6 D6 E5 S6 5A⊃ BCDEF B A& FX IN NO QVNJ	TABLE II.	m-cresol	PY	Et_4NClO_4	0.1	-	-	-	DME Ag/AgCl, 0.1 Me$_4$NCl	216A0	C=0.5-1.5, m=2.87, t=2.70, h=56, MP Aux

TABLE I. Electrochemical Data $C_{21}H_{22}N_2O_2$ DG24

Ref.	C/M	Charact. Potential		Response Const.		Tech.	n	Electrokinetic Data			Products and Identification	Description and Remarks	Code No.
		Value	vs.		Value			Parameter	Value	From			
AA071 0157	-	E_{su}	-0.57	Ag/AgCl	i_{su}	0.60	QE	1.92	$\Delta E_{su/2}$	80	-	-	$C,i_d,\neq$,maleic acid redn.,r	DG16
			-0.83			0.49	-	-		130		-	C,i_a,Tc=1.1,maleic acid redn.,r	
			-1.01			0.41	-	-		166		-	C,i_a,Tc=1.1,maleic acid redn.,r	
			-1.26			0.41	-	-		130		-	C,i_a,Tc=1.1,maleic acid redn.,r	
			-1.70			0.37	-	-		-	-	-	C,i_a,Tc=1.1,maleic acid redn.,r	
AA071 0157	-	E_p	-0.60	Ag/AgCl	$i_p/Cv^{\frac{1}{2}}$	2.27	-	-		-	-	-	$C,i_p/Cv^{\frac{1}{2}}\neq f(v^{\frac{1}{2}})$, maleic acid redn.,r	
AN099 0683	-	$E_{p/2}$	0.80	SCE	-	-	-	-	-	-	-	-	$A,i_p \propto C,r$	DG17 bn65
AN099 0683	-	$E_{p/2}$	0.78	SCE	-	-	-	-	-	-	-	-	$A,i_p \propto C,r$	DG18 bn67
AN099 0683	-	$E_{p/2}$	0.67	SCE	-	-	-	-	-	-	-	-	$A,i_p \propto C,r$	DG19 bn69
AC035 0880	-	$E_{\frac{1}{2}}$	-0.10	SCE	i_ℓ/C	10.0	-	-	-	-	-	-	C,p	DG20
JA094 8471	-	$E_{\frac{1}{2}}$	-1.44 -1.82	SCE	-	-	-	-	αn_a	1.2 1.4	Elog	-	C,p C	DG21
JA094 8471	-	$E_{\frac{1}{2}}$	-1.58 -1.94	SCE	-	-	-	-	αn_a	1.3 1.2	Elog	-	C,p C	DG22
JA094 8471	-	$E_{\frac{1}{2}}$	-1.58 -1.94	SCE	-	-	-	-	αn_a	1.2 1.3	Elog	-	C,p C	DG23
AA072 0169	-	$E_{\frac{1}{2}}$	0.43 -	Ag/AgCl	I i_ℓ	0.388 0.17	-	-	-	-	-	-	$A,\neq$,m-cresolate oxidn.,pK_a=13,r A,i_a,Pr,$i\neq f(C)$	DG24

Code No.	Empirical Formula	Name and WLN Code	Structural Formula	Solvent	Tech.	Medium		μ, M	pH	T, °C	Electrodes	App.	Experimental Parameters
DG25	$C_{21}H_{28}O_3$	16α,17α-oxidoprogesterone T F3 E5 B666 GO PV NUTJ A1 E1 FV1	TABLE II.	DMF	PY	Bu_4NI Et_4NI	0.1 0.01	-	-	25±0.5	DME/MP	OAO	ns
DG26 cm23	$C_{21}H_{29}N_7O_{14}P_2$	nicotinamide adenine dinucleotide,reduced C.A. 58-68-4 T56 BN DN FN HNJ IZ D- BT5OTJ CQ DQ -A&⊃ CD -B&BE E1OPQO&OPO⊃ &OO1- BT5OTJ CQ DQ -A&BE -B&CD E- A T6⊃ N DHJ CVZ &58	TABLE II-3.	H_2O	IL	PHOS TRIS	0.1 ?	-	7.3 8.0 8.0	-	GCE/SCE Pt/SCE CPE/SCE	2AO	C=0.89,v=50, Pt Aux C=0.57,v=5 C=0.52
DG27	$C_{21}H_{30}N_7O_{17}P_3$	nicotinamide adenine dinucleotide phosphate,reduced C.A. 53-57-6 T56 BN DN FN HNJ IZ D- BT5OTJ COPQQO DQ E1OPQO&OPO&O&O1- BT⊃ 5OTJ CQ DQ E- AT6N DHJ CVZ	TABLE II.	H_2O	IL	PHOS	0.1	-	7.3	-	GCE/SCE		C=0.71,v=50, Pt Aux
DG28	$C_{21}H_{32}O$	retinyl methyl ether C.A. 32450-56-9 L6X BUTJ A1 A1 B1U1⊃ Y1&U2U1&U2O1 C1	TABLE II.	DMF 95 DMF	PY PY	Et_4NI Et_4NI	0.04 0.04	-	-	-	DME Ag/AgI DME Ag/AgI	0-0 0-0	m=0.608, t=0.20 m=0.608, t=0.20
DG29	$C_{22}H_{12}$	benzo[g,h,i]perylene C.A. 191-24-2 L666 B6 C6 D6 4ABCD VJ	TABLE II.	MeCN	VY	$NaClO_4$	2.0	-	-	25.0 ±0.1	RPE/SCE	13AO	C=0.030-0.70, A≈0.05,v= 1.7,ω=10,AE
DG30	$C_{22}H_{16}N_2$	1-naphthaldehyde-azine C.A. 2144-00-5 L66J B1UNNU1- BL66J	TABLE II.	EtOH 80	PY	IP		-	3.85 6.25 7.00 9.00 12.25	-	DME/SCE	0-0	ns
DG31	$C_{22}H_{16}N_2$	2-naphthaldehyde-azine C.A. 2144-02-7 L66J C1UNNU1- CL66J	TABLE II.	MeOH 92	PY	Et_4NI	0.02	-	-	-	DME/SCE	0-0	ns
DG32	$C_{11}H_{10}MoO_4$	acetonyl(tricarbonyl)-η-cyclopentadienylmolybdenum L5ØJ Ø-MO-CO&CO&CO&⊃ 1V1 -MO-CO&CO&CO&1V⊃ 1&- ØL5ØJ	$(C_5H_5)Mo(CO)_3$-$CH_2C(O)CH_3$	MG	PY	Bu_4NClO_4	0.1	-	-	22	DME Ag/Ag+	2-0	C=2

TABLE I. Electrochemical Data $\quad C_{22}H_{16}N_2 \quad DG32$

Ref.	C/M	Charact. Potential		Response Const.		n Tech.	n	Electrokinetic Data			Products and Identification	Description and Remarks	Code No.
		Value	vs.		Value			Parameter	Value	From			
AC035 0128	–	$E_{\frac{1}{2}}$	–2.16	SCE	I	4.2	–	–	αn_a	0.79	Elog	–	C,p	DG25
AA078 0271	–	$E_{p/2}$	0.32	SCE	i_p	5F	–	–	–	–	–	–	A,r	DG26 cm23
		E_p	0.60F			0.8F	–	–	–	–	–	–	A,M,r	
			0.75F			0.15F	–	–	–	–	–	–	A,r	
AA078 0271	–	$E_{p/2}$	0.32	SCE	i_p	5F	–	–	–	–	–	–	A,r	DG27
ER003 0538	33a	$E_{\frac{1}{2}}$	–2.07 –2.31 –2.67	SCE	–	–	1:1	2 2 2	–	–	–	–	C,p C C	DG28
ER003 0538	33a	$E_{\frac{1}{2}}$	–2.12 –2.33 –2.47 –2.74	SCE	–	–	1:1	2 1 0.4 0.5	–	–	–	–	C,p C C C	
JA085 2124	–	$E_{\frac{1}{2}}$	1.01 ±0.01	SCE	–	–	–	–	–	–	–	–	A,Av(5),$\Delta E_{\frac{1}{2}} \propto \Delta\beta$,p	DG29
RG035 0015	218 a	$E_{\frac{1}{2}}$	–1.10	SCE	I	6.6	1:1	4	–	–	–	–	C,i_d,p	DG30
			–1.20			6.0		4	–	–	–	–	C,i_d,p	
			–1.27			3.2		2	–	–	–	–	C,i_d,p	
			–1.94			3.1		2	–	–	–	–	C,i_d	
			–1.32			3.4		2	–	–	–	–	C,i_d,p	
			–1.87			3.5		2	–	–	–	–	C,i_d	
			–1.32			2.5		2	–	–	–	–	C,i_d,p	
			–1.90			5.4		–	–	–	–	–	C,i_d	
RG035 0015	218 a	$E_{\frac{1}{2}}$	–1.28 –1.74	SCE	–	–	QE 1:1	2 2	–	–	–	–	C,i_d,p C,i_d	DG31
JA088 5117	–	$E_{\frac{1}{2}}$	–2.0	Ag/Ag$^+$	–	–	QE	1	–	–	–	Mo$^-$ + CH$_3$C(O)CH$_2$, sttd	C,p	DG32

221

Code No.	Empirical Formula	Name and WLN Code	Structural Formula	Solvent	Tech.	Medium		μ, M	pH	T, °C	Electrodes	App.	Experimental Parameters
DG33 bo63	$C_{22}H_{32}O_2$	retinyl acetate C.A. 127-47-9, 7095-40-1 L6UT A B1U1Y&U2U1Y& 42OV1 C C	TABLE II-2.	EtOH 75	IL	H_2SO_4	0.01	-	-	20± 0.1	CPDO SCE	26B0	C=0.005-1.0, v=10, Pt Aux
						BR			2-11		GCE/SCE		C=0.4
DG34	$C_{23}H_{18}N_4$	4-methyl-2,3,6-triphenylpyrazolo[3,4-d]pyrazole C.A. 52887-27-1 L6X BUTJ A1 A1 B1U1 Y1&U2U1Y1&U2OV1 C1	TABLE II.	DMF	PY	Bu_4NI	0.155	-	-	22	DME/MP?	OA?O	t=4.36, h=60
DG35	$C_{24}H_{16}$	5,12-dihydro-5,12-o-benzenonaphthacene C.A. 13395-89-6 L-L E6 C666J DR D-K8J	TABLE II.	DMF	PY	ns		-	-	-	DME?/ SCE?	---	ns
DG36	$C_{24}H_{20}F_6Fe_4O_4P$	tetrakis[carbonyl(-η-cyclopentadienyl)-iron] hexafluorophosphate C.A. 12791-65-0 T4-FE--FE--FE--FE-J ACO A- ØL5ØJ& BCO B- ØL5ØJ& CCO C- ØL5Ø J& DCO D- ØL5ØJ & P -F6	$[(\pi\text{-}C_5H_5)Fe\text{-}(CO)]_4PF_6$	MeCN	QE	Bu_4NPF_6	0.1	-	-	-	Pt/SSCE	---	E_{app}=-1.50
													E_{app}=+1.30
					VR	Bu_4NPF_6	0.1	-	-	22±2	PBE/ SSCE	2--	C=1, v=100
				PrCN	VR	Bu_4NPF_6	0.1	-	-	-	PBE/ SSCE ferrocene	2--	C=1, $E_{\frac{1}{2}}$, ferrocene= 0.37 V vs. SSCE
				CH_2Cl_2	VR	Bu_4NPF_6	0.1	-	-	-	PBE/ SSCE ferrocene	2--	C=1, $E_{\frac{1}{2}}$, ferrocene= 0.46 V vs. SSCE
				DMF	VR	Bu_4NPF_6	0.1	-	-	-	PBE/ SSCE ferrocene	2--	C=1, $E_{\frac{1}{2}}$, ferrocene= 0.43 V vs. SSCE
				DMSO	QE	Bu_4NPF_6	0.1	-	-	-	Pt/SSCE	---	E_{app}=1.20
					VR	Bu_4NPF_6	0.1	-	-	-	PBE/ SSCE ferrocene	2--	C=1, $E_{\frac{1}{2}}$, ferrocene= 0.44 V vs. SSCE
DG37	$C_{24}H_{20}Fe_4O_4$	tetrakis[carbonyl-(η-cyclopentadienyl)-iron(I)] C.A. 12203-87-1 T4-FE--FE--FE--FE-J ACO A- ØL5ØJ& BCO B- ØL5ØJ& CCO C- ØL5Ø J& DCO D- ØL5ØJ	$[(\pi\text{-}C_5H_5)Fe\text{-}(CO)]_4$	CH_2Cl_2	QE	Bu_4NPF_6	0.1	-	-	-	PBE/ SSCE	---	E_{app}=0.80 V
											Pt/SSCE		E_{app}=1.40 V
				MG	PY	Bu_4NClO_4	0.1	-	-	22	DME Ag/Ag+	2-0	C=0.2
DG38	$C_{24}H_{20}N_2$	tetraphenylhydrazine C.A. 632-52-0 RNR&NR&R	$(C_6H_5)_2NN(C_6H_5)_2$	MeCN	VR	$NaClO_4$	0.1	-	-	-	AuBE SCE	2A0	C=0.2-0.9, d=0.15, v=54, MP Aux, PE

TABLE I. Electrochemical Data $C_{24}H_{20}N_2$ DG38

Ref.	C/M	Charact. Potential		Response Const.		Tech.	n	Electrokinetic Data			Products and Identification	Description and Remarks	Code No.
		Value	vs.		Value			Parameter	Value	From			
AN099 0683	-	$E_{p/2}$	0.71	SCE	-	-	sttd	-	-	-	-	-	$A, i_p \propto C, r$	DG33 bo63
		0.74 ±0.02		-	-	-	-	-	-	-	-	$A, E_{p/2} \neq f(pH), r$	
		0.65 ±0.02		-	-	-	-	-	-	-	-	$A, E_{p/2} \neq f(pH), r$	
BJ047 1490	-	$E_{\frac{1}{2}}$	-1.58	MP	i_ℓ/C	4.6	-	-	Elog	52	sttd	-	$C, E_{\frac{1}{2}} \propto LUMO, R?, r$	DG34
BJ048 0416	-	$E_{\frac{1}{2}}$	-2.58	SCE	-	-	-	-	-	-	-	radical anion, ESR	C, p	DG35
JA094 3409	369 a	-	-	-	-	-	QE	2.25	-	-	-	$[(\pi-C_5H_5)Fe(CO)]_4^-$, yellow, ESR	C, p	DG36
		-	-	-	-	-	-	>6	-	-	-	$[(\pi-C_5H_5)Fe(CO)_2(NCCH_3)]^+$, IRS	A, p	
JA094 3409	369 a	E_p	1.04F	SSCE	i_p	28F	-	-	-	-	-	-	C, R	
			0.31F			33F							C, R	
			-1.44F			36F							C, R	
			-1.34F			36F							A, R	
			0.40F			33F							A, R	
			1.12F			28F							A, R	
JA094 3409	369 a	$E_{\frac{1}{2}}$	0.30	SSCE	-	-	-	-	-	-	-	-	C, R, p	
JA094 3409	369 a	$E_{\frac{1}{2}}$	0.38	SSCE	-	-	-	-	-	-	-	-	C, R, p	
JA094 3409	369 a	$E_{\frac{1}{2}}$	0.35	SSCE	-	-	-	-	-	-	-	-	C, R, p	
JA094 3409	369 a	-	-	-	-	-	QE	>7	-	-	-	Fe^{3+}	A, p	
JA094 3409	369 a	$E_{\frac{1}{2}}$	0.35	SSCE	-	-	-	-	-	-	-	-	C, R, p	
JA094 3409	369 a	-	-	-	-	-	QE	1.02	-	-	-	$[(\pi-C_5H_5)Fe(CO)]_4^+$ PF_6, green	A, p	DG37
		-	-	-	-	-	-	>10	-	-	-	$[(\pi-C_5H_5)Fe(CO)]_4^+$ PF_6, UVS, IRS	A, current still at initial value when QE stopped	
JA088 5117	-	$E_{\frac{1}{2}}$	-1.9	Ag/Ag^+	-	-	-	-	-	-	-	-	C, R, i decreases only slightly during QE, p	
JA094 7108	-	XE	0.76	SCE	-	-	sttd	1	ΔE_p	60	sttd	radical cation, ESR	$C, XE=\frac{1}{2}(E_{p,A}+E_{p,C}), p$	DG38
			-		-	-		1		-			C	

223

Code No.	Empirical Formula	Name and WLN Code	Structural Formula	Solvent	Tech.	Medium	μ, M	pH	T, °C	Electrodes	App.	Experimental Parameters
DG39	$C_{24}H_{22}N_4NiO_2$	[b,i]dibenz-6,13-diacetyl-5,12-dimethyl-1,4,8,11-tetra-azacyclotetradeca-2,4,6,9,11,13-hexa-enenickel(II) C.A. 39018-22-9 D5656 C6 N6 1A W A- NI-N IN MN TNJ J1 K01 U1 V01	TABLE II.	MeCN	VY	Bu_4NBF_4 0.05	–	–	25	RPE/SCE	O--	$C \approx 1$
DG40	$C_{24}H_{24}O_6$	2,3,6,7,10,11-hexa-methoxytriphenylene C.A. 808-57-1 L H6 B666J D01 E01 J01 K01 P01 Q01	TABLE II.	CF_3COOH	VA	Bu_4NBF_4 ?	–	–	–	ns	---	$0 \to 1 \to 0$ V, v=150
DG41	$C_{24}H_{24}O_6$	2,3,6,7,10,11-hexa-methoxytriphenylene radical cation C.A. 38032-87-0 L H6 B666J D01 E01 J01 K01 P01 Q01 &8	TABLE II.	CF_3COOH 50 $HFSO_3$ 50	VR	Bu_4NBF_4 ?	–	–	–	ns	---	v=150
DG42 bp53	$C_{24}H_{46}O_4$	didodecanoyl peroxide C.A. 105-74-8 1.1V00V11	$[CH_3(CH_2)_{10}C(O)O]_2$	MeOH 50 C_6H_6 50	PY	LiCl ?	–	–	–	DME/SCE	---	ns
DG43	$C_{25}H_{20}FeO_2Pb$	di-μ-carboxyl-(η-cyclopentadienyl)[triphenylplumbio(III)]-iron(I) C.A. 12132-08-0 L50J Ø-FE-CO&CO&-Pb--R&R&R	$(C_5H_5)Fe(CO)_2$-$Pb(C_6H_5)_3$	MG	PY	Bu_4NClO_4 0.1	–	–	22	DME Ag/Ag+	2-0	C=0.2
DG44	$C_{25}H_{20}FeO_2Sn$	di-μ-carbonyl-(η-cyclopentadienyl)[triphenylstannio(III)]-iron(I) C.A. 12132-09-1 L50J Ø-FE-CO&CO&-SN--R	$(C_5H_5)Fe(CO)_2$-$Sn(C_6H_5)_3$	MG	PY	Bu_4NClO_4 0.1	–	–	22	DME Ag/Ag+	2-0	C=2
DG45	$C_{25}H_{33}FO_7$	9α-fluoro-17α,21-dihydroxy-5-pregnene-3,11,20-trione-3-ethyleneketal-17,-20,21-bismethylenedioxolane L E5 B666 AXXV EXX OX LUTJ A1 BF E1 O-& BT50XOTJ& F-& BT50 XXO EHJ C-& BT50X DOTJ	TABLE II.	EtOH 85 MeOH 5	PY	Bu_4NCl ?	–	–	25	DME/NCE	013A 0	ns
DG46	$C_{25}H_{36}O$	benzhydryl dodecyl ether C.A. 33242-40-9 120YR&R	$(C_6H_5)_2CHO$-$(CH_2)_{11}CH_3$	MeCN	QE	$LiClO_4$ 0.1	–	–	–	Pt Ag/0.1 Ag+	25A-	C=23, A=12.9, E_{app}=1.90, stainless steel Aux
DG47 bp78 cm56	$C_{26}H_{18}$	9,10-diphenylanthracene C.A. 1499-10-1 L C666 BR& IR	TABLE II-2.	DMF	VR	Bu_4NClO_4 0.1	–	–	–	PDE/SCE	3F-	C=0.1-1, v=228

TABLE I. Electrochemical Data $C_{26}H_{18}$ DG47

Ref.	C/M	Charact. Potential Value	vs.	Response Const. Value		Tech.	n	Electrokinetic Data Parameter	Value	From	Products and Identification	Description and Remarks	Code No.
JA094 4529	-	$E_{\frac{1}{2}}$ -1.48 0.94 1.14	SCE	i_ℓ/C	25 24 24	i:i	1 1 1	Tomeš	62 68 73	sttd	-	C,R,r A,Q A,Q	DG39
JA094 4749	-	E_p 0.50F 0.87F 0.77F 0.21F	SCE	i_p	9F 9F 9F 7F	-	-	-	-	-	hexamethoxytriphen-ylene radical cation,ESR hexamethoxytriphen-ylene radical di-cation,ESR - -	A,p A C C	DG40
JA094 4749	-	E_p 0.50F 0.87F 1.34F 1.24F 0.70F 0.34F	SCE	i_p	11F 8F 7.5F 11F 6.5F 8	-	-	-	-	-	- diradical cation triradical cation	A,p A A C C C	DG41
AC035 0880	-	$E_{\frac{1}{2}}$ -0.09	SCE	i_ℓ/C	9.6	-	-	-	-	-	-	C,p	DG42 bp53
JA088 5117	-	$E_{\frac{1}{2}}$ -2.1 -2.0?	Ag/Ag$^+$	-	-	QE	2 -	-	-	-	Fe$^-$ + Pb$^-$,sttd	C,p C	DG43
JA088 5117	-	$E_{\frac{1}{2}}$ -2.6 -2.9	Ag/Ag$^+$	-	-	QE	1 -	-	-	-	Fe$^-$ + Sn,sttd	C,p C,[$(C_6H_5)_3Sn$]$_2$ redn.	DG44
AC034 1440	-	$E_{\frac{1}{2}}$ -2.03	NCE	I	2.3	-	-	-	-	-	-	C,p	DG45
JA094 6812	360 a	-	-	-	-	QE	2.2	-	-	-	benzophenone,70%; dodecanol,70%; NMR,GLC	A,p	DG46
JA094 0691 (JA094 6317)	-	E_p -1.89 1.35	SCE	-	-	-	-	-	-	-	-	C,R,r A,∓	DG47 bp78 cm56

DG48 $C_{26}H_{18}Cl_2$

Code No.	Empirical Formula	Name and WLN Code	Structural Formula	Solvent	Tech.	Medium		μ, M	pH	T, °C	Electrodes	App.	Experimental Parameters
DG48	$C_{26}H_{18}Cl_2$	9,10-dichloro-9,10-dihydro-9,10-diphenylanthracene C.A. 6486-01-7 L C666 BX IXJ BG BR& IG IR	TABLE II.	MeCN	PY	Bu_4NClO_4	0.1	–	–	–	DME Ag/satd $AgNO_3$	5A0	C=0.2
				DMF	VR	Bu_4NClO_4	0.1	–	–	–	Pt Ag/satd $AgNO_3$	5A0	C=0.2, v=500
				THF	VR	Bu_4NClO_4	0.1	–	–	–	Pt Ag/satd $AgNO_3$	5A0	C=0.2, v=500
DG49	$C_{26}H_{20}O$	tetraphenylethylene oxide C.A. 470-35-9 T30XXJ BR& BR& CR& CR	TABLE II.	MeCN	QE	$LiClO_4$	0.1	–	–	–	Pt Ag/0.1 Ag^+	25A-	C=2.86, A=12.9, E_{app}=1.70 V, stainless steel Aux
DG50	$C_{26}H_{44}O_2$	2-methyl-2-(4,8,12-trimethyltridecyl)-6-chromanol C.A. 119-98-2 T66 BOXT&J C1 C3Y1& 3Y1&3Y1&1 HQ	TABLE II.	EtOH 75	IL	H_2SO_4	0.2	–	–	20±0.1	CPDO SCE	26B0	C=0.003-0.7, v=8.3, Pt Aux
DG51	$C_{27}H_{20}O$	1,2,3,3-tetraphenylpropenone C.A. 6333-11-5 RYR&UYR&VR	$(C_6H_5)_2C:C-(C_6H_5)COC_6H_5$	DMF	PY	Bu_4NBF_4	0.50	–	–	25	DME/SCE	23-0	C=1.9-2.1, Pt Aux
DG52	$C_{27}H_{44}O$	cholesta-5,7-dien-3β-ol C.A. 434-16-2 L E5 B666 AX EXTJ A1 E1 FY1&3Y1&1 OQ	TABLE II.	MeOH 75	IL	BR		–	2	20±0.1	GCE/SCE	26B0	C=0.4, v=10, Pt Aux
				MeOH 67 C_6H_6 33	IL	$LiClO_4$	0.05	–	–	20±0.1	GCE/SCE	26B0	v=30, Pt Aux
DG53 bq34	$C_{27}H_{44}O$	vitamin D_3 C.A. 67-97-0 L56 AX FYTJ A1 BY1& 3Y1&1 FU2U- BL6YYTJ AU1 DQ A&D	TABLE II-2.	MeOH 75	IL	BR		–	2	20±0.1	CPDO SCE	26B0	C=0.4, v=10, Pt Aux
				MeOH 67 C_6H_6 33	IL	$LiClO_4$	0.05	–	–	20±0.1	GCE/SCE	26B0	C=ns, v=30, Pt Aux
					VA	$LiClO_4$	0.05	–	–	20±0.1	GCE/SCE	26B0	C=ns, v=10-100, Pt Aux
DG54 cm64	$C_{27}H_{46}O_2$	δ-tocopherol C.A. 119-13-1, 7773-19-5 T66 BOXT&J C C3Y3Y3Y HQ J	TABLE II-3.	EtOH 75	IL	H_2SO_4	0.2	–	–	20±0.1	CPDO SCE	26B0	C=0.003-0.7, v=8.3, Pt Aux
DG55	$C_{28}H_{18}$	dibenzotriptycene L-L G6 E6 C666J DR D- N8J	TABLE II.	DMF	PY	ns		–	–	–	DME?/SCE?		ns
DG56	$C_{28}H_{18}Cl_2N_4$	2,5-bis(4-chlorophenyl)-3,6-diphenylpyrazolo[4,3-d]-pyrazolium anion T55 BNN FNNJ CR DG& DR& GR DG& HR &7/6	TABLE II.	DMF	PY	Bu_4NI	0.155	–	–	22	DME/MP?	0A?0	t=4.36, h=60

TABLE I. Electrochemical Data $C_{28}H_{18}Cl_2N_4$ DG56

Ref.	C/M	Charact. Potential		Response Const.		n Tech.	n	Electrokinetic Data			Products and Identification	Description and Remarks	Code No.
		Value	vs.		Value			Parameter	Value	From			
JA094 9020	-	$E_{\frac{1}{2}}$	-0.2 -1.9 -2.3	SCE	i_ℓ	8F 3F 1.5F	sttd	2 1 1	-	-	-	-	C,r C C	DG48
JA094 9020	-	E_p	-0.4 -0.9 -1.9 -1.8F -0.8F -	SCE	i_p	10F 0.25F 23F 0.3F - 4F	-	-	-	-	-	-	C,Q,r C,≠ C A A A,X	
JA094 9020	-	E_p	-1.50F -1.85F -1.65F	SCE	i_p	22.5F 14F 23F	-	-	-	-	-	-	C,p C A	
JA094 6812	361 a	-	-	-	-	-	QE	3.2	-	-	-	benzophenone,96.5%, NMR,GLC,IRS	A,p	DG49
AN098 0886	-	$E_{p/2}$	0.614	SCE	-	-	-	-	-	-	-	-	A,$i_p \propto C$,r	DG50
JA094 8471	-	$E_{\frac{1}{2}}$	-1.59 -1.7	SCE	-	-	-	-	αn_a -	1.0	Elog	-	C,p C	DG51
AN100 0827	-	$E_{p/2}$	0.84 ±0.02	SCE	-	-	-	-	-	-	-	-	A,$E_{p/2} \neq f(pH)$ for pH=2-11,p	DG52
AN100 0827	-	$E_{p/2}$	1.04	SCE	-	-	-	-	-	-	-	-	A,p	
AN100 0827	-	$E_{p/2}$	0.95 ±0.02	SCE	-	-	-	-	-	-	-	-	A,$E_{p/2} \neq f(pH)$ for pH=2-11,p	DG53 bq34
AN100 0827	-	$E_{p/2}$	1.11	SCE	-	-	-	-	-	-	-	-	A,p	
AN100 0827	-	-	-	-	-	-	-	-	$\beta n_b=$	0.98	$E_p - E_{p/2}$	-	A,≠,i_d,$i_p \propto v^{\frac{1}{2}}$,p	
AN098 0886	-	$E_{p/2}$	0.591	SCE	-	-	-	-	-	-	-	-	A,$i_p \propto C$,r	DG54 cm64
BJ048 0416	-	$E_{\frac{1}{2}}$	-2.65 -2.85	SCE	-	-	-	-	-	-	-	radical anion,ESR	C,P,r? C	DG55
BJ047 1490	-	$E_{\frac{1}{2}}$	-1.10	MP	i_ℓ/C	3.4	-	-	Elog	64	-	-	C,$E_{\frac{1}{2}} \propto$ LUMO,R?,r	DG56

Code No.	Empirical Formula	Name and WLN Code	Structural Formula	Solvent	Tech.	Medium		μ, M	pH	T, °C	Electrodes	App.	Experimental Parameters
DG57 bq77	$C_{28}H_{44}O$	calciferol C.A. 50-14-6 L56 FYTJ A BY&1U1Y&Y FU2U BL6YYTJ AU1 DQ	TABLE II-2.	MeOH 75	IL	BR	-	-	2	20± 0.1	CPDO SCE	26B0	C=0.4,v=10, Pt Aux
				MeOH 67 C_6H_6 33	IL	$LiClO_4$	0.05	-	-	20± 0.1	GCE/SCE	26B0	C=ns,v=30, Pt Aux
DG58 bq78	$C_{28}H_{44}O$	ergosta-5,7,22-trien-3β-ol C.A. 57-87-4 L E5 B666 JU LUTJ A E FY&1U1Y&Y OQ -A&B -B&O	TABLE II-2.	MeOH 75	IL	BR	-	-	2-11	20± 0.1	CPDO SCE	26B0	C=0.4,v=10, Pt Aux
				MeOH 67 C_6H_6 33	IL	$LiClO_4$	0.05	-	-	20± 0.1	GCE/SCE	26B0	C=ns,v=30, Pt Aux
DG59	$C_{28}H_{44}O$	4-[2-(5-hydroxy-2-methylcyclohexen-1-yl)ethenyl]-7a-methyl-3a,6,7,7a-tetrahydroindan C.A. 469-06-7 L56 AX EUTJ A1 BY1& 1U1Y1&Y1&1 F1U1- BL6 AUTJ A1 DQ	TABLE II.	MeOH 67 C_6H_6 33	IL	$LiClO_4$	0.05	-	-	20± 0.1	GCE/SCE	26B0	C=ns,v=30, Pt Aux
DG60	$C_{28}H_{48}O_2$	β-tocopherol C.A. 148-03-8,48223-98-9 T66 BOXT&J C1 C3Y1& 3Y1&3Y1&1 G1 HQ J1	TABLE II.	EtOH 75	IL	H_2SO_4	0.2	-	-	20± 0.1	CPDO SCE	26B0	C=0.003-0.7, v=8.3,Pt Aux
DG61 cm72	$C_{28}H_{48}O_2$	γ-tocopherol C.A. 7616-22-0 T66 BOXT&J C C3Y3Y3Y HQ I J	TABLE II-3.	EtOH 75	IL	H_2SO_4	0.2	-	-	20± 0.1	CPDO SCE	26B0	C=0.003-0.7, v=8.3,Pt Aux
DG62	$C_{28}H_{54}O_4$	ditetradecanoyl peroxide C.A. 3530-28-7 13VOOV13	[$CH_3(CH_2)_{12}C(O)$-O]$_2$	MeOH 50 C_6H_6 50	PY	LiCl	?	-	-	-	DME/SCE	---	ns
DG63 cm75	$C_{29}H_{50}O_2$	α-tocopherol C.A. 59-02-9,10191-41-0,18920-63-3 T66 BOXT&J C C3Y3Y3Y G HQ I J	TABLE II-3.	EtOH 75	IL	H_2SO_4	0.2	-	-	20± 0.1	CPDO SCE	26B0	C=0.003-0.7, v=8.3,Pt Aux
DG64	$C_{30}H_{24}Cl_2N_6Ru$	trisbipyridineruthenium(II) dichloride C.A. 14323-06-9 D B656 GN-RU-NJ H-& HD B656 GN-RU-NJ H -& HD B656 GN-RU-NJ &G &G	TABLE II.	MeCN	VR	Bu_4NBF_4	0.2	-	-	-	Pt Ag wire	---	C=1,v=100
DG65	$C_{30}H_{24}FeN_6O_4S$	tris(bipyridine)-iron(II) sulfate C.A. 14263-81-1 D B656 GN-FE-NJ H-& HD B656 GN-FE-NJ H-& HD B656 GN-FE-NJ &WSW	TABLE II.	H_2O	PY	BOR	?	-	9.6	27	DME/NHE	---	ns

TABLE I. Electrochemical Data $C_{30}H_{24}FeN_6O_4S$ DG65

Ref.	C/M	Charact.	Potential		Response Const.		n Tech.		Electrokinetic Data			Products and Identification	Description and Remarks	Code No.
			Value	vs.		Value			Parameter	Value	From			
AN100 0827	-	$E_{p/2}$	0.95 ±0.02	SCE	-	-	-	-	-	-	-	-	A,$E_{p/2}\neq f(pH)$ for pH=2-11,r	DG57 bq77
AN100 0827	-	$E_{p/2}$	1.11	SCE	-	-	-	-	-	-	-	-	A,r	
AN100 0827	-	$E_{p/2}$	0.84 ±0.03	SCE	-	-	-	-	-	-	-	-	A,$E_{p/2}\neq f(pH)$ for pH=2-11,r	DG58 bq78
AN100 0827	-	$E_{p/2}$	1.04	SCE	-	-	-	-	-	-	-	-	A,r	
AN100 0827	-	$E_{p/2}$	0.89	SCE	-	-	-	-	-	-	-	-	A,r	DG59
AN098 0886	-	$E_{p/2}$	0.520	SCE	-	-	-	-	-	-	-	-	A,$i_p \propto C$,r	DG60
AN098 0886	-	$E_{p/2}$	0.525	SCE	-	-	-	-	-	-	-	-	A,$i_p \propto C$,r	DG61 cm72
AC035 0880	-	$E_{1/2}$	-0.12	SCE	i_ℓ/C	8.2	-	-	-	-	-	-	C,p	DG62
AN098 0886	-	$E_{p/2}$	0.445	SCE	-	-	-	-	-	-	-	-	A,$i_p \propto C$,r	DG63 cm75
JA094 2862	-	E_p	1.63	Ag wire	-	-	i_p/i_p	1	-	-	-	-	C	DG64
			-1.09					1					C	
			-1.27					1				$Ru(bipy)_3^+$	C	
			-1.53					1				$Ru(bipy)_3^0$	C	
			-2.22					≈3				$Ru(bipy)_3^-$	C,probably due to free bipy	
			-1.46					1					A	
			-1.22					1					A	
			-1.03					1					A	
			-1.70					1				$Ru(bipy)_3^{3+}$	A	
JA095 3411	-	$E_{1/2}$	-1.07F	NHE	-	-	-	-	-	-	-	-	C,p	DG65

DG66 $C_{31}H_{46}O_2$

Code No.	Empirical Formula	Name and WLN Code	Structural Formula	Solvent	Tech.	Medium		μ, M	pH	T, °C	Electrodes	App.	Experimental Parameters
DG66	$C_{31}H_{46}O_2$	2-methyl-3-(3,7,11,15-tetramethyl-2-hexadecenyl)-1,4-naphthoquinone C.A. 84-80-0 L66 BV EVJ C1 D2UY1&3Y1&3Y1&3Y1&1	TABLE II.	i-PrOH 75	DI	NH_4Cl	0.06	—	—	25±0.1	DME/SCE	2A0	C=0.006,t=2, ΔE=50,Pt Aux
DG67	$C_{32}H_{12}FeN_8Na_4O_{12}S_4$	tetrasodium phthalocyanineiron-2,9,16,23-tetrasulfonate D656 B-5 B6 B-&5 B5 C5 G6 N6 B&6 K&6/B-0/B-&A&/CJ&/DP&& 6ABBB-B-&C P& AN-FE-N EN NN WN I&N B-N B-NJ ISWO RSWO D&SWO M&SWO &-NA- 4	TABLE II.	H_2O	PY	BOR	?	—	9.6	27	DME/NHE	---	ns
DG68	$C_{32}H_{15}CuN_8NaO_3S$	sodium phthalocyaninecopper(II)-2-sulfonate C.A. 51481-19-7 D656 B-5 B6 B-&5 B6 C5 G6 N6 B&6 K&6 B&6/B-0/B-&A&/CJ&/DP&& GABBB-B-&C P& AN-CU-N EN NN WN I&N B-N B-&NJ ISWO &-NA-	TABLE II.	H_2O	PY	BOR	?	—	9.6	27	DME/NHE	---	ns
DG69	$C_{32}H_{17}N_8NaO_3S$	sodium phthalocyanine-2-sulfonate C.A. 41954-69-2 T-T56 BYNYJ DUN- DT56 BYNJ BUN- DT56 CMJ GSWO BNU- BT56 BYNJ DNU- B-16-J &-NA-	TABLE II.	H_2O	PY	BOR	?	—	9.6	27	DME/NHE	---	ns
DG70	$C_{32}H_{62}O_4$	dihexadecanoyl peroxide C.A. 2697-96-3 15V00V15	$[CH_3(CH_2)_{14}C-(O)O]_2$	MeOH 50 C_6H_6 50	PY	LiCl	?	—	—	—	DME/SCE	---	ns
DG71 cm86	$C_{34}H_{32}ClFeN_4O_4$	protoporphyrin(IX)iron(III)chloride C.A. 15489-47-1 D656 B-5 B6 B-&5 B6 C5/B-K/B-&P/CU/DW& 6ABBB-B-&C W AN-FE-N B-N B-&NJ BG G1 H2VQ L2VQ M1 Q1 R1U1 V1 W1U1	TABLE II.	DMF	PY	Et_4NClO_4	0.1	—	—	25±0.1	DME/SCE	25A0	C=0.4Ap (satd),m=1.08,t=5.8, h=70,Pt Aux
DG72	$C_{34}H_{36}CoN_4O_6$	hematoporphyrincobalt(III) C.A. 29497-66-3 D656 B-5 B6 B-&5 B6 C5/B-K/B-&P/CU/DU& 6ABBB-B-&C W AN-CO-N B-N B-&NJ G1 H2VQ L2VQ M1 Q1 RYQ1 V1 WYQ1	TABLE II.	H_2O	PY	buffer		—	7.5-12.0	—	DME/SCE	0A0	C=1.02
DG73	$C_{36}H_{32}FeN_6O_4$	trans-dicyanoprotoporphyrin(IX)iron-(III) C.A. 16904-05-5 D656 B-5 B6 B-&5 B6 C5/B-K/-&P/CU/DW& 6ABBB-B-&C W AN-FE-N B-N B-&NJ BCN BCN G1 H2VQ L2VQ M1 Q1 R1U1 V1 W1U1	TABLE II.	H_2O	ER	$NaNO_3$	1	—	—	25.0 ±0.1	PDE/SCE	5A0	A=0.224,PE
					VY	$NaNO_3$	1	—	—	25.0 ±0.1	RPE/SCE	3A0	ω=10,PE
				EtOH 30	EE	$NaNO_3$	1	—	—	25.0 ±0.1	PDE/SCE	5A2	A=0.224, τ=10-50 s, PE
					ER	$NaNO_3$	1	—	—	25.0 ±0.1	PDE/SCE	5A2	A=0.224,PE
						$NaNO_3$ NaCN	1 0.5		10.5				C=1.21,A=0.224,τ=0.040 s,PE
													τ=1.85 s
													τ=0.42 s
CONT													

TABLE I. Electrochemical Data $C_{36}H_{32}FeN_6O_4$ (CONT.) DG73

Ref.	C/M	Charact. Potential			Response Const.		Tech.	n	Electrokinetic Data			Products and Identification	Description and Remarks	Code No.
			Value	vs.		Value			Parameter	Value	From			
AN100 0377	–	E_{su}	–0.28F	SCE	i_{su}	0.158F	–	–	–	–	–	–	$C, i_p \propto C$ for $C=$ 0.001–0.01, r	DG66
JA095 3411	–	$E_{\frac{1}{2}}$	–0.97F	NHE	–	–	–	–	–	–	–	–	C,p	DG67
JA095 3411	–	$E_{\frac{1}{2}}$	–1.01F	NHE	–	–	–	–	–	–	–	–	C,p	DG68
JA095 3411	–	$E_{\frac{1}{2}}$	–1.06F	NHE	–	–	–	–	–	–	–	–	C,p	DG69
AC035 0880	–	$E_{\frac{1}{2}}$	–0.10	–	i_ℓ/C	7.7	–	–	–	–	–	–	C,p	DG70
BJ046 3652	–	$E_{\frac{1}{2}}$	–0.28	SCE	–	∞	sttd	1	Tomeš	70	sttd	–	C, i_d, p	DG71 cm86
			–1.20					1		60			C, i_d	
			–1.64					–		–			C	
			–1.84										C	
			–2.4										C	
AA064 0055	–	–	–	–	–	–	–	–	–	–	–	–	O, r?	DG72
JA088 1365	–	$E_{\tau/4}$	–0.420	SCE	–	–	QE	1.01	$E\log t^{\frac{1}{2}}$	60	–	–	C,R,p	DG73
JA088 1365	–	$E_{\frac{1}{2}}$	–0.413	SCE	–	–	QE	1.01	$E\log$	59	–	–	C,R,p	
JA088 1365	–	–	–	–	–	–	–	–	–	–	–	–	C,R,p	
		–	–					–		–			$A, R, E_{0.22}=E_{\tau/4}, C$	
JA088 1365	–	$E_{\tau/4}$	–0.495	SCE	–	–	QE	1.01	$E\log t^{\frac{1}{2}}$	60	–	–	C,R,p	
JA088 1365	–	–	–	SCE	$j\tau^{\frac{1}{2}}/C$	234	–	–	–	–	–	–	$C, j\tau^{\frac{1}{2}}\uparrow$ as $\tau\downarrow$ due to adsorption, p	
		–	–			140	–	–	–	–		–	$C, j\tau^{\frac{1}{2}}\uparrow$ as $\tau\downarrow$ due to adsorption, p	
		–	–			247	–	–	–	–		–	$C, j\tau^{\frac{1}{2}}\uparrow$ as $\tau\downarrow$ due to adsorption	
														CONT

DG73 (CONT.) $C_{36}H_{32}FeN_6O_4$

Code No.	Empirical Formula	Name and WLN Code	Structural Formula	Solvent	Tech.	Medium		μ, M	pH	T, °C	Electrodes	App.	Experimental Parameters
DG73	$C_{36}H_{32}FeN_6O_4$	trans-dicyanoprotoporphyrin(IX) iron-(III)	TABLE II.	EtOH 30	ER	$NaNO_3$ NaCN	1 0.5	-	10.5	25.0 ±0.1	PDE/SCE	5A2	C=4.82, A= 0.224, τ = 0.0125 τ = 3.30
					VY	$NaNO_3$	1	-	-	25.0 ±0.1	RPE/SCE	3A0	ω = 10, PE
DG74	$C_{36}H_{36}CoN_4O_4$	cobalt(II)protoporphyrin(IX) dimethyl ester C.A. 14932-10-6 D656 B-5 B6 B-&5 B6 C5/B-K/B-&P/CU/DW& 6ABBB-B-&C W AN-CO-N B-N B-&NJ G1 H2VO1 L2VO1 M1 Q1 R1U1 V1 W1U1	TABLE II.	DMF	PY	Et_4NClO_4	0.1	-	-	25± 0.1	DME/SCE	25A0	C=0.4Ap (satd), m= 1.08, t=5.8, h=70, Pt Aux
						Et_4NClO_4	0.5						
					HL	Et_4NClO_4	0.5	-	-	-	DME/SCE	25A0	C=0.4Ap (satd), m= 1.08, t=3?, h=70, f=(1-10)E5, Pt Aux
					PA	Et_4NClO_4	0.1	-	-	25± 0.1	DME/SCE	25A0	C=0.4Ap (satd), m= 4.56E-2, t= 345.1, h=50, v=420, Pt Aux
					PV	Et_4NClO_4	0.1	-	-	25± 0.1	DME/SCE	25A0	100Hz, C=0.4Ap (satd), m= 1.08, t=5.8, h=70, Δe=10, Pt Aux
DG75	$C_{36}H_{36}CuN_4O_4$	copper(II)protoporphyrin(IX) dimethyl ester C.A. 15304-60-6 D656 B-5 B6 B-&5 B6 C5/B-K/B-&P/CU/DW& 6ABBB-B-&C W AN-CU-N B-N B-&NJ G1 H2VO1 L2VO1 M1 Q1 R1U1 V1 W1U1	TABLE II.	DMF	PY	Et_4NClO_4	0.1	-	-	25± 0.1	DME/SCE	25A0	C=0.2Ap (satd), m= 1.08, t=5.8, h=70, Pt Aux
						Et_4NClO_4	0.5						
					HL	Et_4NClO_4	0.5	-	-	-	DME/SCE	25A0	C=0. Ap (satd), m= 1.08, t=3?, h=70, f=(1-10)E5, Pt Aux
					PA	Et_4NClO_4	0.1	-	-	25± 0.1	DME/SCE	25A0	C=0.2Ap (satd), m= 4.56E-2, t= 345.1, h=50, v=500, Pt Aux
DG76	$C_{36}H_{36}N_4NiO_4$	nickel(II)protoporphyrin(IX) dimethyl ester C.A. 15304-70-8 D656 B-5 B6 B-&5 B6 C5/B-K/B-&P/CU/DW& 6ABBB-B-&C W AN-NI-N B-N B-&NJ G1 H2HO1 L2VO1 M1 Q1 R1U1 V1 W1U1	TABLE II.	DMF	PY	Et_4NClO_4	0.1	-	-	25± 0.1	DME/SCE	25A0	C=0.4Ap (satd), m= 1.08, t=5.8, h=70, Pt Aux
						Et_4NClO_4	0.5						
					HL	Et_4NClO_4	0.5	-	-	-	DME/SCE	25A0	C=0.4AP (satd), m= 1.08, t=3?, h=70, f=6-10)E5, Pt Aux

TABLE I. Electrochemical Data $C_{36}H_{36}N_4NiO_4$ DG76

Ref.	C/M	Charact. Potential Value	vs.	Response Const.	Value	Tech.	n	Electrokinetic Data Parameter	Value	From	Products and Identification	Description and Remarks	Code No.
JA088 1365	-	-	-	$j\tau^{\frac{1}{2}}/C$	195	-	-	-	-	-	-	$C,j\tau^{\frac{1}{2}}\uparrow$ as $\tau\downarrow$ due to adsorption,p	DG73
	-	-	-		113	-	-	-	-	-	-	$C,j\tau^{\frac{1}{2}}\uparrow$ as $\tau\downarrow$ due to adsorption,p	
JA088 1365	-	$E_{\frac{1}{2}}$ -0.499	SCE	-	-	QE	1.01	Elog	61	-	cyanide hemochrome, QE,spectra	C,R,p	
BJ046 3652	-	$E_{\frac{1}{2}}$ -0.95 -2.06 -2.35	SCE	$i_\ell(u)$	1 0.9 -	-	-	Tomeš	70 60 -	-	-	C,W,i_d,r C,W,i_d C,X	DG74
		-0.96 -2.00			- -				- -			C,p C	
BJ046 3720	-	-	-	-	-	-	-	α $k_{f,h}$ {	0.4 0.64	-	-	C,p	
BJ046 3652	-	E_p -0.97F -1.09F -2.09	SCE	i_p	2F 2.5F 4.0F	-	-	-	-	-	-	C,p C,Mx,i_a C	
BJ046 3652	-	E_{su} -0.96 -1.14	SCE	i_{su}	28F 5.73F	-	-	-	-	-	-	$C,i \propto C, E_{su}=f(C),p$ $C,i=f(C),E_{su} \propto C, i_{desorption}$	
		-2.05			2.0F			ΔE_{su}	100	-		$C,i \propto C, E_{su}=f(C)$	
BJ046 3652	-	$E_{\frac{1}{2}}$ -1.37	SCE	$i_\ell(u)$	1	-	-	Tomeš	60	-	-	$C,i_d,R,i \propto C$ for C= 0.05-0.4, $E_{\frac{1}{2}} \neq f(C)$	DG75
		-1.82			1				60			$C,i_d,R,i \propto C$ for C= 0.05-0.4, $E_{\frac{1}{2}} \neq f(C)$	
		-2.26			1.5				-			$C,\neq,i_d,i \propto C$ for C= 0.05-0.4	
		-1.33 -1.76			- -	-	-	-	-	-	-	C,r C	
BJ046 3720	-	-	-	-	-	-	-	α $k_{f,h}$ { {	0.5 4.1 0.30 0.85	-	-	C,p C	
BJ046 3652	-	E_p -0.81F -1.38F -1.81F -2.31	SCE	i_p	0.33F 1.17F 2.00F 3.83F	-	-	-	-	-	-	C,p C C C	
BJ046 3652	-	$E_{\frac{1}{2}}$ -1.34 -1.89 -2.11 -2.7	SCE	-	-	sttd	1 -	Tomeš	60 60 -	sttd -	-	C,R,p C $C,\neq,X$ $C,\neq,X$	DG76
		-1.28 -1.84			- -	-	-	-	-	-	-	C,p C	
BJ046 3720	-	-	-	-	-	-	-	α $k_{f,h}$ { {	0.55 5.1 0.32 0.76	sttd	-	C,p C	

Code No.	Empirical Formula	Name and WLN Code	Structural Formula	Solvent	Tech.	Medium		μ, M	pH	T, °C	Electrodes	App.	Experimental Parameters
DG77	$C_{36}H_{36}N_4O_4Zn$	zinc(II)protoporphyrin(IX) dimethyl ester C.A. 15304-09-3 D656 B5 B6 B-&5 B6 C5⊃ /B-K/B-&P/CU/DW& GA⊃ BBB-B-&C W AN-ZN-N B-N B-&NJ G1 H2VO1 L2VO1 M1 Q1 R1U1 V1 W1U1	TABLE II.	DMF	PY	Et_4NClO_4	0.1	-	-	25± 0.1	DME/SCE	25AO	C=0.4Ap (satd),m= 1.08,t=5.8, h=70,Pt Aux
						Et_4NClO_4	0.5						
					HL	Et_4NClO_4	0.5	-	-	-	DME/SCE	25AO	C=0.4Ap (satd),m= 1.08,t=3?, h=70,f=(1-10)E5,Pt Aux
DG78	$C_{36}H_{38}ClMnN_4O_5$	manganese(III)chloride protoporphyrin-(IX) dimethyl ester monohydrate C.A. 22357-85-3 D656 B-5 B6 B-&5 B6 C5/B-K/B-&P/CU/DW& 6ABBB-B-&C W AN-MN-N B-N B-&NJ B6 BQH G1 H2VO1 L2VO1 M1 Q1 R1U1 V1 W1U1	TABLE II.	DMF	PY	Et_4NClO_4	0.1	-	-	25± 0.1	DME/SCE	25AO	C=0.4Ap (satd),m= 1.08,t=5.8, h=70,Pt Aux
						Et_4NClO_4	0.5						
					HL	Et_4NClO_4	0.5	-	-	-	DME/SCE	25AO	C=0.4Ap (satd),m= 1.08,t=3?, h=70,f=(1-10)E5,Pt Aux
DG79 cm89	$C_{36}H_{38}N_4O_4$	protoporphyrin(IX) dimethyl ester C.A. 5522-66-7 T-T5YMYJ D2VO1 E1 C⊃ U1- ET5NYJ C1 D2VO1 BU1- BT5MJ C1 D1U1 E1U- BT5NYJ C1 D1U1 E1U- A-16-J	TABLE II.	DMF	PY	Et_4NClO_4	0.1	-	-	25± 0.1	DME/SCE	25AO	C=0.4Ap (satd),m= 1.08,t=5.8, h=70,Pt Aux
						Et_4NClO_4	0.5						
					HL	Et_4NClO_4	0.5	-	-	25± 0.1	DME/SCE	25AO	C=0.4Ap (satd),m= 1.08,t=3?, h=70,f=(1-10)E5,Pt Aux
					PA	Et_4NClO_4	0.1	-	-	25± 0.1	DME/SCE	25AO	C=0.4Ap (satd),m= 4.56E-2,t= 345.1,h=50, v=200,Pt Aux
					PV	Et_4NClO_4	0.1	-	-	25± 0.1	DME/SCE	25AO	C=0.4Ap (satd),m= 1.08,t=5.8, h=70,f=100, Δe=10,Pt Aux
DG80 CONT	$C_{36}H_{42}N_4O_4$	mesoporphyrin(IX) dimethyl ester C.A. 60160-48-7 T-T5YMYJ D2VO1 E1 C⊃ U1- ET5NYJ C1 D2VO1 BU1- BT5MJ C1 D2 E1⊃ U- BT5NYJ C1 D2 C1U- A-16-J	TABLE II.	DMF	PY	Et_4NClO_4	0.1	-	-	25± 0.1	DME/SCE	25AO	C=0.4Ap (satd),m= 1.08,t=5.8, h=70,Pt Aux
						Et_4NClO_4	0.5						

TABLE I. Electrochemical Data $C_{36}H_{42}N_4O_4$ (CONT.) DG80

Ref.	C/M	Charact. Potential		Response Const.		n		Electrokinetic Data			Products and Identification	Description and Remarks	Code No.
		Value	vs.		Value		Tech.	Parameter	Value	From			
BJ046 3652	-	$E_{\frac{1}{2}}$ -1.49	SCE	$i_\ell(u)$	1	-	-	Tomeš	60	sttd	-	$C,R,i_d,i \propto C$ for C= 0.05-0.4,$E_{\frac{1}{2}} \neq f(C)$,p	DG77
		-1.84			1				60			$C,R,i_d,i \propto C$ for C= 0.05-0.4,$E_{\frac{1}{2}} \neq f(C)$	
		-2.30			1.5				-			$C,\not\models,i_d,i \propto C$ for C= 0.05-0.4	
		-1.47 -1.80		-	-	-	-	-	-	-	-	C,p C	
BJ046 3720	-	- -	-	- -	- -	-	-	α $k_{f,h}$	{ 0.43 3.8 { 0.28 0.68	sttd	-	C,p C	
BJ046 3652	-	$E_{\frac{1}{2}}$ -0.37 -1.49 -1.97 -2.4	SCE	-	-	-	-	Tomeš	80 60 - -	sttd	-	C,p C C C	DG78
		- -1.46 -1.93		-	-	-	-	-	-	-	-	C,p C C	
BJ046 3720	-	- -	-	- -	- -	-	-	α $k_{f,h}$	{ 0.48 3.4	sttd	-	C,p	
BJ046 3652	-	$E_{\frac{1}{2}}$ -1.24	SCE	$i_\ell(u)$	1	-	-	Tomeš	60	sttd	-	$C,i \propto C$ for C=0.05- 0.4,$E_{\frac{1}{2}} \neq f(C)$,Tc=0.7, R,p	DG79 cm89
		-1.61			1				60			$C,i \propto C$ for C=0.05- 0.4,$E_{\frac{1}{2}} \neq f(C)$,Tc=0.7, R	
		-2.34			1.8				-			$C,i \propto C$ for C=0.05- 0.4,$\not\models$	
		-1.25 -1.61		-	-	-	-	-	-	-	-	C,p C	
BJ046 3720	-	- -	-	- -	- -	-	-	α $k_{f,h}$	{ 0.50 5.6 { 0.30 1.7	sttd	-	C,p C	
BJ046 3652	-	E_p -0.59F -1.28F	SCE	i_p	0.23F 1.26F	-	-	- $E_{p,A}-E_{p,C}$	- 60	- sttd	-	C,p $C,R,i_p \propto v^{\frac{1}{2}}$	
		-1.63F			1.83F				60			$C,R,i_p \propto v^{\frac{1}{2}}$	
		-2.34			4.1F				-			$C,\not\models$	
BJ046 3652	-	E_{su} -0.95	SCE	i_{su}	0.16F	-	-	-	-	-	-	$C,i_{desorption},i/C= f(C),E_{su}=f(C)$,p	
		-1.24			2.8F			$\Delta E_{su/2}$	90			$C,i \propto C,E_{su} \neq f(C)$,R	
		-1.61			2.43F				90			$C,i \propto C,E_{su} \neq f(C)$,R	
		-2.4			-				-			$C,\not\models$	
BJ046 3652	-	$E_{\frac{1}{2}}$ -1.34	SCE	$i_\ell(u)$	1	sttd	1	Tomeš	60	sttd	-	$C,R,i \propto C$ for C=0.05- 0.4,$E_{\frac{1}{2}} \neq f(C),i_d$,p	DG80
		-1.74			1		1		60			$C,R,i \propto C$ for C=0.05- 0.4,$E_{\frac{1}{2}} \neq f(C),i_d$	
		-2.58			1.8		-		-			$C,i \propto C$ for C=0.05- 0.4,$\not\models,i_d$	
		-1.35 -1.75			0.6F 0.6F	-	-	-	-	-	-	C,p C	CONT

DG80 (CONT.) $C_{36}H_{42}N_4O_4$

Code No.	Empirical Formula	Name and WLN Code	Structural Formula	Solvent	Tech.	Medium		μ, M	pH	T, °C	Electrodes	App.	Experimental Parameters
DG80	$C_{36}H_{42}N_4O_4$	mesoporphyrin(IX) dimethyl ester	TABLE II.	DMF	HL	Et_4NClO_4	0.5	—	—	—	DME/SCE	25AO	C=0.4Ap (satd),m= 1.08,t=3?, h=70,f=(1-10)E5,Pt Aux
					PG	Et_4NClO_4	0.5	—	—	25± 0.1	DME/SCE	25AO	C=0.4Ap (satd),m= 1.08,t=5.8, h=70,ΔE=5.4, f=200,Pt Aux
					RP	Et_4NClO_4	0.5	—	—	25± 0.1	DME/SCE	---	C=0.4Ap (satd),m= 1.08,t=3?, h=70,ΔE=22, f=1.0E6, Pt Aux
DG81	$C_{36}H_{44}AgClN_4O_4$	2,3,7,8,12,13,17,18-octaethylporphinato-silver(III) per-chlorate C.A. 41206-74-0 D656 B-5 B6 B-&5 B6 C5/B-K/B-&P/CU/⊃ DW& 6ABBB-B-&C W AN-AG-N B-N B-&NJ G2 H2 L2 M2 Q2 R2 V2 W2 &G-O4	TABLE II.	C_6H_5CN	VR	Bu_4NClO_4	0.01	—	—	20± 0.1	PBE/SCE	25A-	C=0.1-1,A= 0.043,v=10-1E5,Pt Aux
				DMSO	VR	Bu_4NClO_4	0.01	—	—	20± 0.1	PBE/SCE	25A-	C=0.1-1,A= 0.043,v=10-1E5,Pt Aux
DG82	$C_{36}H_{44}CaN_4$	2,3,7,8,12,13,17,18-octaethylporphinato-calcium(II) C.A. 49606-58-8 D656 B-5 B6 B-&5 B6 C5/B-K/B-&P/CU/DW& 6ABBB-B-&C W AN-CA-N B-N B-&NJ G2 H2 L2 M2 Q2 R2 V2 W2	TABLE II.	PrCN	VR	Bu_4NClO_4	0.1	—	—	20± 0.1	PBE/SCE	25A-	C=0.1-1,A= 0.043,v=10-1E5,Pt Aux
				DMSO	VR	Bu_4NClO_4	0.1	—	—	20± 0.1	PBE/SCE	25A-	C=0.1-1,A= 0.043,v=10-1E5,Pt Aux
DG83	$C_{36}H_{44}CdN_4$	2,3,7,8,12,13,17,18-octaethylporphinato-cadmium(II) C.A. 49661-61-2 D656 B-5 B6 B-&5 B6 C5/B-K/B-&P/CU/DW& 6ABBB-B-&C W AN-CD-N B-N B-&NJ G2 H2 L2 M2 Q2 R2 V2 W2	TABLE II.	PrCN	VR	Bu_4NClO_4	0.1	—	—	20± 0.1	PBE/SCE	25A-	C=0.1-1,A= 0.043,v=10-1E5,Pt Aux
				DMSO	VR	Bu_4NClO_4	0.01	—	—	20± 0.1	PBE/SCE	25A-	C=0.1-1,A= 0.043,v=10-1E5,Pt Aux
DG84	$C_{36}H_{44}CoN_4$	2,3,7,8,12,13,17,18-octaethylporphinato-cobalt(II) C.A. 17632-19-8 D656 B-5 B6 B-&5 B6 C5/B-K/B-&P/CU/DW& 6ABBB-B-&C W AN-CD-N B-N B-&NJ G2 H2 L2 M2 Q2 R2 V2 W2	TABLE II.	PrCN	VR	Bu_4NClO_4	0.1	—	—	20± 0.1	PBE/SCE	25A-	C=0.1-1,A= 0.043,v=10-1E5,Pt Aux
				DMSO	VR	Bu_4NClO_4	0.01	—	—	20± 0.1	PBE/SCE	25A-	C=0.1-1,A= 0.043,v=10-1E5,Pt Aux
DG85	$C_{36}H_{44}CuN_4$	2,3,7,8,12,13,17,18-octaethylporphinato-copper(II) C.A. 14409-63-3 D656 B-5 B6 B-&5 B6 C5/B-K/B-&P/CU/DW& 6ABBB-B-&C W AN-CU-N B-N B-&NJ G2 H2 L2 M2 Q2 R2 V2 W2	TABLE II.	PrCN	VR	Bu_4NClO_4	0.1	—	—	20± 0.1	PBE/SCE	25A-	C=0.1-1,A= 0.043,v=10-1E5,Pt Aux
				DMSO	VR	Bu_4NClO_4	0.01	—	—	20± 0.1	PBE/SCE	25A-	C=0.1-1,A= 0.043,v=10-1E5,Pt Aux

TABLE I. Electrochemical Data $C_{36}H_{44}CuN_4$ DG85

Ref.	C/M	Charact. Potential Value	vs.	Response Const. Value	Tech.	n	Electrokinetic Data Parameter	Value	From	Products and Identification	Description and Remarks	Code No.
BJ046 3720	-	-	-	-	-	-	α $k_{f,h}$	{0.50 3.8} {0.38 1.34}	sttd	-	C,p C	DG80
BJ046 3720	-	E_p	-1.33F -1.75F	SCE	-	-	-	-	-	-	-	C,p C
BJ046 3720	-	E_{rf}	-1.16F -1.72F	SCE	-	-	-	-	-	-	-	C,p C
JA095 5140	-	$E_{0.85p}$	-1.29	SCE	-	-	-	-	-	-	-	C,ligand redn.,R, $E_{\frac{1}{2}},C \propto$ LUMO,r	DG81
JA095 5140	-	$E_{0.85p}$	0.44	SCE	-	-	sttd	1	-	-	-	-	C, redn. of Ag(III) $\rightleftarrows$ Ag(II),R,r
JA095 5140	-	$E_{0.85p}$	0.86 0.50	SCE	-	-	sttd	1 1	-	-	-	-	A,ligand oxidn.,R, $E_{\frac{1}{2}},A \propto$ HOMO,$\Delta E_p=$ 2.25±0.15,r A,ligand oxidn.,R	DG82
JA095 5140	-	$E_{0.85p}$	-1.68	SCE	-	-	sttd	1	-	-	-	-	C,ligand redn.,R, $E_{\frac{1}{2}},C \propto$ LUMO,$\Delta E_p=$ 2.25±0.15,r
JA095 5140	-	$E_{0.85p}$	1.04 0.55	SCE	-	-	sttd	1 1	-	-	-	-	A,ligand oxidn.,R, $E_{\frac{1}{2}},A \propto$ HOMO,$\Delta E_p=$ 2.25±0.15,r A,ligand oxidn.,R	DG83
JA095 5140	-	$E_{0.85p}$	-1.52	SCE	-	-	sttd	1	-	-	-	-	C,ligand redn.,R, $E_{\frac{1}{2}},C \propto$ LUMO,$\Delta E_p=$ 2.25±0.15,r
JA095 5140	-	$E_{0.85p}$	1.00	SCE	-	-	sttd	1	-	-	-	-	A,ligand oxidn.,R, $E_{\frac{1}{2}},A \propto$ HOMO,r	DG84
JA095 5140	-	$E_{0.85p}$	- -1.05	SCE	-	-	sttd	1 1	-	-	-	Co(III) Co(I)	A,$\rightleftarrows$,r C,R
JA095 5140	-	$E_{0.85p}$	1.19 0.79	SCE	-	-	sttd	1 1	-	-	-	-	A,ligand oxidn.,R, $E_{\frac{1}{2}},A \propto$ HOMO,$\Delta E_p=$ 2.25±0.15,r A,ligand oxidn.,R	DG85
JA095 5140	-	$E_{0.85p}$	-1.46	SCE	-	-	sttd	1	-	-	-	-	C,ligand redn.,R, $E_{\frac{1}{2}},C \propto$ LUMO,$\Delta E_p=$ 2.25±0.15,r

Code No.	Empirical Formula	Name and WLN Code	Structural Formula	Solvent	Tech.	Medium	μ, M	pH	T, °C	Electrodes	App.	Experimental Parameters
DG86	$C_{36}H_{44}MgN_4$	2,3,7,8,12,13,17,18-octaethylporphinato-magnesium(II) C.A. 20910-35-4 D656 B-5 B6 B-&5 B6 C5/B-K/B-&P/CU/DW& 6ABBB-B-&C W AN-MG-N B-N B-&NJ G2 H2 L2 M2 Q2 R2 V2 W2	TABLE II.	PrCN	VR	Bu_4NClO_4 0.1	-	-	20±0.1	PBE/SCE	25A-	C=0.1-1, A=0.043, v=10-1E5, Pt Aux
				DMSO	VR	Bu_4NClO_4 0.01	-	-	20±0.1	PBE/SCE	25A-	C=0.1-1, A=0.043, v=10-1E5, Pt Aux
DG87	$C_{36}H_{44}N_4Ni$	2,3,7,8,12,13,17,18-octaethylporphinato-nickel(II) C.A. 24803-99-4 D656 B-5 B6 B-&5 B6 C5/B-K/B-&P/CU/DW& 6ABBB-B-&C W AN-NI-N B-N B-&NJ G2 H2 L2 M2 Q2 R2 V2 W2	TABLE II.	C_6H_5CN	VR	Bu_4NClO_4 0.1	-	-	20±0.1	PBE/SCE	25A-	C=0.1-1, A=0.043, v=10-1E5, Pt Aux
DG88	$C_{36}H_{44}N_4OTi$	2,3,7,8,12,13,17,18-octaethylporphin-atooxotitanium(IV) C.A. 25087-66-5 D656 B-5 B6 B-&5 B6 C5/B-K/B-&P/CU/DW& 6ABBB-B-&C W AN-TI-N B-N B-&NJ BUO G2 H2 L2 M2 Q2 R2 V2 W2	TABLE II.	PrCN	VR	Bu_4NClO_4 0.1	-	-	20±0.1	PBE/SCE	25A-	C=0.1-1, A=0.043, v=10-1E5, Pt Aux
				DMSO	VR	Bu_4NClO_4 0.01	-	-	20±0.1	PBE/SCE	25A-	C=0.1-1, A=0.043, v=10-1E5, Pt Aux
DG89	$C_{36}H_{44}N_4OV$	2,3,7,8,12,13,17,18-octaethylporphin-atooxovanadium(IV) C.A. 27860-55-5 D656 B-5 B6 B-&5 B6 C5/B-K/B-&P/CU/DW& 6ABBB-B-&C W AN-VA-N B-N B-&NJ BUO G2 H2 L2 M2 Q2 R2 V2 W2	TABLE II.	PrCN	VR	Bu_4NClO_4 0.1	-	-	20±0.1	PBE/SCE	25A-	C=0.1-1, A=0.043, v=10-1E5, Pt Aux
				DMSO	VR	Bu_4NClO_4 0.01	-	-	20±0.1	PBE/SCE	25A-	C=0.1-1, A=0.043, v=10-1E5, Pt Aux
DG90	$C_{36}H_{44}N_4Pb$	2,3,7,8,12,13,17,18-octaethylporphinato-lead(II) C.A. 33269-25-9 D656 B-5 B6 B-&5 B6 C5/B-K/B-&P/CU/DW& 6ABBB-B-&C W AN-PB-N B-N B-&NJ G2 H2 L2 M2 Q2 R2 V2 W2	TABLE II.	C_6H_5CN	VR	Bu_4NClO_4 0.1	-	-	20±0.1	PBE/SCE	25A-	C=0.1-1, A=0.043, v=10-1E5, Pt Aux
				DMSO	VR	Bu_4NClO_4 0.01	-	-	20±0.1	PBE/SCE	25A-	C=0.1-1, A=0.043, v=10-1E5, Pt Aux
DG91	$C_{36}H_{44}N_4Pd$	2,3,7,8,12,13,17,18-octaethylporphinato-palladium(II) C.A. 24804-00-0 D656 B-5 B6 B-&5 B6 C5/B-K/B-&P/CU/DW& 6ABBB-B-&C W AN-PD-N B-N B-&NJ G2 H2 L2 M2 Q2 R2 V2 W2	TABLE II.	PrCN	VR	Bu_4NClO_4 0.1	-	-	20±0.1	PBE/SCE	25A-	C=0.1-1, A=0.043, v=10-1E5, Pt Aux
				DMF	VR	Bu_4NClO_4 0.1	-	-	20±0.1	PBE/SCE	25A-	C=0.1-1, A=0.043, v=10-1E5, Pt Aux

TABLE I. Electrochemical Data $C_{36}H_{44}N_4Pd$ DG91

| Ref. | C/M | Charact. Potential | | Response Const. | | Tech. | n | Electrokinetic Data | | | Products and Identification | Description and Remarks | Code No. |
		Value	vs.		Value			Parameter	Value	From			
JA095 5140	-	$E_{0.85p}$	0.77 SCE	-	-	sttd	1	-	-	-	-	A, ligand oxidn., R, $E_{\frac{1}{2}},A \propto HOMO, \Delta E_p = 2.25\pm0.15, r$	DG86
			0.54				1					A, ligand oxidn., R	
JA095 5140	-	$E_{0.85p}$	-1.68 SCE	-	-	sttd	1	-	-	-	-	C, ligand redn., R, $E_{\frac{1}{2}},C \propto LUMO, \Delta E_p = 2.25\pm0.15, r$	
JA095 5140	-	$E_{0.85p}$	0.73 SCE	-	-	sttd	1	-	-	-	uncertain	A, R, $E_{\frac{1}{2}},A \propto HOMO$, $\Delta E_p = 2.25\pm0.15, r$	DG87
			-1.5				-					C, ligand redn., R, $E_{\frac{1}{2}},C \propto LUMO$	
JA095 5140	-	$E_{0.85p}$	1.32 SCE	-	-	sttd	1	-	-	-	-	A, ligand oxidn., R, $E_{\frac{1}{2}},A \propto HOMO, \Delta E_p = 2.25\pm0.15, r$	DG88
			1.03				1					A, ligand oxidn., R	
JA095 5140	-	$E_{0.85p}$	-1.21 SCE	-	-	sttd	1	-	-	-	-	C, ligand redn., R, $E_{\frac{1}{2}},C \propto LUMO, \Delta E_p = 2.25\pm0.15, r$	
			-1.69				1					C, ligand redn., R	
JA095 5140	-	$E_{0.85p}$	1.25 SCE	-	-	sttd	1	-	-	-	-	A, ligand oxidn., R, $E_{\frac{1}{2}},A \propto HOMO, \Delta E_p = 2.25\pm0.15, r$	DG89
			0.96				1					A, ligand oxidn., R	
JA095 5140	-	$E_{0.85p}$	-1.25 SCE	-	-	sttd	1	-	-	-	-	C, ligand redn., R, $E_{\frac{1}{2}},C \propto LUMO, \Delta E_p = 2.25\pm0.15, r$	
			-1.72				1					C, ligand redn., R	
JA095 5140	-	$E_{0.85p}$	0.91 SCE	-	-	sttd	1	-	-	-	-	A, ligand oxidn., R, $E_{\frac{1}{2}},A \propto HOMO, \Delta E_p = 2.25\pm0.15, r$	DG90
			0.65				1					A, ligand oxidn., R	
JA095 5140	-	$E_{0.85p}$	-1.30 SCE	-	-	sttd	1	-	-	-	-	C, ligand redn., R, $E_{\frac{1}{2}},C \propto LUMO, \Delta E_p = 2.25\pm0.15, r$	
JA095 5140	-	$E_{0.85p}$	0.82 SCE	-	-	sttd	1	-	-	-	-	A, ligand oxidn., R, $E_{\frac{1}{2}},A \propto HOMO, \Delta E_p = 2.25\pm0.15, r$	DG91
JA095 5140	-	$E_{0.85p}$	-1.53 SCE	-	-	sttd	1	-	-	-	-	C, ligand redn., R, $E_{\frac{1}{2}},C \propto LUMO, \Delta E_p = 2.25\pm0.15, r$	

Code No.	Empirical Formula	Name and WLN Code	Structural Formula	Solvent	Tech.	Medium	μ, M	pH	T, °C	Electrodes	App.	Experimental Parameters
DG92	$C_{36}H_{44}N_4Zn$	2,3,7,8,12,13,17,18-octaethylporphinatozinc(II) C.A. 17632-18-7 D656 B-5 B6 B-&5 B6 C5/B-K/B-&P/CU/DW& 6ABBB-B-&C W AN-ZN-N B-N B-&NJ G2 H2 L2 M2 Q2 R2 V2 W2	TABLE II.	PrCN	VR	Bu_4NClO_4 0.1	–	–	20±0.1	PBE/SCE	25A-	C=0.1-1,A=0.043,v=10-1E5,Pt Aux
				DMSO	VR	Bu_4NClO_4 0.01	–	–	20±0.1	PBE/SCE	25A-	C=0.1-1,A=0.043,v=10-1E5,Pt Aux
DG93	$C_{36}H_{45}AlN_4O$	2,3,7,8,12,13,17,18-octaethylporphinatohydroxoaluminum(III) C.A. 19529-55-6 D656 B-5 B6 B-&5 B6 C5/B-K/B-&P/CU/DW& 6ABBB-B-&C W AN-AL-N B-N B-&NJ BQ G2 H2 L2 M2 Q2 R2 V2 W2	TABLE II.	PrCN	VR	Bu_4NClO_4 0.1	–	–	20±0.1	PBE/SCE	25A-	C=0.1-1,A=0.043,v=10-1E5,Pt Aux
				DMSO	VR	Bu_4NClO_4 0.01	–	–	20±0.1	PBE/SCE	25A-	C=0.1-1,A=0.043,v=10-1E5,Pt Aux
DG94	$C_{36}H_{45}CrN_4O$	2,3,7,8,12,13,17,18-octaethylporphinatohydroxochromium(III) C.A. 50733-41-0 D656 B-5 B6 B-&5 B6 C5/B-K/B-&P/CU/DW& 6ABBB-B-&C W AN-CR-N B-N B-&NJ BQ G2 H2 L2 M2 Q2 R2 V2 W2	TABLE II.	PrCN	VR	Bu_4NClO_4 0.1	–	–	20±0.1	PBE/SCE	25A-	C=0.1-1,A=0.043,v=10-1E5,Pt Aux
				DMSO	VR	Bu_4NClO_4 0.01	–	–	20±0.1	PBE/SCE	25A-	C=0.1-1,A=0.043,v=10-1E5,Pt Aux
DG95	$C_{36}H_{45}FeN_4O$	2,3,7,8,12,13,17,18-octaethylporphinatohydroxoiron(III) C.A. 24804-01-1 D656 B-5 B6 B-&5 B6 C5/B-K/B-&P/CU/DW& 6ABBB-B-&C W AN-FE-N B-N B-&NJ BQ G2 H2 L2 M2 Q2 R2 V2 W2	TABLE II.	PrCN	VR	Bu_4NClO_4 0.1	–	–	20±0.1	PBE/SCE	25A-	C=0.1-1,A=0.043,v=10-1E5,Pt Aux
				DMSO	VR	Bu_4NClO_4 0.01	–	–	20±0.1	PBE/SCE	25A-	C=0.1-1,A=0.043,v=10-1E5,Pt Aux
DG96	$C_{36}H_{45}GaN_4O$	2,3,7,8,12,13,17,18-octaethylporphinatohydroxogallium(III) C.A. 50733-40-9 D656 B-5 B6 B-&5 B6 C5/B-K/B-&P/CU/DW& 6ABBB-B-&C W AN-GA-N B-N B-&NJ BQ G2 H2 L2 M2 Q2 R2 V2 W2	TABLE II.	PrCN	VR	Bu_4NClO_4 0.1	–	–	20±0.1	PBE/SCE	25A-	C=0.1-1,A=0.043,v=10-1E5,Pt Aux
				DMSO	VR	Bu_4NClO_4 0.01	–	–	20±0.1	PBE/SCE	25A-	C=0.1-1,A=0.043,v=10-1E5,Pt Aux

TABLE I. Electrochemical Data $C_{36}H_{45}GaN_4O$ DG96

Ref.	C/M	Charact. Potential		Response Const.		n Tech.		Electrokinetic Data			Products and Identification	Description and Remarks	Code No.
			Value vs.		Value			Parameter	Value	From			
JA095 5140	-	$E_{0.85p}$	1.02 SCE	-	-	sttd	1	-	-	-	-	A,ligand oxidn.,R, $E_{\frac{1}{2}},A \propto HOMO, \Delta E_p = 2.25\pm0.15, r$	DG92
			0.63				1					A,ligand oxidn.,R	
JA095 5140	-	$E_{0.85p}$	-1.61 SCE	-	-	sttd	1	-	-	-	-	C,ligand redn.,R, $E_{\frac{1}{2}},C \propto LUMO, \Delta E_p = 2.25\pm0.15, r$	
JA095 5140	-	$E_{0.85p}$	1.28 SCE	-	-	sttd	1	-	-	-	-	A,ligand oxidn.,R, $E_{\frac{1}{2}},A \propto HOMO, \Delta E_p = 2.25\pm0.15, r$	DG93
			0.95				1					A,ligand oxidn.,R	
JA095 5140	-	$E_{0.85p}$	-1.31 SCE	-	-	sttd	1	-	-	-	-	C,ligand redn.,R, $E_{\frac{1}{2}},C \propto LUMO, \Delta E_p = 2.25\pm0.15, r$	
JA095 5140	-	$E_{0.85p}$	1.22 SCE	-	-	sttd	1	-	-	-	-	A,ligand oxidn.,R, $E_{\frac{1}{2}},A \propto HOMO, \Delta E_p = 2.25\pm0.15, r$	DG94
			0.99				1					A,ligand oxidn.,R	
JA095 5140	-	$E_{0.85p}$	0.79 SCE	-	-	sttd	1	-	-	-	Cr(IV)	A,R,r	
			-1.14				1				Cr(II)	C,R	
			-1.35				1					C,ligand redn.,R, $E_{\frac{1}{2}},A \propto LUMO, \Delta E_p = 2.25\pm0.15, r$	
JA095 5140	-	$E_{0.85p}$	1.24 SCE	-	-	sttd	1	-	-	-	-	A,ligand oxidn.,R, $E_{\frac{1}{2}},A \propto HOMO, \Delta E_p = 2.25\pm0.15, r$	DG95
			1.00				1					A,ligand oxidn.,R	
JA095 5140	-	$E_{0.85p}$	-0.24 SCE	-	-	sttd	1	-	-	-	Fe(II)	C,R,r	
			-1.33				1					C,ligand redn.,R, $E_{\frac{1}{2}},C \propto LUMO, \Delta E_p = 2.25\pm0.15, r$	
JA095 5140	-	$E_{0.85p}$	1.32 SCE	-	-	sttd	1	-	-	-	-	A,ligand oxidn.,R $E_{\frac{1}{2}},A \propto HOMO, \Delta E_p = 2.25\pm0.15, r$	DG96
			1.01				1					A,ligand oxidn.,R	
JA095 5140	-	$E_{0.85p}$	-1.34 SCE	-	-	sttd	1	-	-	-	-	C,ligand redn.,R, $E_{\frac{1}{2}},C \propto LUMO, \Delta E_p = 2.25\pm0.15, r$	
			-1.80				1					C,ligand redn.,R	

Code No.	Empirical Formula	Name and WLN Code	Structural Formula	Solvent	Tech.	Medium	μ, M	pH	T, °C	Electrodes	App.	Experimental Parameters
DG97	$C_{36}H_{45}InN_4O$	2,3,7,8,12,13,17,18-octaethylporphinato-hydroxoindium(III) C.A. 50733-39-6 D656 B-5 B6 B-&5 B6 C5/B-K/B-&P/CU/DW& 6ABBB-B-&C W AN-IN-N B-N B-&NJ BQ G2 H2 L2 M2 Q2 R2 V2 W2	TABLE II.	PrCN	VR	Bu_4NClO_4 0.1	-	-	20±0.1	PBE/SCE	25A-	C=0.1-1,A=0.043,v=10-1E5,Pt Aux
				DMSO	VR	Bu_4NClO_4 0.01	-	-	20±0.1	PBE/SCE	25A-	C=0.1-1,A=0.043,v=10-1E5,Pt Aux
DG98	$C_{36}H_{45}MnN_4O$	2,3,7,8,12,13,17,18-octaethylporphinato-hydroxomanganese-(III) C.A. 50733-38-5 D656 B-5 B6 B-&5 B6 C5/B-K/B-&P/CU/DW& 6ABBB-B-&C W AN-MN-N B-N B-&NJ BQ G2 H2 L2 M2 Q2 R2 V2 W2	TABLE II.	PrCN	VR	Bu_4NClO_4 0.1	-	-	20±0.1	PBE/SCE	25A-	C=0.1-1,A=0.043,v=10-1E5,Pt Aux
				DMSO	VR	Bu_4NClO_4 0.01	-	-	20±0.1	PBE/SCE	25A-	C=0.1-1,A=0.043,v=10-1E5,Pt Aux
DG99	$C_{36}H_{45}MoN_4O_2$	2,3,7,8,12,13,17,18-octaethylporphinato-hydroxooxomolybdenum(V) C.A. 50733-45-4 D656 B-5 B6 B-&5 B6 C5/B-K/B-&P/CU/DW& 6ABBB-B-&C W AN-MO-N B-N B-&NJ BQ BUO G2 H2 L2 M2 Q2 R2 V2 W2	TABLE II.	PrCN	VR	Bu_4NClO_4 0.1	-	-	20±0.1	PBE/SCE	25A-	C=0.1-1,A=0.043,v=10-1E5,Pt Aux
				DMSO	VR	Bu_4NClO_4 0.01	-	-	20±0.1	PBE/SCE	25A-	C=0.1-1,A=0.043,v=10-1E5,Pt Aux
DH00	$C_{36}H_{45}N_4OSb$	2,3,7,8,12,13,17,18-octaethylporphinato-hydroxoantimony(III) C.A. 50733-37-4 D656 B-5 B6 B-&5 B6 C5/B-K/B-&P/CU/DW& 6ABBB-B-&C W AN-SB-N B-N B-&NJ BQ G2 H2 L2 M2 Q2 R2 V2 W2	TABLE II.	PrCN	VR	Bu_4NClO_4 0.1	-	-	20±0.1	PBE/SCE	25A-	C=0.1-1,A=0.043,v=10-1E5,Pt Aux
				DMSO	VR	Bu_4NClO_4 0.01	-	-	20±0.1	PBE/SCE	25A-	C=0.1-1,A=0.043,v=10-1E5,Pt Aux
DH01	$C_{36}H_{45}N_4OSc$	2,3,7,8,12,13,17,18-octaethylporphinato-hydroxoscandium(III) C.A. 50938-43-7 D656 B-5 B6 B-&5 B6 C5/B-K/B-&P/CU/DW& 6ABBB-B-&C W AN-SC-N B-N B-&NJ BQ G2 H2 L2 M2 Q2 R2 V2 W2	TABLE II.	C_6H_5CN	VR	Bu_4NClO_4 0.1	-	-	20±0.1	PBE/SCE	25A-	C=0.1-1,A=0.043,v=10-1E5,Pt Aux
				DMSO	VR	Bu_4NClO_4 0.01	-	-	20±0.1	PBE/SCE	25A-	C=0.1-1,A=0.043,v=10-1E5,Pt Aux
DH02 CONT	$C_{36}H_{45}N_4OTl$	2,3,7,8,12,13,17,18-octaethylporphinato-hydroxothallium(III) C.A. 58482-63-6 D656 B-5 B6 B-&5 B6 C5/B-K/B-&P/CU/DW& 6ABBB-B-&C W AN-TL-N B-N B-&NJ BQ G2 H2 L2 M2 Q2 R2 V2 W2	TABLE II.	PrCN	VR	Bu_4NClO_4 0.1	-	-	20±0.1	PBE/SCE	25A-	C=0.1-1,A=0.043,v=10-1E5,Pt Aux

TABLE I. Electrochemical Data $C_{36}H_{45}N_4OTl$ (CONT.) DH02

Ref.	C/M	Charact. Potential		Response Const.		n Tech.	Electrokinetic Data			Products and Identification	Description and Remarks	Code No.
		Value	vs.		Value		Parameter	Value	From			
JA095 5140	-	$E_{0.85p}$	1.36	SCE	-	-	sttd 1	-	-	-	-	A,ligand oxidn.,R, $E_{\frac{1}{2}}$,A $\propto$ HOMO, ΔE_p= 2.25±0.15,r	DG97
			1.08				1					A,ligand oxidn.,R	
JA095 5140	-	$E_{0.85p}$	-1.19	SCE	-	-	sttd 1	-	-	-	-	C,ligand redn.,R, $E_{\frac{1}{2}}$,C $\propto$ LUMO, ΔE_p= 2.25±0.15,r	
			-1.59				1					C,ligand redn.,R	
JA095 5140	-	$E_{0.85p}$	>1.4	SCE	-	-	sttd 1	-	-	-	-	A,ligand oxidn.,R, $E_{\frac{1}{2}}$,A $\propto$ HOMO, ΔE_p= 2.25±0.15,r	DG98
			1.12				1					A,ligand oxidn.,R	
JA095 5140	-	$E_{0.85p}$	-0.42	SCE	-	-	sttd 1	-	-	-	Mn(II)	C,R,r	
			-1.61				1					C,ligand redn.,R, $E_{\frac{1}{2}}$,C $\propto$ LUMO, ΔE_p= 2.25±0.15,r	
JA095 5140	-	$E_{0.85p}$	1.43	SCE	-	-	sttd 1	-	-	-	-	A,ligand oxidn.,R, $E_{\frac{1}{2}}$,A $\propto$ HOMO, ΔE_p= 2.25±0.15,r	DG99
JA095 5140	-	$E_{0.85p}$	-0.21	SCE	-	-	sttd 1	-	-	-	Mo(IV)	C,R,r	
			-1.30				1					C,ligand redn.,R, $E_{\frac{1}{2}}$,C $\propto$ LUMO, ΔE_p= 2.25±0.15,r	
			-1.72				1					C,ligand redn.,R	
JA095 5140	-	$E_{0.85p}$	>1.4	SCE	-	-	sttd 1	-	-	-	-	A,ligand oxidn.,R, $E_{\frac{1}{2}}$,A $\propto$ HOMO, ΔE_p= 2.25±0.15,r	DH00
			0.75				1					A,ligand oxidn.,R	
JA095 5140	-	$E_{0.85p}$	-1.07	SCE	-	-	sttd 1	-	-	-	-	C,ligand redn.,R, $E_{\frac{1}{2}}$,C $\propto$ LUMO, ΔE_p= 2.25±0.15,r	
JA095 5140	-	$E_{0.85p}$	1.03	SCE	-	-	sttd 1	-	-	-	-	A,ligand oxidn.,R, $E_{\frac{1}{2}}$,A $\propto$ HOMO, ΔE_p= 2.25±0.15,r	DH01
			0.70				1					A,ligand oxidn.,R	
JA095 5140	-	$E_{0.85p}$	-1.54	SCE	-	-	sttd 1	-	-	-	-	C,ligand redn.,R, $E_{\frac{1}{2}}$,C $\propto$ LUMO, ΔE_p= 2.25±0.15,r	
JA095 5140	-	$E_{0.85p}$	1.31	SCE	-	-	sttd 1	-	-	-	-	A,ligand oxidn.,R, $E_{\frac{1}{2}}$,A $\propto$ HOMO, ΔE_p= 2.25±0.15,r	DH02
			1.00				1					A,ligand oxidn,R	CONT

DH02 (CONT.) $C_{36}H_{45}N_4OT1$

Code No.	Empirical Formula	Name and WLN Code	Structural Formula	Solvent	Tech.	Medium	μ, M	pH	T, °C	Electrodes	App.	Experimental Parameters
DH02	$C_{36}H_{45}N_4OT1$	2,3,7,8,12,13,17,18-octaethylporphinato-hydroxothallium(III)	TABLE II.	DMSO	VR	Bu_4NClO_4 0.01	-	-	20±0.1	PBE/SCE	25A-	C=0.1-1, A=0.043, v=10-1E5, Pt Aux
DH03	$C_{36}H_{46}GeN_4O_2$	2,3,7,8,12,13,17,18-octaethylporphinato-dihydroxogermanium-(IV) C.A. 50936-73-7 D656 B-5 B6 B-&5 B6 C5/B-K/B-&P/CU/DW& 6ABBB-B-&C W AN-GE-N B-N B-&NJ BQ BQ G2 H2 L2 M2 Q2 R2 V2 W2	TABLE II.	PrCN	VR	Bu_4NClO_4 0.1	-	-	20±0.1	PBE/SCE	25A-	C=0.1-1, A=0.043, v=10-1E5, Pt Aux
				DMSO	VR	Bu_4NClO_4 0.01	-	-	20±0.1	PBE/SCE	25A-	C=0.1-1, A=0.043, v=10-1E5, Pt Aux
DH04	$C_{36}H_{46}N_4$	2,3,7,8,12,13,17,18-octaethylporphine C.A. 2683-82-1 T-T5YMYJ D2 E2 CU1-ET5YNJ C2 D2 BU1- E T5MJ C2 D2 F1U- BT5 YNJ C2 D2 E1U- A-16 -J	TABLE II.	PrCN	VR	Bu_4NClO_4 0.1	-	-	20±0.1	PBE/SCE	25A-	C=0.1-1, A=0.043, v=10-1E5, Pt Aux
				DMSO	VR	Bu_4NClO_4 0.01	-	-	20±0.1	PBE/SCE	25A-	C=0.1-1, A=0.043, v=10-1E5, Pt Aux
DH05	$C_{36}H_{46}N_4O_2Si$	2,3,7,8,12,13,17,18-octaethylporphinato-dihydroxosilicon(IV) C.A. 50820-15-0 D656 B-5 B6 B-&5 B6 C5/B-K/B-&P/CU/DW& 6ABBB-B-&C W AN-SI-N B-N B-&NJ BQ BQ G2 H2 L2 M2 Q2 R2 V2 W2	TABLE II.	PrCN	VR	Bu_4NClO_4 0.1	-	-	20±0.1	PBE/SCE	25A-	C=0.1-1, A=0.043, v=10-1E5, Pt Aux
				DMSO	VR	Bu_4NClO_4 0.01	-	-	20±0.1	PBE/SCE	25A-	C=0.1-1, A=0.043, v=10-1E5, Pt Aux
DH06	$D_{36}H_{46}N_4O_2Sn$	2,3,7,8,12,13,17,18-octaethylporphinato-dihydroxotin(IV) C.A. 29008-64-8 D656 B-5 B6 B-&5 B6 C5/B-K/B-&P/CU/DW& 6ABBB-B-&C W AN-SN-N B-N B-&NJ BQ BQ G2 H2 L2 M2 Q2 R2 V2 W2	TABLE II.	PrCN	VR	Bu_4NClO_4 0.1	-	-	20±0.1	PBE/SCE	25A-	C=0.1-1, A=0.043, v=10-1E5, Pt Aux
				DMSO	VR	Bu_4NClO_4 0.01	-	-	20±0.1	PBE/SCE	25A-	C=0.1-1, A=0.043, v=10-1E5, Pt Aux
DH07	$C_{36}H_{48}Cl_2N_4O_8$	diprotonated 2,3,7,8,12,13,17,18-octaethylporphyrin perchlorate C.A. 51319-02-9 T-T5YMYJ D2 E2 CU1- ET5MYJ C2 D2 BU 1- ET5MJ C2 D2 B1U- BT5MYJ C2 D2 E1 U- A-16-J &24 &59 & G-O4 & G-O4	TABLE II.	PrCN	VR	Bu_4NClO_4 0.1	-	-	20±0.1	PBE/SCE	25A-	C=0.1-1, A=0.043, v=10-1E5, Pt Aux
DH08	$C_{36}H_{60}O_2$	retinyl palmitate C.A. 79-81-2 L6X BUTJ A1 A1 B1U1 Y1&U2U1Y1&U2OV15 C1	TABLE II.	EtOH 75	IL	H_2SO_4 0.01	-	-	20±0.1	CPDO SCE	26B0	C=0.005-1.0, v=10, Pt Aux
				MeOH 67 C_6H_6 33	IL	$LiClO_4$ 0.05	-	-	20±0.1	GCE/SCE	26B0	v=30, Pt Aux

TABLE I. Electrochemical Data $C_{36}H_{60}O_2$ (CONT.) DH08

Ref.	C/M	Charact. Potential			Response Const.		n Tech.	n	Electrokinetic Data			Products and Identification	Description and Remarks	Code No.
			Value	vs.		Value			Parameter	Value	From			
JA095 5140	-	$E_{0.85p}$	-1.24	SCE	-	-	sttd	1	-	-	-	-	C,ligand redn.,R, $E_{\frac{1}{2}},C \propto LUMO, \Delta E_p=$ 2.25±0.15,r	DH02
JA095 5140	-	$E_{0.85p}$	1.36	SCE	-	-	sttd	1	-	-	-	-	A,ligand oxidn.,R, $E_{\frac{1}{2}},A \propto HOMO, \Delta E_p=$ 2.25±0.15,r	DH03
			1.09					1					A,ligand oxidn.,R	
JA095 5140	-	$E_{0.85p}$	-1.31	SCE	-	-	sttd	1	-	-	-	-	C,ligand redn., R, $E_{\frac{1}{2}},C \propto LUMO, \Delta E_p=$ 2.25±0.15,r	
JA095 5140	-	$E_{0.85p}$	1.30	SCE	-	-	sttd	1	-	-	-	-	A,ligand oxidn.,R, $E_{\frac{1}{2}},A \propto HOMO, \Delta E_p=$ 2.25±0.15,r	DH04
			0.81					1					A,ligand oxidn.,R	
JA095 5140	-	$E_{0.85p}$	-1.46	SCE	-	-	sttd	1	-	-	-	-	C,ligand redn.,R, $E_{\frac{1}{2}},C \propto LUMO, \Delta E_p=$ 2.25±0.15,r	
			-1.86					1					C,ligand redn.,R	
JA095 5140	-	$E_{0.85p}$	1.19	SCE	-	-	sttd	1	-	-	-	-	A,ligand oxidn.,R, $E_{\frac{1}{2}},A \propto HOMO, \Delta E_p=$ 2.25±0.15,r	DH05
			0.92					1					A,ligand oxidn.,R	
JA095 5140	-	$E_{0.85p}$	-1.35	SCE	-	-	sttd	1	-	-	-	-	C,ligand redn.,R, $E_{\frac{1}{2}},C \propto LUMO, \Delta E_p=$ 2.25±0.15,r	
JA095 5140	-	$E_{0.85p}$	>1.4	SCE	-	-	-	-	-	-	-	-	0,p	DH06
JA095 5140	-	$E_{0.85p}$	-0.90	SCE	-	-	sttd	1	-	-	-	-	C,ligand redn.,R, $E_{\frac{1}{2}},C \propto LUMO, \Delta E_p=$ 2.25±0.15,r	
			-1.30					1					C,ligand redn.,R	
JA095 5140	-	$E_{0.85p}$	1.65	SCE	-	-	sttd	1	-	-	-	-	A,ligand oxidn.,R, $E_{\frac{1}{2}},A \propto HOMO,r$	DH07
AN099 0683	-	$E_{p/2}$	0.85	SCE	-	-	-	-	-	-	-	-	$A, i_p \propto C, X, r$	DH08
AN100 0827	-	$E_{p/2}$ E_p	0.79 1.0F 1.18F	SCE	-	-	-	-	-	-	-	-	A,r A A	
														CONT

Code No.	Empirical Formula	Name and WLN Code	Structural Formula	Solvent	Tech.	Medium		μ, M	pH	T, °C	Electrodes	App.	Experimental Parameters
DH08	$C_{36}H_{60}O_2$	retinyl palmitate	TABLE II.	MeOH 67 C_6H_6 33	VA	$LiClO_4$	0.05	-	-	20±0.1	GCE/SCE	26B0	v=10-100, Pt Aux
DH09	$C_{36}H_{70}O_4$	dioctadecanoyl peroxide C.A. 3273-75-4 17V00V17	$[CH_3(CH_2)_{16}\text{-}C(O)O]_2$	MeOH 50 C_6H_6 50	PY	LiCl	?	-	-	-	DME/SCE	---	ns
DH10	$C_{37}H_{48}N_4$	N-methyl-2,3,7,8,12,13,17,18-octaethyl-porphyrin C.A. 30116-09-7 T-T5YNYJ B1 D2 C2 C- U1- ET5NYJ C2 D2 BU- 1- ET5NJ C2 D2 B1U- BT5NYJ C2 D2 E1U- A- -16-J	TABLE II.	PrCN	VR	Bu_4NClO_4	0.1	-	-	20±0.1	PBE/SCE	25A-	C=0.1-1,A=0.043,v=10-1E5,Pt Aux
				DMSO	VR	Bu_4NClO_4	0.01	-	-	20±0.1	PBE/SCE	25A-	C=0.1-1,A=0.043,v=10-1E5,Pt Aux
DH11	$C_{40}H_{26}$	1,3,6,8-tetraphenyl-pyrene C.A. 13638-82-9 L666 B6 2AB PJ GR& IR& NR& OR	TABLE II.	DMF	VR	Bu_4NClO_4	0.1	-	-	-	PDE/SCE	3F-	C=0.1-1, v=200
DH12	$C_{40}H_{37}FeN_6O_4$	protoporphyrin(IX)-trans-cyanopyridine-iron(III) C.A. 23388-71-8 D656 B-5 B6 B-&5 B6 C5/B-K/B-&P/CU/- DW& 6ABBB-B-&C W AN-FE-N B-N B-&NJ BCN B- AT6NJ& G1 H2VQ L2VQ M1 Q1 R1U1 V1 W1U1	TABLE II.	EtOH 30	ER	$NaNO_3$	1	-	-	25.0±0.1	PDE/SCE	5A0	A=0.224,PE
					VY	$NaNO_3$	1	-	-	25.0±0.1	RPE/SCE	3A0	ω=10,PE
DH13 cm95	$C_{42}H_{28}$	rubrene C.A. 517-51-1 L E6 C666J B&R DR& KR& MR	TABLE II-3.	dioxane 75	PY	Bu_4NI	0.1	-	-	25±0.1	DME/SCE	135A-	i=8-11,m=0.6-0.9
				C_6H_5CN	VA	Bu_4NClO_4	0.1	-	-	20-23	Pt Ag/AgCl	0A0	1.7→-1.7→1.7 V,C=1,A=0.2,v=100
					VR	Bu_4NClO_4	0.1	-	-	-	PDE/SCE	1A2	0.15→-1.40→1.2 V,A=0.2,v=1E3,Pt Aux
					VY	Bu_4NClO_4	0.1	-	-	20-23	Pt Ag/AgCl	0A0	C=1,A=0.2,v=100
				CH_2Cl_2	VR	Bu_4NClO_4	0.2	-	-	-	Pt/SCE	25F-	C=1,v=100
				DMF	VR	Bu_4NClO_4	0.1	-	-	-	Pt/SCE	25A2	1.2→-1.7→1.2 V,v=83.3,A=0.025,Pt Aux
											PDE/SCE	3F-	C=0.1-1,v=249

TABLE I. Electrochemical Data $C_{42}H_{28}$ DH13

Ref.	C/M	Charact. Potential Value	vs.	Response Const.	Value	n Tech.		Electrokinetic Data Parameter	Value	From	Products and Identification	Description and Remarks	Code No.
AN100 0827	-	-	-	-	-	-	-	βn	0.98	E_p- $E_{p/2}$	-	$A, i_p \propto v^{\frac{1}{2}}, \neq, i_d, r$	DH08
AC035 0880	-	$E_{\frac{1}{2}}$	-0.08	-	i_ℓ/C	6.2	-	-	-	-	-	C,p	DH09
JA095 5140	-	$E_{0.85p}$	1.37	SCE	-	-	sttd	1	-	-	-	A,ligand oxidn.,R, $E_{\frac{1}{2}}, A \propto $ HOMO, $\Delta E_p =$ 2.25±0.15,r	DH10
JA095 5140	-	$E_{0.85p}$	-0.73	SCE	-	-	sttd	1	-	-	-	C,ligand redn.,R, $E_{\frac{1}{2}}, C \propto $ LUMO, $\Delta E_p=$ 2.25±0.15,r	
			-1.17					1				C,ligand redn.,R	
JA094 0691	-	E_p	-1.83 1.25	SCE	-	-	-	-	-	-	-	C,R,r A,$\neq$	DH11
JA088 1365	329 k	$E_{\tau/4}$	-0.358	SCE	-	-	QE	1.01	Elogt$^{\frac{1}{2}}$	61	-	-	C,R,p	DH12
JA088 1365	329 k	$E_{\frac{1}{2}}$	-0.360	SCE	-	-	QE	1.01	Elog	61	-	-	C,R,p	
J0025 0611	-	$E_{\frac{1}{2}}$	-1.55 -1.80	SCE	-	-	-	-	-	-	-	C,p C,corrected data from Rec. Trav. Chim., (1955),74, 1525	DH13 cm95
EA013 1197	-	E_p	-1.54F -1.45F 0.96F 1.48F 1.32F 0.88F	Ag/AgCl	-	42F 24F 42F 64F 8F 28F	-	-	-	-	radical anion,stable radical cation, stable dication radical cation, stable -	C,W,r A,W A,W A,W C C	
EA013 1187	-	E_p	-1.39F -1.37F 1.05F 1.06F	SCE	i_p	24F 14F 15F 24F	sttd	1 -	ΔE_p E_p-$E_{p/2}$	66 57	sttd	-	C,W,p A,W C,W A,W,$i_p=f(v^{\frac{1}{2}})$ for v= 50-5000, $E_p \neq f(v)$	
EA013 1197	-	$E_{\frac{1}{2}}$	-1.47	Ag/AgCl	-	-	-	-	-	-	-	-	C,p	
JA094 4872	-	E_p	0.90 0.97	SCE	-	-	-	-	ΔE_p	70 -	sttd	-	C,R,p A,R,E_p calc.	
EA013 1209	-	E_p	-1.5F -1.4 0.84 0.96 -1.48 0.97	SCE	i_p	7F 3F 1F 8F - -	-	-	-	-	rubrene,radical anion,lifetime >> 10 s rubrene,radical cation,lifetime ≈ 5-10 s -	C,W,p A,W C,X A,W C,R,r A,$\neq$	

Code No.	Empirical Formula	Name and WLN Code	Structural Formula	Solvent	Tech.	Medium		μ, M	pH	T, $^\circ$C	Electrodes	App.	Experimental Parameters
DH14	$C_{44}H_{24}CoN_4Na_3O_{12}S_4$	trisodium 5,10,15,-20-tetrakis(4-sulfonatophenyl)porphinatocobalt(III) C.A. 41660-29-1 D656 B-5 B6 B-&5 B6 C5/B-K/B-&P/CU/ DW& 6ABBB-B-&C W AN-CO-N B-N B-&NJ ER DSWO& JR DSWO& OR DSWO& TR DSWO &-NA- 3	TABLE II.	H_2O	PY	BOR	?	-	-	-	DME/SCE	2A-	ns
DH15	$C_{44}H_{24}CoN_4Na_4O_{12}S_4$	tetrasodium 5,10,-15,20-tetrakis(4-sulfonatophenyl)porphinatocobalt(II) C.A. 61004-83-9 D656 B-5 B6 B-&5 B6 C5/B-K/B-&P/CU/ DW& 6ABBB-B-&C W AN-CO-N B-N B-&NJ ER DSWO& JR DSWO& OR DSWO& TR DSWO &-NA- 4	TABLE II.	H_2O	PY	BOR	?	-	-	-	DME/SCE	2A-	ns
DH16	$C_{44}H_{26}N_4Na_4O_{12}S_4$	tetrasodium 5,10,15,-20-tetrakis(4-sulfonatophenyl)porphine C.A. 39050-26-5 T-T5YMYJ CUYR DSWO&- ET5NYJ BUYR DSWO&- ET5MJ BYR DSWO&U- BT5NYJ EYR DSWO&U- A-16-J &-NA- 4	TABLE II.	H_2O	PY	BOR	?	-	9.7	-	DME/SCE	2A-	ns
DH17 bs21	$C_{44}H_{28}CoN_4$	5,10,15,20-tetraphenylporphinatocobalt(II) C.A. 14172-90-8 L656 B-5 B6 B-&6 B6 C5 6ABBB-B-&C W B-CO-J ER& JR& OR& TR	TABLE II-2.	DMF	PY	Bu_4NClO_4	0.2	-	-	-	DME/MP	OAO	C=5.0
DH18 bs31	$C_{44}H_{30}N_4$	5,10,15,20-tetraphenylporphine C.A. 917-23-7 T-T5YMYJ CUYR&- ET5NYJ BUYR&- ET5MJ BYR&- BT5NYJ EYR&- A-16-J	TABLE II-2.	CH_2Cl_2	VR	Bu_4NClO_4	0.2	-	-	-	Pt/SCE	25F-	C=1,v=100
				DMF	VR	Bu_4NClO_4	0.2	-	-	-	Pt/SCE	25F-	C=1,v=100
DH19 CONT	$C_{44}H_{42}ClFeN_6O_4$	trans-dipyridylprotoporphyrin(IX)iron-(III) chloride C.A. 18582-60-0 D656 B-5 B6 B-&5 B6 C5/B-K/B-&P/CU/DW& 6ABBB-B-&C W AN-FE-N B-N B-&NJ B- AT6NJ& B- AT6NJ& G1 H2VQ L2VQ M1 Q1 R1U1 V1 W1U1	TABLE II.	EtOH 10	VY	BOR PYR	0.025 2.5	-	10.0	25.0 ±0.1	RPE/SCE	3AO	C=980,ω=10
				EtOH 30	PY	$NaNO_3$ PYR	1 2.5	-	-	25.0 ±0.1	DME/SCE	3AO	C=2.24
					ER	$NaNO_3$ buffer	1	-	7.9	25.0 ±0.1	PDE/SCE	5AO	A=0.224,PE
					PA	$NaHCO_3$ PYR	0.1 2	-	10.2	25.0 ±0.1	HMDE Ag/AgCl	3AO	C=8.6,v= 129
					VY	$NaNO_3$ buffer	1	-	7.9	25.0 ±0.1	RPE/SCE	3AO	ω=10,PE

TABLE I. Electrochemical Data $C_{44}H_{42}ClFeN_8O_4$ (CONT.) DH19

Ref.	C/M	Charact. Value	Potential vs.	Response Const.	Value	Tech.	n	Electrokinetic Data Parameter	Value	From	Products and Identification	Description and Remarks	Code No.
JA095 2408	-	$E_{\frac{1}{2}}$	-0.65	SCE	-	-	sttd	1	-	-	-	-	C,p	DH14
JA095 2408	-	$E_{\frac{1}{2}}$	-1.10	SCE	-	-	sttd	1	-	-	-	-	C,p	DH15
JA095 2408	-	$E_{\frac{1}{2}}$	-0.87 -1.63	SCE	-	-	sttd	1 1	-	-	-	-	C,p C	DH16
BJ045 2898	-	$E_{\frac{1}{2}}$	-0.82 -1.87	MP	-	-	sttd	1 1	-	-	-	Co(I) tetraphenyl-porphine Co(I) tetraphenyl-porphine radical anion, ESR	C,p C	DH17 bs21
JA094 4872	-	E_p	1.30 1.05 -1.26 -1.66 -1.57 -1.19 1.12 1.36	SCE	i_p	- 14F 18F - - 18F 20F -	-	-	ΔE_p	65 70 70 95 -	sttd	-	C,R,p C,R C,R C,R A,R A,R A,R A,R	DH18 bs31
JA094 4872	-	E_p	1.12 -1.06 -1.00 1.18	SCE	-	-	-	-	ΔE_p	62 60 -	sttd	-	C,R,p C,R A,R A,R	
JA088 1365	-	$E_{\frac{1}{2}}$	-0.16F -0.36F	SCE	i	0.38F 0.59F	Elog!	1 1	Elog Elog $dE_{\frac{1}{2}}/dpH$	60 60 60	sttd sttd plot	-	C,W,R,i_ℓ ↓ as pH ↑,p C,W,R,i_ℓ ↑ as pH ↑	DH19
JA088 1365	-	$E_{\frac{1}{2}}$	-0.128	SCE	D	1.33	-	-	-	-	-	-	C,R,p	
JA088 1365	329 j	$E_{T/4}$	-0.129	SCE	-	-	QE	0.98	Elogt$^{\frac{1}{2}}$	59	-	-	C,R,p	
JA088 1365	-	E_p	-0.18F -0.38F -0.12F	SCE	i_p	1.2F 2.9F 2.0F	-	-	-	-	-	-	C,W,redn. of pyridine hemichrome,p C,W,redn. of hydroxoaquo complex A,W,oxidn. of pyridine hemochrome	
JA088 1365	329 j	$E_{\frac{1}{2}}$	-0.128	SCE	-	-	QE	0.98	Elog	60	-	-	C,R,p	

CONT

DH19 (CONT.) $C_{44}H_{42}ClFeN_6O_4$

Code No.	Empirical Formula	Name and WLN Code	Structural Formula	Solvent	Tech.	Medium		μ, M	pH	T, °C	Electrodes	App.	Experimental Parameters
DH19	$C_{44}H_{42}ClFeN_6O_4$	trans-dipyridylprotoporphyrin(IX)iron(III) chloride	TABLE II.	EtOH 30	VY	$NaNO_3$ buffer	1	–	9.0	25.0 ±0.1	RPE/SCE	3A0	ω = 10, PE
									10.5				
									11.5				
						$NaNO_3$ NaCN PYR	1 3E-4 2	–	–				C=810
DH20	$C_{44}H_{46}ClCoN_6O_6$	trans-dipyridine-hematoporphyrin-cobalt(III) chloride	TABLE II.	PYR 20	PY	CARB	0.05	1.05	8.7	–	DME/SCE	OA0	C=1.02
		D656 B-5 B6 B-&5 B6 C5/B-K/B-&P/CU/ DW& 6ABBB-B-&C W AN-CO-N B-N B-&NJ B- AT6NJ& B- AT6NJ& G1 H2YQ L2VQ M1 Q1 RYQ1 V1 WYQ1 &G			VR	buffer	?	–	8.7	–	PDE/SCE	25-0	C=1.02, v=100, Pt Aux
									12	30±0.5			v=90
DH21	$C_{46}H_{46}ClFeN_6O_4$	trans-di(3-methyl-pyridine)protoporphyriniron(III) chloride C.A.	TABLE II.	EtOH 30	ER	buffer $NaNO_3$	? 1	–	8	25.0 ±0.1	PDE/SCE	5A0	A=0.224, PE
		D656 B-5 B6 B-&5 C6/B-K/B-&P/DW& 6A BBB-B-&C W AN-FE-N B-N B-&NJ B- AT6 NJ C1& B- AT6NJ C1& G1 H2VQ L2VQ M1 Q1 R1U1 V1 W1U1 &G			VY	buffer $NaNO_3$	? 1	–	8	25.0 ±0.1	RPE/SCE	3A0	ω = 10
DH22	$C_{46}H_{46}ClFeN_6O_4$	trans-di(4-methyl-pyridine)protoporphyriniron(III) chloride C.A. 52677-47-1	TABLE II.	EtOH 30	ER	buffer $NaNO_3$	? 1	–	8	25.0 ±0.1	PDE/SCE	5A0	A=0.224, PE
		D656 B-5 B6 B-&5 B6 C5/B-K/B-&P/DW& 6ABBB-B-&C W AN-FE-N B-N B-&NJ B- A T6NJ D1& B- AT6NJ D1& G1 H2VQ L2VQ M1 Q1 R1U1 V1 W1U1 &G			VY	buffer $NaNO_3$	? 1	–	8	25.0 ±0.1	RPE/SCE	3A0	ω = 10, PE
DH23	$C_{46}H_{50}CoN_6O_6$	trans-di(2-methyl-pyridine)hematoporphyrincobalt(III)	TABLE II.	H_2O	PY	buffer 2-methyl-pyridine 2.05	? 2.05	–	8.80	–	DME/SCE	OA0	C=1.02
		D656 B-5 B6 B-&5 C6/B-K/B-&P/CU/DW& 6ABBB-B-&C W AN-CO-N B-N B-&NJ B- A T6NJ B1& B- AT6NJ B1& G1 H2VQ L2VQ M1 Q1 RYQ1 V1 WYQ1											
DH24	$C_{46}H_{50}CoN_6O_6$	trans-di(4-methyl-pyridine)hematoporphyrincobalt(III)	TABLE II.	H_2O	PY	buffer 4-methyl-pyridine 2.05	? 2.05	–	8.80	–	DME/SCE	OA0	C=1.02
		D656 B-5 B6 B-&5 B6 C5/B-K/B-&P/CU/DW& 6ABBB-B-&C W AN-CO-N B-N B-&NJ B- AT6NJ D1& B- AT6NJ D1& G1 H2VQ L2VQ M1 Q1 RYQ1 V1 WYQ1											

TABLE I. Electrochemical Data $C_{46}H_{50}CoN_6O_6$ DH24

Ref.	C/M	Charact. Potential		Response Const.		Tech.	n	Electrokinetic Data			Products and Identification	Description and Remarks	Code No.
		Value	vs.		Value			Parameter	Value	From			
JA088 1365	-	$E_{\frac{1}{2}}$	-0.14F -0.3 to -0.45	SCE	-	-	Elog!	1 1	Elog Elog $dE_{\frac{1}{2}}/dpH$ }	60 60 60	sttd plot	-	C,R,p C,W,R	DH19
	329 k		-0.18F					1	Elog $dE_{\frac{1}{2}}/dpH$ }	60 60	sttd plot	-	C,R,p	
			-0.3 to -0.45					1	}	60 60			C,W,R	
			-0.24F					1	}	60 60		-	C,R,p	
			-0.3 to -0.45					1	}	60 60			C,W,R	
	-		-0.12F		i	19.7F	-	-	-	-	-	-	C,W,redn. of trans-dipyridylprotoporphyriniron(III),p	
			-0.37F				7.5F							C,W;$\log k_1$=5.11, redn. of trans-cyanopyridylprotoporphyriniron(III)
AA064 0055	-	$E_{\frac{1}{2}}$	-0.436	SCE	-	-	QE	1	Elog Tomeš	99 100	-	cobalt(II)hematoporphyrin dipyridine complex,QE	C,$E_{\frac{1}{2}}\neq f(pH)$,$i_d$,$i_\ell/C=$ k for $0.2<C<1.5$	DH20
AA064 0055	-	-	-	-	$i_p/Cv^{\frac{1}{2}}$	0.93	-	-	-	-	-	-	C,$i_{p,A}/i_{p,C}\approx 1.00$, R(?),$i_{p,A}/i_{p,C}=k_1 v^{\frac{1}{2}}+k_2$,$i_p/v^{\frac{1}{2}}=k$,r?	
					-	0.316	-	-	-	-	-	-	C,$i_{p,A}/i_{p,C}=0.96$,r?	
JA088 1365	-	$E_{\tau/4}$	-0.145	SCE	-	-	QE	0.99	Elog$t^{\frac{1}{2}}$	60	-	-	C,R,p	DH21
JA088 1365	-	$E_{\frac{1}{2}}$	-0.147	SCE	-	-	QE	0.99	Elog	59	-	-	C,R,p	
JA088 1365	-	$E_{\tau/4}$	-0.145	SCE	-	-	QE	0.99	Elog$t^{\frac{1}{2}}$	60	-	-	C,R,p	DH22
JA088 1365	-	$E_{\frac{1}{2}}$	-0.147	SCE	-	-	QE	0.99	Elog	59	-	-	C,R,p	
AA064 0055	-	$E_{\frac{1}{2}}$	-0.420	SCE	-	-	-	-	Elog	187	-	-	C,r?	DH23
AA064 0055	-	$E_{\frac{1}{2}}$	-0.425	SCE	-	-	-	-	Elog	65	-	-	C,r?	DH24

Code No.	Empirical Formula	Name and WLN Code	Structural Formula	Solvent	Tech.	Medium	μ, M	pH	T, °C	Electrodes	App.	Experimental Parameters
DH25	$C_{48}H_{50}CoN_6O_8$	trans-di(4-acetyl-pyridine)hematoporphyrincobalt(III) D656 B-5 B6 B-&5 B6 C5/B-K/B-&P/CU/⊃ DW& 6ABBB-B-&C W AN-CO-N B-N B-&NJ B- AT6NJ DV1& B- AT6NJ DV1& G1 H2VQ L2VQ M1 Q1 RYQ1 V1 WYQ1	TABLE II.	H_2O	PY	buffer ? 4-acetyl-pyridine 1.81	-	8.50	-	DME/SCE	OAO	C=1.02
DH26	$C_{50}H_{54}CoN_6O_{10}$	trans-di(ethylnicotinate)hematoporphyrincobalt(III) D656 B-5 B6 B-&5 B6 C5/B-K/B-&P/CU/⊃ DW& 6ABBB-B-&C W AN-CO-N B-N B-&NJ B- AT6NJ DVO2& B- AT6NJ DVO2& G1 H2⊃ VQ L2VQ M1 Q1 RYQ1 V1 WYQ1	TABLE II.	H_2O	PY	buffer ? ethyl nicotinate 1.46	-	8.78	-	DME/SCE	OAO	C=1.02
DH27	$C_{53}H_{83}N_5$	tetrabutylammonium N-methyl-2,3,7,8,-12,13,17,18-octaethylporphyrine T-T5YNYJ B1 D2 E2 CU1- ET5NYJ C2 D2 BU1- ET5MJ C2 D2 B1U- BT5NYJ C2 D2 E1U- A-16-J	TABLE II.	PrCN	VR	Bu_4NClO_4 0.1	-	-	20± 0.1	PBE/SCE	25A-	C=0.1-1,A= 0.043,v=10-1E5,Pt Aux
				DMSO	VR	Bu_4NClO_4 0.01	-	-	20± 0.1	PBE/SCE	25A-	C=0.1-1,A= 0.043,v=10-1E5,Pt Aux
DH28	$C_{54}H_{52}CoN_8O_6$	trans-di(4,4'-bi-pyridine)hematoporphyrincobalt(III) D656 B-5 B6 B-&5 B6 C5/B-K/B-&P/CU/⊃ DW& 6ABBB-B-&C W AN-CO-N B-N B-&NJ B- AT6NJ D- DT6NJ&& B- AT6NJ D- DT6⊃ NJ&& G1 H2VQ L2VQ M1 Q1 RYQ1 V1 WYQ1	TABLE II.	H_2O	PY	buffer ? 4,4'-bipyridine 0.1	-	8.70	-	DME/SCE	OAO	C=1.02
DH29	$C_{56}H_{54}CoN_6O_6$	trans-di(4-phenyl-pyridine)hematoporphyrincobalt(III) D656 B-5 B6 B-&6 B6 C5/B-K/B-&P/CU/DW& 6ABBB-B-&C W AN-CO-N B-N B-&NJ B- A⊃ T6NJ DR&& B- AT6NJ DR&& G1 H2VQ L2VQ M1 Q1 RYQ1 V1 WYQ1	TABLE II.	H_2O	PY	buffer ? 4-phenyl-pyridine 0.1	-	9.46	-	DME/SCE	OAO	C=1.02
DH30	$C_{59}H_{90}O_4$	ubiquinone-10 C.A. 60684-33-5 L6V DVJ B01 C01 E1 F2UY1&/3UY1& 9 1	TABLE II.	H_2O	PY	ns Na dodecyl-sulfate 0.04	-	5.0 6.4 8.6	-	DME/SCE	OAO	m=0.93,t= 3.0,h=75
				EtOH 80	PY	ns	-	5.0 6.4 8.6	-	DME/SCE	OAO	m=0.93,t= 3.0,h=75

TABLE I. Electrochemical Data $C_{59}H_{90}O_4$ DH30

Ref.	C/M	Charact. Potential			Response Const.		n Tech.		Electrokinetic Data			Products and Identification	Description and Remarks	Code No.
			Value	vs.		Value			Parameter	Value	From			
AA064 0055	-	$E_{\frac{1}{2}}$	-0.362	SCE	-	-	-	-	Elog	77	-	-	C,r?	DH25
AA064 0055	-	$E_{\frac{1}{2}}$	-0.291	SCE	-	-	-	-	Elog	93	-	-	C,r?	DH26
JA095 5140	-	$E_{0.85p}$	0.86	SCE	-	-	sttd	1	-	-	-	-	A,ligand oxidn.,R, $E_{\frac{1}{2},A} \propto$ HOMO, ΔE_p= 2.25±0.15,r	DH27
JA095 5140	-	$E_{0.85p}$	-1.37	SCE	-	-	sttd	1	-	-	-	-	C,ligand redn.,R, $E_{\frac{1}{2},C} \propto$ LUMO, ΔE_p= 2.25±0.15,r	
AA064 0055	-	$E_{\frac{1}{2}}$	-0.300	SCE	-	-	-	-	Elog	56	-	-	C,r?	DH28
AA064 0055	-	$E_{\frac{1}{2}}$	-0.384	SCE	-	-	-	-	Elog	68	-	-	C,r?	DH29
BJ048 1354	-	$E_{\frac{1}{2}}$	-0.045	NHE	-	-	sttd	2	$dE_{\frac{1}{2}}/dpH$	8	sttd	-	$C, i_d, i_\ell \propto C, i_\ell \propto h^{\frac{1}{2}}$, $\propto n \neq$ [surfactant],r	DH30
			-0.055		-	-		2		8		-	$C, i_d, i_\ell \propto C, i_\ell \propto h^{\frac{1}{2}}$, $\propto n \neq$ [surfactant],r	
			-0.065		-	-		2		8		-	$C, i_d, i_\ell \propto C, i_\ell \propto h^{\frac{1}{2}}$, $\propto n \neq$ [surfactant],r	
BJ048 1354	-	$E_{\frac{1}{2}}$	0.05	NHE	-	-	sttd	2	$dE_{\frac{1}{2}}/dpH$ p=	60 2	sttd	-	$C, i_d, i_\ell \propto h^{\frac{1}{2}}$,r	
			0.025F		-	-		2		27 1		-	$C, i_d, i_\ell \propto h^{\frac{1}{2}}$,r	
			-0.04		-	-		2		27 1		-	$C, i_d, i_\ell \propto h^{\frac{1}{2}}$,r	

Code No.	Empirical Formula	Name and WLN Code	Structural Formula	Solvent	Tech.	Medium	μ, M	pH	T, °C	Electrodes	App.	Experimental Parameters
DH31		cytochrome C	MW ≈ 13000	H_2O	PY	CITR ?	–	2.50	–	DME/SCE	OAO	C=0.1
												C=0.323
						ACET or CITR?		4.00				C=0.1
						ACET ?		5.0				
								6.00				
						triscacodylate ?		6.05				C=0.0177
												C=0.0355
												C=0.1
												C=0.2045
								7.08				C=0.1
						triscacodylate or BOR?		8.00				
						BOR ?		9.00				C=ns
DH31		cytochrome C	MW ≈ 13000	H_2O	PY	CITR ?	–	2.50	–	DME/SCE	OAO	C=0.1

TABLE I. Electrochemical Data

DH31

Ref.	C/M	Charact. Potential Value	vs.	Response Const.	Value	n Tech.		Electrokinetic Data Parameter	Value	From	Products and Identification	Description and Remarks	Code No.
JA094 8197	-	$E_{\frac{1}{2}}$ -0.40	SCE	i_d/C	0.5	QE at E_{app}= -0.34V	<1	Elog Tomeš	160 120	Plot sttd	-	C, i_a+i_d, r	DH31
		-1.13			4.3		-		60			C	
		-0.595			0.32	-	-		150		-	C, i_a+i_d, r	
		-1.130			0.94				50			C	
		-0.34			1.0-1.2	-	-	Elog	180		-	C, i_a+i_d, r	
		-0.22			1.5	-	-		150			C, i_a+i_d, r	
		-0.12			1.5	-	-		120			C, i_a+i_d, r	
		-0.005			0.85	-	-				-	C, i_a+i_d, r	
		-0.005		D	1.24 1	-	-	$dE_{\frac{1}{2}}/d\log C$ Tomeš	139 90 90		-	$C, i_a+i_d, \Gamma=4E-11$ mol cm^{-2}, r	
		-0.13		i_d/C	1.6	-	-	Elog	120			$C, i_a+i_d, D=0.5$, $i_\ell \propto C$ for $C>0.01$, $i_\ell \propto h^{0.53\pm0.07}, r$	
		-0.160			0.97	-	-	$dE_{\frac{1}{2}}/d\log C$ Tomeš	139 120			C, i_a+i_d, r	
		-0.21			1.4	-	-	Elog	160 -			C, i_a+i_d, r	
		-0.21			0.7	-	-	Elog	100 -			C, i_a+i_d, r	
		-0.42			0.6				100	plot		C	
		-0.41			0.9	-	-		80			C, i_a+i_d, r	

Code No.	Compound oxidized (=R), empirical formula	Compound reduced (=O), empirical formula	Solvent	Medium		μ	pH	T	Electrodes	App.
DH32 dd32 de32	Wurster's Blue $C_{10}H_{16}N_2$	anthracene $C_{14}H_{10}$	DMF	Bu_4NClO_4	0.1	-	-	-	Pt/SCE	23AF2
DH33 dd32 dg47	Wurster's Blue $C_{10}H_{16}N_2$	9,10-diphenylanthracene $C_{26}H_{18}$	DMF	Bu_4NClO_4	0.1	-	-	-	Pt/SCE	23AF2
DH34 dd32 dh10	Wurster's Blue $C_{10}H_{16}N_2$	1,3,6,8-tetraphenyl-pyrene $C_{40}H_{26}$	DMF	Bu_4NClO_4	0.1	-	-	-	Pt/SCE	23AF2
DH35 dd32 dh12	Wurster's Blue $C_{10}H_{16}N_2$	rubrene $C_{42}H_{28}$	DMF	Bu_4NClO_4	0.1	-	-	-	Pt/SCE	23AF2
DH36 dd67	thianthrene $C_{12}H_8S_2$	same	MeCN	Bu_4NClO_4	0.1	-	-	-	Pt/Ag	23AF2
DH37 dd67 de35	thianthrene $C_{12}H_8S_2$	2,5-diphenyl-1,3,4-oxadiazole $C_{14}H_{10}N_2O$	MeCN	Bu_4NClO_4	0.1	-	-	-	Pt/Pt	23AF2
DH38 de04 de35	10-methylphenothia-zine $C_{13}H_{11}NS$	2,5-diphenyl-1,3,4-oxadiazole $C_{14}H_{10}N_2O$	MeCN	Bu_4NClO_4 or Bu_4NBF_4	0.1 0.1	-	-	-	Pt/Ag	23AF2
DH39 de04 de99	10-methylphenothia-zine $C_{13}H_{11}NS$	fluoranthene $C_{16}H_{10}$	DMF	Bu_4NClO_4	0.105	-	-	25	Pt/Pt	3AF
				Bu_4NClO_4	0.1	-	-	-	Pt/SCE	23AF2
DH40 de04 dg64	10-methylphenothia-zine $C_{13}H_{11}NS$	tribipyridineruthenium dichloride $C_{30}H_{24}Cl_2N_6Ru$	MeCN	Bu_4NBF_4	0.2	-	-	-	Pt/Ag	---
DH41 de04 dh17	10-methylphenothia-zine $C_{13}H_{11}NS$	α,β,γ,δ-tetraphenyl-porphine $C_{44}H_{30}N_4$	CH_2Cl_2	Bu_4NClO_4	0.2	-	-	-	Pt/Ag	25F-
			DMF	Bu_4NClO_4	0.1	-	-	-	Pt/Ag	25F-
DH42 de32 de38	anthracene $C_{14}H_{10}$	1,2-dibromo-1,2-di-phenylethane $C_{14}H_{12}Br_2$	DMF	Bu_4NClO_4	0.1	-	-	-	Pt Ag/satd AgNO$_3$	5AO

TABLE I. Electrochemical Data

Exptl. parameters	Ref.	C/M	Type	f	Excitation E$_1$	E$_2$	vs.	Species	Emission Type	Emission Characteristics	Remarks	Code No.
C_R=ns, C_O=ns, A=0.14, Pt Aux(A=13.5)	JA094 0691	351b	□	10	≈0.3	≈-2.0	SCE	$^3O^*$	De	-	π ↑ as magnetic field ↑, r	DH32 dd32 de32
C_R=ns, C_O=ns, A=0.14, Pt Aux(A=13.5)	JA094 0691	351b	□	10	≈0.3	≈-1.9	SCE	$^3O^*$	De		π ↑ as magnetic field ↑, r	DH33 dd32 dg47
C_R=0.23, C_O=0.15, A=0.14, Pt Aux(A=13.5)	JA094 0691	351b	□	10	≈0.3	≈-1.9	SCE	$^3O^*$	De	-	π ↑ as magnetic field ↑, r	DH34 dd32 dh10
C_R=1.5, C_O=1.0, A=0.14, Pt Aux(A=13.5)	JA094 0691	351b	□	10	≈0.3	≈-1.5	SCE	$^3O^*$	De	-	π = max. for magnetic field = 7kG, r	DH35 dd32 dh12
C_R=ns, A=0.17, Pt Aux	JA094 1522	-	□	0.1	1.30	-1.90	Ag	-	-	λ = 460-600 nm, peak at ≈ 500 nm	cmpd. not electroactive at these potentials, π very small, preannihilation ecl, r	DH36 dd67
C_R=7.3, C_O=7.0, A=0.17, Pt Aux	JA094 1522	351d	□	0.2	3.30	-3.30	Pt	$^1R^*$ $^3R^*$	Se De	λ = 430 nm, π has max. value at f=100. λ = 340 nm, π → 0 as f → 500, π$_{340}$/π$_{430}$ = 1/8 and ↓ to 1/20 for f=10	π = f(C), π small for [O] or [R] > 10. π ≠ f(magnetic field), r	DH37 dd67 de35
C_R=2.5, C_O=2.5	JA094 1522	-	-	-	1.0	-2.3	Ag	-	-	λ = 430, 550 nm	π very small, preannihilation ecl of R, r	DH38 de04 de35
C_R=ns, C_O=ns, A=0.08	JA094 6331	-	triple-step	-	0→-1.70→0.77→0			$^3O^*$	-	-	Ø = 8E-5, triplet yield ≈ 30%, r	DH39 de04 de99
C_R=2.40, C_O=3.0, A=0.14, Pt Aux(A=13.5)	JA094 0691	351b	□	10	≈0.9	≈-1.8	SCE	$^3O^*$	De	-	π = max. for magnetic field ≅ 5kG	
C_R=ns, C_O=ns	JA094 2862	374b	□	0.2	1.1	-1.6	Ag	$^3O^*$	-	λ = 610 nm	orange ecl, r	DH40 de04 dg64
C_R=1.0, C_O=0.1	JA094 4872	-	□	0.2	0.84	-1.26	SCE	O	-	λ = 653, 701 nm	π = 4(π for DH64), emission occurs on redn. cycle, r	DH41 de04 dh17
C_R=1.0, C_O=1.0	JA094 4872	-	□	0.2	+1.0	-1.06	SCE	O	-	λ = 662, 702 nm	r	
C_R=ns, C_O=ns	JA094 9020	376a	□ or ∧	-	-	-	-	R	-	-	π very small, r	DH42 de32 de38

Code No.	Compound oxidized (=R), empirical formula	Compound reduced (=O), empirical formula	Solvent	Medium	μ	pH	T	Electrodes	App.
DH43 de35	2,5-diphenyl-1,3,4-oxadiazole $C_{14}H_{10}N_2O$	same	MeCN	Bu_4NClO_4	0.1	-	-	-	Pt/Ag	23AF2
DH44 bi07 de38	pyrene $C_{16}H_{10}$	1,2-dibromo-1,2-diphenylethane $C_{14}H_{12}Br_2$	DMF	Bu_4NClO_4	0.1	-	-	-	Pt Ag/satd $AgNO_3$	5AO
DH45 cl05 de38	1,2-benzanthracene $C_{18}H_{12}$	1,2-dibromo-1,2-diphenylethane $C_{14}H_{12}Br_2$	DMF	Bu_4NClO_4	0.1	-	-	-	Pt Ag/satd $AgNO_3$	5AO
DH46 cl06 de38	naphthacene $C_{18}H_{12}$	1,2-dibromo-1,2-diphenylethane $C_{14}H_{12}Br_2$	DMF	Bu_4NClO_4	0.1	-	-	-	Pt Ag/satd $AgNO_3$	5AO
DH47 cl06 dg48	naphthacene $C_{18}H_{12}$	9,10-dichloro-9,10-diphenylanthracene $C_{26}H_{18}Cl_2$	DMF	Bu_4NClO_4	0.1	-	-	-	Pt Ag/satd $AgNO_3$	5AO
DH48 df80 de99	10-phenylphenothiazine $C_{18}H_{13}NS$	fluoranthene $C_{16}H_{10}$	DMF	Bu_4NClO_4	0.1	-	-	-	PDE Ag/AgCl,satd KCl	---
DH49 cl82 de38	perylene $C_{20}H_{12}$	1,2-dibromo-1,2-diphenylethane $C_{14}H_{12}Br_2$	DMF	Bu_4NClO_4	0.1	-	-	-	Pt Ag/satd $AgNO_3$	5A-
DH50 cl82 dg48	perylene $C_{20}H_{12}$	9,10-dichloro-9,10-dihydro-9,10-diphenylanthracene $C_{26}H_{18}Cl_2$	DMF	Bu_4NClO_4	0.1	-	-	-	Pt Ag/satd $AgNO_3$	5AO
DH51 cm39 de38	coronene $C_{24}H_{12}$	1,2-dibromo-1,2-diphenylethane $C_{14}H_{12}Br_2$	DMF	Bu_4NClO_4	0.1	-	-	-	Pt Ag/satd $AgNO_3$	5AO
DH52 dg47 de38	9,10-diphenylanthracene $C_{26}H_{18}$	1,2-dibromo-1,2-diphenylethane $C_{14}H_{12}Br_2$	DMF	Bu_4NClO_4	0.1	-	-	-	Pt Ag/satd $AgNO_3$	5AO
DH53 dg47	9,10-diphenylanthracene $C_{26}H_{18}$	same	DMF	Bu_4NClO_4	0.1	-	-	-	Pt/SCE	23AF2
				Bu_4NClO_4	0.105	-	-	25	Pt/Pt	3AF-

TABLE I. Electrochemical Data

Exptl. parameters	Ref.	C/M	Type	f	Excitation E_1	E_2	vs.	Species	Type	Emission Characteristics	Remarks	Code No.
C_R=ns,C_O=ns,Pt Aux	JA094 1522	-	□	2	2.2	-2.3	Ag	-	-	-	$\pi \approx 0.01$(π for DH37),$\pi \downarrow$ as $E_1 \to 2.3$ V,r	DH43 de35
C_R=ns,C_O=ns	JA094 9020	376a	□ or ∧	-	-	-	-	R	-	-	π very small,r	DH44 bi07 de38
C_R=ns,C_O=ns	JA094 9020	376a	□ or ∧	-	-	-	-	R	-	-	π very small,r	DH45 c105 de38
C_R=ns,C_O=ns	JA094 9020	376a	□ or ∧	-	-	-	-	R	-	-	π very small,r	DH46 c106 de38
C_R=1.1,C_O=1.0	JA094 9020	375a	-	-	-	-1.9	SCE	R	-	λ = 480,510 nm	$E_{app} \neq f(t)$,r	DH47 c106 dg48
C_R=1.0,C_O=1.0	JA094 4790	-	□	20	0.90	-1.80	Ag/AgCl	$^3O^*$	De	λ = 475 nm, triplet yield = 0.8	emission quenched by trans-stilbene $\emptyset \neq f$[conc.],r	DH48 df80 de99
C_R=ns,C_O=ns	JA094 9020	376a	□ or ∧	-	-	-	-	R	-	-	π medium,r	DH49 c182 de38
C_R=0.05,C_O=0.5	JA094 9020	375a	-	-	>0.5	-2.2	-	R	-	λ = 430,470 nm; $E_{\pi max}$ = -1.90 V vs. SCE	emission at λ = 430 nm is due to O and appears only if [O] >10 [R],r	DH50 c182 dg48
C_R=ns,C_O=ns	JA094 9020	376a	□ or ∧	-	-	-	-	-	-	-	π very small,r	DH51 cm39 de38
C_R=ns,C_O=ns	JA094 9020	376a	□ or ∧	-	-	-	-	R	-	-	π large,r	DH52 dg47 de38
C_R=ns,C_O=ns,A= 0.14,Pt Aux(A=13.5)	JA094 0691	351c	□	10	≈1.4	≈-1.9	SCE	$^1R^*$	Se	-	$\pi \neq f$(magnetic field),r	DH53 dg47
C_R=1.04,A=0.08	JA094 6317	-	triple pot. step	-	0→1.70	-2.3→0	-	$^1R^*$	Se	$\emptyset$ = 1.24E-3	results are anomalous if cathodic step occurs first, $\emptyset \uparrow$ as C $\uparrow$	

Code No.	Compound oxidized (=R), empirical formula	Compound reduced (=O), empirical formula	Solvent	Medium	μ	pH	T	Electrodes	App.
DH54 dg48	9,10-dichloro-9,10-dihydro-9,10-diphenyl-anthracene $C_{26}H_{18}Cl_2$	same	DMF	Bu_4NClO_4	?	-	-	-	Pt Ag/satd $AgNO_3$	5AO
DH55 dg48 dh17	9,10-dichloro-9,10-dihydro-9,10-diphenyl-anthracene $C_{28}H_{18}Cl_2$	α,β,γ,δ-tetraphenyl-porphine $C_{44}H_{30}N_4$	CH_2Cl_2	Bu_4NClO_4	0.1	-	-	-	ns	---
DH56 dg64	trisbipyridineruthenium(II) dichloride $C_{30}H_{24}Cl_2N_6Ru$	same	MeCN	Bu_4NBF_4	0.2	-	-	-	Pt/Ag	---
DH57	phthalocyanine $C_{32}H_{18}N_8$	same	CH_2Cl_2	Bu_4NClO_4	0.1	-	-	-	Pt/Ag	25F-
DH58 dh10	1,3,6,8-tetraphenyl-pyrene $C_{40}H_{26}$	same	DMF	Bu_4NClO_4	0.1	-	-	-	Pt/SCE	23AF2
DH59 dh12 da85	rubrene $C_{42}H_{28}$	p-benzoquinone $C_6H_4O_2$	DMF	Bu_4NClO_4	0.1	-	-	-	Pt/SCE	23AF2
DH60 dh12 de38	rubrene $C_{42}H_{28}$	1,2-dibromo-1,2-diphenylethane $C_{14}H_{12}Br_2$	DMF	Bu_4NClO_4	0.1	-	-	-	Pt Ag/satd $AgNO_3$	5AO
DH61 dh12 dg48	rubrene $C_{42}H_{28}$	9,10-dichloro-9,10-dihydro-9,10-diphenyl-anthracene $C_{26}H_{18}Cl_2$	DMF	Bu_4NClO_4	0.1	-	-	-	Pt Ag/satd $AgNO_3$	5AO
DH62 dh12	rubrene $C_{42}H_{28}$	same	C_6H_5CN	Bu_4NClO_4	0.105	-	-	25	Pt/Pt	3AF-
			DMF	Bu_4NClO_4	0.105	-	-	25	Pt/Pt	3AF
				Bu_4NClO_4	0.1	-	-	-	Pt/SCE	23AF2
DH63 dh12 dh17	rubrene $C_{42}H_{28}$	α,β,γ,δ-tetraphenyl-porphine $C_{44}H_{30}N_4$	CH_2Cl_2	Bu_4NClO_4	0.1	-	-	-	Pt/Ag	25F-
DH64 dh17	α,β,γ,δ-tetraphenyl-porphine $C_{44}H_{30}N_4$	same	CH_2Cl_2	Bu_4NClO_4	0.2				Pt/Ag	25F-
CONT										

TABLE I. Electrochemical Data (CONT.) DH64

Exptl. parameters	Ref.	C/M	Excitation Type	f	E_1	E_2	vs.	Species	Type	Emission Characteristics	Remarks	Code. No.
$C_R=0.2, v=500$	JA094 9020	-	$\wedge$	-	$\geqslant 0.5$	-2.2	SCE	-	-	$\lambda = 430$ nm, $E_{\pi_{max}} = -1.9$ V	ecl in THF is much less intense, r	DH54 dg48
$C_R=1.0, C_0=1.0$	JA094 4872	-	-	-	-2.10	-2.10	SCE	O R	-	$\lambda = 428, 632, 700$ nm	$E \neq f(t)$, r	DH55 dg48 dh17
			$\square$	0.5	0	-1.30	SCE	O	-	$\lambda = 632, 700$ nm	similar λ result if $E_{app} = -1.25$ to -1.65 V $\neq f(t)$, r	
$C_R=ns$	JA094 2862	374a	$\square$	0.2	1.75	-1.60	Ag	-	-	$\lambda = 610$ nm	orange ecl, r	DH56 dg64
$C_R=0.3$	JA094 4872	-	-	1	1.40	-1.70	Ag	-	-	$\lambda = 695$ nm	π very small, r	DH57
$C_R=0.15, A=0.14$, Pt Aux(A=13.5)	JA094 0691	351c	$\square$	10	≈ 1.3	≈ -1.9	SCE	$^1R^*$	Se		$\pi \neq f$(magnetic field), r	DH58 dh10
$C_R=1.0, C_0=2.4, A=0.14$, Pt Aux(A=13.5)	JA094 0691	351b	$\square$	10	≈ 1.0	≈ -0.6	SCE	$^3R^*$	De	-	π = max for magnetic field $\cong$ 4kG, r	DH59 dh12 da85
$C_R=ns, C_0=ns$	JA094 9020	376a	$\square$ or $\wedge$	-	-	-	-	R	-	-	π large, r	DH60 dh12 de38
$C_R=0.3, C_0=0.2$	JA094 9020	375a	-	-	>0.5	-2.2	SCE	R	-	$\lambda = 560$ nm, $E_{\pi_{max}} = -1.5$ and -1.9 V vs. SCE	r	DH61 dh12 dg48
$C_R=0.59-1.25, t_f=1.00$	JA094 6324	-	triple-step	-	2.79(?)	0	$E_{p,C}$	$^3R^*$	-	-	triplet yield of charge transfer $\approx$ 10-30%, $\emptyset = (1.1 \pm 0.1)E-3$, r	DH62 dh12
$C_R=0.38-0.99, A=0.08, t=1.00$	JA094 6324	-	triple pot. step	-	2.8(?)	0	$E_{p,C}$	$^3R^*$	-	-	$\emptyset = (4.92 \pm 0.34)E-4$, r	
$C_R=1.0, A=0.14$, Pt Aux(A=13.5)	JA094 0691	351b	$\square$	10	≈ 1.0	≈ -1.5	SCE	$^3R^*$	De		π=max. for magnetic field=8kG, r	
$C_R=1.0, C_0=0.1$	JA094 4872	351e	$\square$	0.2	0.90	-1.26	SCE	O R	-	$\lambda = 567, 653, 701$ nm	$\pi = 13(\pi$ for DH64), emission occurs on redn. cycle, r	DH63 dh12 dh17
$C_R=0.1$	JA094 4872	351a	$\square$	0.2	1.05	-1.26	SCE	$^1R^*$	-	$\lambda = 655, 705$ nm	$\pi \approx 0.01(\pi$ for DH53), $\pi \neq f(t)$, $\pi = f(C), \pi = f(f)$, r	DH64 dh17
			$\wedge$									CONT

Code No.	Compound oxidized (=R), empirical formula	Compound reduced (=O), empirical formula	Solvent	Medium		μ	pH	T	Electrodes	App.
DH64 dh17	α,β,γ,δ-tetraphenyl-porphine $C_{44}H_{30}N_4$	same	DMF	Bu_4NClO_4	0.1	-	-	-	Pt/Ag	ns
			PCA	Bu_4NClO_4	0.1	-	-	-	Pt/Ag	25F-
	α,β,γ,δ-tetraphenyl-porphine $C_{44}H_{30}N_4$	same	DMF	Bu_4NClO_4						

TABLE I. Electrochemical Data

DH64

Exptl. parameters	Ref.	C/M	Type	f	Excitation E_1	E_2	vs.	Species	Type	Emission Characteristics	Remarks	Code No.
$C_R=3$	JA094 4872	351a	□	0.2	1.12	-1.06	SCE	-	-	$\lambda = 662, 702$ nm	π very small due to instability of radical cation, π↑ if $E_1 = 1.07$ (and $E_2 = -1.06$), r	DH64 dh17
$C_R=0.1$	JA094 4872	351a	□	0.2	1.05	-1.26	SCE	R		$\lambda = 655, 702$ nm	π very small, r	

TABLE II.
STRUCTURAL FORMULAS

This table is a supplement to Table I. It gives the structural formulas of 393 of the 731 compounds included in Table I. The compounds included here are those whose structural formulas could not be unambiguously and clearly represented in line form in the fourth column of Table I. Compounds are listed in the order of their code numbers.

TABLE II. Structural Formulas

DA26	$C_3H_{10}N_6$
DA30	$C_4H_3NO_3$
DA48	C_4H_9NO
DA51	$C_5H_2N_2O_3$
DA52	$C_5H_3NO_3S$
DA56	$C_5H_4N_2O_3$
DA57	$C_5H_4N_2O_3$
DA64	$C_5H_5NO_3$
DA66	$C_5H_5N_3O_2$
DA67	$C_5H_5N_3O_2$
DA68	C_5H_6O
DA69	$C_5H_7ClO_4S_2$
DA70	$C_5H_8NNaS_2$
DA76	$C_5H_{10}N_2O_3S$
DA78	$C_6H_{12}N_2$
DA79	$C_5H_{12}N_2S_2$
DA82	C_6HCl_5S
DA83	$C_6H_4N_2O_3S$
DA87	$C_6H_4O_3S_3$
DA93	$C_6H_5NO_3$
DB01	$C_6H_6AsNO_6$
DB02	$C_6H_6AsNO_7$
DB13	$C_6H_7AsO_5$
DB20	$C_6H_8AsNO_4$

DB24	C_6H_9NO	DB26	$C_6H_{10}NNaS_2$	DB31	$C_6H_{12}N_2$	DB32	$C_6H_{12}N_2$
DB33	$C_6H_{12}N_2$	DB34	$C_6H_{12}N_2$	DB36	$C_6H_{12}N_4$	DB39	$C_6H_{14}N_2$
DB40	$C_6H_{12}N_2$	DB41	$C_6H_{14}N_2O$	DB44	$C_6H_{15}N_3$	DB45	$C_6H_{15}N_3O_2S_2$
DB47	$C_6H_{16}N_4$	DB58	$C_7H_4N_2O_2$	DB69	$C_7H_6O_2$	DB72	$C_7H_7NO_5$
DB74	$C_7H_8ClN_3O_4S_2$	DB76	$C_7H_8N_2O_3$	DB77			$C_7H_8O_3S$
DB79			$C_7H_9NO_3S$	DB80	$C_7H_9NO_3S$	DB81	$C_7H_9NO_3S$

TABLE II. Structural Formulas

DB82	$C_9H_9NO_7$
DB83	$C_7H_{10}BrCl$
DB84	$C_7H_{10}BrCl$
DB85	$C_7H_{10}Cl_2$
DB86	$C_7H_{10}O$
DB88	$C_7H_{10}O$
DB89	$C_7H_{10}O_2$
DB93	$C_7H_{12}N_2$
DB95	$C_7H_{13}N$
DB97	$C_7H_{14}N_2$
DB98	$C_7H_{14}N_2$
DB99	$C_7H_{14}N_2$
DC00	$C_7H_{14}N_2$
DC01	$C_7H_{16}N_2$
DC02	$C_7H_{16}N_2$
DC04	$C_8H_5ClN_2O_2$
DC05	$C_8H_5ClN_2O_2$
DC09	$C_8H_6Br_2$
DC11	$C_8H_6N_2O_2$
DC12	$C_8H_6N_2O_2$
DC23	$C_8H_7NO_3S$
DC34	$C_8H_9NO_4$

DC41	$C_8H_{12}N_2$	DC43	$C_8H_{14}N_2$	DC44	$C_8H_{14}N_2$	DC46	$C_8H_{15}N_3$
DC47	$C_8H_{16}N_2$	DC48	$C_8H_{16}N_2$	DC49	$C_8H_{16}N_2$	DC50	$C_8H_{16}N_2O$
DC51	$C_8H_{16}N_4$	DC52	$C_8H_{18}N_2$	DC53	$C_8H_{18}N_2$	DC54	$C_8H_{18}N_2O$
DC55	$C_8H_{18}N_4$	DC58	$C_9H_5F_6FeS_2$	DC59	$C_9H_5F_6IrS_2$	DC60	$C_9H_5F_6NiS_2$
DC61	$C_9H_5F_6RhS_2$	DC68	$C_9H_8N_2O_3$	DC69	$C_9H_8N_2O_3$	DC70	$C_9H_8N_6$
DC71	$C_9H_8N_6$	DC77	$C_9H_9NO_3S$	DC78	C_9H_{10}

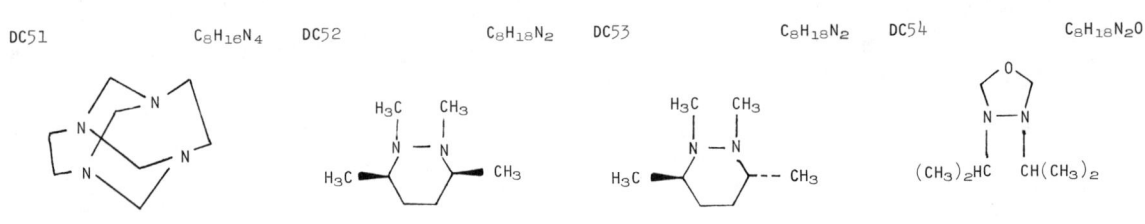

TABLE II. Structural Formulas

DC79 C_9H_{10}

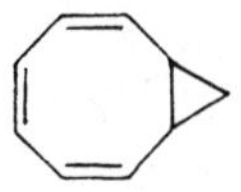

DC85 $C_9H_{11}NO_2$

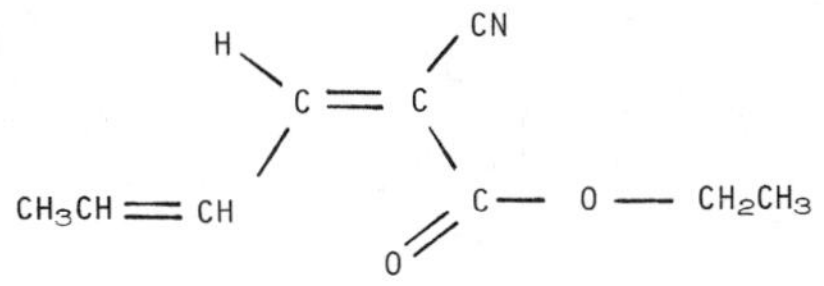

DC87 $C_9H_{13}NO_2$

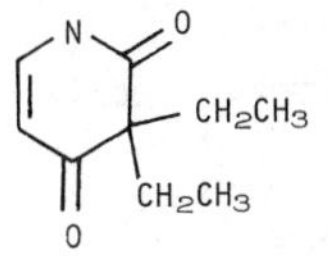

DC89 $C_9H_{14}N_3O_8P$

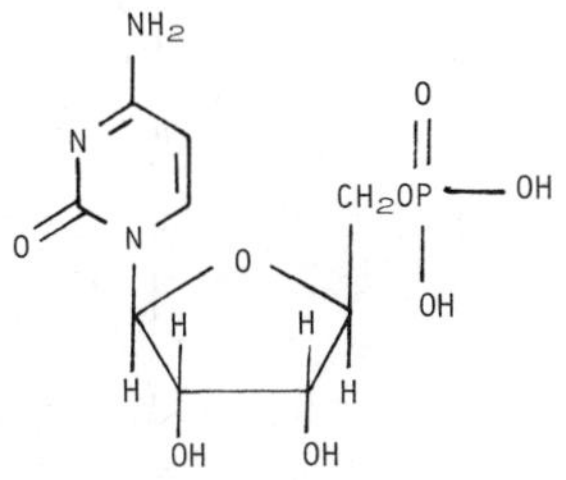

DC90 $C_9H_{15}NO_2$

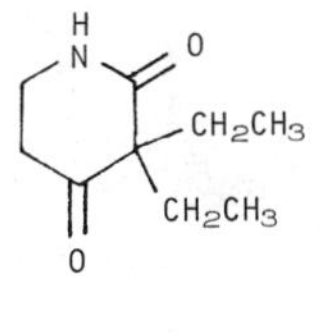

DC91 $C_9H_{16}N_2$

DC92 $C_9H_{17}N_2$

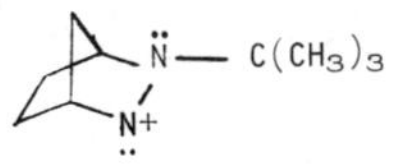

DC93 $C_9H_{18}HgO_4$ — $CH_3COHgCH_2CHO(CH_2)_3CH_3$ with O and OCH_3

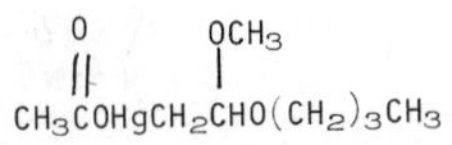

DC94 $C_9H_{18}HgO_4$ — $CH_3COHgCH_2CHOCH_2CH(CH_3)_2$ with O and OCH_3

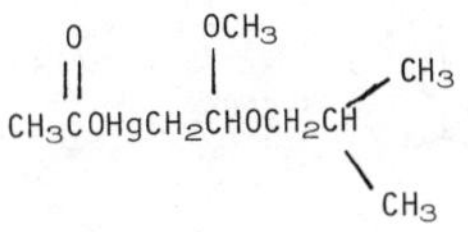

DC96 $C_9H_{18}N_2$

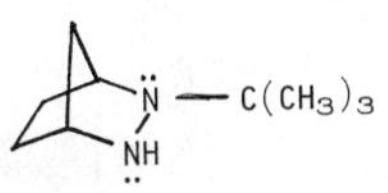

DC97 $C_9H_{20}N_2$

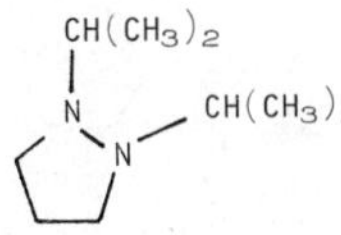

DC98 $C_9H_{20}N_2$

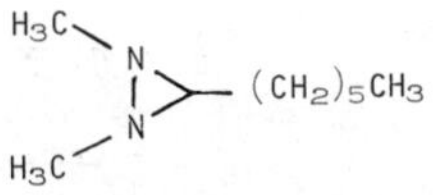

DD02 $C_{10}H_7MnO_4$

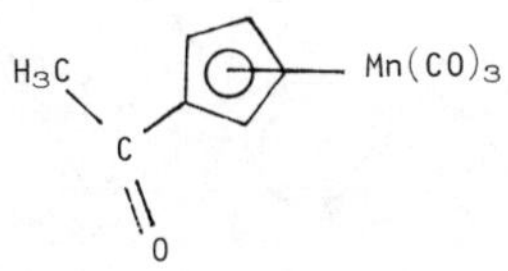

DD04 $C_{10}H_8FeN_4O_2$

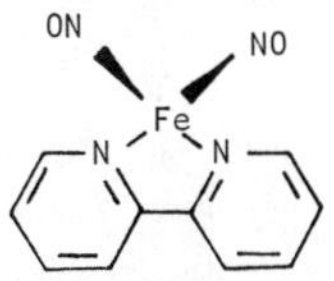

DD06 $C_{10}H_9ClO_4S_2$

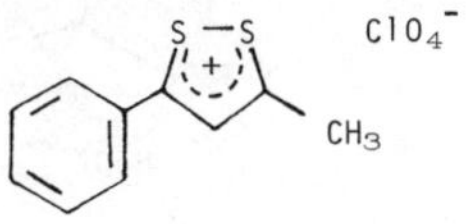

DD15 $C_{10}H_{10}N_2O_2$

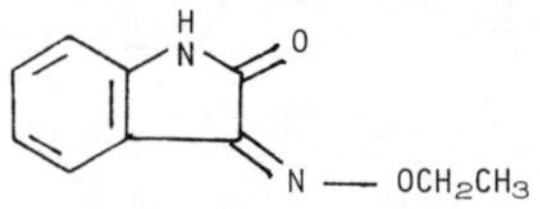

DD19 $C_{10}H_{11}N_3O_2$

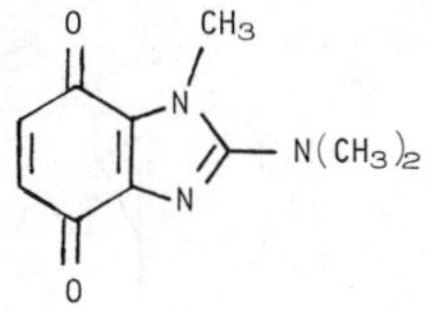

DD20 $C_{10}H_{11}N_3O_2$

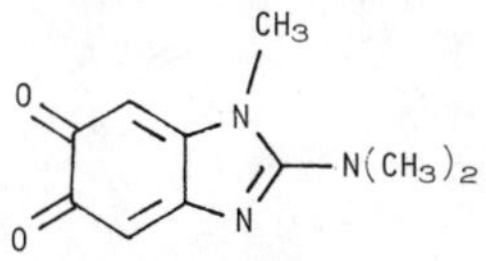

DD21 $C_{10}H_{11}N_3O_2$

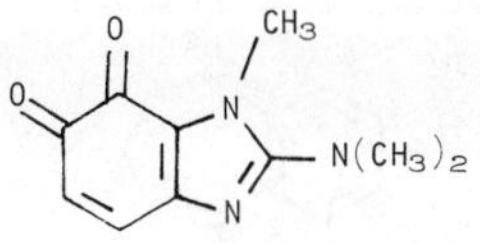

DD22 $C_{10}H_{12}$

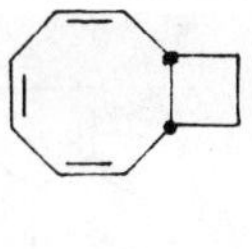

DD23 $C_{10}H_{12}ClNO$

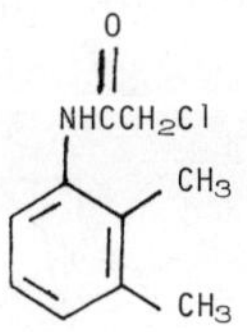

DD24 $C_{10}H_{12}ClNO$

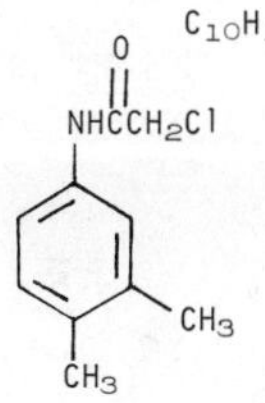

DD25 $C_{10}H_{13}NO$	DD28 $C_{10}H_{14}O$	DD29 $C_{10}H_{14}O$	DD30 $C_{10}H_{14}O_2$
DD31 $C_{10}H_{14}O_4$	DD33 $C_{10}H_{16}N_2O_2S$	DD34 $C_{10}H_{16}O_2$	DD36 $C_{10}H_{17}Br$
DD37 $C_{10}H_{17}NO_2$	DD39 $C_{10}H_{20}N_2$	DD40 $C_{10}H_{20}N_2$	DD41 $C_{10}H_{22}N_2O$
DD42 $C_{10}H_{24}N_2$	DD45 $C_{11}H_8CoN_3O_2$	DD46 $C_{11}H_8FeN_4O_3$	DD47 $C_{11}H_8N_2O$
DD49 $C_{11}H_{10}N_2O_2$	DD52 $C_{11}H_{13}NO_7$	DD54 $C_{11}H_{17}N$	DD56 $C_{11}H_{19}ClHgO$
DD57 $C_{11}H_{19}ClO_4S_2$	DD58 $C_{11}H_{20}N_2$	DD61 $C_{12}H_8Br_2$	DD62 $C_{12}H_8ClN_3OS$

TABLE II. Structural Formulas

DD63	$C_{12}H_8FeN_4O_2$
DD65	$C_{12}H_8N_2O_2$
DD66	$C_{12}H_8S$
DD68	$C_{12}H_9NO$
DD70	$C_{12}H_{10}$
DD78	$C_{12}H_{11}NO$
DD79	$C_{12}H_{11}NO$
DD80	$C_{12}H_{11}N_3O_5$
DD82	$C_{12}H_{12}O$
DD83	$C_{12}H_{14}Br_2N_2$
DD85	$C_{12}H_{16}O$
DD89	$C_{12}H_{19}LiO_2$
DD90	$C_{12}H_{20}O_2$
DD92	$C_{12}H_{20}O_2$
DD93	$C_{12}H_{21}ClHgO$
DD94	$C_{12}H_{22}N_2$
DD95	$C_{12}H_{22}N_2O$
DD96	$C_{12}H_{28}FeN_4$
DD97	$C_{13}H_6Cl_6O_2$
DD98	$C_{13}H_8CoN_3O_2$
DE00	$C_{13}H_{10}N_2O_4$
DE01	$C_{13}H_{10}N_2O_4$
DE03	$C_{13}H_{11}ClN_4O$

DE05 $C_{13}H_{12}$	DE06 $C_{13}H_{12}Cl_2O_4$	DE08 $C_{13}H_{12}N_4O_2S$	DE09 $C_{13}H_{12}N_4O_2S$
DE12 $C_{13}H_{14}Br_2N_2$	DE15 $C_{13}H_{16}O$	DE17 $C_{13}H_{18}O$	DE18 $C_{13}H_{22}N_2$
DE19 $C_{13}H_{24}N_2$	DE23 $C_{14}H_8N_2$	DE24 $C_{14}H_8N_2NiO_2$	DE25 $C_{14}H_8N_2O_4$
DE29 $C_{14}H_8O_8S_2$	DE30 $C_{14}H_9ClN_6$	DE33 $C_{14}H_{10}ClN_3O$	
DE36 $C_{14}H_{10}N_6$	DE38 $C_{14}H_{12}Br_2$	DE40 $C_{14}H_{12}O$	
DE41 $C_{14}H_{13}N_5O_2$	DE47 $C_{14}H_{15}N_5O_2$	DE48 $C_{14}H_{15}N_5O_2$	DE49 $C_{14}H_{16}Br_2N_2$

TABLE II. Structural Formulas

DE50	$C_{14}H_{16}Br_2N_2$	DE51	$C_{14}H_{16}N_6$	DE52	$C_{14}H_{16}O$
DE53	$C_{14}H_{18}BCuF_4N_4$	DE54	$C_{14}H_{18}BF_4N_4Ni$	DE55	$C_{14}H_{18}CuN_4$
DE56	$C_{14}H_{18}FeO$	DE57	$C_{14}H_{18}N_2$	DE58	$C_{14}H_{18}N_4Ni$
DE60	$C_{14}H_{20}O_8$	DE61	$C_{14}H_{22}CuN_4$	DE62	$C_{14}H_{22}N_4Ni$
DE63	$C_{16}H_{22}N_4Zn$	DE64	$C_{14}H_{22}N_6$	DE65	$C_{14}H_{24}O_2$
DE66	$C_{14}H_{25}ClHgO$	DE67	$C_{14}H_{25}NO_2$	DE68	$C_{14}H_{32}N_4$

DE71 $C_{15}H_{10}BrClO_4S_2$	DE73 $C_{15}H_{10}Cl_2N_2O$	DE74 $C_{15}H_{10}Cl_2N_2O_2$	DE76 $C_{15}H_{10}I_2N_2O$
DE77 $C_{15}H_{10}N_4O_6$	DE79 $C_{15}H_{11}BrN_2O$	DE81 $C_{15}H_{11}ClN_2O_2$	DE82 $C_{15}H_{11}ClO_4S_2$
DE83 $C_{15}H_{11}FN_2O$	DE90 $C_{15}H_{13}ClN_2$	DE91 $C_{15}H_{13}ClN_4O$	DE92 $C_{15}H_{14}ClN_3$
DE93 $C_{15}H_{14}N_4O_2$	DF05 $C_{16}H_{12}$	DF06 $C_{16}H_{13}ClN_2O$	DF07 $C_{16}H_{13}ClN_2O$
DF08 $C_{16}H_{13}ClN_2O$	DF09 $C_{16}H_{13}ClN_2O$	DF10 $C_{16}H_{13}ClN_2O_2$	DF11 $C_{16}H_{13}ClN_2O_2$
DF12 $C_{16}H_{13}ClN_2O_4$	DF13 $C_{16}H_{13}ClO_4S_2$	DF14 $C_{16}H_{13}ClO_5S_2$	

TABLE II. Structural Formulas

DF15	$C_{16}H_{13}Cl_3N_2$	DF16	$C_{16}H_{13}Cl_3N_2$	DF17	$C_{16}H_{13}FIN_3O$	DF18	$C_{16}H_{13}NO$
DF20	$C_{16}H_{14}BrN_3O$	DF21	$C_{16}H_{14}ClIN_2$	DF23	$C_{16}H_{14}ClN_3O_2$	DF24	$C_{16}H_{14}ClN_3O_2$
DF25	$C_{16}H_{14}N_2O$	DF26	$C_{16}H_{14}N_2O$	DF27	$C_{16}H_{14}N_2OS$	DF28	$C_{16}H_{14}N_2O_2$
DF29	$C_{16}H_{15}ClN_2$	DF30	$C_{16}H_{15}ClN_2O$	DF31	$C_{16}H_{15}ClN_2O_2S$	DF32	$C_{16}H_{15}NO$
DF33	$C_{16}H_{15}N_3O$	DF34	$C_{16}H_{15}N_3O_2$	DF35	$C_{16}H_{16}ClN_3$	DF37	$C_{16}H_{16}N_2$
DF39	$C_{16}H_{22}Fe$	DF40	$C_{16}H_{22}N_8Ni$	DF41	$C_{16}H_{28}FeN_4$	DF42	$C_{16}H_{28}FeN_4$

DF43	$C_{16}H_{28}FeN_4$	DF44	$C_{16}H_{28}FN_4$	DF45	$C_{16}H_{30}FeN_4$	DF46 $C_{16}H_{30}FeN_4$
DF47	$C_{16}H_{32}FeN_4$	DF49	$C_{16}H_{36}FeN_4$	DF50	$C_{17}H_{10}FeO_3$	
DF51	$C_{17}H_{13}ClN_3O$	DF53	$C_{17}H_{14}ClN_3$	DF54	$C_{17}H_{14}N_2O$	DF55 $C_{17}H_{14}N_2O_2$
DF56	$C_{17}H_{15}ClN_2O$	DF57	$C_{17}H_{15}ClN_2O_2$	DF58	$C_{17}H_{15}ClO_4S_2$	
DF59	$C_{17}H_{15}ClO_6S_2$	DF60	$C_{17}H_{15}NO_3$	DF61	$C_{17}H_{15}N_3O_2$	
DF63	$C_{17}H_{16}ClN_3O$	DF64	$C_{17}H_{16}ClN_3O$	DF65	$C_{17}H_{16}ClN_3O$	DF66 $C_{17}H_{16}ClN_3O$

TABLE II. Structural Formulas

DF67 $C_{17}H_{16}ClN_3O$	DF69 $C_{17}H_{17}ClN_2OS$	DF70 $C_{17}H_{17}ClN_2S$	DF71 $C_{17}H_{17}N_3O$
DF72 $C_{17}H_{17}N_3OS$	DF76 $C_{18}H_{12}N_2O_2S_2$		DF77 $C_{18}H_{13}NO_2S$
DF78 $C_{18}H_{13}NO_3S$	DF79 $C_{18}H_{13}NO_3S$	DF80 $C_{18}H_{13}NS$	DF81 $C_{18}H_{14}N_4Ni$
DF82 $C_{18}H_{15}AlCl_4O$	DF85 $C_{18}H_{15}IO$	DF86 $C_{18}H_{16}ClN_3O_2$	DF87 $C_{18}H_{16}FN_3O_2$
DF88 $C_{18}H_{16}N_4$	DF89 $C_{18}H_{16}N_4$	DF91	$C_{18}H_{17}Mo_2O_6P$
DF92 $C_{18}H_{17}N_3O_3$	DF93 $C_{18}H_{18}ClNO_5S_2$	DF94	$C_{18}H_{21}BrO_2$

DF95 $C_{18}H_{21}BrO_2$ DF96 $C_{18}H_{22}N_4O_5S_2$

DF97 $C_{18}H_{24}N_5O_{13}P$ DF99 $C_{18}H_{26}N_8Ni$

DG02 $C_{19}H_{14}N_2$

DG03 $C_{19}H_{18}ClN_3O_2$ DG04 $C_{19}H_{20}ClN_3O$

DG05 $C_{19}H_{25}N_8O_{12}P$ DG06 $C_{20}H_{10}Br_2O_5$

DG07 $C_{20}H_{10}Cl_2O_5$

TABLE II. Structural Formulas

| DG08 | $C_{20}H_{10}I_2O_5$ | DG09 | $C_{20}H_{12}N_2Na_2O_7S_2$ | DG10 | $C_{20}H_{15}NO$ |

| DG11 | $C_{20}H_{18}N_4Ni$ | DG12 | $C_{20}H_{20}N_2$ | DG13 | $C_{20}H_{20}N_2O$ | DG15 | $C_{20}H_{22}N_2$ |

| DG16 | $C_{20}H_{23}ClN_2O_4$ | DG22 | $C_{21}H_{16}O$ | DG23 | $C_{21}H_{16}O$ |

| DG24 | $C_{21}H_{22}N_2O_2$ | DG25 | $C_{21}H_{28}O_3$ | DG27 | $C_{21}H_{30}N_7O_{17}P_3$ |

| DG28 | $C_{21}H_{32}O$ |

| DG29 | $C_{22}H_{12}$ | DG30 | $C_{22}H_{16}N_2$ | DG31 | $C_{22}H_{16}N_2$ |

DG34 $C_{23}H_{18}N_4$	DG35 $C_{24}H_{16}$	DG39 $C_{24}H_{22}N_4NiO_2$
DG40 $C_{24}H_{24}O_6$	DG41 $C_{24}H_{24}O_6$	DG45 $C_{25}H_{33}FO_7$
DG48 $C_{26}H_{18}Cl_2$	DG49 $C_{26}H_{20}O$	DG50 $C_{26}H_{44}O_2$
DG52 $C_{27}H_{44}O$	DG55 $C_{28}H_{18}$	
DG56 $C_{28}H_{18}Cl_2N_4$	DG59 $C_{28}H_{44}O$	
DG60 $C_{28}H_{48}O_2$		

TABLE II. Structural Formulas

DG64 $C_{30}H_{24}Cl_2N_6Ru$ DG65 $C_{30}H_{24}FeN_6O_4S$ DG66 $C_{31}H_{46}O_2$

DG67 $C_{32}H_{12}FeN_8Na_4O_{12}S_4$ DG68 $C_{32}H_{15}CuN_8NaO_3S$

DG69 $C_{32}H_{17}N_8NaO_3S$ DG71 $C_{34}H_{32}ClFeN_4O_4$

DG72 $C_{34}H_{36}CoN_4O_6$ DG73 $C_{36}H_{32}FeN_6O_4$

DG74 $C_{36}H_{36}CoN_4O_4$ DG75 $C_{36}H_{36}CuN_4O_4$

DG76 $C_{36}H_{36}N_4NiO_4$ DG77 $C_{36}H_{36}N_4O_4Zn$

TABLE II. Structural Formulas

DG78 $C_{36}H_{38}ClMnN_4O_5$

DG79 $C_{36}H_{38}N_4O_4$

DG80 $C_{36}H_{42}N_4O_4$

DG81 $C_{36}H_{44}AgClN_4O_4$

DG82 $C_{36}H_{44}CaN_4$

DG83 $C_{36}H_{44}CdN_4$

DG84 　　　　　　　　　$C_{36}H_{44}CoN_4$　　DG85　　　　　　　　　$C_{36}H_{44}CuN_4$

DG86 　　　　　　　　　$C_{36}H_{44}MgN_4$　　DG87　　　　　　　　　$C_{36}H_{44}N_4Ni$

DG88 　　　　　　　　　$C_{36}H_{44}N_4OTi$　　DG89　　　　　　　　　$C_{36}H_{44}N_4OV$

TABLE II. Structural Formulas

DG90 $C_{36}H_{44}N_4Pb$ DG91 $C_{36}H_{44}N_4Pd$

DG92 $C_{36}H_{44}N_4Zn$ DG93 $C_{36}H_{45}AlN_4O$

DG94 $C_{36}H_{45}CrN_4O$ DG95 $C_{36}H_{45}FeN_4O$

DG96　　　　　　　　　　　$C_{36}H_{45}GaN_4O$　　DG97　　　　　　　　　　　$C_{36}H_{45}InN_4O$

DG98　　　　　　　　　　　$C_{36}H_{45}MnN_4O$　　DG99　　　　　　　　　　　$C_{36}H_{45}MoN_4O_2$

DH00　　　　　　　　　　　$C_{36}H_{45}N_4OSb$　　DH01　　　　　　　　　　　$C_{36}H_{45}N_4OSc$

TABLE II. Structural Formulas

DH02 $C_{36}H_{45}N_4OTl$

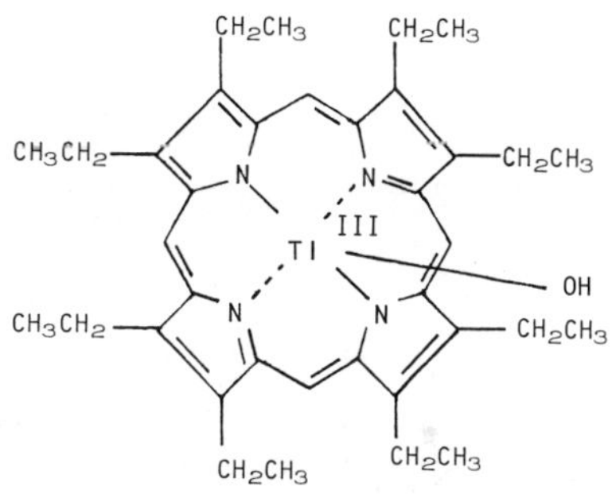

DH03 $C_{36}H_{46}GeN_4O_2$

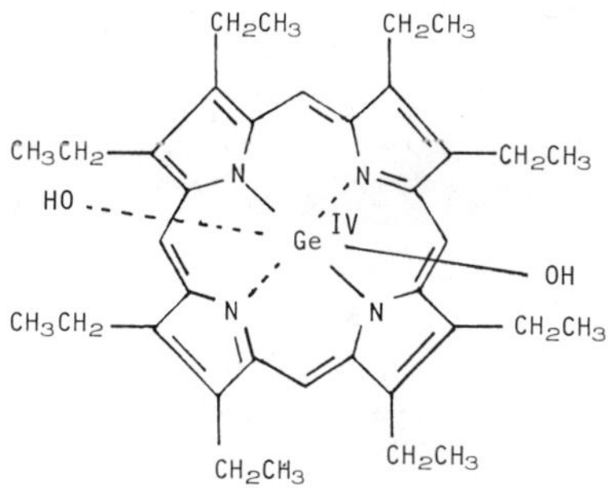

DH04 $C_{36}H_{46}N_4$

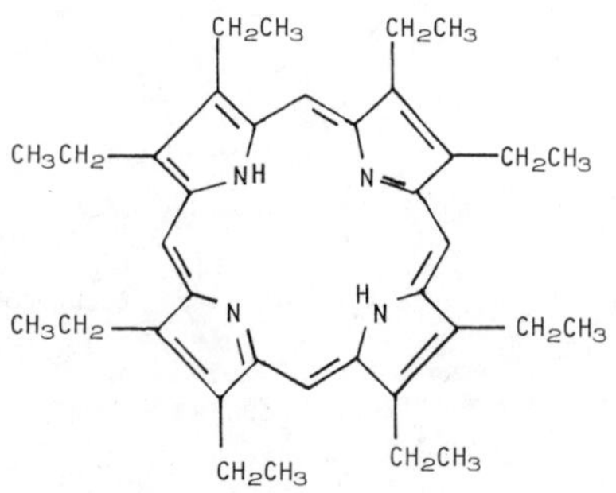

DH05 $C_{36}H_{46}N_4O_2Si$

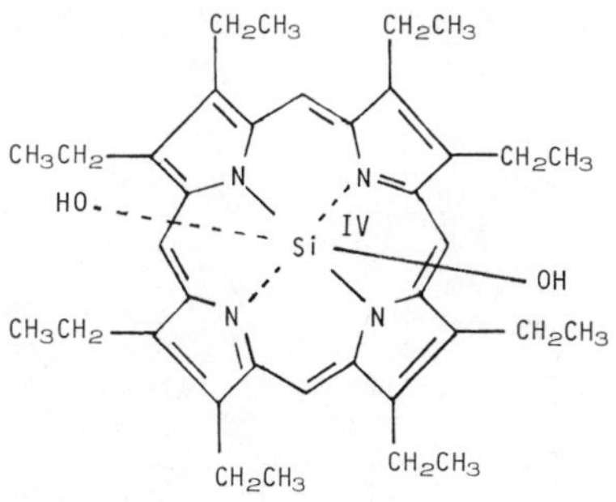

DH06 $C_{36}H_{46}N_4O_2Sn$

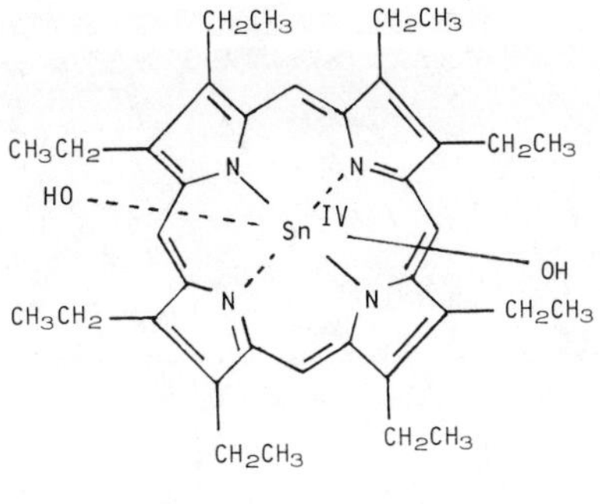

DH07 $C_{36}H_{48}Cl_2N_4O_8$

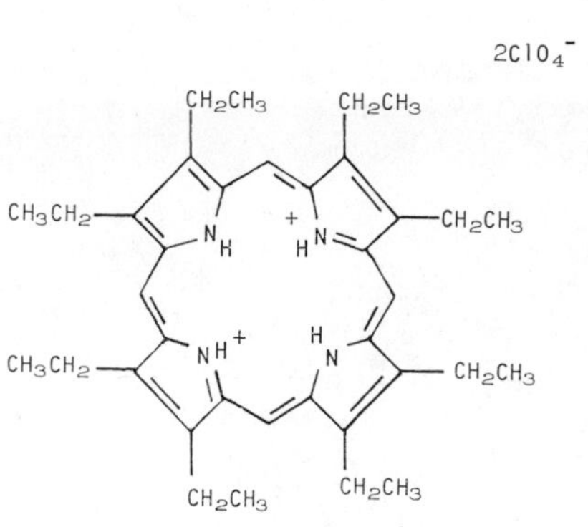

$2ClO_4^-$

DH08 $C_{36}H_{60}O_2$

DH10 $C_{37}H_{48}N_4$ DH11 $C_{40}H_{26}$

DH12 $C_{40}H_{37}FeN_6O_4$ DH14 $C_{44}H_{24}CoN_4Na_3O_{12}S_4$

TABLE II. Structural Formulas

DH15 $C_{44}H_{24}CoN_4Na_4O_{12}S_4$ DH16 $C_{44}H_{26}N_4Na_4O_{12}S_4$

DH18 $C_{44}H_{30}N_4$ DH19 $C_{44}H_{42}ClFeN_6O_4$

DH20 $C_{44}H_{46}ClCoN_6O_6$ DH21 $C_{46}H_{46}ClFeN_6O_4$

DH22 $C_{46}H_{46}ClFeN_6O_4$ CH23 $C_{46}H_{50}CoN_6O_6$

DH24 $C_{46}H_{50}CoN_6O_6$ DH25 $C_{48}H_{50}CoN_6O_8$

DH26 $C_{50}H_{54}CoN_6O_{10}$ DH27 $C_{53}H_{83}N_5$

TABLE II. Structural Formulas

DH28 $C_{54}H_{52}CoN_8O_6$

DH29 $C_{56}H_{54}CoN_6O_6$

DH30 $C_{59}H_{90}O_4$

TABLE III.
COURSES AND MECHANISMS OF HALF-REACTIONS

This table is a supplement to Table I. It gives equations for the electron-transfer steps, and for the chemical reactions associated with them, that are involved in the half-reactions undergone by the compounds listed in Table I. It also gives values of the rate and equilibrium constants of these processes whenever those values appeared in the original article.

References to this table appear in Column 15 ("C/M") of Table I in the form "42b". Such an entry signifies that this table provides equations for the steps by which the half-reaction proceeds, and that it may also contain information about one or more physical constants associated with these steps. If Column 15 of Table I is blank, there is no additional information to be found in Table III.

The general nature of the mechanism is indicated by the Arabic number in the reference in Column 15 of Table I. For oxamide (code number DA06) the entry in that column is "359a" and hence the Arabic number is "359". Under this number in Table III it is possible to find the mechanism for the reduction of oxamide, and often also closely related mechanisms or groups of mechanisms and information about other compounds that are reduced by the same mechanism or a closely related one. Neither the "359" nor the "a" has any electrochemical or chemical significance. These Arabic numbers appear consecutively in Table III and are centered on its pages. For the number "359" the relevant information can be found on page 342.

To prevent repetition, related mechanisms are frequently grouped together. Each individual mechanism is given by a set of several equations and is denoted by a combination of the Arabic number and a letter, e.g., "359a". Following this reference number, which is also given in Column 15, Table I, are the numbers of the equations that were proposed for the mechanism. For example, "359a: 359-1 to -4" means that the individual steps for oxamide are given by the equations numbered 359-1, 359-2, 359-3, and 359-4. The chemical equations for individual steps follow. In another example "71d: 71-1, -2, and -9" means that the individual steps are given by the equations numbered 71-1, 71-2, and 71-9.

On the same line in Table III, opposite the number and letter which define each particular mechanism (e.g., 359a or 71d) the code numbers (referring to Table I) are given for the compounds for which this mechanism was proposed. For mechanism 359a the code number DA06 is the only one given: this means that oxamide (DA06) is the only one of the compounds dealt with in this volume that follows this mechanism. For mechanism 71d two code numbers (DD38 and DE67) are given: this means that mechanism 71d was proposed for the compounds having the code numbers DD38 and DE67.

The chemical equations are followed by a summary of the rate and equilibrium constants that pertain to it and that appear in the chemical equations (e.g., the rate constant k_4 in the case of mechanism 359a). Such constants are given here only for compounds that appear in Table I of the present volume (but not for compounds dealt with in Volumes I through III).

For some compounds the proposed course or mechanism of the electrode process can vary according to the composition of the supporting electrolyte, the nature of the electrode used, or other experimental conditions. This is so for N-nitrosopiperidine (code number DA75), for which the entry "82d" appears twice in Column 15 in Table I for values obtained at pH 2, while the entry "82b" appears against the value measured at pH 10. This indicates that different mechanisms operate in acidic and alkaline media. To simplify orientation among processes in varying media, brief characterizations are given in Table III following the numbers of steps involved in the given mechanism. For example, "82b: 82-4 (basic)" means that in basic media (e.g., at pH 10) a four-electron reduction of the unprotonated form of the nitrosoamine takes place. Alternatively, "82d: 81-1, -2, -5, -6, and -12 (0 < pH < 7)" means that N-nitrosopiperidine at pH 2 is adsorbed, protonated in the adsorbed state, and reduced in two successive two-electron steps. The conjugate base of the nitrosoamine, represented as a carbanion, is not reducible. Information can be also found under mechanism 82 about the true value of pK_a (0 ± 2) and the pH-value at which waves of the acid and base are equal height (denoted as pK_2, equal to about 7 for N-nitrosopiperidine).

To illustrate the situation when related mechanisms are collected under a single number, examples appearing under number 82 in Table III can be again used. For N-nitrososarcosinamide (code number DA24) the entries "82a" and "82c" appear in Column 15 of Table I. Inspection of the page in Table III dealing with mechanism 82 indicates that the reduction of N-nitrososarcosinamide in acidic media takes place in one four-electron step involving the surface-protonated species, but that at pH-values greater than 8 the unprotonated form is reduced by the transfer of two electrons to each molecule of N-nitrosoamine. A carbanion is formed in a competitive reaction.

Inspection of entries under number 82 indicates that mechanisms given by various combinations of individual steps apply to a number of N-nitrosoamines. By using the same individual step in a number of mechanisms space is saved in Table III.

For some compounds bearing two or more electroactive groups, more than one mechanism applies. This can be indicated either by multiple entries in Column 15 or Table I, or, more frequently, by a reference following the enumeration of the reaction steps. For example, under mechanism number 130 we find that the reduction of 5-nitro-2-(thiocyanatomethyl)furan (code number DA83) occurs by a mechanism given as "130h: 130-1 and -2 as one step, -13, -14, followed by 126b". This means that reduction of the nitro group occurs in one four-electron step, followed by protonation of the hydroxylamino group and its two-electron reduction. The step denoted by 126b refers to the reduction of the thiocyanato group.

Many of the mechanisms that are cited in Table I have already appeared in Volumes II and III. If the equations appearing in such a mechanism were applicable without change to a compound in Volume IV, the mechanism appears in Table III without the equations, which may be found in Volume II or III. This is exemplified by mechanism 14. However, if new equations were needed, all the equations, including those previously given in Volumes II and III, are given here. This is exemplified by mechanism 65, where steps 65-1 through 65-13 were given in Volume II, and steps 65-14 through 65-16 were given in Volume III. A few mechanisms that appeared in previous volumes have been replaced by modified versions. In such cases under the number of the new mechanism (e.g., 117) there is a note "To replace old 7" or "To replace old 120 and 121". Some mechanisms that appeared in Volumes II and III are not applicable to any of the compounds in this volume. These have been omitted from Table III in this volume, as exemplified by the absence of mechanisms 5 and 6.

Since closely related compounds are often (though not always) reduced or oxidized in similar ways, there are many cases in which groups or substituents that do not affect the course or mechanism of a half-reaction are represented by symbols like "R", "R^1", "R^2", "R^3", or "Ar". The symbol "X" is often used to denote a halogen substituent. Roman numerals are used to denote intermediates; the Roman numeral in each equation representing the consumption of an intermediate is followed by parentheses containing the number of the equation in which that intermediate was formed. This facilitates finding the formula of the intermediate and following its course through the mechanism.

Unless otherwise indicated in a line or table following the last equation, "R" or "R^1" denotes an alkyl, and "Ar" an aryl, group. We have attempted to show enough of each molecular structure to enable the reader to decipher the chemical significance of each equation without undue difficulty. In cases where a reaction step is restricted to compounds with certain groups R or X, this is indicated under the Roman numeral indicating the intermediate. Hence the dimerization in step 113-13 in mechanism 113 is represented by an equation reading

$$2 \text{ IX } (113\text{-}10) \rightarrow$$
$$\text{where } R^1 = H$$

to denote that the step applies only to radicals having structure IX, which appears in equation 113-10, if the group R^1 is a hydrogen atom.

Most of the rate or equilibrium constants given in this Table were deduced from the electrochemical data; the original literature must of course be consulted for details. A few values that were obtained by spectrophotometric or other nonelectrochemical techniques are also included to permit comparison with, or to aid in the interpretation of, the accompanying electrochemical results.

TABLE III. Courses and Mechanisms of Half-Reactions

4

Mechanism:
4a: 4-3 to -5

4b: 4-1 and -2

Proposed for Compound:
-

DA01

$$C(NO_2)_4 + 2e \rightarrow C(NO_2)_3^- + NO_2^- \qquad 4\text{-}1$$

$$C(NO_2)_3^- + H^+ \rightleftarrows CH(NO_2)_3 \qquad 4\text{-}2$$

$$CH(NO_2)_3 + 8H_2O + 12e \rightarrow HONHC(:NH)NHOH + 12OH^- \qquad 4\text{-}3$$

$$HONHC(:NH)NHOH \xrightarrow[\text{slow}]{OH^-} H_2NC(:NH)NO + H_2O \qquad 4\text{-}4$$

$$H_2NC(:NH)NO + 2H^+ + 2e \rightarrow H_2NC(:NH)NHOH \qquad 4\text{-}5$$

7

Mechanism:
7b: 7-4 to 7-7

Proposed for Compound:
DA63

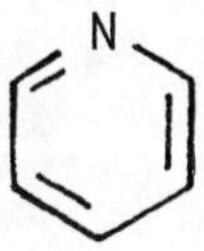

pyridine (I) + HA $\rightleftarrows$ pyridinium (II) + A⁻   7-1

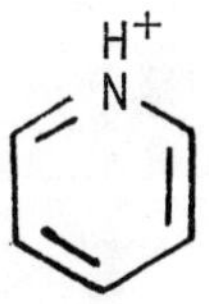

II (7-1) + e $\rightleftarrows$ (III)   7-2

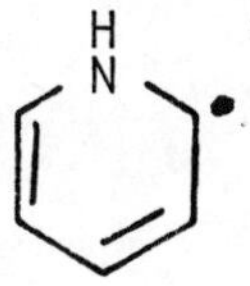

III (7-2) → inactive product   7-3

I (7-1) + e $\rightleftarrows$ (IV)   7-4

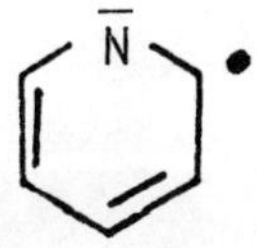

IV (7-4) + HA → III (7-2) + A⁻   7-5

III (7-2) + e → 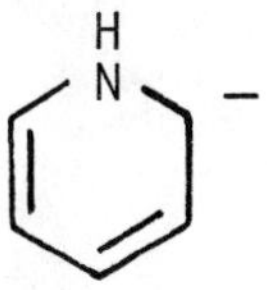 (V)   7-6

V (7-6) + HA → [2H-pyridine structure with NH and H] + A⁻   7-7

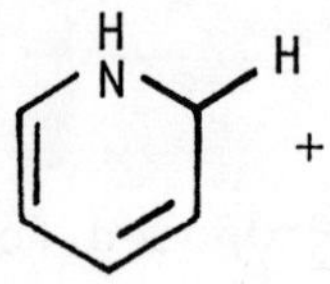

8

Mechanism:

8d: 8-5; -6 and -7 as one step, and -11 to -13

8e: 8-10; -6 and -7 as one step, and -11 to -13

8f: 8-5; -6 and -7 as one step, and -8

Proposed for Compound:

DA13

DA14

DA88, DD08, DD09, DD10, DD11, DD12, DD13, DD18, DD50, DD51, DD84

$RHgX \rightleftarrows RHg^+ + X^-$ 8-1

$RHg^+ + e \rightleftarrows RHg\cdot$ 8-2

$RHg\cdot + e \rightarrow R^- + Hg$ 8-3

$R^- + H^+ \rightleftarrows RH$ 8-4

$RHgX + e \rightleftarrows RHg\cdot + X^-$ 8-5

$2RHg\cdot \xrightleftharpoons{fast} (RHg)_2$ 8-6

$(RHg)_2 \rightleftarrows Hg + R_2Hg$ 8-7

$RHg\cdot + H^+ + e \rightarrow RH + Hg$ 8-8

$R^- + H_2O \rightarrow RH + OH^-$ 8-9

$RX + Hg + e \rightarrow RHg\cdot + X^-$ 8-10

$RHg\cdot + e \rightarrow RHg^-$ 8-11

$RHg^- + RX \rightarrow R_2Hg + X^-$ 8-12

$RHg^- + H^+ \xrightarrow{?} RH + Hg$ 8-13

$R_2Hg + 2H^+ + 2e \rightarrow 2RH + Hg$ 8-14

$R = NCCH_2CH_2-, Ar-, Ar\underset{I}{C}HCOOC_2H_5$

$X = I, Cl, Br$

TABLE III. Courses and Mechanisms of Half-Reactions

9

Mechanism:
9a: 9-1 to -4

Proposed for Compound:
DA03

14

Mechanism:
14a: 14-1 and -2

14b: 14-1 to -4

14c: 14-1 to -6

Proposed for Compound:
DB91, DC16, DC17, DC18, DC20, DC27, DC81, DD23, DD24

DA29, DC19, DC74, DC75

DC10, DC66, DC67

19

Mechanism:
19a: 19-2 and -5 to -7 (basic)

19b: 19-3 and -4 (acidic)

19c: 19-1, -2 and -8 to -12 (at PGE)

Proposed for Compound:
-

-

DA05

$HOOCCOOH \rightleftarrows HOOCCOO^- + H^+$	19-1
$HOOCCOO^- \rightleftarrows {}^-OOCCOO^- + H^+$	19-2
$HOOCCOOH \rightarrow (HOOCCOOH)_{ads}$	19-3
$(HOOCCOOH)_{ads} \rightarrow 2CO_2 + 2H^+ + 2e$	19-4
$HOOCCOO^- \rightarrow (\cdot OOCCOO\cdot)_{ads} + H^+ + 2e$	19-5
${}^-OOCCOO^- \rightarrow (\cdot OOCCOO\cdot)_{ads} + 2e$	19-6
$(\cdot OOCCOO\cdot)_{ads} \rightleftarrows 2CO_2$	19-7
${}^-OOCCOO^- \rightarrow {}^-OOCCOO\cdot + e$	19-8
${}^-OOCCOO\cdot \rightarrow \cdot OOCCOO\cdot + e$	19-9
$\cdot OOCCOO\cdot \rightarrow 2CO_2$	19-10
$HOOCCOO^- \rightarrow HOOCCOO\cdot + e$	19-11
$HOOCCOO\cdot \rightarrow \cdot OOCCOO\cdot + H^+ + e$	19-12

21

Mechanism:
21b: 21-1, -2, and -4

Proposed for Compound:
DC57

23

Mechanism:
23d: 23-7

Proposed for Compound:
DA28, DA37, DB25, DG14

$$R = -COOCH_3,\ -CCl:CCl_2,\ -CH_2COCH_3,\ C_6H_5\overset{|}{C}HCOOCH_3$$

33

Mechanism:
33a: 33-1

Proposed for Compound:
DG28

$R^1CH:CHCH_2$ = retinyl, R^2 = methyl

43

Mechanism:
43a: 43-1 and -2

Proposed for Compound:
DA21, DA41, DA46, DC29, DC83

R^1 = vinyl

R^2 = methyl, vinyl, ethyl, phenyl, benzyl

65

Mechanism:
65d: 65-1 to -3, -8 and -9

Proposed for Compound:
DD02

65r: 65-10, -12 and -17

DE02

TABLE III. Courses and Mechanisms of Half-Reactions

$$ArC(:O)R + H^+ \underset{}{\overset{K_1}{\rightleftarrows}} ArC(:OH)R^+ \qquad 65\text{-}1$$

$$ArC(:OH)R^+ + e \rightleftarrows Ar\dot{C}(OH)R \qquad 65\text{-}2$$

$$2Ar\dot{C}(OH)R \rightarrow \text{pinacol} \qquad 65\text{-}3$$

$$Ar\dot{C}(OH)R + [ArCOR]^{\bar{\cdot}} \rightarrow RArC(O^-)C(OH)ArR \qquad 65\text{-}4$$

$$RArC(O^-)C(OH)ArR + H^+ \underset{}{\overset{k_5}{\rightleftarrows}} RArC(OH)C(OH)ArR \qquad 65\text{-}5$$

$$Ar\dot{C}(OH)R + Hg \rightarrow \text{organomercuric compounds} \qquad 65\text{-}6$$

$$Ar\dot{C}(OH)R + \text{solvent} \rightarrow \text{products} \qquad 65\text{-}7$$

$$Ar\dot{C}(OH)R + e \rightarrow ArC(OH)R^- \qquad 65\text{-}8$$

$$ArC(OH)R^- + H^+ \rightleftarrows ArCHOHR \qquad 65\text{-}9$$

$$ArC(:O)R + e \rightarrow [ArCOR]^{\bar{\cdot}} \qquad 65\text{-}10$$

$$[ArCOR]^{\bar{\cdot}} + H^+ \underset{}{\overset{K_{11}}{\rightleftarrows}} Ar\dot{C}(OH)R \qquad 65\text{-}11$$

$$[ArCOR]^{\bar{\cdot}} + e \rightarrow ArCOR^{2-} \qquad 65\text{-}12$$

$$ArCOR^{2-} + 2H^+ \rightleftarrows ArCH(OH)R \qquad 65\text{-}13$$

$$2[ArCOAr]^{\bar{\cdot}} + (CH_3CO)_2O \rightarrow 2CH_3COO\dot{C}Ar_2 + O^{2-} \qquad 65\text{-}14$$

$$CH_3COO\dot{C}Ar_2 + e \rightarrow CH_3COO\bar{C}Ar_2 \qquad 65\text{-}15$$

$$2CH_3COO\bar{C}Ar_2 + (CH_3CO)_2O \rightarrow 2CH_3COOC(COCH_3)Ar_2 + O^{2-} \qquad 65\text{-}16$$

$$2[ArCOR]^{\bar{\cdot}} \rightleftarrows ArCOR + ArCOR^{2-} \qquad 65\text{-}17$$

Ar = C_6H_5 or cyclopentadienyltricarbonylmanganese

R = CH_3, C_6H_5,

Code No.	K_{17} at 50°C
DE02	10^{-12}

69

Mechanism:	Proposed for Compound:
69b: 69-1 to -4 and -6 to -8	DB65, DB71, DE39
69c: 69-1 and -9	DC13
69d: 69-1 to -3 and -6	DC14

$O_2NC_6H_4CH_2X + e \rightarrow [O_2NC_6H_4CH_2X]^{\bar{\cdot}}$ 69-1

$[O_2NC_6H_4CH_2X]^{\bar{\cdot}} \xrightarrow{k_2} O_2NC_6H_4CH_2^{\cdot} + X^-$ 69-2

$O_2NC_6H_4CH_2^{\cdot} + HSo \rightarrow O_2NC_6H_4CH_3 + So^{\cdot}$ 69-3

$O_2NC_6H_4CH_3 + e \rightarrow [O_2NC_6H_4CH_3]^{\bar{\cdot}}$ 69-4

$[O_2NC_6H_4CH_3]^{\bar{\cdot}} + O_2NC_6H_4CH_2X \rightarrow O_2NC_6H_4CH_3 + [O_2NC_6H_4CH_2X]^{\bar{\cdot}}$ 69-5

$2 O_2NC_6H_4CH_2^{\cdot} \rightarrow (O_2NC_6H_4CH_2)_2$ 69-6

$(O_2NC_6H_4CH_2)_2 + e \rightarrow (O_2NC_6H_4CH_2)_2^{\bar{\cdot}}$ 69-7

$(O_2NC_6H_4CH_2)_2^{\bar{\cdot}} + e \rightarrow (O_2NC_6H_4CH_2)_2^{2-}$ 69-8

$[O_2NC_6H_4CH_2X]^{\bar{\cdot}} \rightarrow O_2NC_6H_4\bar{C}HX + \tfrac{1}{2}H_2$ 69-9

X = CN, 4-NO$_2$

    Cl, 4-NO$_2$

71

Mechanism:	Proposed for Compound:
71a: 71-1 to -4	DC85
71d: 71-1, -2, and -9	DD38, DE67
71e: 71-1 and -10 (anhyd. DMF)	DA62
71f: 71-1, -8, and -11 (H$_2$O)	DA62

$R^1CH{:}CR^2R^3 + e \rightarrow (R^1CHCR^2R^3)^{\bar{\cdot}}$ 71-1

$(R^1CHCR^2R^3)^{\bar{\cdot}} + e \rightarrow (R^1CHCR^2R^3)^{2-}$ 71-2

$(R^1CHCR^2R^3)^{\bar{\cdot}} + \text{excess } R^1CH{:}CR^2R^3 \rightarrow (R^1CHCR^2R^3)_2^{\bar{\cdot}}$ 71-3

$(R^1CHCR^2R^3)^{\bar{\cdot}} + BH \rightarrow R^1CHCR^2R^3H\cdot + B^-$ 71-4

$(R^1CHCR^2R^3)^{\bar{\cdot}} + CO_2 \rightarrow {}^-O_2CCHR^1\dot{C}R^2R^3$ 71-5

${}^-O_2CCHR^1\dot{C}R^2R^3 + e \rightarrow {}^-O_2CCHR^1\bar{C}R^2R^3$ 71-6

${}^-O_2CCHR^1\bar{C}R^2R^3 + CO_2 \rightarrow {}^-O_2CCHR^1CR^2R^3CO_2^-$ 71-7

$2R^1CH{:}CR^2R^3{}^{\bar{\cdot}} \xrightarrow{k_8} \begin{array}{c} R^1\bar{C}HCR^2R^3 \\ | \\ R^1\underline{C}HCR^2R^3 \end{array}$ (I)   71-8

TABLE III. Courses and Mechanisms of Half-Reactions

$(R^1CHCR^2R^3)^{2-} + 2H^+ \xrightarrow{fast} R^1CH_2CHR^2R^3$ 71-9

$(R^1CHCR^2R^3)^{\cdot} + n\ R^1CH{:}CR^2R^3 \rightarrow \left(\begin{matrix} \ \ \ \ R^2 \\ CH-C \\ R^1\ \ \ R^3 \end{matrix}\right)^{\cdot}_{n+1}$ 71-10

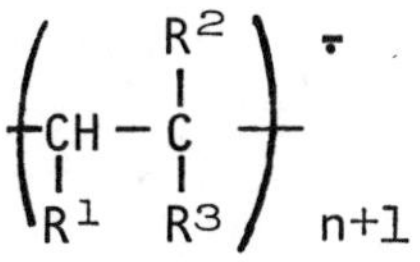

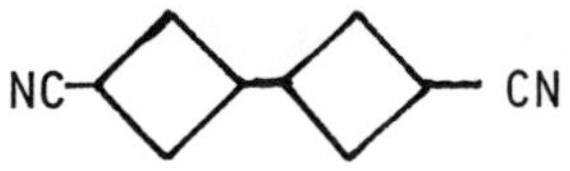

$\begin{matrix} NC\bar{C}HCHCH{=}CH_2\ (Ia{-}71{-}8) \\ | \\ NC\underline{C}HCHCH{=}CH_2 \end{matrix} + 2H^+ \rightarrow NC-\square\square-CN$ 71-11

R^1 = H or CN

R^2 = CH_3 or H

R^3 = -C(O)OCH$_2$CH$_2$N(piperidine with C$_2$H$_5$ groups) , -C(O)OCH$_2$CH$_2$N(C$_2$H$_5$)$_2$ or -CH=CH$_2$

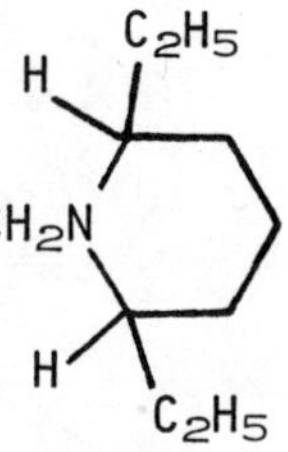

72

Mechanism:
72c: 72-1 and -5 to -8

Proposed for Compound:
DB48, DB49, DB50, DB51, DB52, DB53, DB57

$XC_6H_4CN + e \rightleftarrows XC_6H_4CN^{\cdot}$ 72-1

$2XC_6H_4CN^{\cdot} + 2H^+ \rightarrow XC_6H_4CN +$ [cyclohexadiene with X and CN]   72-2

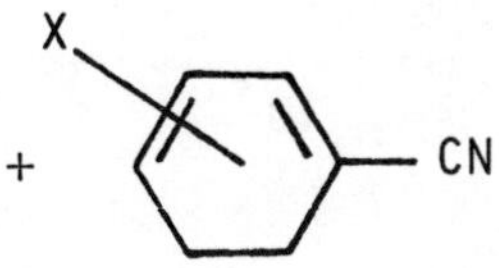

$XC_6H_4CN^{\cdot} + H^+ \rightarrow$ [cyclohexadienyl radical with X, H, CN] (I)   72-3

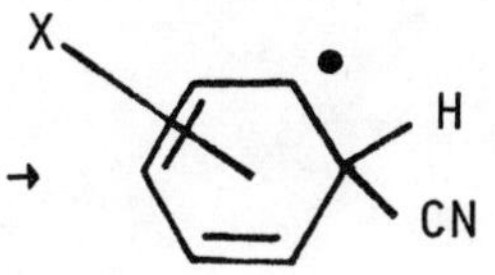

$I\,(72{-}3) + e \rightarrow XC_6H_5 + CN^-$ 72-4

$$XC_6H_4CN^{\cdot -} \xrightarrow{fast} C_6H_4CN\cdot + X^- \qquad 72\text{-}5$$

$$C_6H_4CN\cdot + CN^- \rightarrow NCC_6H_4CN^{\cdot -} \qquad 72\text{-}6$$

$$C_6H_4CN\cdot + e \rightarrow C_6H_4CN^- \qquad 72\text{-}7$$

$$C_6H_4CN^- + HSo \rightarrow C_6H_5CN + So^- \qquad 72\text{-}8$$

X = 2-Br, 3-Br, 4-Br, 2-Cl, 3-Cl, 4-Cl, 4-I

74

Mechanism: Proposed for Compound:

74c: 74-7 to -10 (MeCN)   DA31

74d: 74-4, -5, and -11 to -16 (pH < 2.5)   DA31

74e: 74-17 and -14 to -16 (pH > 2.5)   DA31

pyrazine (I) + H$^+$ + e → protonated pyrazinyl radical (II)   74-1

II(74-1) + H$^+$ + e → 1,4-dihydropyrazine (III)   74-2

$$I(74\text{-}1) + III(74\text{-}2) \underset{}{\overset{K_3}{\rightleftarrows}} 2\,II(74\text{-}1) \qquad 74\text{-}3$$

monoprotonated pyrazine (IV) ⇌ I(74-1) + H$^+$ pK$_4$ 74-4

TABLE III. Courses and Mechanisms of Half-Reactions

(V) ⇌ IV(74-4) + H⁺   pK₅   74-5

V(74-5) + H⁺ + 2e ⇌ (VI)   74-6

I(74-1) + e ⇌ (VII)   74-7

VII(74-7) + HA $\xrightarrow{k_8}$ II(74-1) + A⁻   74-8

II(74-1) + e → (VIII)   74-9

VIII(74-9) + HA → III(74-2) + A⁻   74-10

V(74-5) + e ⇌ (IX)   74-11

IX(74-11) + H⁺ + e ⇌ VI(74-6)   74-12

VI(74-6) + I(74-1) + H⁺ → 2 IX(74-11)   74-13

VI(74-6) ⇌ III(74-2) + H⁺   74-14

III(74-2) ⇌ (X)   74-15

X (74-15) + H_2O → $H_2NCH:CHNHCH_2CHO$ (Hydrolysis is possible) 74-16

I (74-1) + $3H^+$ + 2e → VI (74-6) 74-17

Code No.	HA	$k_8, dm^3mol^{-1}s^{-1}$	k_5
DA31	H_2O	22 ± 8	
	1M $HClO_4$		100

75

Mechanism:
75c: 75-1 and -5 to -7

Proposed for Compound:
DA12, DA32, DA33

$R + e \rightleftarrows R^{\bullet -}$ 75-1

$R^{\bullet -} + R \rightarrow RR^{\bullet -}$ 75-2

$R + 2e \rightarrow R^{2-}$ 75-3

$R^{2-} + R \rightarrow 2R^{\bullet -}$ 75-4

$R^{\bullet -} + HA \xrightarrow{k_5} RH\cdot + A^-$ 75-5

$RH\cdot + e \rightleftarrows RH^-$ 75-6

$RH^- + HA \rightarrow RH_2 + A^-$ 75-7

R = 1,3,5-triazine, pyridazine or pyrimidine

Code No.	HA	$k_5, dm^3mol^{-1}s^{-1}$
DA32	H_2O	30 ± 7

TABLE III. Courses and Mechanisms of Half-Reactions

82

Mechanism:　　　　　　　　　　　　　　　　　　　　　　Proposed for Compound:

82a: 82-1 to -3 (acidic)　　　　　　　　　　　　　　　DA24, DB75, DD74, DE42

82b: 82-4 (basic)　　　　　　　　　　　　　　　　　　DA10, DA75, DA76, DB42, DD95

82c: 82-10, -4, and -12 (pH > 8)　　　　　　　　　　　DA24, DE42

82d: 82-1, -2, -5, and -6 ($0 < pH < 7$)　　　　　　　DA10, DA75, DA76, DB42, DD95

82e: 82-11 ($9 < pH < 12$)　　　　　　　　　　　　　　DB75, DD74

$R^1R^2NNO(I) \rightleftarrows R^1R^2NNO_{ads}$ (II)   82-1

$R^1R^2\overset{H^+}{\frown}NNO_{ads}$ (III) $\rightleftarrows$ II(82-1) + H^+ K_2 82-2

III(82-2) + $3H^+$ + 4e $\overset{k_3}{\rightarrow}$ $R^1R^2NNH_2$ (IV) + H_2O 82-3

2 I(82-1) + $3H_2O$ + 4e → $2R^1R^2NH$ (V) + N_2O + $4OH^-$ 82-4

III(82-2) + H^+ + 2e → R^1R^2NNHOH (VI)   82-5

VI(82-5) + $2H^+$ + 2e → IV(82-3) + H_2O 82-6

$R^1R^2NNH_2OH^+$ (VII) $\rightleftarrows$ VI(82-5) + H^+ K_7 82-7

VII(82-7) + H^+ + 2e → IV(82-3) + H_2O 82-8

IV(82-3) + $2H^+$ + 2e → decomposition   82-9

$R^3CH_2\overset{R^1}{\underset{|}{N}}NO$ (Ia) $\rightleftarrows$ $R^3\overset{R^1}{\underset{|}{\bar{C}}}HNNO$ + H^+ K_{10} 82-10

I (82-1) + $2H_2O$ + 2e → R^1R^2NNHOH + $2OH^-$ 82-11

$R^3\overset{R^1}{\underset{|}{\bar{C}}}HNNO$ + ne $\not\rightarrow$ 82-12

R^1 = CH_3, $(CH_3)_2CH$, cycloC_6H_{11}, C_6H_5, $C_6H_5CH_2$

R^2 = CH_3, thiolane, $(CH_3)_2CH$, cyclo-C_6H_{11}, $C_6H_5CH_2$, CH_2CONH_2, C_6H_5, NH_2, OC_2H_5,
    $C(NH)NHNO_2$

R^3 = C_6H_5, NH_2CO

R^1 and R^2 = piperidine

Code No.	pK_2(UVS,PY)	pK_2'(PY)	pK_{10}(UVS,PY)
DA10	0±2	≈7	
DA24	0	8.5	12.75
DA75	0±2	≈7	
DA76	0±2	≈7	
DB42	0±2	≈7	
DB75	0.25	7.4	
DD74	-2.0	7	
DD95	0±2	≈7	
DE42	0.7	7	13.7

89

Mechanism:
89c: 89-4 to -6, -2 and -7 to -9

Proposed for Compound:
DA36, DC88, DC89, DF97, DF98, DG05

$R^1N\text{-pyrimidinone}\text{-}R$ (I) + 2e + 2H$^+$ → $R^1N\text{-dihydropyrimidinone}\text{-}R,H$ (II)   89-1

II (89-1) → $R^1N\text{-pyrimidinone}$ (III) + HR   89-2

2 III (89-2) + e + H$^+$ → R^1N–dimer–NR^1 (decomposes) (IV)   89-3

I (89-1) + H$^+$ ⇌ $R^1N\text{-pyrimidinone-}R$ protonated (V)   89-4

V (89-4) + 2e → $R^1N\text{-}R$ (VI)   89-5

VI (89-5) + H$^+$ → II (89-1)   89-6

TABLE III. Courses and Mechanisms of Half-Reactions

III(89-2) + H⁺ ⇌ [structure VII: pyrimidinone cation with R¹N, NH⁺, O] (VII)   89-7

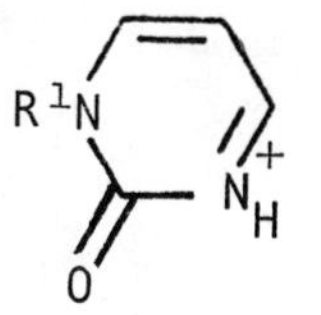

VII(89-7) + e → [structure VIII: radical with R¹N, NH, O] (VIII)   89-8

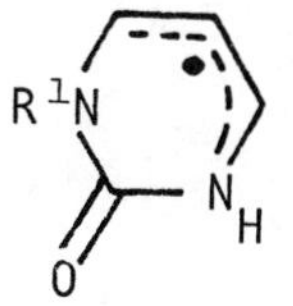

2VIII(89-8) → IV (89-3)   89-9

R = NH₂, R¹ = H, ribofuranosyl or substituted ribofuranosyl unit.

Code No.	k_2, s^{-1}	$k_5, cm\ s^{-1}$	k_6, s^{-1}
DA36	10	0.1	5×10^4

92

Mechanism:
92c: 92-4 and -5 as one step

Proposed for Compound:
DA15

R = CH₂:CH

105

Mechanism:
105c: 105-2 and -5

Proposed for Compound:
DB15

CRC Handbook Series in Organic Electrochemistry

113
(To replace old 113)

Mechanism:	Proposed for Compound:
113a: 113-1 to -3 (at PGE)	AE04
113b: 113-3 (at PGE)	AE02
113c: 113-1 to -5 (1 < pH < 3)	AE03
113d: 113-10 (3 < pH < 5)	AE03
113e: 113-10 and 11 as one step, -12, and -2 to -5 (5 < pH < 5.8)	AE03
113f: 113-3 to -6	product of 117(AD65,AD66, AE01)AD95,AD96,AE13
113g: 113-1, -7 to -9, -5 and -6	AE13,AE17,AG96,AG99,AH38, AK97,AT12,AU02,BB28
113h: 113-10, -13, (-11 and -12 as one step), and -14 (at DME, pH < 5)	AE02
113i: 113-10, -13 and -14 (at DME, pH > 5 or at PGE)	AE02
113j: 113-17, 18, 13, 19, 20, -11, and -12 (neutral)	DA59
113k: 113-15 and -16, -11 and -12 as one step (acidic)	DA59
113l: 113-15, -21 to -24 and -16, -11 and -12 as one step (HClO$_4$)	DA59

(I) + 2H$^+$ + 2e → (II)   113-1

II(113-1) → (III) + R^{1}H   113-2

TABLE III. *Courses and Mechanisms of Half-Reactions*

III(113-2) + 2H⁺ + 2e → (IV)   113-3

IV(113-3) + 2H⁺ + 2e → (V)   113-4

V(113-4) + H_2O → (VI)   113-5
where R^2 = H

VI(113-5) $\xrightarrow{H_2SO_4}$ + H_2CO 113-6

II(113-1) + 2H⁺ + 2e → (VII)   113-7
where R^2 = H

VII(113-7) $\xrightarrow{slow}$ (VIII) + R^1H 113-8

VIII(113-8) + 2H⁺ + 2e → V(113-4)   113-9
where R^2 = H

I (113-1) + H$^+$ + e → (IX)   113-10

IX (113-10) + e → (X)   113-11

X (113-11) + H$^+$ → II (113-1)   113-12

2 IX (113-10) →
where R^1 = H   (XI)   113-13

XI (113-13) + 2H$^+$ + 2e $\overset{?}{\rightarrow}$ 2 II (113-1)   113-14

(XII) ⇄ I (113-1) + H$^+$ 113-15

XII (113-15) + e → IX (113-10)   113-16

I (113-1) + e → (XIII)   113-17

I (113-1) + XIII (113-17) → IX (113-10) +   113-18

TABLE III. Courses and Mechanisms of Half-Reactions

XIII (113-17) + e → (XIV) 113-19

XIV (113-19) + 2H⁺ → II (113-1) 113-20

(XV) ⇌ XII (113-15) + H⁺ 113-21

(XVI) ⇌ II (113-1) + H⁺ 113-22

XVI (113-22) + H⁺ + 2e → V (113-4) 113-23

XV (113-21) + H⁺ + 2e → XVI (113-22) 113-24

R^1 = H, SH, NH_2, NR_2, OCH_3, OH

R^2 = H, SH

116

Mechanism:
116a: 116-1 to -3

Proposed for Compound:
DB64, DC06, DC07

Code No.	RX	k_1, dm³mol⁻¹s⁻¹
DB64	EtBr	15
DC06	MeI	0.04
DC07	MeI	0.67

117
(To replace old 117)

Mechanism:		Proposed for Compound:
117a:	117-1 to -15 followed by 113f ($0 < pH < 4$)	AE01
117b:	117-1 to -11, (-12 and -13 as one step), -14, -16, -17, and -18 followed by 113f ($4 < pH < 5$)	AE01
117c:	117-1 and -14 as one step, -16, -17, and -18 followed by 113f ($6 < pH < 7$)	AE01
117d:	117-1, -2, -10, and -11 followed by 113-4 to -6 ($7 < pH < 11$)	AE01
117e:	117-1 and -2 ($pH > 11$)	AE01
117f:	117-1 to -7 ($2 < pH < 3$)	AD65, AD66
117g:	117-1 to -7, -20, and -21 followed by 113-9, -5, and -6 ($3 < pH < 6$)	AD65, AD66
117h:	117-1 to -4 and -19 to -24 followed by 113f ($7 < pH < 9$)	AD65, AD66
117i:	117-22 to -24 ($9 < pH < 13$)	AD65, AD66

(I) + 2H⁺ + 2e → (II)   117-1

II(117-1) $\xrightarrow{slow}$ (III) + H$_2$SO$_3$ 117-2

II(117-1) $\xrightarrow[slow]{very}$ (IV) + H$_2$O   117-3

IV (117-3) + 2H⁺ + 2e → (V)   117-4

TABLE III. Courses and Mechanisms of Half-Reactions

V (117-4) $\xrightarrow{\text{very slow}}$ [structure VI: purine with SOH at C6, R at C2] (VI) + H₂O   117-5

VI (117-5) + 2H⁺ + 2e⁻ → [structure VII: dihydro purine with H, SOH at C6, R at C2] (VII)   117-6

VII (117-6) $\xrightarrow{\text{very slow}}$ [structure VIII: purine with SH at C6, R at C2] (VIII) + H₂O   117-7

VIII (117-7) + 2H⁺ + 2e⁻ → [structure IX: dihydro purine with H, SH at C6, R at C2] (IX)   117-8

IX (117-8) → III (117-2) + H₂S   117-9

III (117-2) + 4H⁺ + 4e⁻ → [structure X: tetrahydro purine with R at C2] (X)   117-10

X (117-10) $\xrightarrow{\text{slow}}$ [structure XI: dihydropurine] (XI) + H₂SO₃   117-11

where R = HO₃S

315

X(117-10) $\xrightarrow{\text{slow}}$ [XII structure] (XII) + H_2O 117-12
where R = HO_3S

XII(117-12) + $6H^+$ + 6e → XI (117-11) + H_2S + $2H_2O$ 117-13

II(117-1) + $2H^+$ + 2e → [XIII structure] (XIII)   117-14

XIII(117-14) → [XIV structure] (XIV) + $2H_2SO_3$ 117-15
where R = HO_3S

XIII(117-14) → [XV structure] (XV) + $2H_2O$ 117-16
where R = HO_3S

XV(117-16) + $4H^+$ + 4e → [XVI structure] (XVI)   117-17

XV(117-17) $\xrightarrow{\text{slow}}$ XIV(117-15) + $2H_2SO_2$ 117-18
where R = H

V(117-4) → XIV (117-15) + H_2SO_2 117-19
where R = H

V(117-4) + $2H^+$ + 2e → [XVII structure] (XVII)   117-20

TABLE III. Courses and Mechanisms of Half-Reactions

XVII(117-20) $\xrightarrow{\text{slow}}$ [purine with H, R at 2-position] + H_2SO_2 117-21

IV (117-3) $\rightleftarrows$ [purine with SO_2^-] (XIX) + $2H^+$ 117-22
where R = H

XIX(117-22) + $2H^+$ + 2e → [dihydropurine with H, SO_2^-] (XX)   117-23

XX (117-23) $\xrightarrow{\text{slow}}$ [purine] + HSO_2^- 117-24

R = SO_3H, H

119

Mechanism:	Proposed for Compound:
119a: 119-1 to -10	DD30
119c: 119-2 and -3 (proton donor present)	DD76
119h: 119-6, -11, -15, and -5	DD76
119i: 119-20, 21, -4 and -5	DD76

$HOC_6H_4OH \rightleftarrows 2H^+ + {}^-OC_6H_4O^-$ 119-1

[benzoquinone] (I) + H^+ → [protonated benzoquinone OH^+] (II)   119-2

II (119-2) + e → [hydroquinone with OH top, O• bottom] (III) 119-3

III (119-3) + e → [hydroquinone dianion with OH top, O⁻ bottom] (IV) 119-4

IV (119-4) + H⁺ ⇌ [hydroquinone with OH top, OH bottom] (V) 119-5

I (119-2) + e → [semiquinone radical anion with O• top, O⁻ bottom] (VI) 119-6

V (119-5) → [radical cation with OH•⁺ top, OH bottom] (VII) + e 119-7

VII (119-7) → [phenoxyl radical with O• top, OH bottom] (VIII) + H⁺ 119-8

2 VIII (119-8) $\xrightarrow{fast}$ hemiketal, reaction partly heterogeneous 119-9

hemiketal $\xrightarrow{slow}$ I (119-1) + V (119-5) 119-10

TABLE III. Courses and Mechanisms of Half-Reactions

VI (119-6) + e → [benzene ring with two O⁻ groups para] (IX)   119-11

VII (119-7) → [benzene ring with two OH⁺ groups para] (X) + e   119-12

II (119-2) + 2e + H⁺ → V (119-5)   119-13

I (119-2) + 2e + 2H⁺ → V (119-5)   119-14

IX (119-11) + H⁺ ⇌ IV (119-4)   119-15

[naphthoquinone with OH, OH, SO₃⁻] → [naphthoquinone with OH, O⁻, SO₃⁻] (XI) + H⁺   119-16

XI (119-16) → [naphthoquinone with =O, SO₃⁻] + H⁺ + 2e   119-17

II (119-2) + H⁺ ⇌ X (119-12)   119-18

VI (119-6) + H⁺ ⇌ III (119-3)   119-19

I (119-2) + V (119-5) ⇌ [two quinone rings hydrogen-bonded O----H—O] (XII)   119-20

319

XII(119-20) + e → III(119-3) + IV(119-4)   119-21

Code No.	$K_{20} dm^3 mol^{-1}$
DD76	0.083 ± 0.004

120
(To replace old 120 and 121)

Mechanism: | Proposed for Compound:
120a: 120-1 to -3 (at PGE) | AE04

120b: 120-1 and -4 (ACET, pH 2.3) | AE03

120c: 120-1, -5, and -6 (NH_3, pH 9) | AE03

120d: 120-7 | AE03

120e: 120-8 and -9 (at PGE) | AE02

2 (I) → (II) + $2H^+$ + 2e   120-1

II(120-1) →
where R = SH
+ $2H^+$ + 2e   120-2

I(120-1) + $6H_2O$ → + $12H^+$ + 12e   120-3

TABLE III. Courses and Mechanisms of Half-Reactions

II(120-1) + (O₂?) $\xrightarrow{\text{1M HOAc}}$ [structure: purine-S-S(=O)-purine]

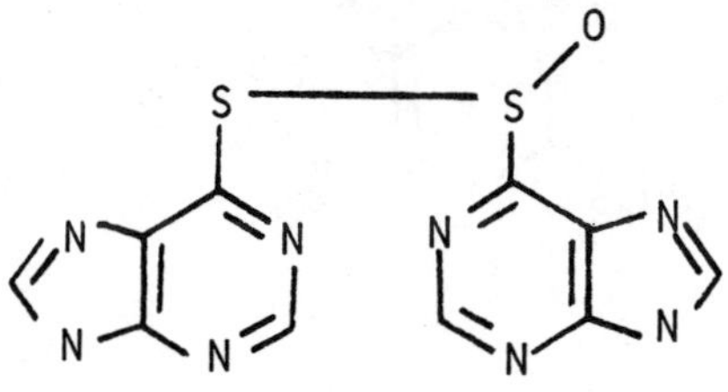

where R = H

or [structure: purine-S-S(=O)₂-purine]   120-4

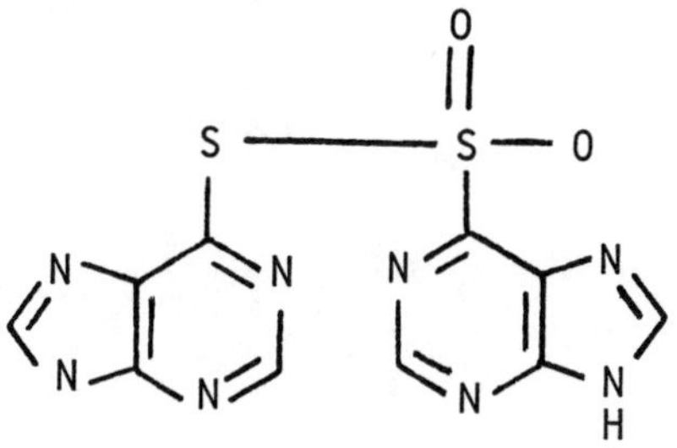

II(120-1) $\xrightarrow[\substack{NH_3 \\ H_2O \\ O_2?}]{pH\ 9}$ I(120-1) + [purine-SO₂H] (III)

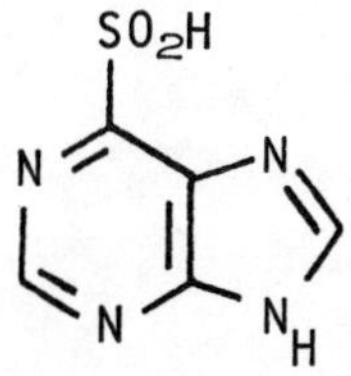

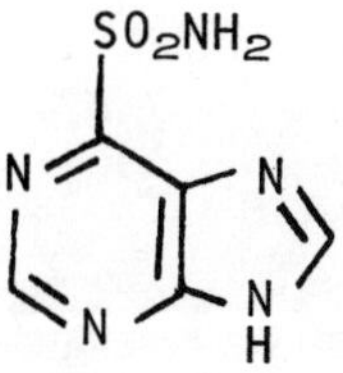

\+ [purine-SO₂NH₂] (IV)(unbalanced)   120-5

I(120-1) $\xrightarrow[\substack{NH_3 \\ H_2O \\ O_2?}]{pH\ 9}$ III(120-5) + IV(120-5) + H⁺ + e +

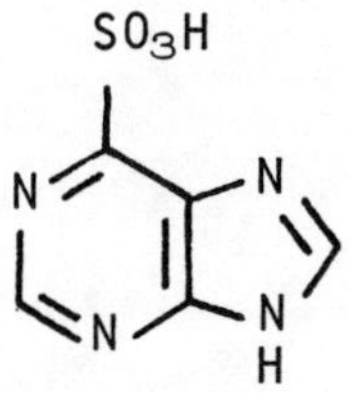

[purine-SO₃H] (V)(unbalanced)   120-6

I(120-1) + 2H₂O $\xrightarrow[\text{CARB}]{pH\ 9}$ V(120-6) + 6H⁺ + 6e   120-7

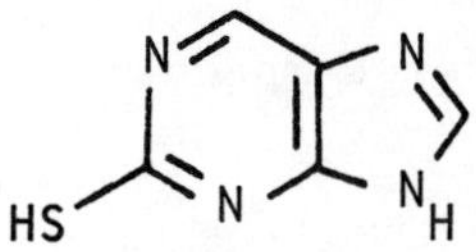

[HS-purine] (VI) → [purine-S-S-purine] (VII)

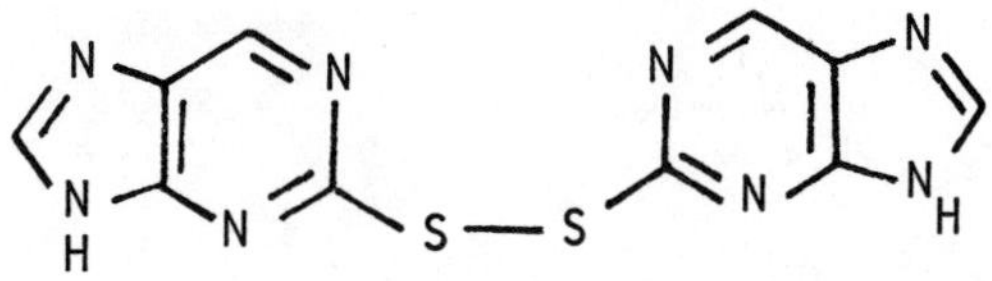

\+ 2H⁺ + 2e   120-8

VI (120-8) + 3H₂O → [structure: 2-sulfo-purine-like, HO₃S-C=N, fused imidazole with NH] + 6H⁺ + 6e   120-9

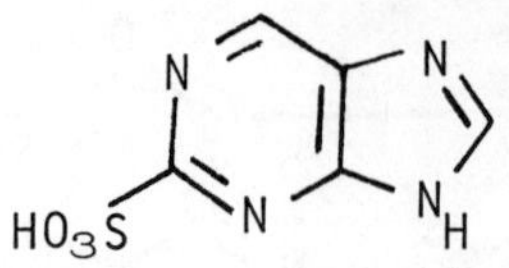

126

Mechanism:
126b: 126-1, -4, and -6

Proposed for Compound:
product of 130k (DA83), DC23, DC77

$$ArSCN + e \rightleftarrows ArSCN^{\overline{\cdot}}$$ 126-1

$$ArSCN^{\overline{\cdot}} + e \rightarrow ArSCN^{2-}$$ 126-2

$$ArSCN^{2-} \xrightarrow{fast} ArS^- + CN^-$$ 126-3

$$ArSCN^{\overline{\cdot}} \xrightarrow{slow} ArS^{\cdot} + CN^-$$ 126-4

$$ArS^{\cdot} + e \rightarrow ArS^-$$ 126-5

$$ArS^{\cdot} + H^+ + e \rightarrow ArSH$$ 126-6

130

Mechanism:
130b: 130-1 and -4

Proposed for Compound:
DB63

130g: 130-1 and -2

DB62

130k: 130-1 and 2 as one step, -13, -14, followed by 126b (pH < 4)

DA83

135

Mechanism:
135a: 135-1, -2, and -5 (pH < 9)

Proposed for Compound:
DC93, DC94

135b: 135-1, and -3 to -5 (pH > 9)

DC93, DC94

TABLE III. Courses and Mechanisms of Half-Reactions

$R^1CH(OMe)CHR^2HgX$

Code No.	R^1	R^2	X
DC93	butoxy	H	CH_3COO
DC94	isobutoxy	H	CH_3COO

141

Mechanism: Proposed for Compound:

141f: 141-7 and -8   DF76

141g: 141-9 to -14 and -2 (pH < 7)   DB35

141h: 141-9, -10, -13, and -14 (pH > 7)   DB35

141i: 141-16 (pH < 7)   DB09

141j: 141-2 and -15 (pH > 7)   DB09

$RS^- + Hg \rightleftarrows RSHg + e$	141-1
$RSH \rightleftarrows RS^- + H^+$ K_2	141-2
$2RSHg \rightarrow Hg(SR)_2 + Hg$	141-3
$Hg(SR)_2 \rightleftarrows RSSR + Hg$	141-4
$RSSR + Hg \rightleftarrows (RS)_2Hg_{ads}$	141-5
$(RS)_2Hg_{ads} + 2e + 2H^+ \rightleftarrows 2RSH + Hg$	141-6
$R^1SSR^2 + e \rightarrow (R^1SSR^2)^{\bar{\cdot}}$	141-7
$(R^1SSR^2)^{\bar{\cdot}} + e \rightleftarrows R^1S^- + R^2S^-$ (protonation possible)	141-8
$2Hg + RSSR \rightarrow Hg_2(SR)_2$	141-9
$Hg_2(SR)_2 + 2H^+ + 2e \rightarrow 2Hg + 2RSH$	141-10
$R^1SSR^2 + e \rightarrow R^1S^- + R^2S^{\cdot}$	141-11
$R^2S^{\cdot} + e \rightarrow R^2S^-$	141-12
$R^1SSR^2 + 2H^+ + e \rightarrow R^1SH + R^2SH^+$	141-13
$R^2SH^+ + e \rightarrow R^2SH$	141-14
$2RS^- + Hg \rightleftarrows (RS)_2Hg + 2e$	141-15
$RSH + Hg \rightleftarrows RSHg + H^+ + e$	141-16

$R = C_6H_5$, $HOOCCH(NH_2)CH_2$,

156

Mechanism: | Proposed for Compound:
156g: 156-18 to -21 (acidic) | DC72
156h: 156-1, -2, -22, -20 and -21 (basic) | DC72

$R^1CH:CHCOR^2 + e \rightarrow R^1\dot{C}HCH:C(O^-)R^2$ 156-1

$R^1\dot{C}HCH:C(O^-)R^2 + H^+ \rightleftarrows R^1\dot{C}HCH:C(OH)R^2$ 156-2

$2R^1\dot{C}HCH:C(OH)R^2 \rightarrow$ (racemic structure) (racemic) 156-3

$2R^1\dot{C}HCH:C(OH)R^2 \rightarrow R^1CH_2CH_2COR^2 + R^1CH:CHCOR^2$ 156-4

$R^1\dot{C}HCH:C(O^-)R^2 + CO_2 \rightarrow {}^-O_2CCHR^1\dot{C}HCOR^2$ 156-5

${}^-O_2CCHR^1\dot{C}HCOR^2 + e \rightarrow {}^-O_2CCHR^1CH:C(O^-)R^2$ 156-6

${}^-O_2CCHR^1CH:C(O^-)R^2 + CO_2 \rightarrow {}^-O_2CCHR^1CH(CO_2^-)COR^2$ 156-7

$CO_2 + e \rightarrow CO_2^{\cdot -}$ 156-8

$CO_2^{\cdot -} + R^1CH:CHCOR^2 \rightarrow {}^-O_2CCHR^1\dot{C}HCOR^2$ 156-9

$2R^1\dot{C}HCH:C(O^-)R^2 \rightarrow$

$\begin{array}{c} H \\ | \\ R^1-C-CH=C(O^-)R^2 \\ | \\ R^1-C-CH=C(O^-)R^2 \\ | \\ H \end{array}$ (I) 156-10

$I(156\text{-}10) + 2CO_2 \rightarrow$

$\begin{array}{cc} H & CO_2^- \\ | & | \\ R^1-C-CH-COR^2 \\ | \\ R^1-C-CH-COR^2 \\ | & | \\ H & CO_2^- \end{array}$ (II) 156-11

TABLE III. Courses and Mechanisms of Half-Reactions

$2^-O_2CCHR^1\dot{C}HCOR^2 \rightarrow$ II(156-11)   156-12

$R^1\dot{C}HCH:C(O^-)R^2 + R^1CH:CHCOR^2 \rightarrow$

$$\begin{array}{c} H \\ | \\ R^1-C-\dot{C}HCOR^2 \\ | \\ R^1-C-\underline{C}HCOR^2 \\ | \\ H \end{array} \quad (III)$$

156-13

III(156-13) + $CO_2 \rightarrow$

$$\begin{array}{c} H \\ | \\ R^1-C-\dot{C}HCOR^2 \\ | \\ R^1-C-CHCOR^2 \\ | \quad | \\ H \quad CO_2^- \end{array} \quad (IV)$$

156-14

IV(156-14) + e $\rightarrow$

$$\begin{array}{c} H \\ | \\ R^1-C-\bar{C}HCOR^2 \\ | \\ R^1-C-CHCOR^2 \\ | \quad | \\ H \quad CO_2^- \end{array} \quad (V)$$

156-15

V(156-15) + $CO_2 \rightarrow$ II(156-11)   156-16

III(156-13) + $R^1\dot{C}HCH:C(O^-)R^2 \rightleftarrows$ I(156-10) + $RCH:CHCOR^2$ 156-17

$\overbrace{R^1CH:CHCOR^2}^{H^+} \rightleftarrows R^1CH:CHCOR^2 + H^+$ 156-18

$\overbrace{R^1CH:CHCOR^2}^{H^+} + e \rightarrow R^1CH_2\dot{C}HCOR^2$ 156-19

$R^1CH_2\dot{C}HCOR^2 + e \rightarrow R^1CH_2\overbrace{CHC}^{-}-R^2$
                                          $\|$
                                          O
156-20

$R^1CH_2CH_2COR^2 \rightleftarrows R^1CH_2\overbrace{CHCR^2}^{-} + H^+$
                                          $\|$
                                          O
156-21

$R^1\dot{C}HCH:C(OH)R^2 \rightleftarrows R^1CH_2\dot{C}HCOR^2$ 156-22

$R^1 = C_6H_5, R^2 = H$

174

Mechanism:
174b: 174-3 and -2

Proposed for Compound:
DD93, DE66

$$RHgCl + H^+ + e \to RHg\cdot + HCl \qquad 174\text{-}1$$

$$RHg\cdot + H^+ + e \to RH + Hg \qquad 174\text{-}2$$

$$RHgCl + e \to RHg\cdot + Cl^- \qquad 174\text{-}3$$

$$R = (CH_2)n \begin{Bmatrix} CHOC_2H_5 \\ | \\ C- \\ \| \\ CH \end{Bmatrix} \quad \text{where } n = 8, 10$$

198

Mechanism:
198d: 198-1 to -3

Proposed for Compound:
DD69, DD72

198e: 198-9 to -12 (weak acid present) — DD69, DD72

198f: 198-9, -11, -7 and -8 (strong acid present) — DD69, DD72, DE95

198g: 198-9, -13 to -16, -11, -7 and -8 (strong acid present) — DE07, DE46

198h: 198-17, -9, -18, -11, -12, -19, and -20 — DD75

$$Ar^1N{:}NAr^2 + e \rightleftarrows [Ar^1N{:}NAr^2]^{\cdot -} \qquad 198\text{-}1$$

$$[Ar^1N{:}NAr^2]^{\cdot -} + e \rightleftarrows [Ar^1N{:}NAr^2]^{2-} \qquad 198\text{-}2$$

$$[Ar^1N{:}NAr^2]^{2-} + HSo \to Ar^1N{:}NAr^2H^- + So^- \qquad 198\text{-}3$$

$$Ar^1N{:}NAr^2H^- \to Ar^1N{:}NAr^2 + 2e + H^+ \qquad 198\text{-}4$$

$$[Ar^1N{:}NAr^2]^{2-} + O_2 \to Ar^1N{:}NAr^2 + (OH^-?) \qquad 198\text{-}5$$

$$[Ar^1N{:}NAr^2]^{\cdot -} + HQ \to Ar^1NH\dot{N}Ar^2 + Q^- \qquad 198\text{-}6$$

$$Ar^1NH\dot{N}Ar^2 + e \to Ar^1NH\overline{N}Ar^2 \qquad 198\text{-}7$$

TABLE III. Courses and Mechanisms of Half-Reactions

$Ar^1NH\bar{N}Ar^2 + HA \rightarrow Ar^1NHNHAr^2 + A^-$ 198-8

$Ar^1N(H^+):NAr^2\ (I) \rightleftarrows Ar^1N:NAr^2 + H^+$ 198-9

$Ar^1N:NAr^2 + HA \rightleftarrows Ar^1N(H^+):NAr^2 + A^-$ 198-10

$Ar^1N(H^+):NAr^2 + e \xrightarrow{slow} Ar^1NH\dot{N}Ar^2$ 198-11

$Ar^1NH\dot{N}Ar^2 + e + HA \xrightarrow{fast} Ar^1NHNHAr^2 + A^-$ 198-12

$Ar^1N(H^+):NC_6H_4XH^+ \rightleftarrows Ar^1N(H^+):NC_6H_4X\ (Ia - 198-9) + H^+$ 198-13

$Ar^1N:NC_6H_4XH^+ \rightleftarrows Ar^1N:NC_6H_4X + H^+$ 198-14

$Ar^1N(H^+):NC_6H_4XH^+ + e \xrightarrow{slow} Ar^1NH\dot{N}C_6H_4XH^+$ 198-15

$Ar^1NH\dot{N}C_6H_4XH^+ + H^+ + e \xrightarrow{fast} Ar^1NHNHC_6H_4XH^+$ 198-16

$Ar^1N:NC_6H_4YH \rightleftarrows Ar^1N:NC_6H_4Y^- + H^+$ 198-17

$Ar^1N(H^+):NC_6H_4Y^- \rightleftarrows Ar^1N:NC_6H_4Y^- + H^+$ 198-18

$Ar^1N(H^+):NC_6H_4Y^- + e \xrightarrow{slow} Ar^1NH\dot{N}C_6H_4Y^-$ 198-19

$Ar^1NH\dot{N}C_6H_4Y^- + H^+ + e \xrightarrow{fast} Ar^1NHNHC_6H_4Y^-$ 198-20

$Ar = C_6H_5,\ 4-O_2NC_6H_4$

$HQ = $ hydroquinone

$ArX = 4-CH_3OC_6H_4,\ 4-(CH_3)_2NC_6H_4,\ 4-(CH_3)_3N^+C_6H_4$

$ArYH = 4-HO_3SC_6H_4$

218

Mechanism:　　　　　　　　　　　　　　　　　　　　　　Proposed for Compound:
218a:　218-1 and -2　　　　　　　　　　　　　　　　　　DG30, DG31

219

Mechanism:　　　　　　　　　　　　　　　　　　　　　　Proposed for Compound:
219b:　219-3 and -6 $(0 < pH < 3)$　　　　　　　　　　　DF12

219c:　219-3, -4, and -7 $(3 < pH < 5)$　　　　　　　　　DF12

219d: 219-4 and -8 (5 < pH < 12)   DF12

219e: 219-4, -5, -9, and -10 (pH > 12)   DF12

219f: 219-11, -12, and -14 (pH < 4)   DE74, DE81

219g: 219-12 and -15 (6 < pH < 12)   DE74, DE81

219h: 219-13 and -16 (10 < pH 13)   DE74, DE81

[Structure] + $2H^+$ + $2e$ → (I) + H_2O 219-1

$I(219-1)$ + $2H^+$ + $2e$ → (II)   219-2

[Structure III] ⇌ [Structure IV] + H^+ pK_3 219-3

$IV(219-3)$ ⇌ (V) + H^+ 219-4

TABLE III. Courses and Mechanisms of Half-Reactions

V(219-4) ⇌ [structure VI] (VI) + H⁺   219-5

where R² = H

III(219-3) + H⁺ + 2e → [structure VII] (VII)   219-6

IV(219-3) + 2H⁺ + 2e → VII(219-6)   219-7

V(219-4) + 2H⁺ + 2e → [structure]   219-8

VI(219-5) + H⁺ + e → [structure VIII] (VIII)   219-9

VIII(219-9) + H⁺ + e → [structure]   219-10

[Structure IX with OH₂⁺] (IX) ⇌ [Structure X with OH] (X) + H⁺   219-11

X(219-11) ⇌ [Structure XI with OH] (XI) + H⁺   219-12

XI(219-12) ⇌ [Structure XII with O⁻] (XII) + H⁺   219-13

IX(219-11) + 2H⁺ + 5e → [reduced structure]⁻ + H_2O 219-14

X(219-11) + 3H⁺ + 4e → [reduced structure] + H_2O 219-15

TABLE III. Courses and Mechanisms of Half-Reactions

XI(219-12) + 2H$^+$ + 2e → [structure: benzodiazepinone with substituents R^1, R^2, R^4, and OH]

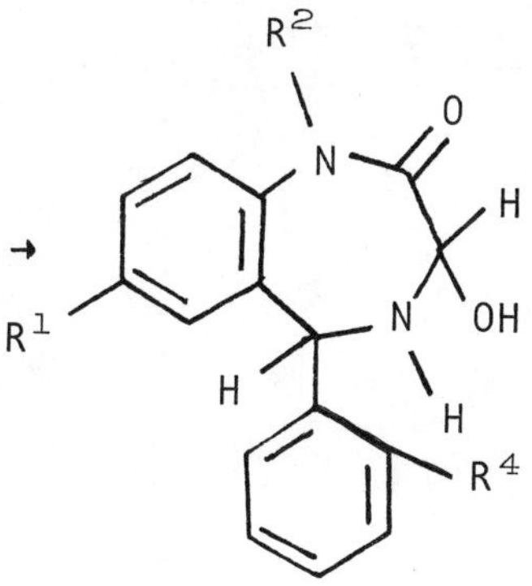

219-16

$R^1 = Cl$

$R^2 = H$

$R^3 = -COOH, OH$

$R^4 = H, Cl$

Code No.	Technique	pK$_3$	pK$_3$'	pK$_4$	pK$_4$'	pK$_5$
DF12	Spectrophotometry	3.5	–	–	–	12.5
	Polarography	3.0	7	5.0	9.0	12.0

Code No.	Technique	pK$_{11}$	pK$_{11}$'	pK$_{12}$	pK$_{12}$'	pK$_{13}$	pK$_{13}$'
DE74	Spectrophotometry	–	–	1.8	–	11.5	–
	Polarography	< 2	≈ 5	≈ 2	≈ 11	–	> 14
DE81	Spectrophotometry	–	–	2.0	–	11.6	–
	Polarography	< 2	≈ 5	≈ 2	≈ 11	–	> 14

241

Mechanism:
241a: 241-1 to -7

Proposed for Compound:
DF90

Mechanism: | Proposed for Compound:
277d: 277-13 and -14 (acidic) | DB04

277e: 277-5 and -4 (basic) | DB04

[Structure: protonated pyridine-3-carboxamide (I) ⇌ pyridine-3-carboxamide (II) + H⁺]   277-1

I (277-1) + e ⇌ [dihydropyridinyl radical structure] (III)   277-2

2 III (277-2) → [bis-dihydropyridine dimer structure] (IV)   277-3

III (277-2) + H⁺ + e → [dihydropyridine carboxamide structure] (V)   277-4

II (277-1) + H⁺ + e ⇌ III (277-2)   277-5

IV (277-3) → II (277-1) + 2H⁺ + 2e   277-6

II (277-1) + e ⇌ [pyridinyl radical anion structure] (VI)   277-7

TABLE III. Courses and Mechanisms of Half-Reactions

2 VI(277-7) $\xrightarrow{k_8}$ [dimer structure: HRN-C(=O)-pyridine-CH-CH-pyridine-C(=O)-NRH] (VII)   277-8

VII(277-8) + 2H$^+$ → IV(277-3)   277-9

VI(277-7) + e → [pyridine-C(=O)-NRH]$^{2-}$ (VIII)   277-10

VIII(277-10) + 2H$^+$ → V(277-4)   277-11

V(277-4) → II(277-1) + 2H$^+$ + 2e   277-12

II(277-1) + 2H$^+$ + e → [dihydropyridine radical-C(=O)-NH$_2$] (IX)   277-13
where R = H

IX(277-13) + H$^+$ + e → [dihydropyridine-C(=O)-NH$_2$]   277-14

314

Mechanism: Proposed for Compound:
314b: 314-10 to -13   DC38

[phenol with R^1, R^2, R^3 substituents] → [phenoxonium with R^1, R^2, R^3] (I) + H$^+$ + 2e   314-1

$$I(314\text{-}1) \leftrightarrow \text{[II]} \leftrightarrow \text{[III]} \qquad 314\text{-}2$$

$$\left.\begin{array}{c} I(314\text{-}1) \\ \updownarrow \\ II(314\text{-}2) \\ \updownarrow \\ III(314\text{-}2) \end{array}\right\} + H_2O \rightarrow \text{(IV)} + H^+ \qquad 314\text{-}3$$

$$IV(314\text{-}3) + H^+ \rightarrow \text{(V)} + (R^3)^+ \qquad 314\text{-}4$$

$$V(314\text{-}4) \rightarrow \text{[quinone]} + 2H^+ + 2e \qquad 314\text{-}5$$

$$(CH_3)_3C^+ + H_2O + CH_3CN \rightarrow (CH_3)CNHCOCH_3 + H^+ \qquad 314\text{-}6$$

For $R^3 = (CH_3)_3C$

$$\left.\begin{array}{c} I(314\text{-}1) \\ \updownarrow \\ II(314\text{-}2) \\ \updownarrow \\ III(314\text{-}2) \end{array}\right\} + \text{pyridine} \rightarrow \text{[pyridinium adduct]} \qquad 314\text{-}7$$

$$\left.\begin{array}{c} I(314\text{-}1) \\ \updownarrow \\ II(314\text{-}2) \\ \updownarrow \\ III(314\text{-}2) \end{array}\right\} + \text{pyridine} \rightarrow \text{(VI)} + (R^2)^+ \qquad 314\text{-}8$$

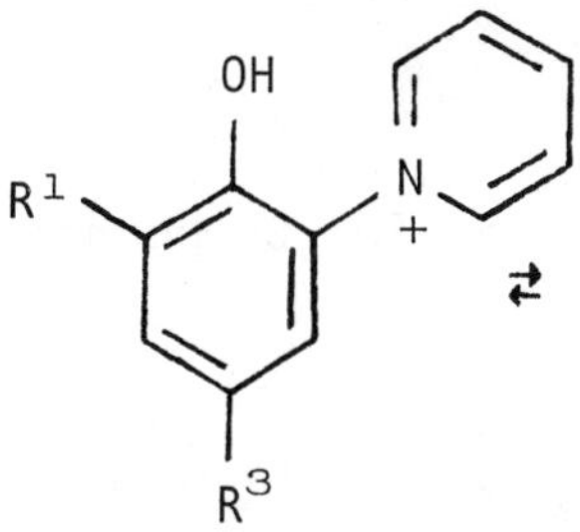

TABLE III. Courses and Mechanisms of Half-Reactions

[Structure: 2-R¹-6-(pyridinium-N-yl)-4-R³-phenol] ⇌ VI(314-8) + H⁺   314-9

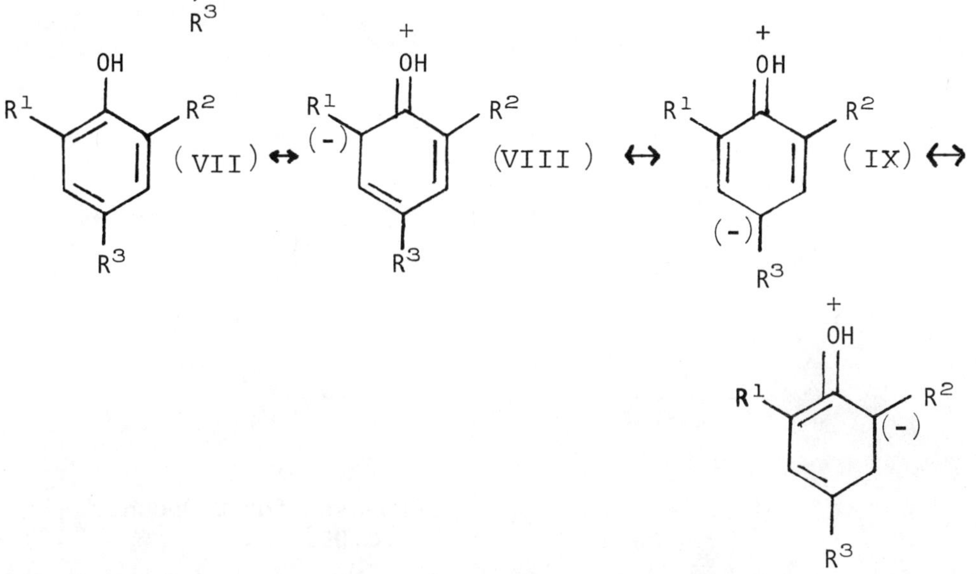

(VII) ↔ (VIII) ↔ (IX) ↔ (X)   314-10

$$\begin{matrix} VIII\,(314\text{-}10) \\ \updownarrow \\ IX\,(314\text{-}10) \\ \updownarrow \\ X\,(314\text{-}10) \end{matrix} \quad \rightarrow \quad \left. \begin{matrix} I\,(314\text{-}1) \\ \updownarrow \\ II\,(314\text{-}2) \\ \updownarrow \\ III\,(314\text{-}2) \end{matrix} \right\} + H^+ + 2e \quad\quad 314\text{-}11$$

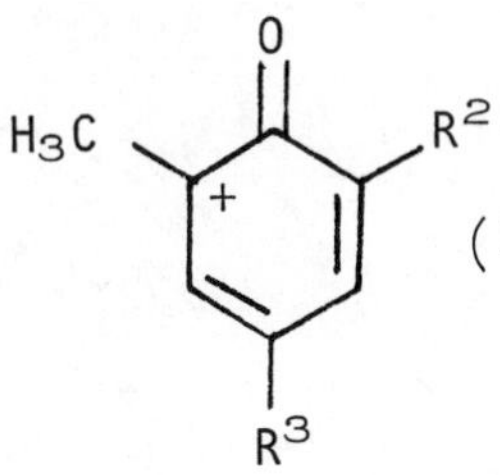

(III)(where R¹ = CH₃) (314-2) 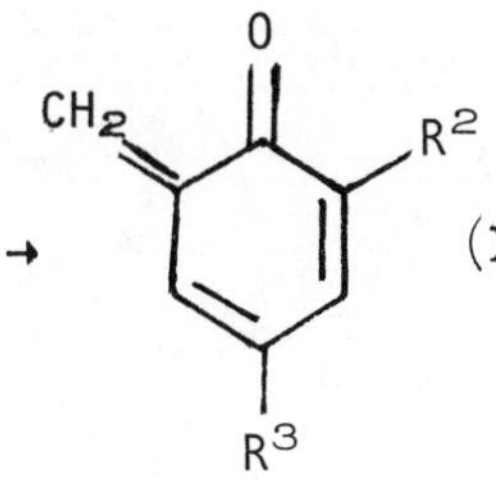 → (XI) + H⁺   314-12

2 XI(314-12) → [bis-phenol stilbene: HO-C₆H₂(R²)(R³)-CH=CH-C₆H₂(R²)(R³)-OH]   314-13

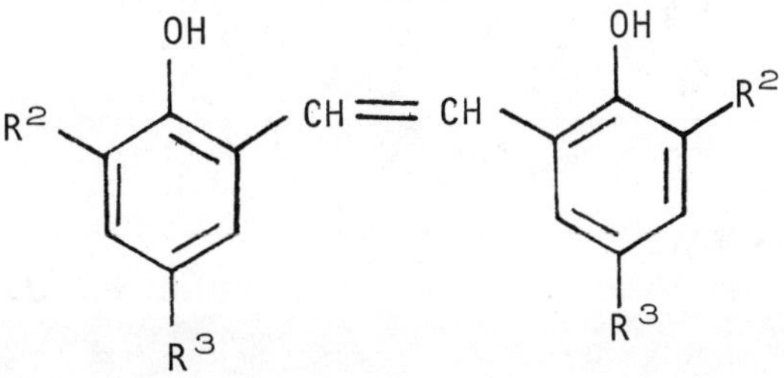

$R^1 = CH_3$

$R^2 = CH_3$

$R^3 = H$

329

Mechanism:
329j: 329-2 (7.5 < pH < 8.3)

329k: 329-8 and -20 (pH > 10.5)

P = porphyrin

S = pyridine

M = Fe

X = CN^-, OH^-

Proposed for Compound:
DH19

DH12, DH19

335

Mechanism:
335c: 335-1 to -3

335d: 335-5

335e: 335-5 to -8

Proposed for Compound:
DB28, DD87

DD86

DD88

$$Hg + RH \rightarrow HgR^+ + H^+ + 2e \qquad 335\text{-}1$$

$$RH + HgR^+ \underset{}{\overset{slow}{\rightleftharpoons}} HgR_2 + H^+ \qquad 335\text{-}2$$

$$Hg + HgR_2 \rightarrow 2HgR^+ + 2e \qquad 335\text{-}3$$

$$RH + e \rightarrow R^- + \tfrac{1}{2}H_2 \qquad 335\text{-}4$$

$$MR_2 \rightarrow M^{2+} + 2R^+ + 4e \qquad 335\text{-}5$$

$$MR_2 + e \rightleftharpoons MR_2^{\cdot -} \qquad 335\text{-}6$$

$$MR_2^{\cdot -} + e \rightleftharpoons MR_2^{2-} \qquad 335\text{-}7$$

$$MR_2^{\cdot -} \xrightarrow[slow]{rearrangement} \text{electroinactive form} \qquad 335\text{-}8$$

M = Cd, Zn

RH = O-ethyl thioacetothioacetate

TABLE III. Courses and Mechanisms of Half-Reactions

351

Mechanism:
351a: 351-1, -2, -3, -6, and -12

351b: 351-1 to -12

351c: 351-1, -2, -13, and -12

351d: 351-1, -2, -3, -6, and -13 to -18

351e: 351-1 to -6 and -20 to -23

Proposed for Compound:
DH64

DH32, DH33, DH34, DH35, DH39, DH59, DH62

DH53, DH58

DH37

DH63

$O \rightarrow O^+ + e$	351-1
$R + e \rightarrow R^-$	351-2
$O^+ + R^- \rightarrow O + {}^3R^*$	351-3
$O^+ + {}^3R^* \rightarrow O^+ + R$	351-4
$R^- + {}^3R^* \rightarrow R^- + R$	351-5
$2{}^3R^* \rightarrow {}^1R^* + R$	351-6
$O^+ + {}^1R^* \rightarrow O^+ + R$	351-7
$R^- + {}^1R^* \rightarrow R^- + R$	351-8
${}^3R^* \rightarrow R$	351-9
${}^1R^* \rightarrow R$	351-10
${}^1R^* \rightarrow {}^3R^*$	351-11
${}^1R^* \rightarrow R + h\nu$	351-12
$O^+ + R^- \rightarrow O + {}^1R^*$	351-13
$O^+ + R^- \rightarrow {}^1O^* + R$	351-14
${}^1O^* \rightarrow O + h\nu$	351-15
$O^+ + R^- \xrightarrow{?} O + R + h\nu$	351-16
$2{}^3R^* \xrightarrow{?} 2R + h\nu$	351-17
${}^1O^* \rightarrow {}^3O^*$	351-18
${}^3O^* + R \rightarrow O + {}^3R^*$	351-19
$O^+ + R^- \rightarrow {}^3O^* + R$	351-20

$2\,^3O^* \rightarrow O + {}^1O^*$ 351-21

${}^1O^* + R \rightarrow O + {}^1R^*$ 351-22

${}^3R^* + O \rightarrow R + {}^3O^*$ 351-23

Code No.	O	R
DH32	anthracene	Wursters Blue
DH33	9,10-diphenylanthracene	Wursters Blue
DH34	1,3,6,8-tetraphenylpyrene	Wursters Blue
DH35	rubrene	Wursters Blue
DH37	2,5-diphenyl-1,3,4-oxadiazole	thianthrene
DH39	fluoranthene	10-methylphenothiazine
DH53	9,10-diphenylanthracene	9,10-diphenylanthracene
DH58	1,3,6,8-tetraphenylpyrene	1,3,6,8-tetraphenylpyrene
DH59	p-benzoquinone	rubrene
DH62	rubrene	rubrene
DH63	α,β,γ,δ-tetraphenylporphine	rubrene
DH64	α,β,γ,δ-tetraphenylporphine	rubrene

355

Mechanism:
355a: 355-1 and -2
355b: 355-1 to -3

Proposed for Compound:
DC62, DC65, DC76
DC63

$ArC(O)OCH_2CX_3 + 2e \rightarrow ArC(O)OCH_2CX_2^- + X^-$ 355-1

$ArC(O)OCH_2CX_2^- + H^+ \rightarrow ArC(O)OH + CH_2{:}CX_2$ 355-2

$ArC(O)OCH_2CX_2^- + H^+ \rightarrow ArC(O)OCH_2CHX_2$ 355-3

$X = Br, Cl, I, H$

$Ar = C_6H_5$

TABLE III. Courses and Mechanisms of Half-Reactions

356

Mechanism:
356a: 356-1, -3, and -4 (pH ≤ 7)

356b: 356-1, -2, -5, and -6 (pH ≥ 10)

356c: 356-1, -2, and -5 to -7 (7 < pH < 8)

Proposed for Compound:
DG06, DG07

DG06, DG07

DG08

[structure] ⇌ [structure] (I) + H⁺   356-1

I(356-1) ⇌ [structure] (II) + H⁺   356-2

I(356-1) + 2e + 2H⁺ → [structure] (III)   356-3

III(356-3) + 2e + 2H⁺ → [structure]   356-4

II(356-2) + e → (IV)   356-5

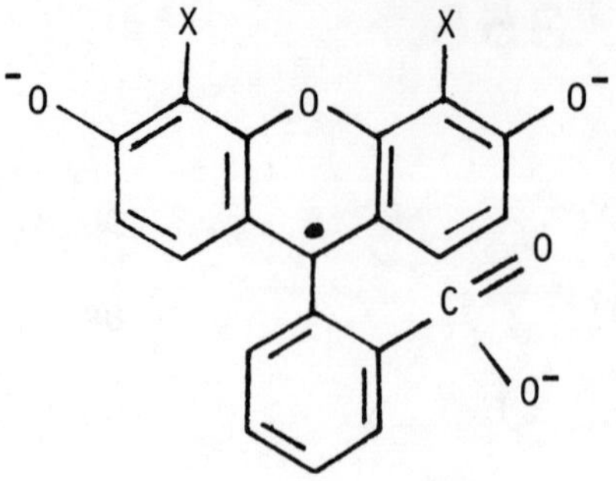

IV(356-5) + e + H⁺ → (V)   356-6

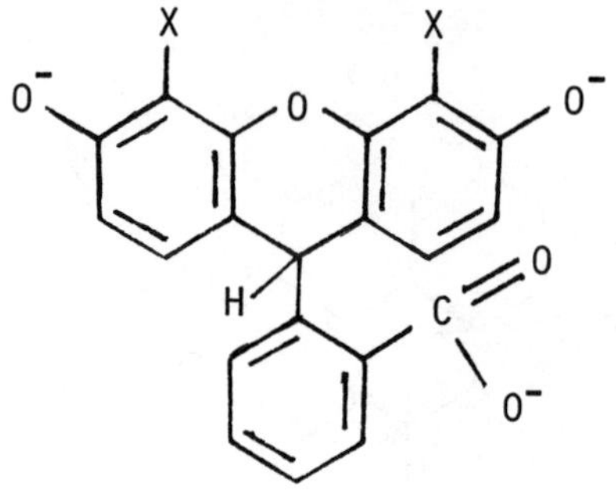

V(356-6) + 2e + 2H⁺ →   356-7

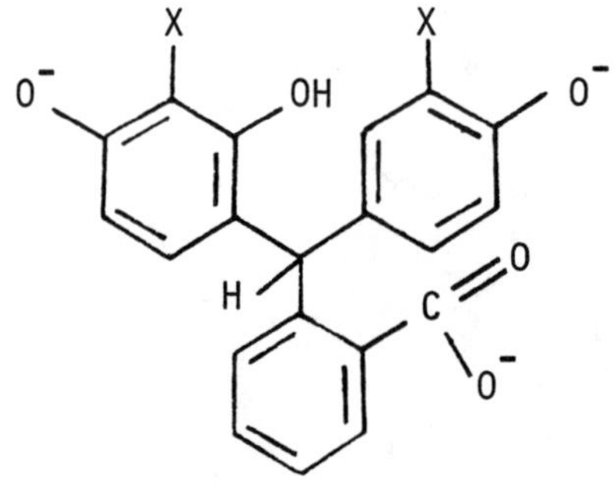

X = Br, Cl, I

357

Mechanism:
357a: 357-1 to -8

Proposed for Compound:
DB92

$BrC(CH_3)_2COC(CH_3)_2Br + 2e \rightarrow$ (I) + Br⁻   357-1

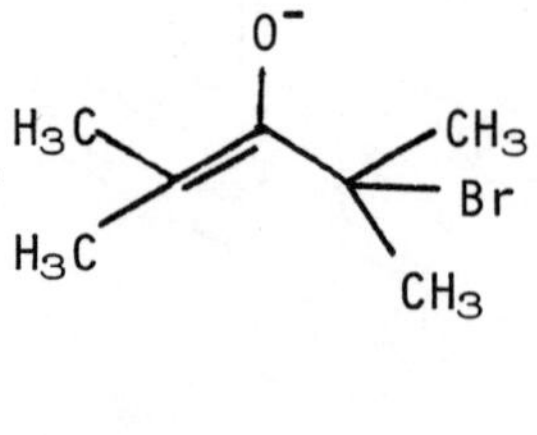

I (357-1) + HSo → (II) + HSō   357-2

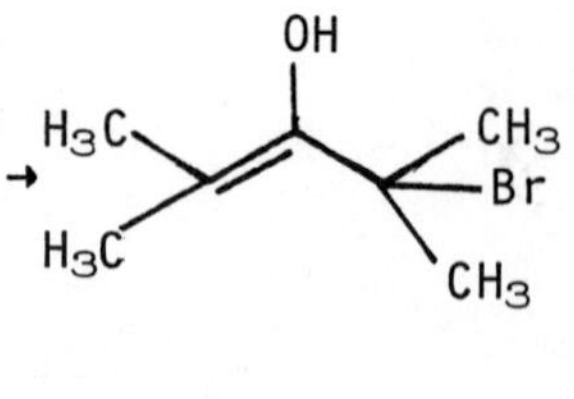

TABLE III. Courses and Mechanisms of Half-Reactions

II(357-2) → [(CH3)2C(OH)-C+(CH3)2] (III) + Br⁻   357-3

III(357-3) + HSo → (CH3)2CH-C(=O)-C(CH3)2-So + H⁺   357-4

III(357-3) → (CH3)2CH-C(=O)-C(CH3)=CH2 (IV) + H⁺   357-5

I(357-1) → [(CH3)2C(Oδ⁻)-Cδ+(CH3)(CH2)(CH3)] (V) + Br⁻   357-6

V(357-6) + furan → [Diels-Alder adduct] and [(CH3)2CH-C(=O)-C(CH3)(cyclopentadienyl)CH3]   357-7

V(357-6) → IV(357-5)   357-8

358

Mechanism: Proposed for Compound:
358a: 358-1 and -2   DE78

2-phenyl-1,3-indandione + 2e + 2H⁺ —fast→ [2-phenyl-3-hydroxy-indan-1-one] (I) ads   358-1

I(358-1) + 2e + 2H⁺ → 2-phenyl-indane-1,2-diol (1,3-diol)   358-2

359

Mechanism:
359a: 359-1 to -4

Proposed for Compound:
DA06

$$H_2NC(:O)C(:O)NH_2 + 2e \rightarrow H_2NC(O^-):C(O^-)NH_2 \qquad 359\text{-}1$$

$$H_2NC(O^-):C(O^-)NH_2 \leftrightarrow H_2NC(:O)\bar{C}(O^-)NH_2 \qquad 359\text{-}2$$

$$H_2NC(:O)\bar{C}(O^-)NH_2 + 2H^+ \rightleftarrows H_2NC(:O)CH(:O) + NH_3 \qquad 359\text{-}3$$

$$H_2NC(:O)CH(:O) + 2H^+ + 2e \xrightarrow{k_4} H_2NC(:O)CH_2OH \qquad 359\text{-}4$$

Code No.	Medium	pH	k_4, s^{-1}, VA	from QE	PY
DA06	MB	7	50	30	—
	BOR	10.0	10	13	6.5
	Cl$^-$/OH$^-$	11.6	1	2	.1

360

Mechanism:
360a: 360-1 to -4 (MeCN)

Proposed for Compound:
DE43, DG46

360b: 360-1 to -5 (MeCN)

DC37, DC82, DD53, DE17, DE44, DE96

360c: 360-1, -2, -6, and -7 (MeCN)

DD27, DE94

$$C_6H_5\underset{R}{\overset{|}{C}}HOR' \rightarrow \left[C_6H_5\underset{R}{\overset{|}{C}}HOR'\right]^{\pm} (I) + e \qquad 360\text{-}1$$

$$I\ (360\text{-}1) \rightarrow C_6H_5\underset{R}{\overset{|}{\overset{+}{C}}}OR'\ (II) + H^+ + e \qquad 360\text{-}2$$

TABLE III. Courses and Mechanisms of Half-Reactions

$$\text{II}(360\text{-}2) + H_2O \rightarrow C_6H_5\underset{R}{\overset{\overset{OH}{|}}{C}}OR'\ (\text{III}) + H^+ \qquad 360\text{-}3$$

$$\text{III}(360\text{-}3) \rightarrow C_6H_5\overset{\overset{O}{\|}}{C}R + R'OH \qquad 360\text{-}4$$

$$\text{III}(360\text{-}3) + H_2O \rightarrow C_6H_5C\overset{O}{\underset{OH}{\diagdown}} + 2e + 2H^+ + R'OH \qquad 360\text{-}5$$

where R = H

$$\text{II}(360\text{-}2) \rightarrow C_6H_5\underset{OR'}{\overset{|}{C}}=CH_2\ (\text{IV}) + H^+ \qquad 360\text{-}6$$

where R = CH$_3$

$$\text{IV}(360\text{-}6) \rightarrow \text{products} + ne \qquad 360\text{-}7$$

R = H, CH$_3$, C$_6$H$_5$

R' = CH$_3$, C$_2$H$_5$, C(CH$_3$)$_3$, C$_6$H$_5$CH$_2$, CH$_3$CO, cyclo-C$_6$H$_{11}$, C$_8$H$_{17}$, or C$_{12}$H$_{25}$

361

Mechanism:
361a: 361-1 to -4

Proposed for Compound:
DG49

$$Ar_2C\underset{O}{-}CAr_2 \rightarrow \left[Ar_2C\underset{O}{-}CAr_2\right]^{+\cdot}\ (\text{I}) + e \qquad 361\text{-}1$$

$$\text{I}(361\text{-}1) \rightarrow Ar_2C-\overset{+}{\overset{\cdot}{O}}-CAr_2 \qquad 361\text{-}2$$

$$Ar_2C-\overset{+}{\overset{\cdot}{O}}-CAr_2 + 2H_2O \rightarrow Ar_2C(OH)OC(OH)Ar_2 + 2H^+ + e \qquad 361\text{-}3$$

$$Ar_2C(OH)OC(OH)Ar_2 \rightarrow 2\ Ar_2C{:}O + H_2O \qquad 361\text{-}4$$

Ar = C$_6$H$_5$

362

Mechanism:	Proposed for Compound:
362a: 362-1 and -3 to -7	DB85
362b: 362-1 and -3 to -10 (SR$_2$ present)	DB85
362c: 362-2 to -7	DB83, DB84
362d: 362-2 to -10 (SR$_2$ present)	DB84

[norbornane with CCl$_2$] + e → [norbornane with C(•)Cl] (I) + Cl$^-$ 362-1

[norbornane with CClBr] or [norbornane with CBrCl] + e → I(362-1) + Br$^-$ 362-2

I(362-1) + e → [norbornane with C(:)$^-$Cl] (II)   362-3

II(362-3) ↔ [norbornane with C(:)Cl]$^-$ (III)   362-4

II(362-3) or III(362-4) + H$^+$ → [norbornane with CHCl]   362-5

TABLE III. Courses and Mechanisms of Half-Reactions

II (362-3) or   III(362-4) → [norbornyl carbene]   (IV) + Cl⁻   362-6

IV(362-6) → [nortricyclene] + ½H₂   362-7

[norbornyl-Cl(X)] + R₂S → [norbornyl-SR₂⁺] (V) + XCl   362-8

V(362-8) + H⁺ → [norbornyl-H, SR₂⁺] (VI)   362-9

VI(362-9) + H⁺ + 2e → [norbornane] + R₂S   362-10

X = Br, Cl;   R₂S = C₆H₅SCH₂COOH, [2-methylthio-3-methyl-phenol with SCH₃, CH₃, OH substituents]

Conc. of Compound = 0.02 M   Products, %

Code No.	Medium	Proton Donor	conc.	nortricyclene	endo-2-chloronorbornane	norbornane
DB83 (E_{app} = -1.1 V)	Et_4NBr, 0.1 M	H_2O	0.066	51	49	
			0.225	38	62	
			1.000	23	77	
DB84 (E_{app} = -1.1 V)	Et_4NBr, 0.1 M	H_2O	0.010	58	42	
			0.075	53	47	
			0.110	42	58	
			0.350	30	70	
			1.000	22	78	
			5.000	20	80	
		Phenol, 1.0 M + H_2O	0.015	39	61	
			0.100	41	59	
			0.500	28	72	
			1.000	23	77	
		CH_3OH, 1.0 M + H_2O	0.03–0.04	27	73	
		CH_3COOH, 1.0 M + H_2O	0.03–0.04	32	68	
		2,4,5-trimethyl-phenol, 0.1 M + H_2O	0.03–0.04	50	50	
		hydroquinone, 1.0 M + H_2O	0.03–0.04	40	60	
		n-octyl sulfide, 0.2 M + H_2O	0.040	72	28	
			0.150	54	46	
			0.385	47	53	
		$C_6H_5SCH_2COOH$	0.2–1.0	10	54	36
		4-methylthio-3-methylphenol	0.2	22	72	6
			1.0	18	72	5
		n-hexylsulfide, 0.2 M + phenol	1.000	23	42	35

TABLE III. Courses and Mechanisms of Half-Reactions

Conc. of Compound = 0.02 M Products, % (cont.)

Code No.	Medium	Proton Donor	conc.	nortricyclene	endo-2-chloronorbornane	norbornane
DB84 (cont.)	Et$_4$NI, 0.1 M	H$_2$O	0.150	26	74	
	Et$_4$NClO$_4$, 0.1 M	H$_2$O	0.200	38	62	
	Et$_3$NHBr, 0.1 M	H$_2$O	0.150	8	92	
	LiBr, 0.1 M	H$_2$O	0.075	52	58	
			0.350	35	65	
			1.000	25	75	
		Phenol, 1.0 M +	0.017	13	87	
		H$_2$O	0.155	11	89	
			0.500	10	90	
			1.000	11	89	
DB85 (E_{app} = -1.8)	Et$_4$NBr, 0.1 M	H$_2$O	0.016	63	37	
			0.075	52	48	
		n-hexyl sulfide		23	42	0

363

Mechanism:
363a: 363-1 to -5

Proposed for Compound:
DC92, DC96

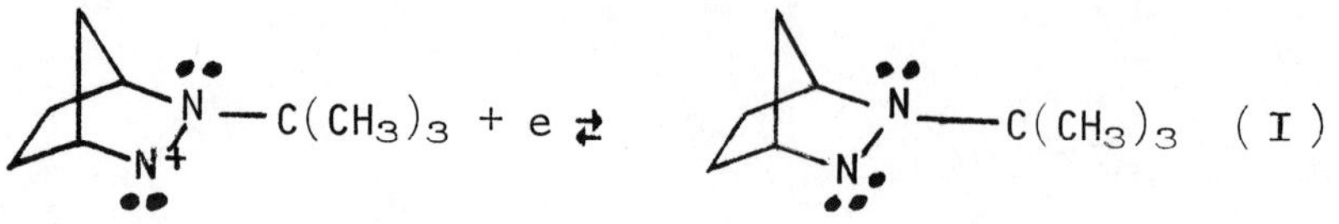

[structure: bicyclic N-N⁺ with —C(CH₃)₃] + e ⇌ [structure: bicyclic N-N with —C(CH₃)₃] (I)   363-1

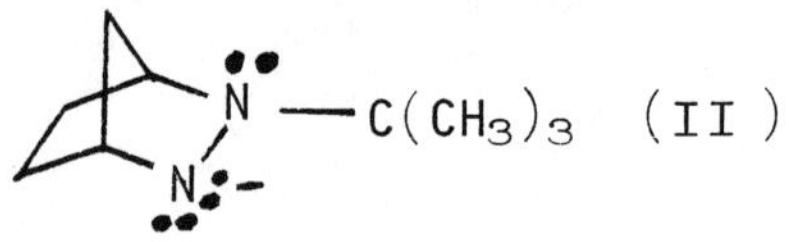

I (363-1) + e ⇌ [structure: bicyclic N-N⁻ —C(CH₃)₃] (II)   363-2

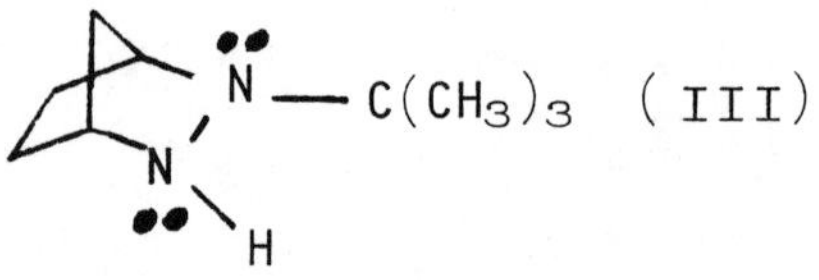

II (363-2) + H⁺ ⇌ [structure: bicyclic N—C(CH₃)₃, N-H] (III)   363-3

I (363-1) + H⁺ ⇌ [structure: bicyclic N—C(CH₃)₃, N⁺•-H] (IV)   363-4

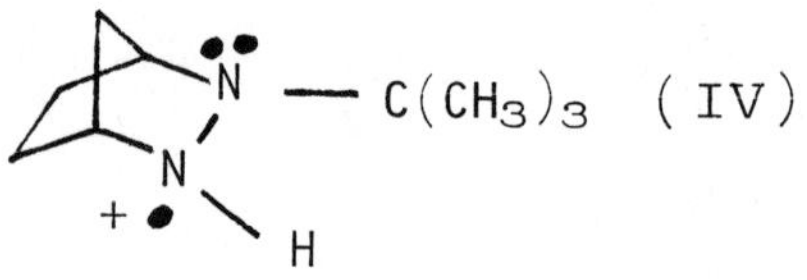

IV (363-4) + e ⇌ III (363-3)   363-5

364

Mechanism:
364a: 364-1 to -3

364b: 364-3

Proposed for Compound:
DD49

DD05

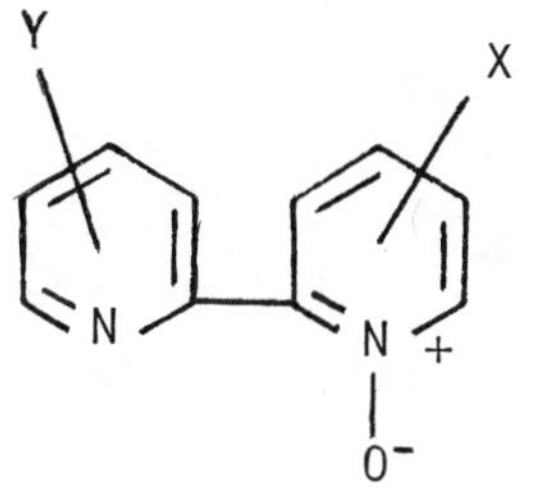

TABLE III. Courses and Mechanisms of Half-Reactions

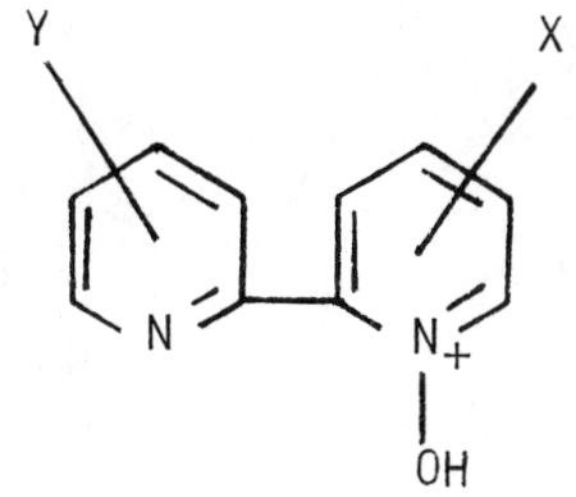

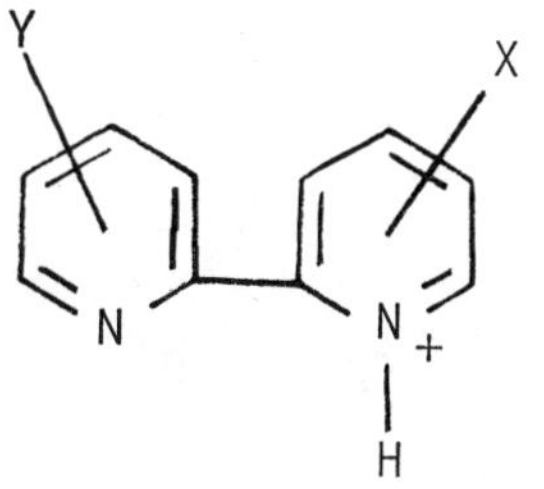

364-1: [bipyridine N-oxide] + H⁺ → (I) [protonated N-OH bipyridinium]

$$I\ (364\text{-}1) + 2H^+ + 2e \rightarrow (II) + H_2O$$

364-2

$$II\ (364\text{-}2) + 2H^+ + 2e \rightarrow$$ [reduced dihydro-bipyridinium]

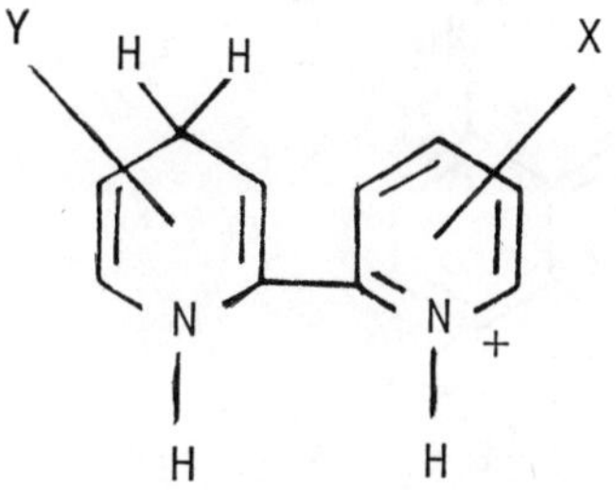

364-3

X = 4-OCH$_3$, H

Y = H

365

Mechanism:	Proposed for Compound:
365a: 365-1 to -3 (pH < 3)	DE00
365b: 365-2, -4, and -5 (3 < pH < 9)	DE00

(I) ⇌ (II) + H⁺   365-1

II(365-1) ⇌ (III) + H⁺   365-2

I(365-1) + 2H⁺ + 4e → (IV) + 2H₂O   365-3

II(365-1) + H⁺ + 2e → (V) + H₂O   365-4

V(365-4) + 2H⁺ + 2e → IV(365-3) + H₂O   365-5

X = 6-OCH₃

TABLE III. Courses and Mechanisms of Half-Reactions

366

Mechanism:
366a: 366-1, -6, and -3

366b: 366-1 (aprotic)

366c: 366-1 to -5 (aprotic with H₂O present)

Proposed for Compound:
DC24

DD70, DF05

DD70, DF05

[cyclooctatetraene numbered 1-8] + e ⇌ [cyclooctatetraene radical anion] (I)   366-1

I(366-1) + HSo → [cyclooctatetraenyl radical with H] (II) + So⁻   366-2

II(366-2) + e ⇌ [cyclooctatetraenyl anion with H] (III)   366-3

III(366-3) + HSo → [cyclooctatriene with 2H] (IV) + So⁻   366-4

IV(366-4) + e ⇌ [cyclooctatriene radical anion]   366-5

I(366-1) + Bu₄N⁺ → II(366-2) + Bu₃N + CH₃CH:CHCH₃   366-6

Code No.	
DC24	1-8 = H
DD70	1,2 = benzo, 3-8 = H
DF05	1,2 and 5,6 = benzo, 3,4,7,8 = H

HSo = H₂O

367

Mechanism:
367a: 367-1 to -3 (micro scale)

Proposed for Compound:
DA97, DA98, DA99, DB00, DB01, DB02, DB11, DB12, DB13, DB18, DB19, DB20, DB78

367b: 367-1, -2, -5, and -6 (macro scale)

DA97, DA98, DA99, DB00, DB01, DB02, DB11, DB12, DB13, DB18, DB19, DB20, DB78

367c: 367-2 and -3 (micro scale)

DB10

367d: 367-2 and -4 (macro scale)

DB10

$$ArAs(OH)_2(=O) + 2H^+ + 2e \rightarrow ArAs(OH)_2(OH) + H_2O \qquad 367\text{-}1$$

$$ArAs(OH)_2 + 4H^+ + 4e \rightarrow ArAsH_2 + 2H_2O \qquad 367\text{-}2$$

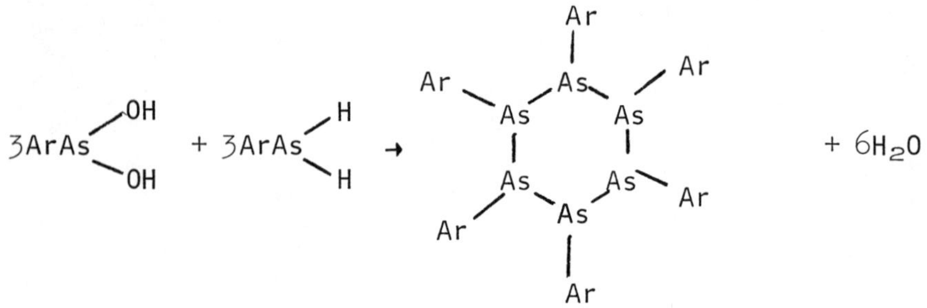

$$3ArAs(OH)_2 + 3ArAsH_2 \rightarrow (ArAs)_6\text{ ring} + 6H_2O \qquad 367\text{-}3$$

$$ArAsH_2 + 2ArAs(OH)_2 \rightarrow \tfrac{1}{x}[(ArAs)_3O]_x + 3H_2O \qquad 367\text{-}4$$

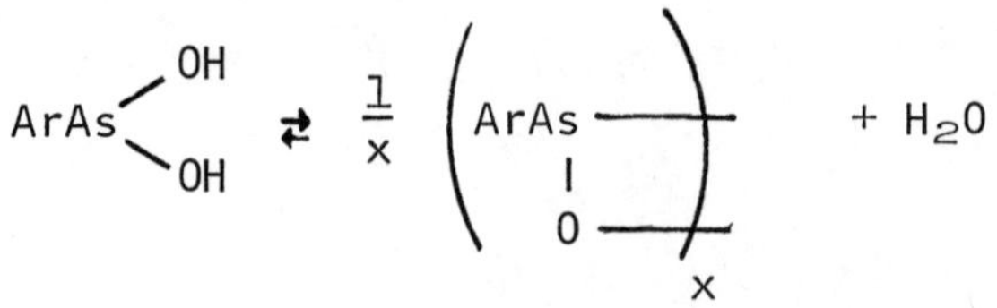

$$ArAs(OH)_2 \rightleftharpoons \tfrac{1}{x}\left(ArAs{-}O{-}\right)_x + H_2O \qquad 367\text{-}5$$

TABLE III. Courses and Mechanisms of Half-Reactions

$$\text{ArAs}\genfrac{}{}{0pt}{}{\diagup H}{\diagdown H} + (N-1)\ \text{ArAs}\genfrac{}{}{0pt}{}{\diagup OH}{\diagdown OH} \rightarrow \frac{1}{y}\left[(\text{ArAs})_N O_{N-2}\right]_y\ yN\ H_2O\ (\text{N depends on conditions}) \qquad 367\text{-}6$$

368

Mechanism:
368a: 368-1 to -9

Proposed for Compound:
DD43, DD44, DE70

$$\left[(B_9C_5H_{15})_2Ni^{II}\right]^{2-}\ (\text{I}) \rightleftarrows \left[(B_9C_5H_{15})_2Ni^{III}\right]^{-}\ (\text{II}) + e \qquad 368\text{-}1$$

$$\text{II}(368\text{-}1) \rightleftarrows \left[(B_9C_5H_{15})_2Ni^{IV}\right]\ (\text{III}) + e \qquad 368\text{-}2$$

$$2\,\text{II}(368\text{-}1) + \tfrac{1}{2}O_2 \rightarrow 2\left[(B_9C_5H_{15})_2Ni^{IV}\right]_{\text{orange}}\ (\text{IV}) + O^{2-} \qquad 368\text{-}3$$

$$\text{III}(368\text{-}2) \xrightarrow[0°C]{k_4} \text{IV}(368\text{-}3) \qquad 368\text{-}4$$

$$\text{IV}(368\text{-}3) + e \rightleftarrows \left[(B_9C_5H_{15})_2Ni^{III}\right]^{-}_{\text{orange}}\ (\text{V}) \qquad 368\text{-}5$$

$$\text{V}(368\text{-}5) + e \rightleftarrows \left[(B_9C_5H_{15})_2Ni^{II}\right]^{2-}_{\text{orange}} \qquad 368\text{-}6$$

$$\text{IV}(368\text{-}3) \xrightarrow[150°C]{k_7} \left[(B_9C_5H_{15})_2Ni^{IV}\right]_{\text{yellow}}\ (\text{VI}) \qquad 368\text{-}7$$

$$\text{VI}(368\text{-}7) + e \rightarrow \left[(B_9C_5H_{15})_2Ni^{III}\right]^{-}_{\text{yellow}}\ (\text{VII}) \qquad 368\text{-}8$$

$$\text{VII}(368\text{-}8) + e \rightarrow \left[(B_9C_5H_{15})_2Ni^{II}\right]^{2-}_{\text{yellow}} \qquad 368\text{-}9$$

369

Mechanism:
369a: 369-1 to -7 (nonaqueous)

Proposed for Compound:
DG36, DG37

$[(\pi\text{-}C_5H_5)Fe(CO)]_4$ (I) $\rightleftarrows$ $[(\pi\text{-}C_5H_5)Fe(CO)]_4^+$ (II) + e   369-1

I (369-1) + e $\rightleftarrows$ $[(\pi\text{-}C_5H_5)Fe(CO)]_4^-$ (III)   369-2

II (369-1) + III (369-2) $\rightarrow$ 2I (369-1)   369-3

II (369-1) $\rightleftarrows$ $[(\pi\text{-}C_5H_5)Fe(CO)]_4^{2+}$ (IV) + e   369-4

IV (369-4) + 2So $\rightarrow$ 2$[(\pi\text{-}C_5H_5)Fe(CO)_2So]^+$ (V) (36%) + 2e + Fe^{3+} (30%) + Fe^{2+} (6%) + $2C_5H_5^-$ 369-5

IV (369-4) $\rightarrow$ $2(\pi\text{-}C_5H_5)Fe^+$ + $[(\pi\text{-}C_5H_5)Fe(CO)_2]_2$ (VI)   369-6

VI (369-6) + 2So $\rightarrow$ 2 V(374-5) + 2e   369-7

II (369-1) + 2So $\rightarrow$ 2 V(374-5) + $2Fe^{3+}$ + $2C_5H_5^-$ + 3e   369-8

So = solvent

370

Mechanism:
370a: 370-1 to -5

Proposed for Compound:
DD57, DE71, DE82, DF13, DF14, DF58, DF59, DF93

[structure with S—S ring, Ar substituents, cation] + e $\rightleftarrows$ [radical structure] (I)   370-1

[dimeric structure with Ar, S—S groups] $\rightleftarrows$ 2 I (370-1)   370-2

I (370-1) + e $\rightarrow$ [structure] (II)   370-3

TABLE III. Courses and Mechanisms of Half-Reactions

$$II(370\text{-}3) \xrightarrow{\text{fast}} \underset{Ar}{\overset{S}{\diagdown}}=\underset{Ar}{\overset{S^-}{\diagup}} \quad (III) \qquad 370\text{-}4$$

$$III(370\text{-}4) \rightarrow I(370\text{-}1) + e \qquad 370\text{-}5$$

Ar = CH_3, t-Bu, C_6H_5, 4-$CH_3C_6H_4$, 4-$CH_3OC_6H_4$, 4-BrC_6H_4, 4-$(CH_3)_2NC_6H_4$

Code No.	T°C	25	20	15	10	5	0	-5
DE82	$K_2 \times 10^5$, $dm^3 mol^{-1}$	57	37	25	16	9.7	5.7	3.2

Code No.	$\dfrac{K_2}{K_2 \text{ for DE82}}$, in CH_2Cl_2 at room temp.	K_2, $dm^3 mol^{-1}$ at 25°C	ΔH, kcal mol^{-1}	ΔS, cal deg^{-1} mol^{-1}
DD57	—	8.2×10^{-2}	13.4	40
DE71	1.1	—	—	—
DE82	1	5.7×10^{-4}	16.0	39.8
DF13	0.55	—	—	—
DF14	0.13	—	—	—
DF58	0.54	—	—	—
DF59	0.06	5.5×10^{-5}	10.4	20
DF93	0.26	—	—	—

371

Mechanism:	Proposed for Compound:
371a: 371-1 to -3 (M=Pt) or 371-1, -4, and -5 (M=Pt) or 371-1, -6, and -7 (M=Pt oxide film) or 371-1, -8, -5 and -7 (M=Pt oxide film)(sulfate or perchlorate medium)	DD66, DD77
371b: 371-1, -6, and -7 (M=Cl⁻ adsorbed on Pt) or 371-9 to -11 and -7 (chloride medium)	DD77

$Ar_2S + M \rightleftarrows Ar_2\overset{+\cdot}{S}M + e$ 371-1

$Ar_2\overset{+\cdot}{S}M + H_2O \rightleftarrows Ar_2\overset{OH}{\underset{|}{S}}M\cdot + H^+$ 371-2

$Ar_2\overset{OH}{\underset{|}{S}}M\cdot \rightarrow Ar_2SO + H^+ + M + e$ 371-3

$2Ar_2\overset{+\cdot}{S}M \xrightarrow{slow} Ar_2S^{2+} + Ar_2S + 2M$ 371-4

$Ar_2S^{2+} + H_2O \rightleftarrows Ar_2SO + 2H^+$ 371-5

$Ar_2\overset{+\cdot}{S}M + H_2O \rightarrow Ar_2\overset{+}{S}OH + M + H^+ + e$ 371-6

$Ar_2\overset{+}{S}OH \rightleftarrows Ar_2SO + H^+$ 371-7

$Ar_2\overset{+\cdot}{S}M \rightarrow Ar_2S^{2+} + M + e$ 371-8

$Cl^- + Pt \rightleftarrows ClPt + e$ 371-9

$ClPt + Ar_2S \rightleftarrows Ar_2\overset{+\cdot}{S}Cl + Pt + e$ 371-10

$Ar_2\overset{+\cdot}{S}Cl + H_2O \rightarrow Ar_2\overset{+}{S}OH + H^+ + Cl^-$ 371-11

$Ar_2S = (C_6H_5)_2S$, (dibenzothiophene)

372

Mechanism:
372a: 372-1 to -3 (MeCN)

372b: 372-1 to -3 and -4 to -7 or -8 to -11 (MeOH)

Proposed for Compound:
DA45, DC25, DC80

DA45, DC25, DC80

$R^1R^2C{:}CH_2 \rightarrow (R^1R^2C{:}CH_2)_{ads}$ 372-1

$(R^1R^2C{:}CH_2)_{ads} \rightarrow (R^1R^2CCH_2)^{+\cdot} + e$ 372-2

$(R^1R^2CCH_2)^{+\cdot} \rightarrow R^1R^2CCH_2^{2+} + e$ 372-3

$R^1R^2CCH_2^{2+} + 2R^3OH \rightarrow R^1-\underset{\underset{OR^3}{|}}{\overset{\overset{R^2}{|}}{C}}-CH_2OR^3 + 2H^+$ 372-4

TABLE III. Courses and Mechanisms of Half-Reactions

$$R^1R^2CCH_2^{2+} + R^1R^2C:CH_2 \rightarrow (R^1R^2\overset{+}{C}CH_2)_2 \qquad 372\text{-}5$$

$$(R^1R^2\overset{+}{C}CH_2)_2 + 2R^3OH \rightarrow (R^1-\underset{OR^3}{\overset{R^2}{\underset{|}{\overset{|}{C}}}}-CH_2)_2 + 2H^+ \qquad 372\text{-}6$$

$$(R^1R^2\overset{+}{C}CH_2)_2 \rightarrow (R^1R^2C:CH)_2 + 2H^+ \qquad 372\text{-}7$$

$$(R^1R^2CCH_2)^{\cdot +} + R^3OH \rightarrow R^1R^2\dot{C}CH_2OR^3 + H^+ \qquad 372\text{-}8$$

$$R^1R^2\dot{C}CH_2OR^3 \rightarrow R^1R^2\overset{+}{C}CH_2OR^3 + e \qquad 372\text{-}9$$

$$R^1R^2\overset{+}{C}CH_2OR^3 + R^3OH \rightarrow R^1-\underset{OR^3}{\overset{R^2}{\underset{|}{\overset{|}{C}}}}CH_2OR^3 + H^+ \qquad 372\text{-}10$$

$$(R^1R^2CCH_2)^{\cdot +} + R^1R^2C:CH_2 \rightarrow R^1R^2\dot{C}CH_2CH_2\overset{+}{C}R^1R^2 \qquad 372\text{-}11$$

$R^1 = H, CH_3$

$R^2 = OC_2H_5, C_6H_5$

$R^3 = CH_3$

373

Mechanism:	Proposed for Compound:
373a: 373-1 to -4 (pH < 4)	DE06
373b: 373-1, -2, and -5 to -7 (4 < pH < 13)	DE06
373c: 373-8 to -14 (11 < pH < 14, 2% EtOH)	DE06
373d: 373-8, -13, and -14 (pH > 14, 2% EtOH)	DE06

$$HOOCCH_2O\text{-}Ar\text{-}\underset{\underset{\underbrace{CH_2}_{H^+}}{\|}}{C}OCC_2H_5 \text{ (I)} \underset{pK_1}{\rightleftharpoons} HOOCCH_2O\text{-}Ar\text{-}\underset{\underset{CH_2}{\|}}{C}OCC_2H_5 \text{ (II)} + H^+ \qquad 373\text{-}1$$

$$II \text{ (373-1)} \rightleftarrows {}^-OOCCH_2O\text{-}Ar\text{-}\underset{\underset{CH_2}{\|}}{C}OCC_2H_5 \text{ (III)} + H^+ \qquad 373\text{-}2$$

I (373-1) + e → HOOCCH$_2$O—Ar—COCHC$_2$H$_5$ (IV)   373-3
                                                       |
                                                      •CH$_2$

IV (373-3) + e + H$^+$ → HOOCCH$_2$O—ArCOCHC$_2$H$_5$ 373-4
                                              |
                                             CH$_3$

III (373-2) + 2e → $^-$OOCCH$_2$O—Ar—C═CC$_2$H$_5$ (V)   373-5
                                   | |
                                  O$^-$ CH$_2^-$

V (373-5) + 2H$^+$ $\underset{pK_6}{\rightleftarrows}$ $^-$OOCCH$_2$O—Ar—C═CC$_2$H$_5$ (VI)   373-6
                                                              | |
                                                             OH CH$_3$

V (373-5) + 2H$^+$ ⇌ $^-$OOCCH$_2$O—ArC—CHC$_2$H$_5$ (VII)   373-7
                                   ‖ |
                                   O CH$_3$

III (373-2) + e → [$^-$OOCCH$_2$O—ArCOCC$_2$H$_5$]$^{\cdot -}$ (VIII)   373-8
                                        ‖
                                       CH$_2$

VIII (373-8) + H$^+$ ⇌ $^-$OOCCH$_2$O—ArCOĊC$_2$H$_5$ (IX)   373-9
                                            |
                                           CH$_3$

IX (373-9) + e → $^-$OOCCH$_2$OArCO$\bar{\text{C}}$C$_2$H$_5$ (X)   373-10
                                        |
                                       CH$_3$

X (373-10) + H$^+$ ⇌ VI (373-6)   373-11

X (373-10) + H$^+$ ⇌ VII (373-7)   373-12

VII (373-7) + e → $^-$OOCCH$_2$OArĊOCC$_2$H$_5$ 373-13
                                    |
                                   CH$_3$

VIII (373-8) + e → V (373-5)   373-14

Ar = (2,3-dichloro-1,4-disubstituted benzene ring with Cl, Cl at positions 2,3)

Code No.	Solvent	pK_1	pK_1'	pK_6
DE06	60% EtOH	4.6	5.7	10.0

TABLE III. Courses and Mechanisms of Half-Reactions

374

Mechanism:
374a: 374-1 to -5
374b: 374-2, -6, and -7

Proposed for Compound:
DH56
DH40

$Ru(bipy)_3^{2+} \rightleftarrows Ru(bipy)_3^{3+} + e$ 374-1

$Ru(bipy)_3^{2+} + e \rightleftarrows Ru(bipy)_3^{+}$ 374-2

$Ru(bipy)_3^{+} + e \rightleftarrows Ru(bipy)_3$ 374-3

$Ru(bipy)_3 + e \rightleftarrows Ru(bipy)_3^{-}$ 374-4

$Ru(bipy)_3^{+} + Ru(bipy)_3^{3+} \rightarrow Ru(bipy)_3^{2+*} + Ru(bipy)_3^{2+}$ 374-5

[N-methylphenothiazine (I)] → [N-methylphenothiazine radical cation (I)]$^{+\cdot}$ + e   374-6

$I(374\text{-}6) + Ru(bipy)_3^{+} \rightarrow Ru(bipy)_3^{2+*} + I(374\text{-}6)$ 374-7

$Ru(bipy)_3$ and $Ru(bipy)_3^{-}$ as well as $Ru(bipy)_3^{+}$ can react to produce electrochemiluminescence

375

Mechanism:
375a: 375-1 to -6

Proposed for Compound:
DH47, DH50 DH61

[9,10-dichloro-9,10-diphenyl-9,10-dihydroanthracene] (I) + 2e → [9,10-diphenylanthracene] (II) + 2Cl^{-} 375-1

II (375-1) + e → [structure of 9,10-diphenylanthracene radical anion] (III)   375-2

III(375-2) + I(375-1) → II(375-1) + [structure of 9-chloro-9,10-diphenyl-9,10-dihydroanthracene] (IV) + Cl⁻   375-3

$Ar + e \rightarrow Ar^{\cdot -}$ 375-4

$Ar^{\cdot -} + IV(375\text{-}3) \rightarrow {}^3Ar^* + II(375\text{-}1) + Cl^-$ 375-5

${}^3Ar^* \rightarrow \rightarrow \rightarrow Ar + h\nu$ 375-6

Ar = naphthacene, perylene, rubrene

376

Mechanism:
376a: 376-1 to -8

Proposed for Compound:
DH42, DH44, DH45, DH46, DH49, DH51, DH52, DH60

$Ar + e \rightarrow Ar^{\cdot -}$ 376-1

$Ar^{\cdot -} + C_6H_5CHBrCHBrC_6H_5 \rightleftarrows Ar + [C_6H_5CHBrCHBrC_6H_5]^{\cdot -}$ 376-2

$[C_6H_5CHBrCHBrC_6H_5]^{\cdot -} \rightarrow [C_6H_5CHCHBrC_6H_5]^{\cdot} + Br^-$ 376-3

$Ar^{\cdot -} + [C_6H_5CHCHBrC_6H_5]^{\cdot} \rightleftarrows {}^1Ar^* + [C_6H_5CHCHBrC_6H_5]^-$ 376-4

$Ar + [C_6H_5CHCHBrC_6H_5]^{\cdot} \rightleftarrows Ar^{\cdot +} + [C_6H_5CHCHBrC_6H_5]^-$ 376-5

$[C_6H_5CHCHBrC_6H_5]^- \rightarrow C_6H_5CH\!:\!CHC_6H_5 + Br^-$ 376-6

$Ar^{\cdot -} + Ar^{\cdot +} \rightarrow {}^1Ar^* + Ar$ 376-7

${}^1Ar^* \rightarrow Ar + h\nu$ 376-8

Ar = anthracene, pyrene, 1,2-benzanthracene, naphthacene, perylene, coronene, rubrene

TABLE IV.
COMPOUNDS INCLUDED IN TABLE I

This index is a list of the compounds that appear in Table I. It contains the names that are used in Table I and also, for most of the compounds, one or more synonyms as well as a few common trivial names. All these names are arranged in alphabetical order, and each is followed by the code number assigned to the compound in Table I.

Most of the entries that have code numbers between DH32 and DH64 pertain to mixtures containing two electroactive substances, of which one is reduced while the other is oxidized. Each of these electroactive substances is indexed here.

TABLE IV. Compounds Included in Table I

Acenaphthylene, DD60

6-Acetamido-2-phenylchromanone, DF60

6-Acetamido-2-phenyl-4-oxochroman, DF60

7-Acetamido-2,3-dihydro-5-phenyl-1H-1,4-benzodiazepin-2-one, DF61

(Acetato-0)(2-butoxy-2-methoxyethyl)-mercury, DC93

Acetonyl(tricarbonyl)-η-cyclopentadienyl-molybdenum, DG32

7-Acetoxy-2-methylamino-5-phenyl-3H-1,4-benzodiazepine 4-N-oxide, DF92

5-Acetoxy-2-pentanone, DB94

7-Acetylamido-2,3-dihydro-5-phenyl-1H-1,4-benzodiazepin-2-one, DF61

7-Acetylamido-1,2,3-trihydro-5-phenyl-1,4-benzodiazepin-2-one, DF61

7-Acetylbenzo[c]phenothiazine 12-dioxide, DF78

12-Acetylbenzo[a]phenothiazine 7-dioxide, DF79

12-Acetylbenzo[a]phenothiazine 7-oxide, DF77

1-Acetyl-4-tert-butyl-1-cyclohexene, DD90

1-Acetyl-7-chloro-2,3-dihydro-5-phenyl-1H-1,4-benzodiazepine, DF56

Acetyl-η-cyclopentadienyltricarbonyl-manganese(II), DD02

7-Acetyl-1,3-dihydro-5-phenyl-2H-1,4-benzodiazepin-2-one, DF55

7-Acetyl-5-(2-fluorophenyl)-2-methyl-amino-3H-1,4-benzodiazepine 4-N-oxide, DF87

2-Acetylmethylamino-7-chloro-5-phenyl-1,4-benzodiazepine 4-N-oxide, DF86

2-Acetylmethylamino-7-chloro-5-phenyl-3H-1,4-benzodiazepine 4-N-oxide, DF86

2-Acetyl-5-nitrofuran, DA95

2-Acetyl-5-nitrooxacyclopentadiene, DA95

5-Acetyloxy-2-pentanone, DB94

7-Acetyl-1,2,3-trihydro-5-phenyl-1,4-benzodiazepin-2-one, DF55

Acrolein, DA15

Acrylaldehyde, DA15

Acrylamide, DA18

Acrylonitrile, DA11

Alizarine Red S, DE22

Allopurinol, DA60

2-Aminobenzenearsonic acid, DB18

4-Aminobenzenearsonic acid, DB19

1-Aminobutane, DA49

α-(N-Aminochloroacetyl)toluene, DC81

4-Aminochlorobenzene, DB03

1-Amino-7-chloro-2,3-dihydro-5-phenyl-1H-1,4-benzodiazepine, DE92

2-Amino-7-chloro-5-phenyl-3H-1,4-benzodiazepin 4-N-oxide, DE85

1-Amino-7-chloro-1,2,3-trihydro-5-phenyl-1,4-benzodiazepine, DE92

4-Aminohydroxybenzene, DB15

3-Amino-4-hydroxybenzenearsonic acid, DB20

3-Amino-4-hydroxyphenylarsonic acid, DB20

4-Amino-2-hydroxypyrimidine, DA36

1-Amino-2-mercaptobenzene, DB16

1-Amino-4-mercaptobenzene, DB17

2-Amino-3-mercaptopropanoic acid, DA23

2-Amino-3-mercaptopropionic acid, DA23

2-Amino-4-methylbenzenesulfonic acid, DB79

2-Amino-5-methylbenzenesulfonic acid, DB80

5-Amino-2-methylbenzenesulfonic acid, DB81

4-Aminomethyl-5-hydroxymethyl-2-methylpyridin-3-ol, DC42

4-(Aminomethyl)-5-hydroxy-6-methyl-3-pyridinemethanol, DC42

2-Aminonaphthalene, DD07

1-Amino-3-nitrobenzene, DB05

1-Amino-4-nitrobenzene, DB06

2-Amino-3-nitropyridine, DA66

2-Amino-5-nitropyridine, DA67

4-Amino-2-oxo-1,2-dihydropyrimidine, DA36

4-Aminophenol, DB15

2-Aminophenylarsonic acid, DB18

4-Aminophenylarsonic acid, DB19

5-(2-Aminophenyl)-7-chloro-2,3-dihydro-1-methyl-1H-1,4-benzodiazepine, DF35

4-Amino-2(1)pyrimidone, DA36

4-Amino-1-β-D-ribofuranosyl-2-(1H)-pyrimidinone, DC88

4-Amino-1-β-D-ribofuranosyl-2-(1H)-pyrimidinone-2'-monophosphate, DC89

α-Amino-β-thiolpropionic acid, DA23

2-Aminothiophenol, DB16

4-Aminothiophenol, DB17

Ammonium 2-carboxy-1-pyrrolidinecarbodithioate, DB45

Ammonium 1-pyrrolidinecarbodithiolate, DA79

Ammonium pyrrolidine-1-dithiocarboxylate, DA79

Ammonium tetramethylenedithiocarbamate 2-carboxylate, DB45

Ammonium-N,N-trimethylenedithiocarbamate, DA79

3-(4-Anisyl)-5-(4-dimethylaminophenyl)-1,2-dithiolylium perchlorate, DF93

1-(4-Anisyl)-2-phenyldiazene, DE07

3-(4-Anisyl)-5-phenyl-1,2-dithiolylium perchlorate, DF14

Anthracene, DE32, DH32, DH42

9,10-Anthracenedione-1,5-disulfonic acid, DE28

9,10-Anthracenedione-2,6-disulfonic acid, DE29

9,10-Anthracenedione-1-sulfonic acid, DE26

9,10-Anthracenedione-2-sulfonic acid, DE27

(Anthracene)tricarbonyliron(0), DF50

Anthraquinone-1,5-disulfonic acid, DE28

Anthraquinone-2,6-disulfonic acid, DE29

9,10-Anthraquinone-1,5-disulfonic acid, DE28

9,10-Anthraquinone-2,6-disulfonic acid, DE29

Anthraquinone-1-sulfonic acid, DE26

Anthraquinone-2-sulfonic acid, DE27

9,10-Anthraquinone-1-sulfonic acid, DE26

9,10-Anthraquinone-2-sulfonic acid, DE27

9,10-Anthraquinone-1-sulfonic acid, sodium salt, DE21

Ascorbic acid, DB23

Ativan, DE74

Axerophthol, DG19

Azabenzene, DA63

1-Azabicyclo[2.2.2]octane, DB95

Azobenzene, DD72

Azobenzene-4-sulfonic acid, DD75

Benzaldehyde, DB66

1,2-Benzanthracene, DH45

Benzene, DA96

Benzeneacetaldehyde, DC28

Benzenearsonic acid, DB11

1,4-Benzenedicarboxylic acid dinitrile, DC03

1,4-Benzenediol, DB07

Benzenethiol, DB09

Benzhydryl dodecyl ether, DG46

Benzhydryl methyl ether, DE43

Benzimidazole-4,7-quinone, DB58

Benzo[b]bicyclo[6.1.0]nona-2,4,6-triene, DE05

2,3-Benzobicyclo[6.1.0]nona-2,4,6-triene, DE05

Benzocyclooctatetraene, DD70

Benzoic acid, DB68

Benzonitrile, DB61

Benzo[g,h,i]perylene, DG29

1,12-Benzoperylene, DG29

Benzophenone, DE02

o-Benzoquinone, DA85

p-Benzoquinone, DA86, DH59

1,2-Benzoquinone, DA85

1,4-Benzoquinone, DA86, DH59

Benzotriptycene, DG35

Benzyl acetate, DC82

Benzyl tert-butyl ether, DD53

N-Benzylchloracetamide, DC81

Benzyl cyclohexyl ether, DE17

Benzyl ether, DE44

Benzyl methyl ether, DC37

Benzyloxycarbonylmethyl(dimethyl)(2-dimethylaminoethyl)ammonium chloride, DE97

Benzyl 3,3,6-trimethyl-3,6-diaza-3-iumheptanoate chloride, DE97

Benzyl vinyl sulfone, DC83

TABLE IV. Compounds Included in Table I

2,2'-Biazine, DD05

cis-Bicyclo[6.2.0]deca-2,4,6-triene, DD22

cis-Bicyclo[6.1.0]nona-2,4,6-triene, DC78

trans-Bicyclo]6.1.0]nona-2,4,6-triene, DC79

4,4'-Biphenyldinitrile, DE23

4-Biphenylyl(phenyl)diazomethane DG02

2,2'-Bipyridine, DD05

2,2'-Bipyridylium-1,1'-ethylene-5 5'-dimethyl bromide, DE49

1,1'-Bipyrrolidine, DC47

Bis(α-carbethoxybenzyl)mercury, DG14

1,1-Bis(carbethoxy)-2,2-dimethyl-ethene, DD35

2,5-Bis(4-chlorophenyl)-3,6-diphenyl-pyrazolo[4,3-d]pyrazolium anion, DG56

2,5-Bis(4-chlorophenyl)-3,6-diphenyl-1,2,4,5-tetraazapentalene, DG56

Bis[η-cyclopentadienyldicarbonyliron(I)], DE34

Bis[cyclopentadienyldicarbonyliron(I)], dimethyltin(II), DF36

Bis[cyclopentadienyldicarbonyl-ruthenium(I), DE37

Biscyclopentadienyliron(II), DD14

Bis(cyclopentadienylnickel(I))acetylene, DD81

Bis[π-[η-cyclopentadienylnickel(I)]]-acetylene, DD81

Bis[cyclopentadienyltricarbonyl-chromium(I)], DF01

Bis[η-cyclopentadienyltricarbonyl-chromium(I)mercury (II), DF00

Bis[π-cyclopentadienyltricarbonyl-chromium(I)mercury (II), DF00

Bis[η-cyclopentadienyltricarbonyl-molybdenum(I)], DF03

Bis[π-cyclopentadienyltricarbonyl-molybdenum(I)], DF03

Bis[η-cyclopentadienyltricarbonyl-molybdenum(I)mercury(0), DF02

Bis[π-cyclopentadienyltricarbonyl-molybdenum(I)mercury(0), DF02

Bis[dicarbonyl-(η-cyclopentadienyl)-iron(I)], DE34

Bis[dicarbonyl-η-cyclopentadienyliron(I)]-dimethylstannane(II), DF36

Bis[dicarbonyl-η-cyclopentadienyliron-(I)]dimethyltin(II), DF36

Bis[dicarbonyl-(η-cyclopentadienyl)-ruthenium(I)], DE37

1,4-Bis(dimethylamino)benzene, DD32, DH32, DH33, DH34, DH35

Bis(dimethylamino)methane, DA81

3,7-Bis(dimethylamino)phenothiazin-5-ium chloride, DF38

1,6-Bis(dimethylamino)pyrene, DG12

Bis(O-ethyl thioacetothioacetato)-cadmium(II), DD86

Bis(O-ethyl thioacetothioacetato)-mercury(II), DD87

Bis(O-ethyl thioacetothioacetato)-zinc(II), DD88

Bis(ethyl 3-thionobutyrato)mercury-(II), DD87

Bishomotetraazaadamantane, DC51

3,8-Bis(1-hydroxyethyl)-2,7,12,18-tetramethyl-13,17-bis(3-propionic acid)porphinato-trans-bis(4-acetyl pyridine)cobalt(III), DH25

3,8-Bis(1-hydroxyethyl)-2,7,12,18-tetramethyl-13,17-bis(3-propionic acid)porphinato-bis(4,4'-bi-pyridine)cobalt(III), DH28

3,8-Bis(1-hydroxyethyl)-2,7,12,18-tetramethyl-13,17-bis(3-propionic acid)porphinato-bis(ethyl nicotinate)-cobalt(III), DH26

3,8-Bis(1-hydroxyethyl)-2,7,12,18-tetramethyl-13,17-bis(3-propionic acid)porphinato-bis(2-methyl pyridine)cobalt(III), DH23

3,8-Bis(1-hydroxyethyl)-2,7,12,18-tetramethyl-13,17-bis(3-propionic acid)porphinato-bis(4-methyl-pyridine)cobalt(III), DH24

3,8-Bis(1-hydroxyethyl)-2,7,12,18-tetramethyl-13,17-bis(3-propionic acid)porphinato-bis(4-phenyl-pyridine)cobalt(III), DH29

3,8-Bis(1-hydroxyethyl)-2,7,12,18-tetramethyl-13,17-bis(3-propionic acid)porphinato-trans-(di-pyridine)cobalt(III) chloride, DH20

3,8-Bis(1-hydroxyethyl)-2,7,12,18-tetramethyl-13,17-di(3-propionic acid)porphinatocobalt(III), DG72

13,17-Bis(2-methoxycarbonylethyl)-2,7,12,18-tetramethyl-3,8-divinyl-porphinatoaquamanganese(III) chloride, DG78

13,17-Bis(2-methoxycarbonylethyl)-2,7,12,18-tetramethyl-3,8-divinyl-porphinatocobalt(II), DG74

13,17-Bis(2-methoxycarbonylethyl)-2,7,12,18-tetramethyl-3,8-divinyl-porphinatocopper(II), DG75

13,17-Bis(2-methoxycarbonylethyl)-2,7,12,18-tetramethyl-3,8-divinyl-porphinatonickel(II), DG76

13,17-Bis(2-methoxycarbonylethyl)-2,7,12,18-tetramethyl-3,8-divinyl-porphinatozinc(II), DG77

13,17-Bis(2-methoxycarbonylethyl)-2,7,12,18-tetramethyl-3,8-divinyl-porphine, DG79

Bis(methoxycarbonyl)mercury(II), DA37

3,5-Bis(4-methoxyphenyl)-1,2-dithio-1-ium perchlorate, DF59

Bis methylformate mercury(II), DA37

3,5-Bis(4-methylphenyl-1,2-dithiol-1-ium perchlorate, DF58

1,2-Bis(4-nitrophenyl)ethane, DE39

5,5-Bis(3-nitrophenyl)-2,4-imidazolidinedione, DE77

Bis(2-oxopropyl)mercury, DB25

Bis(2-quinolyl)disulfide di-N-oxide, DF76

Bis[(tricarbonyl)(η-cyclopentadienyl)-chromium(I)], DF01

Bis[(tricarbonyl)(η-cyclopentadienyl)-chromium(1)mercury(0), DF00

Bis[(tricarbonyl)(η-cyclopentadienyl)-molybdo(I)], DF03

Bis[(tricarbonyl)(η-cyclopentadienyl)-molybdo(I)mercury(0), DF02

Bis[(tricarbonyl-η-cyclopentadienyl-trifluoromethyl)molybdenum], DF75

Bis(trichloroethenyl)mercury(II), DA28

Bis(3,4,6-trichloro-2-hydroxyphenyl)-methane, DD97

Bis(trichlorovinyl)mercury(II), DA28

Bis(trifluoromethylcyclopentadienyl-tricarboxylmolybdenum), DF75

Bis($_\pi$-μ-1,2-trimethylene-1,2-dicarbollyl)nickel(IV)(orange form), DD43

Bis($_\pi$-μ-1,2-trimethylene-1,2-dicarbollyl)nickel(IV)(yellow form), DD44

Bis($_\pi$-μ-1,2-trimethylene-1,2-dicarbaundecaborane(9)nickel(IV), (orange form), DD43

Bis[$_\pi$-μ-1,2-trimethylene-1,2-dicarbaundecaboran(9)nickel(IV), (yellow form), DD44

2-Bromobenzonitrile, DB48

3-Bromobenzonitrile, DB49

4-Bromobenzonitrile, DB50

4-Bromobornane, DD36

4-Bromocamphane, DD36

3'-Bromo-2-chloroacetanilide, DC16

4'-Bromo-2-chloroacetanilide, DC17

2-exo-Bromo-2-endo-chlorobicyclo-[2.2.1]heptane, DB83

2-exo-Bromo-2-endo-chloro-norbornane, DB83

3-Bromo-2-chloro-3-phenylpropanoic acid, DC64

β-Bromo-α-chloro-β-phenylpropionic acid, DC64

7-Bromo-1,3-dihydro-5-phenyl-2H-1,4-benzodiazepin-2-one, DE79

5-Bromo-3,6-dinitropseudocumene, DC73

5-Bromo-3,6-dinitro-1,2,4-trimethyl-benzene, DC73

16α-Bromo-3-hydroxyestra-1,3,5(10)-trien-17-one, DF94

16β-Bromo-3-hydroxyestra-1,3,5(10)-trien-17-one, DF95

7-Bromo-2-methylamino-5-phenyl-3H-1,4-benzodiazepine 4-N-oxide, DF20

N-(3-Bromophenyl)chloroacetamide, DC16

N-(4-Bromophenyl)chloroacetamide, DC17

3-(4-Bromophenyl)-5-phenyl-1,2-dithiolylium perchlorate, DE71

7-Bromo-1,2,3-trihydro-5-phenyl-1,4-benzodiazepin-1-one, DE79

trans-2-Butenal, DA38

trans-2-Butenenitrile, DA35

trans-2-Buten-1-nitrile, DA35

trans-2-Butenonitrile, DA35

2-Butoxy-2-methoxyethylmercuric acetate, DC93

tert-Butoxyphenylmethane, DD53

α-(tert-Butoxy)toluene, DD53

n-Butylamine, DA49

TABLE IV. Compounds Included in Table I

2-Butylamino-7-chloro-5-phenyl-1,4-benzodiazepine 4-N-oxide, DG04

2-Butylamino-7-chloro-5-phenyl-3H-1,4-benzodiazepine 4-N-oxide, DG04

2-tert-Butyl-2-azonia-3-aza-2-norborene ion, DC92

4-(tert-Butyl)-1-cyanocyclohexene, DD54

4-tert-Butyl-1-cyano-1-cyclohexene, DD54

4-tert-Butyl-1-cyclohexen-1-nitrile, DD54

4-tert-Butycyclohexen-1-yl methyl ketone, DD90

4-tert-Butycyclohexylideneacetic acid, lithium salt, DD89

2-tert-Butyl-2,3-diazabicyclo[2.2.1]-heptane, DC96

2-tert-Butyl-2,3-diaza-3-eniumbicyclo-[2.2.1]heptane ion, DC92

2-tert-Butyl-2,3-diazanorbornane, DC96

2-tert-Butyl-3-methyl-2,3-diazabicyclo-[2.2.1]heptane, DD39

2-tert-Butyl-3-methyl-2,3-diazanorbornane, DD39

tert-Butyl peroxydodecanoate, DF48

t-Butyl peroxymyristate, DG01

tert-Butyl peroxynonanoate, DE20

t-Butyl peroxytetradecanoate, DG01

tert-Butyl trans-styryl ketone, DE14

Calciferol, DG57

2-Carbamoyl-7-chloro-2,3-dihydro-1-methyl-5-phenyl-1H-1,4-benzodiazepine, DF63

2-Carbamoyl-7-chloro-1-methyl-1,2,3-trihydro-5-phenyl-1,4-benzodiazepine, DF63

3-Carbamoylpyridine, DB04

2-Carbethoxy-2,4-hexadienoate nitrile, DC85

2-Carbethoxy-5-nitrofuran, DB72

1-Carbmethoxy-4-tert-butyl-1-cyclohexene, DD92

5-Carbmethoxyfurylthiocyanate, DC23

(4-Carbmethoxy-5-methyl-2-furyl)-thiocyanate, DC77

5-Carbmethoxy-2-thiofurylalcohol, DB77

Carbontetrachloride, DA00

Carbonyl(nitrosyl)-2,2'-bipyridine-cobalt, DD45

Carbonyl(nitrosyl)-1,10-phenanthrolinecobalt, DD98

3-Carboxy-7-chloro-2,2-dihydroxy-5-phenyl-2H-1,4-benzodiazepine, DF12

3-Carboxy-7-chloro-1,2,3-trihydro-2,2-dihydroxy-5-phenyl-1,4-benzodiazepine, DF12

1-Carboxyethyl nitrate, DA19

2-Carboxy-5-nitrofuran, DA55

9-(2-Carboxyphenyl)-4,5-dibromo-6-hydroxy-3H-xanthen-3-one, DG06

9-(2-Carboxyphenyl)-4,5-dichloro-6-hydroxy-3H-xanthen-3-one, DG07

9-(2-Carboxyphenyl)-4,5-diiodo-6-hydroxy-3H-xanthen-3-one, DG08

3-Carboxy-2-pyridone, DA93

3-Carboxy-1,4-thipyrone-2,6-dithiol, DA87

Catechol dimethyl ether, DC39

Chalcone, DE88

Chlorazapam, DF12

7-Chloro-1-acetamido-2,3-dihydro-5-phenyl-1H-benzodiazepine, DF64

Chloroacetanilide, DC27

7-Chloro-1-acetylamino-2,3-dihydro-5-phenyl-1H-1,4-benzodiazepine, DF64

7-Chloro-1(N-acetyl)amino-1,2 3-trihydro-5-phenyl-1,4-benzodiazepine, DF64

N-(Chloroacetyl)aniline, DC27

N-(Chloroacetyl)benzylamine, DC81

N-(Chloroacetyl)-3-bromoaniline, DC16

N-(Chloroacetyl)-4-bromoaniline, DC17

N-(Chloroacetyl)-3-chloroaniline, DC20

N-(Chloroacetyl)-2,3-dimethylaniline, DD23

N-(Chloroacetyl)-3,4-dimethylaniline, DD24

N-(Chloroacetyl)-4-iodoaniline, DC18

4-Chloroaniline, DB03

4-Chlorobenzaldehyde, DB60

4-Chlorobenzenearsonic acid, DA97

2-Chlorobenzonitrile, DB51

3-Chlorobenzonitrile, DB52

4-Chlorobenzonitrile, DB53

2-exo-Chloro-2-endo-bromobicyclo-[2.2.1]-heptane, DB84

2-exo-Chloro-2-endo-bromonorbornane, DB84

6-Chloro-3-chloromethyl-3,4-dihydro-2-methyl-7-sulfamoyl-1,2,4-benzothiadiazine 1,1-dioxide, DC84

7-Chloro-5-(2-chlorophenyl)-2,3-dihydro-1H-1,4-diazepin-2-one, DE73

2-Chloro-5-(2-chlorophenyl)-2,3-dihydro-3-hydroxy-1H-1,4-benzodiazepin-2-one, DE74

7-Chloro-2-cyano-2,3-dihydro-1-methyl-5-phenyl-1H-1,4-benzodiazepine, DF53

7-Chloro-2-cyano-1-methyl-1,2,3-trihydro-5-phenyl-1,4-benzodiazepine, DF53

1-Chloro-1,3-dibromopropane, DA16

7-Chloro-5-(2,3-dichlorophenyl)-2,3-dihydro-1-methyl-1H-1,4-benzodiazepine, DF15

7-Chloro-5-(2,6-dichlorophenyl)-2,3-dihydro-1-methyl-1H-1,4-benzodiazepine, DF16

6-Chloro-3,4-dihydro-2H-1,2,4-benzothiadiazine-7-sulfonamide 1,1-dioxide, DB74

7-Chloro-1,3-dihydro-5-(2-fluorophenyl)-2H-1,4-benzodiazepin-2-one, DE72

7-Chloro-1,3-dihydro-2-hydrazino-5-phenyl-2H-1,4-benzodiazepine 4-N-oxide, DE91

7-Chloro-2,3-dihydro-2-hydroxy-1-methyl-5-phenyl-1H-1,4-benzodiazepine, DF30

7-Chloro-1,3-dihydro-3-hydroxy-5-phenyl-2H-1,4-benzodiazepin-2-one, DE81

7-Chloro-1,3-dihydro-5-(1H-indol-1-yl)-2H-1,4-benzodiazepin-2-one, DF51

7-Chloro-2,3-dihydro-1-methanoyl-5-phenyl-1H-1,4-benzodiazepine, DF06

7-Chloro-1,3-dihydro-5-(2-methoxyphenyl)-2H-1,4-benzodiazepin-2-one, DF10

7-Chloro-2,3-dihydro-1-(N-methylcarbamoyl)-5-phenyl-1H-1,4-benzodiazepine, DF65

7-Chloro-2,3-dihydro-1-methylformyl-5-phenyl-1H-1,4-benzodiazepine, DF56

7-Chloro-2,3-dihydro-1-methyl-5-(2-methylmercaptophenyl)-1H-1,4-benzodiazepine, DF70

7-Chloro-2,3-dihydro-1-methyl-5-(2-methylsulfinylphenyl)-1H-1,4-benzodiazepine, DF69

7-Chloro-2,3-dihydro-1-methyl-5-(2-nitrophenyl)-1H-1,4-benzodiazepine, DF23

7-Chloro-2,3-dihydro-N-methyl-5-phenyl-1H-1,4-benzodiazepin-1-carboxamide, DF65

7-Chloro-2,3-dihydro-1-methyl-5-phenyl-1H-1,4-benzodiazepin-2-carboxamide, DF63

7-Chloro-2,3-dihydro-1-methyl-5-phenyl-1H-1,4-benzodiazepin-2-carboxylic acid, DF57

7-Chloro-2,3-dihydro-1-methyl-5-phenyl-1H-1,4-benzodiazepine, DF29

7-Chloro-2,3-dihydro-1-methyl-5-phenyl-1H-1,4-benzodiazepin-2-ol, DF30

7-Chloro-1,3-dihydro-1-methyl-5-phenyl-2H-1,4-benzodiazepin-2-one, DF09

7-Chloro-1,3-dihydro-5-(2-methylphenyl)-2H-1,4-benzodiazepin-2-one, DF07

7-Chloro-1,3-dihydro-5-(1-methyl-1H-pyrazol-5-yl)-2H-1,4-benzodiazepin-2-one, DE03

7-Chloro-2,3-dihydro-1-methylsulfonyl-5-phenyl-1H-1,4-benzodiazepine, DF31

7-Chloro-2,3-dihydro-5-phenyl-1H-1,4-benzodiazepin-1-amine, DE92

7-Chloro-2,3-dihydro-5-phenyl-1H-1,4-benzodiazepin-1-carbaldehyde, DF06

7-Chloro-2,3-dihydro-5-phenyl-1H-1,4-benzodiazepine, DE90

7-Chloro-1,3-dihydro-5-phenyl-2H-1,4-benzodiazepin-2-one, DE80

7-Chloro-1,3-dihydro-5-(2-pyridinyl)-2H-1,4-benzodiazepin-2-one, DE33

6-Chloro-3,4-dihydro-7-sulfamoyl-2H-1,2,4-benzothiadiazine 1,1-dioxide, DB74

7-Chloro-1,3-dihydro-5-(2-thiazolyl)-2H-1,4-benzodiazepin-2-one, DD62

7-Chloro-2-dimethylamino-5-phenyl-3H-1,4-benzodiazepine 4-N-oxide, DF66

TABLE IV. Compounds Included in Table I

7-Chloro-2-dimethylamino-5-phenyl-1,4-benzodiazepine 4-N-oxide, DF66

2-[4-Chloro-α-(2-dimethylamminium-ethyl)benzyl]pyridine maleate, DG16

7-Chloro-2-ethylamino-5-phenyl-2H-1,4-benzodiazepine 4-N-oxide, DF67

7-Chloro-2-ethylamino-5-phenyl-1,4-benzodiazepine 4-N-oxide, DF67

7-Chloro-2-(ethylformyl)methylamino-5-phenyl-1,4-benzodiazepine 4-N-oxide, DG03

7-Chloro-5-(2-fluorophenyl)-1,3-dihydro-4-benzodiazepin-2-one, DE72

7-Chloro-2-hydrazino-1,2,3-trihydro-5-phenyl-1,4-benzodiazepine 4-N-oxide, DE91

7-Chloro-2-hydroxy-1-methyl-1,2,3-trihydro-5-phenyl-1,4-benzodiazepine, DF30

7-Chloro-3-hydroxy-5-phenyl-1,3-dihydro-2H-1,4-benzodiazepin-2-one, DE81

7-Chloro-5-(4-hydroxyphenyl)-2-methylamino-3H-1,4-benzodiazepine 4-N-oxide, DF24

cis-2-Chloromercury-3-ethoxycyclononene, DD55

cis-2-Chloromercury-3-methoxycyclodecene, DD56

trans-2-Chloromercury-3-methoxycyclotridecene, DE66

trans-2-Chloromercury-3-methoxycycloundecene, DD93

Chloro(11-methoxy-1-cycloundecen-1-yl)mercury, DD93

7-Chloro-2-methoxy-5-phenyl-3H-1,4-benzodiazepine 4-N-oxide, DF11

7-Chloro-2-(N-methylacetamido)-5-phenyl-3H-1,4-benzodiazepine 4-N-oxide, DF86

7-Chloro-2-methylamino-5-phenyl-3H-1,4-benzodiazepine 4-N-oxide, DF22

7-Chloro-1-methylformyl-1,2,3-trihydro-5-phenyl-1,4-benzodiazepine, DF56

4-Chloromethylnitrobenzene, DB65

7-Chloro-N-methyl-5-phenyl-3H-1,4-benzodiazepin-2-carboximidic acid ethyl ester 4-N-oxide, DG03

7-Chloro-N-methyl-5-phenyl-3H-1,4-benzodiazepin-2-carboximidic acid methyl ester 4-N-oxide, DF86

7-Chloro-2-methyl-5-phenyl-3H-1,4-benzodiazepine 4-N-oxide, DF08

7-Chloro-2[N-methylpropionamido)-5-phenyl-3H-1,4-benzodiazepine 4-N-oxide, DG03

7-Chloro-1-methylsulfonyl-1,2,3-trihydro-5-phenyl-1,4-benzodiazepine, DF31

7-Chloro-1-methyl-1,2,3-trihydro-5-phenyl-1,4-benzodiazepin-2-carboxylic acid, DF57

7-Chloro-1-methyl-1,2,3-trihydro-5-phenyl-2H-1,4-benzodiazepine, DF29

α-Chloro-4-nitrotoluene, DB65

Chlorpheniramine maleate, DG16

4-Chlorophenylarsonic acid, DA97

N-(3-Chlorophenyl)chloroacetamide, DC20

5-(2-Chlorophenyl)-2,3-dihydro-7-iodo-1-methyl-1H-1,4-benzodiazepine, DF21

γ-(4-Chlorophenyl)-N,N-dimethyl-2-pyridinepropanammonium maleate, DG16

2-(4-Chlorophenyl)-5-phenyl-1,2,3,4,5,6-hexazapentalene, DE30

5-(4-Chlorophenyl)-2-phenyl-1,2,3-triazolo[4,5-d]triazolium anion DE30

[3-(4-Chlorophenyl)-3-(2-pyridyl-1-propyl]dimethylammonium maleate DG16

5-(2-Chlorophenyl)-1,2,3-trihydro-7-iodo-1-methyl-1H-1,4-benzodiazepine, DF21

7-Chloro-5-(2-thiazolyl)-1,4-benzodiazepin-2-one, DD62

7-Chloro-1,2,3-trihydro-5-(1H-indol-1-yl)-1,4-benzodiazepin-2-one, DF51

7-Chloro-1,2,3-trihydro-5-(2-methoxyphenyl)-1,4-benzodiazepin-2-one, DF10

7-Chloro-1,2,3-trihydro-2-methyl-5-(2-methylphenyl)-1,4-benzodiazepin-2-one, DF07

7-Chloro-1,2,3-trihydro-1-methyl-2'-nitro-5-phenyl-1,4-benzodiazepine, DF23

369

7-Chloro-1,2,3-trihydro-5-phenyl-1,4-benzodiazepine, DE90

7-Chloro-1,2,3-trihydro-5-phenyl-1,4-benzodiazepin-1-al, DF06

7-Chloro-1,2,3-trihydro-5-phenyl-1,4-benzodiazepin-2-one, DE80

Cholecalciferol, DG53

Cholesta-5,7-dien-3β-ol, DG52

Cinnamaldehyde, DC72

Cobalt(III)hematoporphyrin, DG73

Cobalt(III)hematoporphyrin di-(4-acetylpyridine) complex, DH25

Cobalt(III)hematoporphyrin di-(4,4'-bipyridine) complex, DH28

Cobalt(II)protoporphyrin(IX) dimethyl ester, DG74

Cobalt(III)hematoporphyrin di-(ethyl nicotinate) complex, DH26

Cobalt(III)hematoporphrin di-(4-phenylpyridine) complex, DH29

Cobalt(III)hematoporphyrin di-(2-picoline) complex, DH23

Cobalt(III)hematoporphyrin di-(4-picoline) complex, DH24

Cobalt(III)hematoporphyrin di-pyridine complex, DH20

Coenzyme Q_{10}, DH30

Copper(II)protoporphrin(IX) dimethyl ester, DG75

Coronene, DH51

CpC, DF98

CpG, DG05

CpU, DF97

m-Cresol, sodium salt, DB73

Crotonaldehyde, DA38

Crotonitrile, DA35

Cyanide hemichrome, DG72

Cyanobenzene, DB61

2-Cyanobromobenzene, DB48

3-Cyanobromobenzene, DB49

4-Cyanobromobenzene, DB50

1-Cyano-1,3-butadiene, DA62

2-Cyanochlorobenzene, DB51

3-Cyanochlorobenzene, DB52

4-Cyanochlorobenzene, DB53

(2-Cyanoethyl)iodomercury, DA13

2-Cyanoethylmercuric iodide, DA13

4-Cyanoiodobenzene, DB57

2-Cyano-5-nitrofuran, DA51

trans-1-Cyanopropene, DA35

Cyanopyridyl hemichrome, DH12

4-Cyanotoluene, DC21

1,2-Cyclohexadiendione, DA85

Cyclohexatriene, DA96

2-Cyclohexenaldehyde, DB86

2-Cyclohexenecarbaldehyde, DB86

Cyclohexylhydroxylamine, DB38

(Cyclohexyloxy)methylbenzene, DE17

2-Cyclohexyl-5-phenyl-1,2,3,4,5,6-tetraazapentalene, DE51

2-Cyclohexyl-5-phenyl-1,2,3-triazolo-[4,5,d]-1,2,3-triazolium anion, DE51

Cyclooctatetraene, DC24

1,3,5-Cyclooctatriene, DC35

Cyclopentadienyldicarbonyliron(I)-triphenyllead(0), DG43

Cyclopentadienyldicarbonyliron(I)-triphenyltin(III), DG44

Cyclopentadienyldicarbonylmolybdenum(I)dimethyltin(II)cyclopentadienyltricarbonylmolybdenum(I), DF68

π-Cyclopentadienyldicarbonylmolydenum-(I)trimethyltin(III), DD26

η-Cyclopentadienylperfluorodimethylethylene-1,2-disulfidoiridium(III), DC59

η-Cyclopentadienylperfluorodimethylethylene-1,2-disulfidoiron(III), DC58

η-Cyclopentadienylperfluorodimethylethylene-1,2-disulfidonickel(III), DC60

η-Cyclopentadienylperfluorodimethylethylene-1,2-disulfidorhodium(III), DC61

η-Cyclopentadienyltrifluoromethyltricarbonylmolybdenum(0), DC56

6,6-Cyclopentamethylene-1,5-diazabicyclo[3.1.0]hexane, DC91

9,9-Cyclopentamethylene-1,5-diazabicyclo[3.3.1]nonane, DD94

TABLE IV. Compounds Included in Table I

8,8-Cyclopentamethylene-1,5-diazabicyclo[3.2.1]octane, DD58

2,2-Cyclopentamethylene-1,3-diazatricyclo[3.2.2.2^{3,7}]decane, DE18

2-Cyclopentenone, DA68

Cysteine, DA23

L-Cystine, DB35

Cytidine, DC88

Cytidine monophosphate, DC89

Cytidylic acid, DC89

Cytidylyl (3'→5') cytidine, DF98

Cytidylyl (3'→5') guanidine, DG05

Cytidylyl (3'→5') uridine, DF97

Cytochrome C, DH31

Cytosine, DA36

Dabco, DB31

[1,2,3,4,6,10,11,12,13,15-Decahydro-7,16-dimethyldibenz[f,m][1,2,4,5,8,9,11,12]-octaazacyclotetradecinato(2-)-N^6,N^9,N^{15},N^{18}]nickel, DF40

7-Dehydrocholesterol, DG52

N-Desalkylflurazepam, DE72

N-Desmethylchloridiazepoxide, DE85

Diacetonylmercury, DB25

trans-Di[4-acetylpyridine)hematoporphyrincobalt(III), DH25

Diammonium proline-N-dithiocarboxylate, DB45

3,5-Di(4-anisyl)-1,2-dithiolylium perchlorate, DF59

1,3-Diazaadamantane, DC43

1,6-Diazabicyclo[4.4.0]deca-3,8-diene, DC41

1,6-Diazabicyclo[4.4.0]decane, DC48

1,5-Diazabicyclo]3.3.1]nonane, DB98

1,5-Diazabicyclo[3.2.2]nonane, DB97

1,4-Diazabicyclo]2.2.2]octane, DB31

1,5-Diazabicyclo[3.2.1]octane, DB32

1,5-Diazabicyclo[3.3.0]octane, DB33

2,2'-Diazabiphenyl, DD05

4,5-Diazaphenanthrene, DD64

1,3-Diazatricyclo[3.3.1.1^{3,7}]decane, DC43

Diazepam, DF09

1,2-Diazine, DA32

1,3-Diazine, DA33

1,4-Diazine, DA31

α-Diazomethyl 3-trifluoromethylphenyl ketone, DC57

5,9-Diazonia-5,6,7-trihydrodibenzo[a,c]azepine bromide, DE12

[b,i]Dibenz-6,13-diacetyl-5,12-dimethyl-1,4,8,11-tetraazacyclotetradeca-2,4,6,9,11,13-hexaenenickel(II), DG39

[b,i]Dibenz-5,12-dimethyl-1,4,8,11-tetraazacyclotetradeca-2,4,6,9,11,13,-hexaenenickel(II), DG11

sym-Dibenzocycloactatetraene, DF05

Dibenzo[a,e]cycloactatetraene, DF05

Dibenzo[b,i]-1,4,8,11-tetraaza-2,5,7,9,12,14-cyclotetradecahexaenenickel(II), DF81

Dibenzothiophene, DD66

Dibenzo[b,d]thiophene, DD66

Dibenzotriptycene, DG55

[b,i]Dibenz-1,4,8,11-tetraazacyclodeca-2,4,6,9,11,13,hexaenenickel(II), DF81

Dibenzyl-N-nitrosoamine, DE42

trans-Di-(4,4'-bipyridine)hematoporphyrincobalt(III), DH28

1,2-Dibromoacenaphthene, DD61

1,2-Dibromobenzocyclobutene, DC09

1,3-Dibromo-1-chloropropane, DA16

1,2-Dibromo-1,2-dihydroacenaphthylene, DD61

2,4-Dibromo-2,4-dimethyl-3-pentanone, DB92

1,2-Dibromo-1,2-diphenylethane, DE38, DH42, DH44, DH45, DH46, DH49, DH51, DH52, DH60

(1,2-Dibromoethyl)benzene, DC26

4',5'-Dibromofluorescein, DG06

1,2-Dibromo(phenyl)ethane, DC26

Dibromostyrene, DC26

2,3-Dibromothiole, DA29

2,3-Dibromothiophene, DA29

3,5-Di(tert-butyl)-1,2-dithiolylium perchlorate, DD57

3,4-Di-tert-butyl-1,3,4-oxadiazolidine, DD41

[Dicarbonyl-η-cyclopentadienylmolybdenum(I)][tricarbonyl-η-cyclopentadienylmolybdenum(I)]dimethyltin(II), DF68

Dicarbonyl-[η-cyclopentadienyl)(trimethylstannio(III)]molybdenum(I), DD26

Dicarbonyl-(η-cyclopentadienyl)[triphenylplumbio(III)]iron(I), DG43

Dicarbonyl-(η-cyclopentadienyl)[triphenylstannio(III)]iron(I), DG44

Dicarbonyl-1,10-phenanthrolinenickel(0), DE24

2,2-Dichloroacetanilide, DC19

2,3'-Dichloroacetanilide, DC20

N-(Dichloroacetyl)aniline, DC19

N-(Dichloroacetyl)-3-methylaniline, DC74

N-(Dichloroacetyl)-4-methylaniline, DC75

2,2-Dichlorobicyclo[2.2.1]heptane, DB85

9,10-Dichloro-9,10-dihydro-9,10-diphenylanthracene, DG48, DH47, DH50, DH54, DH55, DH61

2,2-Dichloroethyl benzoate, DC65

4',5'-Dichlorofluorescein, DG07

[2,3-Dichloro-4-(2-methylenebutyryl)-phenoxy]acetic acid, DE06

[2,3-Dichloro-4-(2-methylene-1-oxobutyl)phenoxy]acetic acid, DE06

2,2-Dichloronorbornane, DB85

2',7-Dichloro-5-phenyl-2,3-dihydro-1,4-benzodiazepin-2-one, DE73

1,4-Dicyanobenzene, DC03

4,4'-Dicyanobiphenyl, DE23

trans-Dicyanoprotoporphyrin(IX)iron-(III), DG72

[f,m]Dicyclohexane-3,10-diethyl-1,2,4,5,8,9,11,12-octaazacyclotetradeca-2,5,7,9,12,14-hexanenickel(II), DF99

[f,m]Dicyclohexane-3,10-dimethyl-1,2-4,5,8,9,11,12-octaazacyclotetradeca-2,5,7,9,12,14-hexaenenickel, DF40

2,5-Di(cyclohexyl)-1,2,3,4,5,6-hexaazapentalene, DE64

Dicyclohexyl-N-nitrosoamine, DD95

2,5-Dicyclohexyl-1,2,3-triazolo[4,5-d]-1,2,3-triazolium anion, DE64

Dicyclopentadienyliron(II), DD14

Dicysteine, DB35

Didecanoyl peroxide, DG20

Didodecanoyl peroxide, DG42

Diethenyl sulfone, DA41

Diethenyl sulfoxide, DA39

Diethylaminoethyl methacrylate, DD38

7,14-Diethyl-bis(1,2-cyclohexylenyl-[c,j]-1,2,5,6,8,9,12,13-octaaza-2,4,7,9,11,14-cyclotetradecahexaenenickel(II), DF99

3,8-Diethyl-13,17-bis(2-methoxycarbonylethyl)-2,7,12,18-tetramethylporphine, DG80

7,16-Diethyl-1,2,3,4,6,10,11,12,13,15-decahydrodibenz[f,m][1,2,4,5,8,9,11,12-octaazacyclotetradecinato(2-)-N^6,N^9,N^{15},N^{18}]nickel(II), DF99

3,3-Diethyl-2,4-dioxopiperidine, DC90

3,3-Diethyl-2,4-dioxotetrahydropyridine, DC87

2,2-Diethyl-5,8-ethanoperhydropyrazolo[1,2-a]pyridazine, DE19

2,2-Diethylhexahydro-5,9-ethano-1H-pyrazolo-[1,2-α]-pyridazine, DE19

Diethyl isopropylidenemalonate, DD35

3,3-Diethyl-5-methyl-2,4-dioxopiperidine, DD37

3,3-Diethyl-5-methyl-2,4-piperidinedione, DD37

trans-Di(ethylnicotinate)hematoporphyrincobalt(III), DH26

3,4-Diethyl-1,3,4-oxadiazolidine, DB41

3,3-Diethyl-2,4-piperidinedione, DC90

1,2-Diethylpyrazolidine, DC01

3,3-Diethyl-2,4-(1H,3H)pyridinedione, DC87

N,N'-Diethyltrimethylenehydrazine, DC01

Dihexadecanoyl peroxide, DG70

5,12-Dihydro-5,12-o-benzenonaphthacene- DG35

cis-1a,9b-Dihydro-1H-benzo[a]cyclopropa[c]cyclooctene, DE05

[1,1'-(7,16-Dihydro-6,15-dimethyl-dibenzo[b,i]-[1,4,8,11]tetraazacyclotetradecine-7,16-diyl)bis-[ethanoato](2-)N^5,N^9,N^{14},N^{18}]-nickel(II), DG39

3,4-Dihydro-2,8-dimethyl-2-(4,8,12-trimethyltridecyl-2H-1-benzopyran-6-ol, DG54

TABLE IV. Compounds Included in Table I

1,3-Dihydro-7-fluoro-5-phenyl-2H-1,4-benzodiazepin-2-one, DE83

7-(1,3-Dihydro-2-isoindolyl)-7-azabicyclo[4.1.0]heptane, DE57

2,3-Dihydro-5-(5-isothiazolyl)-1-methyl-7-nitro-1H-1,4-benzodiazepine, DE08

1,3-Dihydro-7-methoxy-5-phenyl-2H-1,4-benzodiazepin-2-one, DF28

2,3-Dihydro-1-methyl-5-(1-methyl-1H-imidazol-2-yl)-7-nitro-1H-1,4-benzodiazepine, DE47

2,3-Dihydro-1-methyl-5-(1-methyl-1H-pyrazol-5-yl)-7-nitro-1H-1,4-benzodiazepine, DE48

2,3-Dihydro-1-methyl-7-nitro-5-phenyl-1H-1,4-benzodiazepine, DF34

2,3-Dihydro-1-methyl-7-nitro-5-(4-pyridyl)-1H-1,4-benzodiazepine, DE93

2,3-Dihydro-1-methyl-7-nitro-5-(2-pyrimidinyl)-1H-1,4-benzodiazepine, DE41

2,3-Dihydro-1-methyl-7-nitro-5-(2-thiazolyl)-1H-1,4-benzodiazepine, DE09

3,4-Dihydro-2-methyl-4-oxo-3-o-tolyl-quinazoline, DF25

2,3-Dihydro-1-methyl-5-phenyl-1H-1,4-benzodiazepine, DF37

1,3-Dihydro-7-methyl-5-phenyl-2H-1,4-benzodiazepin-2-one, DF26

1,3-Dihydro-7-methylthio-5-phenyl-2H-1,4-benzodiazepin-2-one, DF27

3,4-Dihydro-2-methyl-2-(4,8,12-trimethyltridecyl)-2H-benzo[b]pyran-6-ol, DG50

1,3-Dihydro-7-nitro-5-phenyl-2H-1,4-benzodiazepin-2-one, DE84

1,2-Dihydro-2-oxo-1-B-D-ribofuranosyl-3-pyridinecarboxylic acid, DD52

1,3-Dihydro-5-phenyl-2H-1,4-benzodiazepin-2-one, DE87

1,3-Dihydro-5-phenyl-7-trifluoromethyl-2H-1,4-benzodiazepin-2-one, DF04

3,4-Dihydro-2,5,7,8-tetramethyl-2-(4,8,12-trimethyltridecyl)-2H-1-benzopyran-6-ole, DG63

3,4-Dihydro-2,5,8-trimethyl-2-(4,8,12-trimethyltridecyl)-2H-1-benzopyran-6-ol, DG60

3,4-Dihydro-2,7,8-trimethyl-2-(4,8,12-trimethyltridecyl)-2H-1-benzopyran-6-ol, DG61

Dihydroxo(phenyl)arsenic, DB10

1,2-Dihydroxy-9,10-anthraquinone-3-sulfonic acid, sodium salt, DE22

1,3-Dihydroxybenzene, DB08

1,4-Dihydroxybenzene, DB07

2,4-Dihydroxybenzenearsonic acid, DB13

2,4-Dihydroxy-5-nitrobenzenearsonic acid, DB02

2,4-Dihydroxy-5-nitrophenylarsonic acid, DB02

2,4-Dihydroxyphenylarsonic acid, DB13

7,9-Diiodo-1,3-dihydro-5-phenyl-2H-1,4-benzodiazepin-2-one, DE76

4',5'-Diiodofluorescein, DG08

7,9-Diiodo-5-phenyl-2,3-dihydro-1,4-benzodiazepin-2-one, DE76

1,2-Diisopropyl-1,2-diazacyclopentane, DC97

1,2-Diisopropyl-1,2-diazolidine, DC97

Diisopropyl-N-nitrosoamine, DB42

3,4-Diisopropyl-1,3,4-oxadiazolidine, DC54

1,2-Diisopropylpyrazolidine, DC97

2,6-Dimercapto-4-oxo-4H-thiopyran-3-carboxylic acid, DA87

2,6-Dimercapto-1,4-thiopyrone-3-carboxylic acid, DA87

1,2-Dimethoxybenzene, DC39

5,6-Dimethoxy-2-methyl-3-(3,7,11,15,19,23,27,31,35,39-decamethyl-2,6,10,14,18,22,26,30,34,38-tetracontadecaenyl)-1,4-benzoquinone, DH30

x-Dimethylaminoazobenzene, DE45

4-(Dimethylamino)azobenzene, DE46

x-Dimethylaminodiphenyldiazene, DE45

4-(Dimethylaminodiphenyldiazene, DE46

x-Dimethylaminophenylazobenzene, DE45

4-Dimethylaminophenylazobenzene, DE46

3-[4-(Dimethylamino)phenyl]-5-(4-methoxyphenyl)-1,2-dithiol-1-ium perchlorate, DF93

1-(4-Dimethylaminophenyl)-2-phenyldiazene, DE46

1,3-Dimethylbenzimidazolone-4,7-quinone, DC68

1,3-Dimethylbenzimidazolone-5,6-quinone, DC69

1,1'-Dimethyl-4,4'-bipyridylium bromide, DD83

7,14-Dimethyl-bis(1,2-cyclohexylenyl-[c,j])-1,2,5,6,8,9,12,13-octaaza-2,4,7,9,11,14-cyclotetradecahexaene-nickel(II), DF40

Dimethyl cis-butendioate, DB22

Dimethyl trans-butendioate, DB21

(3,3-Dimethyl-2-butyl)ferrocene, DF39

2',3'-Dimethyl-2-chloroacetanilide, DD23

3',4'-Dimethyl-2-chloroacetanilide, DD24

2,5-trans-Dimethyl-3,4-cyclotetramethyleneoxa-3,4-diazolidine, DC50

2,3-Dimethyl-2,3-diazabicyclo-[2.2.1]-heptane, DB99

2,3-Dimethyl-2,3-diazabicyclo[2.2.1]-hept-5-ene, DB93

2,3-Dimethyl-2,3-diazabicyclo-[2.2.2]-octa-5-ene, DC44

2,3-Dimethyl-2,3-diazabicyclo[2.2.2]-octane, DC49

1,4-Dimethyl-1,4-diazacyclohexane, DB39

1,2-Dimethyl-1,2-diaza-4-cyclohexene, DB34

1,4-Dimethyl-1,4-diazaspiro[4.5]-decane, DD40

2,7-Dimethyl-9,10-dihydro-8a,10a-diazoniaphenanthrene bromide, DE49

3,6-Dimethyl-9,10-dihydro-8a,10a-diazoniaphenanthrene bromide, DE50

Dimethyl-2(N,N-dimethyl)aminoethyl-(benzoxycarbonyl)methylammonium chloride, DE97

1,3-Dimethyl-4,7-dioxobenzimidazolone, DC68

1,3-Dimethyl-5,6-dioxobenzimidazolone, DC69

7,8-Dimethyl-2,4-dioxo-10(D-ribo-2,3,4,5-tetrahydroxypentyl)-2,3,4,10 tetrahydrobenzo[g]pteridine, DF74

1,4-Dimethyl-3,6-diphenylpyrazolo-[4,5-d]pyrazole, DF88

2,4-Dimethyl-3,6-diphenylpyrazolo-[3,4-d]pyrazole, DF89

1,4-Dimethyl-3,6-diphenyl-1,2,4,5-tetraazapentalene, DF88

1,5-Dimethyl-3,6-diphenyl-1,2,4,5-tetraazapentalene, DF89

3,5-Dimethyl-1,2-dithiolylium perchlorate, DA69

4,4-Dimethyl-1,1'-ethano-2,2'-bipyridylium bromide, DE50

5,5'-Dimethyl-1,1'-ethano-2,2'-bipyridylium bromide, DE49

Dimethyl cis-1,2-ethylenedicarboxylate, DB22

Dimethyl trans-1,2-ethylenedicarboxylate, DB21

Dimethyl fumarate, DB21

2,5-Dimethyl-1,2,3,4,5,6-hexaazabicyclo[2.2.1]heptane, DA26

1,4-Dimethylhexahydropyrazine, DB39

1,2-Dimethylhexahydropyridazine, DB40

1,2-Dimethyl-3-hexyl-1,2-diazacyclopropane, DC98

1,2-Dimethyl-3-n-hexyldiaziridine, DC98

Dimethyl maleate, DB22

3,8-Dimethyl-2-methoxyazocine, DD25

Dimethyl(4-methylphenyl)amine, DC86

Dimethyl-N-nitrosoamine, DA10

2,6-Dimethylphenol, DC38

N-(2,3-Dimethylphenyl)chloroacetamide, DD23

N-(3,4-Dimethylphenyl)chloroacetamide, DD24

5,5-Dimethyl-3-phenyl-2-cyclohexenone, DE52

trans-4,6-Dimethyl-1-phenyl-1-penten-3-one, DE14

N,N'-Dimethylpiperazine, DB39

N,N'-Dimethylpyrazolidine, DA78

1,2-Dimethylpyrazolidine, DA78

trans-Di(2-methylpyridine)hematoporphyrincobalt(III), DH23

trans-Di(4-methylpyridine)hematoporphyrincobalt(III), DH24

trans-Di(3-methylpyridine)protoporphyriniron(III) chloride, DH21

trans-Di(4-methylpyridine)protoporphyriniron(III) chloride, DH22

7,8-Dimethyl-10-ribitylisoalloxazine, DF74

TABLE IV. Compounds Included in Table I

2,3-Dimethyl-1,4,8,11-tetraazacyclo-tetradecaneiron(II) dication, DD96

1,2-Dimethyl-1,2,3,6-tetrahydropyridazine, DB34

7,8-Dimethyl-10-(D-ribo-2,3,4,5-tetrahydroxypentyl)isoalloxazine, DF74

5,8-Dimethyltocol, DG60

7,8-Dimethyltocol, DG61

N,N-Dimethyl-4-toluidine, DC86

Dimethyl(p-tolyl)amine, DC86

3,7-Dimethyl-9-(2,6,6-trimethyl-1-cyclohexene-1-yl)-2,4,6,8-nonatetraen-1-al, DG17

3,7-Dimethyl-9-(2,6,6-trimethyl-1-cyclohexen-1-yl)-2,4,6,8-nonatetraen-1-ol, DG19

3,7-Dimethyl-9-(2,6,6-trimethyl-1-cyclohexen-1-yl)-2,4,6,8-nonatetraen-1-yl acetate, DG33

3,7-Dimethyl-9-(2,6,6-trimethyl-1-cyclohexen-1-yl)-2,4,6,8-nonatetranoic acid, DG18

3,7-Dimethyl-9-(2,6,6-trimethyl-1-cyclohexen-1-yl)-2,4,6,8-nonatrien-1-yl palmitate, DH08

9,10-Dinitroanthracene, DE25

1,3-Dinitrobenzene, DA84

4,4'-Dinitrobibenzyl, DE39

3,6-Dinitro-5-bromo-1,2,4-trimethylbenzene, DC73

3,3'-Dinitrodiphenylhydantoin, DE77

Dinitrosyl-2,2'-bipyridineiron, DD04

Dinitrosyl-1,10-phenanthrolineiron, DD63

Dinitrosyldi(2-pyridyl ketone)iron, DD46

Dinitrosyl[di-(2-pyridyl) ketone]iron, DD46

Dinonanoyl peroxide, DG00

Dioctadecanoyl peroxide, DH09

4,7-Dioxobenzimidazole, DB58

2,4-Dioxo-3,3-diethyl-5-methylpiperidine, DD37

2-(2,6-Dioxo-3-piperidinyl)-1H-isoindole-1,3(2H)-dione, DE01

N-(2,6-Dioxo-3-piperidyl)phthalimide, DE01

β,β-Diphenylacrolein, DE89

9,10-Diphenylanthracene, DG47, DH33, DH52, DH53

cis-1,2-Diphenyl-2-benzoylethene, DG22

trans-1,2-Diphenyl-2-benzoylethene, DG23

2,2-Diphenyl-1-benzoylethene, DG21

1,4-Diphenyl-1,3-butadiene, DF19

Diphenylbutadiyne, DE98

1,3-Diphenyl-5,7-diazatricyclo[3.3.1.$1^{3,7}$]decane, DG15

5,7-Diphenyl-1,3-diazatricyclo[3.3.1.$1^{3,7}$]decane, DG15

1,3-Diphenyl-5,7-diazatricyclo[3.3.1.$1^{3,7}$]decan-2-one, DG13

5,7-Diphenyl-1,3-diazatricyclo[3.3.1.$1^{3,7}$]decan-6-one, DG13

Diphenyldiazene, DD72

3,5-Diphenyl-1,2-dithiolylium perchlorate, DE82

Diphenylene disulfide, DD67, DH36, DH37

1,1-Diphenyl-2-formylethene, DE89

N,N'-Diphenylguanidien, DE11

1,3-Diphenylguanidine, DE11

2,5-Diphenyl-1,2,3,4,5,6-hexaazapentalene, DE36

Diphenyl ketone, DE02

2,6-Diphenyl-4-methylpyrylium tetrachloroaluminate, DF82

Diphenyl-N-nitrosamine, DD74

2,5-Diphenyloxadiazole, DE35, DH37, DH38, DH43

2,5-Diphenyl-1,3,4-oxadiazole, DE35, DH37, DH38, DH43

trans-2,3-Diphenyloxirane, DE40

3,3-Diphenylprop-2-eneal, DE89

1,3-Diphenylpropenone, DE88

trans-1,3-Diphenylpropenone, DE88

trans-Di(4-phenylpyridine)hematoporphyrincobalt(III), DH29

Diphenyl sulfide, DD77

Diphenylthiocarbazone, DE10

Diprotonated 2,3,7,8,12,13,17,18-octaethylporphyrin perchlorate, DH07

trans-Dipyridinehematoporphyrincobalt(III) chloride, DH20

2,2'-Dipyridyl, DD05

Di-2-pyridyl ketone, DD47

trans-Dipyridylprotoporphyrin(IX)iron-(III) chloride, DH19

Distyryl, DF19

Ditetradecanoyl peroxide, DG62

3,3'-Dithiobis(2-aminopropanoic acid), DB35

2,2'-Dithiodiquinoline 1,1'-dioxide, DF76

Dithizon, DE10

3,5-Di(4-tolyl)-1,2-dithiolylium perchlorate, DF58

Divinyl sulfone, DA41

Divinyl sulfoxide, DA39

1,1'-[(Dodecyloxy)methylene]bisbenzene, DG46

Doriden, DE13

Durohydroquinone, DD30

Ergocalciferol, DG57

Ergosta-5,7,22-trien-3β-ol DG58

Ergosterin, DG58

Ergosterol, DG58

Ethacrynic acid, DE06

Ethanediamide, DA06

Ethanedioic acid, DA05

Ethenyl ethyl ether, DA45

1-Ethenyl-2-oxopyrrolidine, DB24

1-Ethenyl-2-pyrrolidinone, DB24

N-Ethenyl-2-pyrrolidone, DB24

Ethenylsulfonylbenzene, DC29

cis-9-Ethoxy-1-cyclononen-1-yl-chloromercury, DD55

2-Ethoxy-2-cyclopentenone, DB89

Ethoxyethene, DA45

1-Ethoxy-1-phenylethane, DD27

Ethyl 2-bromomercurio-2-(4-isopropylphenyl)acetate, DE16

Ethyl α-bromomercury(2-bromophenyl)-acetate, DD11

Ethyl-α-bromomercury(3-bromophenyl)-acetate, DD12

Ethyl-α-bromomercury(4-bromophenyl)-acetate, DD13

Ethyl 2-bromomercury-2-(2-bromophenyl)-ethanoate, DD11

Ethyl 2-bromomercury-2-(3-bromophenyl)-ethanoate, DD12

Ethyl 2-bromomercury-2-(4-bromophenyl)-ethanoate, DD13

Ethyl α-bromomercury(3-tert-butylphenyl)acetate, DE59

Ethyl 2-bromomercury-2-(3-tert-butylphenyl)ethanoate, DE59

Ethyl α-bromomercury(4-chlorophenyl)-acetate, DD08

Ethyl 2-bromomercury-2-(4-chlorophenyl)ethanoate, DD08

Ethyl 2-bromomercury(4-ethylphenyl)-acetate, DD84

Ethyl 2-bromomercury-2-(4-ethylphenyl)ethanoate, DD84

Ethyl α-bromomercury(4-fluorophenyl)acetate, DD09

Ethyl 2-bromomercury-2-(4-fluorophenyl)ethanoate, DD09

Ethyl α-bromomercury(4-iodophenyl)-acetate, DD10

Ethyl 2-bromomercury-2-(4-iodophenyl)-ethanoate, DD10

Ethyl α-bromomercury(2-methylphenyl)-acetate, DD50

Ethyl α-bromomercury(3-methylphenyl)-acetate, DD51

Ethyl 2-bromomercury-2-(2-methylphenyl)ethanoate, DD50

Ethyl 2-bromomercury-2-(3-methylphenyl)ethanoate, DD51

Ethyl α-bromomercuryphenylacetate, DD18

Ethyl 2-bromomercury-2-phenylethanoate, DD18

Ethyl 4-tert-butylcyclohexylidene acetate, DE65

5-Ethyl-5-(sec-butyl)-2-thiobarbituric acid, DD33

Ethyl 2-cyano-2,4-hexadienoate, DC85

Ethyl α-cyano-β-propenylacrylate, DC85

Ethyl 2-cyano-3-propenylpropenoate, DC85

Ethylene bis(3-mercaptopropionate), DC45

1,1'-Ethylene-4,4'-dimethyl 2,2'-bipyridylium bromide, DE50

1,4-Ethylenepiperidine, DB95

Ethyl ethenyl oxide, DA45

TABLE IV. Compounds Included in Table I

Ethyl ethenyl sulfone, DA46

Ethyl N-methyl-N-nitrosocarbamate, DA44

2-(2'-Ethyl-6'-methylpiperidyl)ethyl methacrylate, DE67

5-Ethyl-5-(1-methylpropyl)-4,6-dioxo-2-thioxohexahydropyrimidine, DD33

5-Ethyl-5(1-methylpropyl)-2-thiobarbituric acid, DD33

Ethyl 5-nitro-2-furoate, DB72

5-Ethyl-5-(3-nitrophenyl)barbituric acid, DD80

5-Ethyl-5-(3-nitrophenyl)-2,4,6-(1H,3H,5H)pyrimidinetrione, DD80

5-Ethyl-5-(3-nitrophenyl)-2,4,6-trioxoperhydropyrimidine, DD80

3-Ethyl-3-phenyl-2,6-diketopiperidine, DE13

3-Ethyl-3-phenyl-2,6-dioxopiperidine, DE13

Ethyl 1-phenylethyl ether, DD27

2-Ethyl-2-phenylglutarimide, DE13

3-Ethyl-3-phenyl-2,6-piperidinedione, DE13

(Ethylsulfonyl)ethene, DA46

O-Ethyl thioacetothioacetate, DB28

Ethyl 3-thioketothiobutanoate, DB28

Ethyl vinyl ether, DA45

Ethyl vinyl sulfone, DA46

Ethylxanthic acid potassium salt, DA17

Fast Red E, DG09

Ferrocene, DD14

Fluoranthene, DE99, DH39, DH48

9-Fluorenone, DD99

9α-Fluoro-17α,21-dihydroxy-5-pregnene-3,11,20-trione-3-ethyleneketal-17,20,21-bismethylenedioxolane, DG45

5-(2-Fluorophenyl)-7-iodo-2-methylamino-3H-1,4-benzodiazepine 4-N-oxide, DF17

7-Fluoro-1,2,3-trihydro-5-phenyl-1,4-benzodiazepin-2-one, DE83

Formaldehyde sodium sulfoxylate, DA04

Formylbenzene, DB66

1-Formyl-4-chlorobenzene, DB60

3-Formylcyclohexene, DB86

2-Formyl-1-hydroxybenzene, DB70

1-Formyl-1-methylethyl nitrate, DA42

1-Formyl-3-nitrobenzene, DB62

1-Formyl-4-nitrobenzene, DB63

2-Formyl-5-nitrofuran, DA54

2-Formyl-4-nitrothiophene, DA52

2-Formyl-5-nitrothiophene, DA53

2-Formylphenol, DB70

D-Glucose, DB37

Glutethimide, DE13

Glycol dimercaptopropionate, DC45

Guaiacol glyceryl ether, DD31

Guaiacuran, DD31

Hematoporphyrincobalt(III), DG73

Hemin, DG71

3-Heptyn-2-one, DB87

Hexachlorobuta-1,3-diene, DA27

3,3',4,4',6,6'-Hexachloro-2,2'-dihydroxydiphenylmethane, DD97

Hexachlorophene, DD97

2,4a,5,6,7,8-Hexahydro-4a,8-dimethyl-naphthalen-2-one, DD85

Hexahydro-1,2,4,5-tetrakis(1-methylethyl)-1,2,4,5-tetrazine, DE68

2,3,6,7,10,11-Hexamethoxytriphenylene, DG40

2,3,6,7,10,11-Hexamethoxytriphenylene, radical cation, DG41

Hexamethylenetetramine, DB36

5,7,7,12,14,14-Hexamethyl-1,4,8,11-tetraaza-4,11-cyclotetradeca-dieniron(II) dication, DF47

5,5,7,12,12,14-Hexamethyl-1,4,8,11-tetraazacyclotetradecaneiron(II) dication bis[tetrafluoroborate (1-)], DF49

5,5,7,12,12,14-Hexamethyl-1,4,8,11-tetraaza-1,3,7,10-cyclotetradeca-tetraeneiron(II) dication, DF41

5,5,7,12,12,14-Hexamethyl-1,4,8,11-tetraaza-1,3,8,10-cyclotetradeca-tetraeniron(II) dication, DF42

5,5,7,12,14,14-Hexamethyl-1,4,8,11-
  tetraaza-1,3,7,11-cyclotetradeca-
  tetraeneiron(II) dication, DF43

5,7,7,12,14,14-Hexamethyl-1,4,8,11-
  tetraaza-1,4,8,11-cyclotetradeca-
  tetraeneiron(II) dication, DF44

5,7,7,12,14,14-Hexamethyl-1,4,8,11-
  tetraaza-1,3,8-cyclotetradeca-
  trieneiron(II) dication, DF45

5,7,7,12,14,14-Hexamethyl-1,4,8,11-
  tetraaza-1,4,11-cyclotetradeca-
  trieneiron(II) dication, DF46

Hexamine, DB36

2-Hexynoic acid, lithium salt, DB14

μ-Hydrido-μ-dimethylphosphinobis-
  [(cyclopentadiene)tricarbonylmolyb-
  denum(I)], DF91

μ-Hydrido-μ-dimethylphosphinobis[tri-
  carbonyl(η-cyclopentadienyl)molyb-
  denum(I)], DF91

Hydrochlorothiazide, DB74

Hydroquinol, DB07

Hydroquinone, DB07

4-Hydroxyaniline, DB15

4-Hydroxyazobenzene, DD73

2-Hydroxy-1,1'-azonaphthalene-4',6-
  disulfonic acid disodium salt, DG09

2-Hydroxybenzaldehyde, DB70

4-Hydroxybenzenearsonic acid, DB12

1-Hydroxybutylferrocene, DE56

2-Hydroxy-3-carboxypyridine, DA93

4-Hydroxydiphenyldiazene, DD73

3β-Hydroxy-5,7,22-ergostatriene, DG58

3-Hydroxy-5-(hydroxymethyl)-2-methyl-
  isonicotinaldehyde, DC33

3-Hydroxy-5-(hydroxymethyl)-2-methyl-
  isonicotinaldehyde-5-phosphate, DC36

3-Hydroxy-5-(hydroxymethyl)-2-methyl-
  isonicotinic acid, DC34

3-Hydroxy-5-(hydroxymethyl)-2-methyl-
  4-pyridinecarboxylic acid, DC34

Hydroxymethanesulfinic acid sodium salt,
  DA04

4-[2-(5-Hydroxy-2-methylcyclohexen-1-
  yl)ethenyl]-7a-methyl-3a,6,7,7a-
  tetrahydroindan, DG59

N-(2-Hydroxymethyl-3,4-dihydroxy-5-
  tetrahydrofuryl)-3-carboxypyridine-
  (1H)2-one, DD52

2-Hydroxymethyl-5-nitrofuran, DA65

5-Hydroxy-6-methyl-3,4-pyridine-
  dimethanol, DC40

2-Hydroxynicotinic acid, DA93

1-Hydroxy-4-nitrobenzene, DA92

4-Hydroxynitrobenzene, DA94

4-Hydroxy-3-nitrobenzenearsonic acid,
  DB01

4-Hydroxy-3-nitrophenylarsonic acid,
  DB01

2-Hydroxy-3-nitropyridine, DA56

2-Hydroxy-5-nitropyridine, DA57

4-Hydroxyphenlyarsonic acid, DB12

4-Hydroxyphenylazobenzene, DD73

2-Hydroxyphenylmercury chloride, DA88

1-(4-Hydroxyphenyl)-2-phenyldiazene,
  DD73

4-Hydroxypyrazolo[3,4-d]pyrimidine,
  DA60

2-Hydroxy-3-pyridinecarboxylic acid,
  DA93

Hydroxytriphenylstannane, DF90

Hydroxytriphenyltin, DF90

8-Hydroxyxanthine, DA61

7-Imidazo[4,5-d]pyrimidine, DA59

1-H-Indole-2,3-dione 3-(O-ethyloxime),
  DD15

4-Iodobenzonitrile, DB57

4'-Iodo-2-chloroacetanilide, DC18

1-Iodo-2-cyanoethane, DA14

2-Iodoethyl benzoate, DC76

3-Iodomercuripropanonitrile, DA13

N-(4-Iodophenyl)chloroacetamide, DC18

3-Iodopropanonitrile, DA14

β-Iodopropionitrile, DA14

3-Iodopropionitrile, DA14

Iron(III)chloride protoporphrin(IX),
  DG71

Isatin 3-(O-ethyloxime), DD15

Isatin-3-oxime ethyl ether, DD15

3-Isobutoxy-5,5-dimethyl-2-cyclo-
  hexen-1-one, DD91

TABLE IV. Compounds Included in Table I

2-Isobutoxy-2-methoxyethylmercuric acetate, DC94

Isopropylmagnesium bromide, DA22

Isotachysterol, DG59

Lactoflavin, DF74

Lauroyl peroxide, DG42

Lithium 4-tert-butylcyclohexylideneacetate, DD89

Lithium 2-hexynoate, DB14

Lorazepam, DE74

Maleic acid, DG16

Manganese(III)chloride protoporphyrin-(IX) dimethyl ester monohydrate, DG78

Menadione, DD48

Mercaptoacetic acid, DA07

β-Mercaptoalanine, DA23

2-Mercaptoaniline, DB16

4-Mercaptoaniline, DB17

3-Mercapto-1,5-diphenylformazan, DE10

Mercuric bismethylformate, DA37

Mercuric trichloroethylene, DA28

Mercury, chloro methyl, DA03

Mercury phenylmercaptide, DD71

Mesoporphyrin(IX) dimethyl ester, DG80

Metadiazine, DA33

Methanamine, DB36

2-Methanoyl-5-nitrofuran, DA54

Methaqualone, DF25

4-Methoxyazobenzene, DE07

2-Methoxybenz[b]azocine, DD78

2-Methoxy-1-benzazocine, DD78

4-Methoxybenz[d]azocine, DD79

4-Methoxy-3-benzazocine, DD79

4-Methoxy-2,2'-bipyridyl-1-oxide, DD49

1-Methoxycarbonyl-4-tert-butylcyclohexene, DD92

cis-10-Methoxy-1-cyclodecen-1-yl-chloromercury, DD56

6-Methoxydibenz[b,f]azocine, DF18

6-Methoxy-11,12-dihydrodibenz[b,f]-azocine, DF32

1-Methoxy-3,7-dimethyl-9-(2,6,6-trimethyl-1-cyclohexenyl)-2,4,6,8-nonatetraene, DG28

6-Methoxy-1-hydroxyphenazine-5,10-dioxide, DE00

(Methoxymethyl)benzene, DC37

6-Methoxy-1-phenazinol-5,10-dioxide, DE00

3-(2-Methoxyphenoxy)-1,2-propanediol, DD31

1-(4-Methoxyphenyl)-2-phenyldiazene, DE07

3-(4-Methoxyphenyl)-5-phenyl-1,2-dithiol-1-ium perchlorate, DF14

6-Methoxytribenz[b,d,f]azocine, DG10

2-Methoxy-3,5,5-trimethyl-2-cyclohexen-1-one, DD34

Methyclothiazide, DC84

Methyl acetylenecarboxylate, DA34

β-Methylacrolein, DA38

Methyl acrylate, DA40

Methyl allylacetate, DB29

2-Methylamino-7-methylmercapto-5-phenyl-3H-1,4-benzodiazepine 4-N-oxide, DF72

2-Methylamino-7-methylthio-5-phenyl-1,4-benzodiazepine 4-N-oxide, DF72

2-Methylamino-5-phenyl-3H-1,4-benzodiazepine-2-carboxylic acid methyl ester 4-N-oxide, DF92

2-Methylamino-5-phenyl-3H-1,4-benzodiazepine 4-N-oxide, DF33

2-Methylamino-5-phenyl-1,4-benzodiazepine 4-N-oxide, DF33

4-Methylbenzenearsonic acid, DB78

1-Methylbenzimidazole-4,7-quinone, DC11

1-Methylbenzimidazole-6,7-quinone, DC12

4-Methylbenzonitrile, DC21

2-Methyl-7,8-benzophenazine, DF52

Methyl-1,4-benzoquinone, DB69

Methyl benzyl ether, DC37

Methyl benzyl oxide, DC37

Methyl 4-bromo-5-oxohexanoate, DB91

Methyl trans-2-butenoate, DA73

Methyl 3-butenoate, DA72

Methyl 4-tert-butyl-1-cyclohexenoate, DD92

(5-Methyl-4-carbmethoxy-2-furyl)thiocyanate, DC77

1-Methyl-2-chlorobenzimidazole-4,7-quinone, DC04

1-Methyl-2-chlorobenzimidazole-6,7-quinone, DC05

1-Methyl-2-chloro-4,7-dioxobenzimidazole, DC04

1-Methyl-2-chloro-6,7-dioxobenzimidazole, DC05

Methyl cinnamate, DD16

Methyl trans-cinnamate, DD17

Methyl crotonate, DA73

5-Methyl-2-cyclohexenone, DB88

3'-Methyl-2,2-dichloroacetanilide, DC74

4'-Methyl-2,2-dichloroacetanilide, DC75

1-Methyl-2-dimethylaminobenzimidazole-4,7-quinone, DD19

1-Methyl-2-dimethylaminobenzimidazole-5,6-quinone, DD20

1-Methyl-2-dimethylaminobenzimidazole-6,7-quinone, DD21

1-Methyl-2-dimethylamino-4,7-dioxobenzimidazole, DD19

1-Methyl-2-dimethylamino-5,6-dioxobenzimidazole, DD20

1-Methyl-2-dimethylamino-6,7-dioxobenzimidazole, DD21

1-Methyl-4,7-dioxobenzimidazole, DC11

1-Methyl-6,7-dioxobenzimidazole, DC12

1-Methyl-4,9-dioxo-(2,3-d)naphthoimidazole, DD65

2-Methyl-4,6-diphenylpyrylium iodide, DF85

2,2'-Methylenebis(3,4,6-trichlorophenol), DD97

Methylene Blue, DF38

Methyl ethenyl sulfone, DA21

Methyl 2-hexynoate, DB90

2-Methyl-3-hydroxy-4-aminomethyl-5-hydroxymethylpyridine, DC42

2-Methyl-3-hydroxy-4,5-bis(hydroxymethyl)pyridine, DC40

2-Methyl-3-hydroxy-4-carboxy-5-hydroxymethylpyridine, DC34

2-Methyl-3-hydroxy-4-formyl-5-hydroxymethylpyridine, DC33

9a-Methyl-4-[2-(3-hydroxy-6-methylenecyclohexylidene)ethylidene]-1-(1,5-dimethylhexyl)perhydroindene, DG53

Methyl 5-(mercaptomethyl)-2-furoate, DB77

Methylmercuric chloride, DA03

7-Methyl-2-methylamino-5-phenyl-1,4-benzodiazepin 4-N-oxide, DF71

7-Methyl-2-methylamino-5-phenyl-3H-1,4-benzodiazepin 4-N-oxide, DF71

1-Methyl-5-(1-methyl-2-imidazolyl)-7-nitro-(2H,3H)-1,4-benzodiazepine, DE47

1-Methyl-5-(1-methyl-5-pyrazolyl)-7-nitro-(2H,3H)-1,4-benzodiazepine, DE48

Methyl 2-methyl-5-(thiocyanatomethyl)-3-furoate, DC77

1-Methyl-1H-naphth[2,3-d]imidazol-4,7-dione, DD65

1-Methyl(2,3-d)naphthoimidazole-4,9-quinone, DD65

2-Methyl-1,4-naphthoquinone, DD48

2-Methyl-2-nitratopropanal, DA42

2-Methyl-nitratopropanoic acid, DA43

4-Methyl-2-nitroaniline, DB76

4-Methylnitrobenzene, DB71

2-Methyl-5-nitrofuran, DA64

1-Methyl-7-nitro-5(2-pyrimidyl)-(2H,3H)-1,4-benzodiazepine, DE41

2-(N-Methyl-N-nitrosoamino)acetamide, DA24

N-Methyl-N-nitrosoaniline, DB75

N-Methyl-N-nitrosoglycinamide, DA24

N-Methyl-N-nitroso-N'-nitroguanidine, DA09

N-Methyl-N-nitrosophenylamine, DB75

N-Methyl-N-nitrosourea, DA08

1-Methyl-7-nitro-1,2,3-trihydro-5-phenyl-1,4-benzodiazepine, DF34

N-Methyl-2,3,7,8,12,13,17,18-octaethylporphrin, DH10

23-Methyl-1,2,7,8,12,13,17,18-octaethylporphine, DH10

2-Methylol-5-nitrofuran, DA65

Methyl 4-pentenoate, DB29

3-Methyl-3-penten-2-one, DB27

TABLE IV. Compounds Included in Table I

10-Methylphenothiazine, DE04, DH38, DH39, DH40, DH41

Methyl β-phenylacrylate, DD16

Methyl trans-β-phenylacrylate, DD17

4-Methylphenylarsonic acid, DB78

1-Methylphenylazo-2-hydroxynaphthalene, DF54

1-[(2-Methylphenylazo]-2-naphthalenol, DF54

2-Methyl-5-phenylazonia-1,2,3-triazolo-[4,5-d]-1,2,3-triazole anion, DC71

1-Methyl-5-phenylazonia-1,2,3-triazolo-[4,5-d]-1,2,3-triazole, DC70

N-(3-Methylphenyl)dichloroacetamide, DC74

N-(4-Methylphenyl)dichloroacetamide, DC75

3-Methyl-5-phenyl-1,2-dithiolylium perchlorate, DD06

1-Methyl-1-phenylethylene, DC80

2-Methyl-5-phenyl-1,2,3,4,5,6-hexaazapentalene, DC71

3-(4-Methylphenyl)-5-phenyl-1,2-dithiol-1-ium perchlorate, DF13

Methyl 3-Phenylpropenoate, DD16

Methyl trans-3-phenylpropenoate, DD17

1-Methyl-5-phenyl-1,2,3,4,5,6-tetraazapentalene, DC70

N-(3-Methylphenyl)trichloroacetamide, DC66

N-(4-Methylphenyl)trichloroacetamide, DC67

1-Methyl-5-phenyl-1,2,3-trihydro-1,4-benzodiazepine, DF37

2-Methyl-3-phytyl-1,4-naphthoquinone, DG66

Methyl propargylate, DA34

Methyl 2-propenoate, DA40

Methyl propiolate, DA34

3-(2-Methylpropoxy-5,5-dimethyl-2-cyclohexen-1-one, DD91

2-(2-Methylpropoxy)-2-methoxyethylmercuric acetate, DC94

Methyl propynoate, DA34

Methylquinone, DB69

Methyl retinol, DG28

α-Methylstyrene, DC80

(Methylsulfonyl)ethene, DA21

2-Methyl-3-(3,7,11,15-tetramethyl-2-hexadecenyl)-1,4-naphthoquinone, DG66

Methyl 5-(thiocyanatomethyl)-2-furoate, DC23

Methylthionine chloride, DF38

8-Methyltocol, DG54

2-Methyl-3-o-tolyl-4-(3H)-quinazolinone, DF25

2-Methyl-3-(o-tolyl)-4-quinazolone, DF25

1-Methyl-3,5,7-triazaadamantane, DC46

1-Methyl-3,5,7-triazatricyclo-[3.3.1.1^{3,7}]-decane, DC46

3'-Methyl-2,2,2-trichloroacetanilide, DC66

4'-Methyl-2,2,2-trichloroacetanilide, DC67

1-Methyl-1,2,3-trihydro-7-nitro-5-(4-pyridyl)-1,4-benzodiazepine, DE93

7-Methyl-1,2,3-trihydro-5-phenyl-1,4-benzodiazepin-2-one, DF26

2-Methyl-2-(4,8,12-trimethyltridecyl)-6-chromanol, DG50

4-Methyl-2,3,6-triphenylpyrazolo[3,4-d]-pyrazole, DG34

1-Methyl-3,5,6-triphenylpyrazolo[4,5-c]-pyrazole, DG34

1-Methyl-3,5,6,-triphenyl-1,2,4,5-tetraazapentalene, DG34

Methyl vinylacetate, DA72

Methyl vinyl sulfone, DA21

Methyprylon, DD37

Miazine, DA33

Morpholine, DA48

Myristoyl peroxide, DG62

NADH, DG26

NADPH, DG27

Naphthacene, DH46, DH47

1-Naphthaldehydeazine, DG30

2-Naphthaldehydeazine, DG31

Naphthalene, DD03

1,2-Naphthalenedione, DD00

1,4-Naphthalenedione, DD01

1,2-Naphthoquinone, DD00

1 4-Naphthoquinone, DD01

2-Naphthylamine, DD07

Naphthyl-1,1'-azo-2-naphthol-4',6-disulfonic acid, disodium salt, DG09

Niacinamide, DB04

Nickel(II)protoporphrin(IX) dimethyl ester, DG76

Nicotinamide, DB04

Nicotinamide adenine dinucleotide, phosphate, reduced, DG27

Nicotinamide adenine dinucleotide, reduced, DG26

Nicotinic acid, DA90

Nicotinic acid amide, DB04

2-Nitratoisobutanoic acid, DA43

2-Nitratoisobutyraldehyde, DA42

2-Nitratoisobutyric acid, DA43

2-Nitratopropanoic acid, DA19

α-Nitratopropionic acid, DA19

Nitrazepam, DE84

9-Nitroanthracene, DE31

3-Nitroaniline, DB05

4-Nitroaniline, DB06

4-Nitroazobenzene, DD69

3-Nitrobenzaldehyde, DB62

4-Nitrobenzaldehyde, DB63

Nitrobenzene, DA91

2-Nitrobenzenearsonic acid, DA98

3-Nitrobenzenearsonic acid, DA99

4-Nitrobenzenearsonic acid, DB00

4-Nitrobenzontrile, DB59

2-Nitrobenzotrifluoride, DB54

3-Nitrobenzotrifluoride, DB55

4-Nitrobenzotrifluoride, DB56

4-Nitrobenzyl chloride, DB65

4-Nitrobenzylcyanide, DC13

4-Nitrobenzyl thiocyanate, DC14

4-Nitro-α-cyanotoluene, DC13

Nitrocyclohexane, DB30

5-Nitro-2,4-dihydroxyphenylarsonic acid, DB02

4-Nitrodiphenyldiazene, DD69

(2-Nitroethyl)benzene, DC31

Nitroform, DA02

4-Nitro-2-formylthiphene, DA52

5-Nitro-2-formylthiophene, DA53

5-Nitro-2-furaldehyde, DA54

5-Nitro-2-furaldehyde diacetate, DB82

2-Nitrofuran, DA30

5-Nitro-2-furancarbaldehyde, DA54

5-Nitro-2-furancarbaldehyde diacetate, DB82

5-Nitrofuran-2-carboxaldehyde diacetate, DB82

5-Nitro-2-furancarboxylonitrile, DA51

5-Nitrofurfural, DA54

5-Nitrofurfuryl alcohol, DA65

5-Nitro-2-furoic acid, DA55

5-Nitro-2-furonitrile, DA51

5-Nitro-2-furylcarbonitrile, DA51

(5-Nitro-2-furyl)methanol, DA65

2'-Nitro-1'-hydroxyethylbenzene, DC32

3-Nitro-4-hydroxyphenylarsonic acid, DB01

1-Nitro-2-hydroxy-2-phenylethane, DC32

2-Nitrooxole, DA30

3-Nitrophenobarbital, DD80

4-Nitrophenol, DA94

4-Nitrophenylacetonitrile, DC13

2-Nitrophenylarsonic acid, DA98

3-Nitrophenylarsonic acid, DA99

4-Nitrophenylarsonic acid, DB00

1-Nitro-2-phenylethene, DC22

4-Nitrophenylphenyldiazene, DD69

3-Nitro-2-pyridone, DA56

5-Nitro-2-pyridone, DA57

Nitrosobenzene, DA89

N-Nitrosodimethylamine, DA10

N-Nitrosodiphenylamine, DD74

N-Nitroso-3-methylaminosulfolane, DA76

N-Nitroso-3-methylaminotetrahydrothiophene 1,1-dioxide, DA76

3-(N-Nitrosomethylamino)thiolane, DA76

4-Nitrosophenol, DA92

N-Nitrosopiperidine, DA75

1-Nitrosopiperidine, DA75

N-Nitrososarcosamide, DA24

N-Nitrososarcosinamide, DA24

β-Nitrostyrene, DC22

TABLE IV. Compounds Included in Table I

4-Nitro-2-thienylcarboxaldehyde, DA52

5-Nitro-2-thienylcarboxaldehyde, DA53

5-Nitro-2-(thiocyanatomethyl)furan, DA83

4-Nitro-2-thiophenaldehyde, DA52

5-Nitro-2-thiophenaldehyde, DA53

4-Nitrotoluene, DB71

2-Nitro(trifluoromethyl)benzene, DB54

3-Nitro(trifluoromethyl)benzene, DB55

4-Nitro(trifluoromethyl)benzene, DB56

2-Nitro-α,α,α-trifluorotoluene, DB54

3-Nitro-α,α,α-trifluorotoluene, DB55

4-Nitro-α,α,α-trifluorotoluene, DB56

2,3,7,8,12,13,17,18-Octaethylporphinatocadmium(II), DG83

2,3,7,8,12,13,17,18-Octaethylporphinatocalcium(II), DG82

2,3,7,8,12,13,17,18-Octaethylporphinatocobalt(II), DG84

2,3,7,8,12,13,17,18-Octaethylporphinatocopper(II), DG85

2,3,7,8,12,13,17,18-Octaethylporphinatodihydroxogermanium(IV), DH03

2,3,7,8,12,13,17,18-Octaethylporphinatodihydroxosilicon(IV), DH05

2,3,7,8,12,13,17,18-Octaethylporphinatodihydroxotin(IV), DH06

2,3,7,8,12,13,17,18-Octaethylporphinatohydroxoaluminum(III), DG93

2,3,7,8,12,13,17,18-Octaethylporphinatohydroxoantimony(III), DH00

2,3,7,8,12,13,17,18-Octaethylporphinatohydroxochromium(III), DG94

2,3,7,8,12,13,17,18-Octaethylporphinatohydroxogallium(III), DG96

2,3,7,8,12,13,17,18-Octaethylporphinatohydroxoindium, DG97

2,3,7,8,12,13,17,18-Octaethylporphinatohydroxoiron(III), DG95

2,3,7,8,12,13,17,18-Octaethylporphinatohydroxomanganese(III), DG98

2,3,7,8,12,13,17,18-Octaethylporphinatohydroxooxomolybdenum(V), DG99

2,3,7,8,12,13,17,18-Octaethylporphinatohydroxoscandium(III), DH01

2,3,7,8,12,13,17,18-Octaethylporphinatohydroxothallium(III), DH02

2,3,7,8,12,13,17,18-Octaethylporphinatolead(II), DG90

2,3,7,8,12,13,17,18-Octaethylporphinatomagnesium(II), DG86

2,3,7,8,12,13,17,18-Octaethylporphinatonickel(II), DG87

2,3,7,8,12,13,17,18-Octaethylporphinatooxotitanium(IV), DG89

2,3,7,8,12,13,17,18-Octaethylporphinatooxovanadium(IV), DG89

2,3,7,8,12,13,17,18-Octaethylporphinatopalladium(II), DG91

2,3,7,8,12,13,17,18-Octaethylporphinatosilver(III) perchlorate, DG81

2,3,7,8,12,13,17,18-Octaethylporphinatozinc(II), DG92

2,3,7,8,12,13,17,18-Octaethylporphine, DH04

1,2,3,4,4a,5,6,7-Octahydronaphthalen-1-one, DD28

1,2,3,4,5,6,7,8-Octahydronaphthalen-1-one, DD29

1,2,3,4,4a,5,6,7-Octahydronaphthalone, DD28

2-Octyl benzyl ether, DE96

Oizine, DA32

Orange OT, DF54

Orotic acid, DA58

Orthodiazine, DA32

Oxalamide, DA06

Oxalic acid, DA05

Oxamide, DA06

Oxazepam, DE81

16α,17α-Oxido-4-pregnene-3,20-dione, DG25

16α,17α-Oxidoprogesterone, DG25

2-Oxo-3-indolidineethoxyazane, DD15

4-Oxo-1-pentyl acetate, DB94

4,4'-Oxybis[benzenesulfonyl(isopropylidenehydrazide)], DF96

1,1'-[Oxybis(methylene)]bisbenzene, DE44

4,4'-Oxydibenzenesulfonic acid bis-(isopropylidenehydrazide), DF96

Palmitoyl peroxide, DG70

Paradiazine, DA31

Pentacarbonyl[dicarbonyl-π-cyclopentadienylferrio(0)]manganese(0), DD59

Pentacarbonylmanganese(0)cyclopentatriendicarbonyliron(I), DD59

Pentacarbonylmanganese(0)[dicarbonyl-η-cyclopentadienyliron(I)], DD59

Pentachloromercaptobenzene, DA82

Pentachlorothiophenol, DA82

2,4-Pentadienenitrile, DA62

9,9-Pentamethylene-1,5-diazatricyclo-[3.3.1.1^{3,7}]decane, DE18

trans-3-Penten-2-one, DA71

Perchlorobutadiene, DA27

Perhydro-1,4-oxazine, DA48

Perylene, DH49, DH50

o-Phenanthroline, DD64

1,10-Phenanthroline, DD64

4,5-Phenanthroline, DD64

Phenindione, DE78

Phenoxazine, DD68

Phenylacetaldehyde, DC28

Phenylacetaldoxime, DC30

Phenylacetylene, DC08

β-Phenylacrolein, DC72

Phenyl arsenoxide, DB10

Phenyl arsenoxide (dehydrated form), DB10

Phenylarsonic acid, DB11

Phenylarsonous acid, DB10

trans-2-Phenyl-1-benzoylethene, DE88

β-Phenylchalcone, DG21

N-Phenylchloroacetamide, DC27

Phenyl cyanide, DB61

2-Phenyl-2-cyclohexenone, DD82

Phenyldiazenecarbothioic acid 2-phenylhydrazide, DE10

N-Phenyldichloroacetamide, DC19

2-Phenyl-1,3-diketohydrindene, DE78

trans-1-Phenyl-4,4-dimethyl-1-penten-3-one, DE14

4-Phenyldiphenyldiazomethane, DG02

Phenyl ethenyl sulfone, DC29

1-Phenylethyl benzoate, DE94

Phenylethylene, DC25

1-Phenylethyl ethyl ether, DD27

Phenylethyne, DC08

Phenylglyoxalic acid, DC15

3-Phenyl-3-hepten-2-one, DE15

2-Phenyl-1,3-indandione, DE78

2-Phenyl-1H-indene-1,3(2H)-dione, DE78

Phenylmercaptan, DB09

1-Phenyl-1-methylethylene, DC80

1-Phenyl-2-nitroethane, DC31

1-Phenyl-2-nitroethanol, DC32

Phenyloxoacetic acid, DC15

Phenyloxoethanoic acid, DC15

Phenyloxylic acid, DC15

10-Phenylphenothiazine, DF80, DH48

3-Phenylpropenal, DC72

Phenylsulfonylethene, DC29

(Phenylthio)benzene, DD77

S-Phenyl thiophenol, DD77

3-Phenyl-5-(4-tolyl)-1,2-dithiolylium, perchlorate, DF13

N-Phenyltrichloroacetamide, DC10

5-Phenyl-1,2,3-trihydro-1,4-benzodiazepin-2-one, DE87

Phenyl vinyl sulfone, DC29

3-(N-Phthalimido)glutarimide, DE01

Phthalocyanine, DH57

Phytomenadione, DG66

β-Picoline hemichrome chloride, DH21

γ-Picoline hemichrome chloride, DH22

Potassium ethyldithiocarbonate, DA17

Potassium ethylxanthate, DA17

Potassium ethylxanthogenate, DA17

Proline-N-dithiocarboxylic acid diammonium salt, DB45

1-Propanearsonic acid, DA25

1,1'-Propano-2,2'-bipyridylium bromide, DE12

2-Propenal, DA15

Propenamide, DA18

Propenoic acid amide, DA18

(1-Propenyl-2-carbethoxy)-2-cyanoethylene, DC85

Propylarsonic acid, DA25

TABLE IV. Compounds Included in Table I

Protoporphyrin(IX) trans-cyanopyridine-iron(III), DH12

Protoporphyrin(IX) dimethyl ester, DG79

Protoporphyrin(X) dimethyl ester cobalt-(II), DG74

Protoporphyrin(IX) dimethyl ester copper(II), DG75

Protoporphine(IX) dimethyl ester, DG79

Protoporphyrin(IX) iron(III)chloride, DG71

Provitamin D_3, DG52

Purine, DA59

Purine-2,6,8-triol, DA61

Pyrazine, DA31

1H-Pyrazolo[3,4-d]pyrmidin-4-ol, DA60

Pyridazine, DA32

Pyridine, DA63

3-Pyridinecarboxylic acid, DA90

(Pyridine hemichrome) chloride, DH19

Pyridoxal, DC33

Pyridoxal-5-monophosphoric acid ester, DC36

Pyridoxal 5-phosphate, DC36

4-Pyridoxic acid, DC34

Pyridoxine, DC40

Pyridoxol, DC40

Pyrimidine, DA33

Pyrithyldione, DC87

Pyrocatechol dimethyl ether, DC39

Pyrene, DH44

Pyridoxamine, DC42

Quinhydrone, DD76

Quinone, DA86, DH59

Quinone oxmine, DA92

Quinuclidine, DB95

Resorcinol, DB08

Retinal, DG17

Retinene, DG17

Retinoic acid, DG18

Retinol, DG19

Retinyl acetate, DG34

Retinyl methyl ether, DG28

Retinyl palmitate, DH08

Riboflavin, DF74

1-β-D-Ribofuranosylcytosine, DC88

Rubrene, DH13, DH35, DH59, DH60, DH61

Salicyaldehyde, DB70

9,10-Secocholesta-5,7,10(19)-trien-3β-ol, DG53

9,10-Secoergosta-5,7,10(19),22-tetraen-3β-ol, DG57

9,10-Secoergosta-5(10),6,8(14),22-tetraen-3β-ol, DG59

Sedulon, DC90

Serenid-D, DE81

Sodium 9,10-anthracenedione-1-sulfonate, DE21

Sodium 1-anthraquinonesulfonate, DE21

Sodium 9,10-anthraquinone-1-sulfonate, DE21

Sodium cyclopentadienetricarbonyl-molybdate(0), DC07

Sodium cyclopentadienyldicarbonyl-ruthenate(0), DB64

Sodium (cyclopentadienyl)tricarbonyl-chromate(0), DC06

Sodium dibutyldithiocarbamate, DC95

Sodium diethyldithiocarbamate, DA74

Sodium diethyldithiocarboxylate, DA74

Sodium 1,2-dihydroxy-9,10-anthracene-dione 3-sulfonate, DE22

Sodium 1,2-dihydroxyanthraquinone-3-sulfonate, DE22

Sodium 1,2-dihydroxy-9,10-anthraqui-none-3-sulfonate, DE22

Sodium dimethyldithiocarbamate, DA20

Sodium 7,8-dimethyl-10-ribitylisoal-loxazine 5'-phosphate, DF73

Sodium dipropyldithiocarbamate, DB96

Sodium flavin mononucleotide, DF73

Sodium formaldehydesulfoxylate, DA04

Sodium hydroxymethanesulfinate, DA04

Sodium 2-methylphenoxide, DB73

Sodium pentamethylenedithiocarbamate, DB26

Sodium phthalocyaninecopper(II)-2-sulfonate, DG68

Sodium phthalocyanine-2-sulfonate, DG69

Sodium 1-piperidinecarbodithioate, DG26

Sodium 1-piperidinecarbodithiocarboxylate, DB26

Sodium 1-pyrrolidinecarbodithioate, DA70

Sodium 1-pyrrolidinedithiocarboxylate, DA70

Sodium riboflavin phosphate, DF73

Sodium riboflavin 5'-phosphate, DF73

Sodium tetramethylenedithiocarbamate, DA70

Sodium vitamin B_2 phosphate, DF73

Spiro(cyclohexane-1,9'-[1,5]diazabicyclo[3.3.1]nonane, DD94

Spiro[cyclohexane-1,8'-[1,5]diazabicyclo[3.2.1]octane, DD58

Spiro[cyclohexane-1,2'-[1,3]diazatricyclo[3.3.1.1^{3,7}]decane, DE18

Stearoyl peroxide, DH09

Stesolid, DF09

trans-Stilbene oxide, DE40

Strychnidin-10-one, DG24

Strychnine, DG24

Styrene, DC25

Styrenedibromide, DE38, DH42, DH44, DH45, DH46, DH49, DH51, DH52, DH60

2-Sulfophthalocyanine sodium salt, DG69

Terephthalonitrile, DC03

Tetraazaadamantane, DB36

1,3,5,7-Tetraazatricyclo[3.3.1.1^{3,7}]decane, DB36

1,3,6,8-Tetraazatricyclo[4.4.1.1^{3,8}]dodecane, DC51

Tetrabutylammonium N-methyl-2,3,7,8-12,13,17,18-octaethylporphyrin, DH27

1,1,2,2-Tetra(carbethoxy)ethene, DE60

Tetrachloromethane, DA00

Tetraethyl ethenetetracarboxylic acid, DE60

1,2,4,5-Tetraethylperhydro-s-tetrazine, DD42

5,6,7,8-Tetrahydro-4a,8-dimethyl-2(4aH)naphthalenone, DD85

Tetrahydro-1,3-dimethyl-1H,3H-[1,3,4]-oxadiazolo[3,4-a]pyridazine DC50

1,2,3,6-Tetrahydro-2,6-dioxo-4-pyrimidinecarboxylic acid, DA58

Tetrahydro-N-methyl-N-Nitroso-3-thiophenamine 1,1-dioxide, DA76

Tetrahydro-1,4-oxazine, DA48

Tetrahydro-2H-1,4-oxazine, DA48

1,2,4,5-Tetraisopropylperhydro-s-tetrazine, DE68

Tetrakis[carbonyl-(η-cyclopentadienyl)-iron(I)], DG37

Tetrakis[carbonyl-(η-cyclopentadienyl)-iron] hexafluorophosphate, DG36

Tetrakis(π-cyclopentadienyliron(I)-carbonyl), DG37

Tetrakis(π-cyclopentadienyliron carbonyl) hexafluorophosphate, DG36

N,N,N',N'-Tetramethylaminomethanamidine, DA80

Tetramethylammonium bis[π-μ-trimethylene-(3)-1,2-dicarbaundecaboranyl-(9)]cobaltate(III), DE69

Tetramethylammonium bis[π-μ-1,2-trimethylene-(3)-1,2-dicarbaundecaboranyl(9)nickelate(III), DE70

Tetramethylammonium bis[π-μ-1,2-trimethylene-(3)-1,2-dicarbolly]-cobaltate(III), DE69

Tetramethylammonium bis[π-μ-1,2-trimethylene-(3)-1,2-dicarbolly]-nickelate(III), DE70

2,7,12,18-Tetramethyl-13,17-bis(3-propionic acid)-3,8-divinylporphinatobis(3-methylpyridine)iron-(III) chloride, DH21

2,7,12,18-Tetramethyl-13,17-bis(3-propionic acid)-3,8-divinylporphinatobis(4-methylpyridine)iron-(III) chloride, DH22

2,7,12,18-Tetramethyl-13,17-bis(3-propionic acid)-3,8-divinylporphinato-trans-(cyanopyridine)iron-(III), DH12

Tetramethyldiazane, DA50

2,3,5,6-Tetramethyl-1,4-dihydroxybenzene, DD30

TABLE IV. Compounds Included in Table I

2,7,12,18-Tetramethyl-13,17-di(3-propionic acid)-3,8-divinylporphinatodicyanoiron(III), DG73

2,7,12,18-Tetramethyl-13,17-di(3-propionic acid)-3,8-divinylporphinatodipyridineiron(III) chloride, DH19

2,7,12,18-Tetramethyl-13,17-di(3-propionic acid)-3,8-divinylporphinatoiron(III) chloride, DG71

1,3,5,8-Tetramethyl-2,4-divinylporphine-6,7-dipropionic acid ferrichloride, DG71

N,N,N',N'-Tetramethylethylenediamine, DB46

s-Tetramethylguanidine, DA80

1,2-cis-3,6-Tetramethylhexahydropyridazine, DC52

1,2-trans-3,6-Tetramethylhexahydropyridazine, DC53

Tetramethylhydrazine, DA50

2,3,5,6-Tetramethylhydroquinone, DD30

N,N,N',N'-Tetramethylmethylenediamine, DA81

1,2-cis-3,6-Tetramethylperhydropyridazine, DC52

1,2-trans-3,6-Tetramethylperhydropyridazine, DC53

1,2,4,5-Tetramethylperhydro-s-tetrazine, DB47

N,N,N',N'-Tetramethyl-4-phenylenediamine, DD32, DH32, DH33, DH34, DH35

N,N,N',N'-Tetramethyl-1,6-pyrenediamine, DG12

2,4,6,8-Tetramethyl-2,4,6,8-tetraazabicyclo[3.3.0]octane, DC55

5,7,12,14-Tetramethyl-1,4,8,11-tetraazacyclotetradeca-2,5,7,9,12,14-hexaenecopper(II), DE55

5,7,12,14-Tetramethyl-1,4,8,11-tetraazacyclotetradeca-2,4,7,9,12,14-hexaenecopper(I) tetrafluoroborate, DE53

5,7,12,14-Tetramethyl-1,4,8,11-tetraazacyclotetradeca-2,4,7,9,12,14-hexaenenickel(II), DE58

5,7,12,14-Tetramethyl-1,4,8,11-tetraazacyclotetradeca-2,5,7,9,12,14-hexaenenickel(I) tetrafluoroborate, DE54

5,7,12,14-Tetramethyl-1,4,8,11-tetraazacyclotetradeca-5,7,12,14-tetraenecopper(II), DE61

5,7,12,14-Tetramethyl-1,4,8,11-tetraazacyclotetradeca-5,7,12,14-tetraenenickel(II), DE62

5,7,12,14-Tetramethyl-1,4,8,11-tetraazacyclotetradeca-5,7,12,14-tetraenezinc(II), DE63

Tetranitromethane, DA01

Tetraphenylethylene oxide, DG49

Tetraphenylhydrazine, DG38

5,6,11,12-Tetraphenylnaphthacene, DH13, DH35, DH59, DH60, DH61

Tetraphenyloxirane, DG49

5,10,15,20-Tetraphenylporphinatocobalt(II), DH17

$\alpha,\beta,\gamma,\delta$-Tetraphenylporphine, DH18, DH41, DH55, DH63, DH64

5,10,15,20-Tetraphenylporphine, DH18, DH41, DH55, DH63, DH64

$\alpha,\beta,\gamma,\delta$-Tetraphenylporphinecobalt(II)- DH17

1,2,3,3-Tetraphenylpropenone, DG51

1,3,6,8-Tetraphenylpyrene, DH11, DH34, DH58

Tetrasodium iron phthalocyanine-4,4',4'',4'''-tetrasulfonate, DG67

Tetrasodium phthalocyanineiron-2,9,16,23-tetrasulfonate, DG67

Tetrasodium 5,10,15,20-tetrakis(4-sulfonatophenyl)porphine, DH16

Tetrasodium meso-tetra(p-sulfonatophenyl)porphinatocobalt(II), DH15

Tetrasodium 5,10,15,20-tetra(4-sulfonatophenyl)porphinatocobalt(II), DH15

Tetrasodium meso-tetra(4-sulfonatophenyl)porphine, DH16

Thalidomide, DE01

Thianthracene, DD67, DH36, DH37

Thianthrene, DD67, DH36, DH37

Thiobenzoic acid, DB67

4-Thiocyanatomethylnitrobenzene, DC14

Thioglycolic acid, DA07

Thiophenol, DB09

Tocol, DG50

α-Tocopherol, DG63

β-Tocopherol, DG60

γ-Tocopherol, DG61

δ-Tocopherol, DG54

3-Toluidine-4-sulfonic acid, DB79

4-Toluidine-2-sulfonic acid, DB81

4-Toluidine-3-sulfonic acid, DB80

4-Tolunitrile, DC21

1-o-Tolyl-azo-2-naphthol, DF54

α-Tolylcyclopentadienyltricarbonyl-
molybdenum(II), DE86

s-Triazine, DA12

1,3,5-Triazine, DA12

2,2,2-Tribromoethyl benzoate, DC62

Tricarbonyl-η-cyclopentadienylmolyb-
denum(I)dicarbonyl-η-cyclopentadienyl-
iron(II), DE75

Tricarbonyl-η-cyclopentadienylmolyb-
denum(I)dicarbonylcyclopentadienyl-
iron(II), DE75

Trichloroacetanilide, DC10

N-(Trichloroacetyl)aniline, DC10

N-(Trichloroacetyl)-3-methyl-
aniline, DC66

N-(Trichloroacetyl)-4-methyl-
aniline, DC67

2,2,2-Trichloroethyl benzoate, DC63

Triethylamine, DB43

Triethylenediamine, DB31

3-Trifluoromethyl-ω-diazoacetophenone,
DC57

7-Trifluoromethyl-1,2,3-trihydro-5-
phenyl-1,4-benzodiazepin-2-one, DF04

1,2,3-Trihydro-7-methoxy-5-phenyl-1,4-
benzodiazepin-2-one, DF28

1,2,3-Trihydro-7-methylthio-5-phenyl-
1,4-benzodiazepin-2-one, DF27

1,2,3-Trihydro-7-nitro-5-phenyl-1,4-
benzodiazepin-2-one, DE84

4-Trimethylammonioazobenzene perchlorate, DE95

4-Trimethyl(azobenzene)phenylammonium
perchlorate, DE95

1,1'-Trimethylene-2,2'-bipyridylium
bromide, DE12

1,2,3-Trimethylperhydropyridazine,
DC02

1,3,5-Trimethylperhydro-1,3,5-tri-
azine, DB44

1,2,2-Trimethyl-1-propyl)ferrocene,
DF39

1,2,3-Trimethyl-1,2,3,6-tetrahydro-
pyridazine, DC00

5,7,8-Trimethyltocol, DG63

2,5,8-Trimethyl-2-(4,8,12-trimethyl-
tridecyl)-6-chromanol, DG60

Trinitromethane, DA02

Triphenylarsine oxide, DF83

1,1,3-Triphenylpropenone, DG21

cis-1,2,3-Triphenyl-2-propenone, DG22

trans-1,2,3-Triphenyl-2-propenone,
DG23

Triphenyltin hydroxide, DF90

Tris(bipyridine)iron(II) sulfate, DG65

Trisbipyridineruthenium(II) dichloride,
DG64, DH40, DH56

Tris(carbonyl-η-cyclopentadienylcobalt), DF84

Tris(π-cyclopentadienylcarbonylcobalt),
DF84

Tris(η-cyclopentadienyl)dicarbonyl-
trinickel, DF62

Trisodium 5,10 15,20-tetrakis(4-
sulfonatophenyl)porphinatocobalt-
(III), DH14

Trisodium meso-tetra(p-sulfonatophenyl)porphinatocobalt(III), DH14

Tris[tricarbonylcobalt(I)]methyl
chloride, DC99

Ubiquinone-10, DH30

Uracil-6-carbonylic acid, DA58

Uric acid, DA61

Valium, DF09

Veratrole, DC39

Vinylacrylonitrile, DA62

Vinylbenzene, DC25

1-Vinyl-2-pyrrolidone, DB24

Vitamin A, DG19

Vitamin A acetate, DG33

Vitamin A aldehyde, DG17

Vitamin A palmitate, DH08

Vitamin B_2, DF74

Vitamin B_6, DC40

Vitamin D_2, DG57

Vitamin D_3, DG53

Vitamin K_1, DG66

Vitamin K_3, DD48

Vival, DF09

TABLE IV. Compounds Included in Table I

Wood sugar, DA77

Wurster's Blue, DD32, DH32, DH33, DH34, DH35

2,6-Xylenol, DC38

D-Xylose, DA77

Zinc(II)protoporphyrin(IX) dimethyl ester, DG77

TABLE V.
FUNCTIONAL-GROUP INDEX

This index divides the 731 compounds that appear in Table I into 107 groups and subgroups of chemically related compounds. It is preceded by a list of these groups which begins on the following page.

So that none of the final categories would contain more than about 50 compounds, many main groups have been divided and some have been subdivided. Frequently the division was made, first according to the molecular frame (e.g., alicyclic, aliphatic, aromatic or benzenoid, and heterocyclic compounds) and then according to the number of substituents present, but for halogenated compounds the first division was according to the kind of halogen atom, and organometallic ones were divided according to the metallic atom.

In each category there appear, in alphabetical order, the names of the compounds, together with the code numbers assigned to them in Table I, in which the functional group named is, or may be, electroactive; but compounds in which that functional group is present but not electroactive are omitted. Compounds for which the products of the half-reaction are unknown are listed under each of the groups they contain.

TABLE V. Functional-Group Index

ACETYLENES, see HYDROCARBONS and UNSATURATED COMPOUNDS and their SALTS

ACIDS, CARBOXYLIC (see also KETONES and UNSATURATED COMPOUNDS)
    Aliphatic
    Aromatic
    Heterocyclic and Others

ACID SALTS, see ACIDS, CARBOXYLIC

ACIDS, SULFONIC, see SULFUR COMPOUNDS

ALCOHOLS, see HYDROXY COMPOUNDS

ALDEHYDES (see also UNSATURATED COMPOUNDS)
    Aliphatic and Alicyclic
    Aromatic
    Heterocyclic

AMIDES
    Aliphatic, not Aryl-substituted
    Aliphatic, Aryl-substituted
    Aliphatic, Heterocyclic-substituted
    Heterocyclic, benzodiapinones
    Heterocyclic, Others
    Others

AMINES (see also NITROSO COMPOUNDS)
    Aliphatic and Alicyclic
    Anilines
        N-Unsubstituted
        N-Substituted
    Aromatic, Others
    Cyclic
    Heterocyclic, Benzodiazepinones
    Heterocyclic, Others
    Quaternary

AMINE SALTS, see AMINES

AMINO ACIDS, see ACIDS, CARBOXYLIC

ANHYDRIDES, CARBOXYLIC

ARSENIC, see PHOSPHORUS

AZINES, see IMINES

AZO COMPOUNDS

BENZOQUINONES, see QUINONES

CARBAMATES, see ESTERS, NONCARBOXYLIC

CARBOHYDRATES and SUGARS

CARBOXYLIC ACIDS, see ACIDS, CARBOXYLIC

DIAZO COMPOUNDS

DYES

EPOXIDES, see ETHERS

ESTERS, CARBOXYLIC

ESTERS, NONCARBOXYLIC
    Carbamates, Thiocarbamates and Xanthates
    Nitrates
    Phosphates
    Thioacetates

ETHERS and EPOXIDES

GUANIDINES

HALOGEN COMPOUNDS
    Aliphatic and Alicyclic
        Bromo
        Chloro
        Fluoro
        Iodo
        Mixed Halogens
    Aromatic
        Bromo
        Chloro
        Fluoro
        Iodo
    Heterocyclic, Benzodiazepinones
        Bromo
        Chloro
        Fluoro
        Iodo
    Heterocyclic, Others
    Steroids

HETEROCYCLIC COMPOUNDS
    Nitrogen
        Pyridines
        One Ring-Nitrogen Atom, Monocyclic (except Pyridines)
        One Ring-Nitrogen Atom, Polycyclic
        Two Ring-Nitrogen Atoms, Monocyclic
        Benzodiazepines
        Two Ring-Nitrogen Atoms, Polyclic (except Benzodiazepines)
        Three Ring-Nitrogen Atoms
        Four Ring-Nitrogen Atoms, Tetraazacyclotetradecanes
        Four Ring-Nitrogens, Porphines
        Four Ring-Nitrogens, Others
        Six or more Ring-Nitrogens
    Oxygen
    Sulfur

HYDRAZINES and HYDRAZIDES

HYDROCARBONS
    Aromatic
        One or more rings, Uncondensed

Two Condensed Rings
  Three or more Condensed Rings
  Aromatic Compounds in Complexes (see also ORGANOMETALLICS)

HYDROXY COMPOUNDS (see also QUINONES)
  Aliphatic and Alicyclic
  Aromatic
    Phenols
    Two or more Rings
  Heterocyclic
    Carbohydrates
    Others
  Hydroxylamines

HYDROXY SALTS, see HYDROXY COMPOUNDS

IMIDES, see HETEROCYCLIC COMPOUNDS

IMINES and OTHER AZOMETHINES (see also OXIMES)
  Aliphatic and Alicyclic
  Aromatic
  Cyclic

ISOTHIOCYANATES and THIOCYANATES

KETONES (see also HETEROCYCLIC COMPOUNDS; QUINONES and HYDROQUINONES; and UNSATURATED COMPOUNDS
  Dialkyl Ketones
  Aryl-Alkyl Ketones
  Alkyl-Heterocyclic Ring Ketones
  Diaryl Ketones
  Diheterocyclic Ring Ketones
  Alicyclic Ketones
  Heterocyclic Ketones

LACTAMS, see LACTOMES

LACTONES and LACTAMS (see also HETEROCYCLIC COMPOUNDS)

METHOXIMES, see OXIMES

NITRILES
  Aliphatic and Alicyclic
  Benzonitriles
  Heterocyclic

NITRO COMPOUNDS
  Aliphatic and Alicyclic
  Nitrobenzenes
  Nitroaromatics with Condensed Rings
  Nitro Derivaties of N-Heterocycles
  Nitro Derivatives of O-Heterocycles
  Nitro Derivatives of S-Heterocycles

NITROSO and NITROSYL COMPOUNDS

ORGANOMETALLICS
  Al
  Ag
  B
  Ca
  Cd
  Co
  Cr
  Cu
  Fe
  Ga
  Ge
  Hg
  In
  Ir
  Mg
  Mn
  Mo
  Ni
  Pb
  Pd
  Rh
  Ru
  Sb
  Sc
  Si
  Sn
  Ti
  Tl
  V
  Zn

N-OXIDES

OXIMES and ALKOXIMES

PHENOLS, see HYDROXY COMPOUNDS

PEROXIDES, HYDROPEROXIDES and PERACIDS

PHOSPHORUS and ARSENIC COMPOUNDS (see also ESTERS, NONCARBOXYLIC, and SULFUR COMPOUNDS)

QUARTERNARY COMPOUNDS, see AMINES

QUINONES and HYDROQUINONES (see also HYDROXY COMPOUNDS)
  Benzoquinones
  Polycyclic Quinones
  Heterocyclic Quinones

SUGARS, see CARBOHYDRATES

SULFATES, see ESTERS, NONCARBOXYLIC

SULFIDES, see SULFUR COMPOUNDS

SULFONAMIDES, see SULFUR COMPOUNDS

SULFONATES, see SULFUR COMPOUNDS

TABLE V. Functional-Group Index

SULFONES, see SULFUR COMPOUNDS

SULFONIC ACIDS, see SULFUR COMPOUNDS

SULFUR COMPOUNDS (see also HETERO-
  CYCLIC COMPOUNDS, ORGANOMETALLICS,
  and ESTERS, NONCARBOXYLIC)
  Thiols
  Sulfides, Thioethers, and Sulfonium
    Ions
  Disulfides
  Dithiolium Compounds
  Thiocarbamates, Dithiocarbamates and
    Xanthates
  Thiones, Thioamides, Thiobarbiturates,
    Thiopyrimidines, and Thiopurines
  Sulfonamides
  Sulfuric and Thiosulfuric Acid
    Derivatives, Sulfonates, and
    Sulfinates
  Organometallic Complexes
  Sulfoxides, Sulfones and S-dioxides

THIOCYANATES, see ISOTHIOCYANATES

THIOLS, see SULFUR COMPOUNDS

THIOUREAS, see UREAS

UNSATURATED COMPOUNDS (see also
  HYDROCARBONS, and QUINONES and
  HYDROQUINONES)
  Ethylenic
    Double bond Isolated from Nonhydro-
      carbon Functional Groups
    α,β-Unsaturated Aldehydes
    α,β-Unsaturated Ketones
    α,β-Unsaturated Sulfones and Sulf-
      oxides
    α,β-Unsaturated Acids, Thioacids,
      Esters and Amides
    α,β-Unsaturated Nitriles
    Unsaturated Amines, Halogens,
      Alcohols, Ethers, and Nitro-
      compounds (see also
      IMINES
  Alicyclic
    Steroids
  Heterocyclic Compounds (Nonaromatic)
    Porphines
  Acetylenic

UREAS and THIOUREAS (see also HETERO-
  CYCLIC COMPOUNDS and KETONES, Heterocyclic)

ACETYLENES, see HYDROCARBONS and UNSATURATED COMPOUNDS

ACIDS, CARBOXYLIC and their SALTS (see also HYDROGEN-ION REDUCTION, KETONES, and UNSATURATED COMPOUNDS)

Aliphatic

3-bromo-2-chloro-3-phenylpropanoic acid, DC64

tert-butyl peroxydodecanoate, DF48

[3-(4-chlorophenyl)-3-(2-pyridyl-1-propyl]dimethylammonium maleate, DG16

cysteine, DA23

L-cystine, DB35

trans-di(4-acetylpyridine)hematoporphyrincobalt(III), DH25

trans-di(4,4'-bipyridine)hematoporphyrincobalt(III), DH28

[2,3-dichloro-4-(2-methylene-1-oxobutyl)phenoxy]acetic acid, DE06

trans-dicyanoprotoporphyrin(IX)-iron(III), DG72

didecanoyl peroxide, DG20

didodecanoyl peroxide, DG42

trans-di(ethylnicotinate)hematoporphyrincobalt(III), DH26

dihexadecanoyl peroxide, DG70

trans-di(2-methylpyridine)hematoporphyrincobalt(III), DH23

trans-di(4-methylpyridine)hematoporphyrincobalt(III), DH24

trans-di(3-methylpyridine)protoporphyriniron(III) chloride, DH21

trans-di(4-methylpyridine)protoporphyriniron(III) chloride, DH22

dinonanoyl peroxide, DG00

dioctadecanoyl peroxide, DH09

trans-di(4-phenylpyridine)hematoporphyrincobalt(III), DH29

trans-dipyridinehematoporphyrincobalt(III) chloride, DH20

trans-dipyridylprotoporphyrin(IX)-iron(III) chloride, DH19

ditetradecanoyl peroxide, DG62

hematoporphyrincobalt(III), DG73

lithium 4-tert-butylcyclohexylideneacetate, DD89

lithium 2-hexynoate, DB14

2-methyl-2-nitratopropanoic acid, DA43

2-nitratopropanoic acid, DA19

oxalic acid, DA05

phenylglyoxalic acid, DC15

protoporphyrin(IX)-trans-cyanopyridineiron(III), DH12

protoporphyrin(IX)iron(III) chloride, DG71

retinoic acid, DG18

thioglycolic acid, DA07

Aromatic

benzoic acid, DB68

9-(2-carboxyphenyl)-4,5-dibromo-6-hydroxy-3H-xanthen-2-one, DG06

9-(2-carboxyphenyl)-4,5-dichloro-6-hydroxy-3H-xanthen-3-one, DG07

9-(2-carboxyphenyl)-4,5-diiodo-6-hydroxy-3H-xanthen-3-one, DG08

thiobenzoic acid, DB67

Heterocyclic and others

Chlorazepam, DF12

7-chloro-2,3-dihydro-1-methyl-5-phenyl-1H-1,4-benzodiazepin-2-carboxylic acid, DF57

diammonium proline-N-dithiocarboxylate, DB45

2,6-dimercapto-1,4-thiopyrone-3-carboxylic acid, DA87

3-hydroxy-5-(hydroxymethyl)-2-methyl-4-pyridinecarboxylic acid, DC34

N-(2-hydroxymethyl-3,4-dihydroxy-5-tetrahydrofuryl)-3-carboxypyridine-(1H)2-one, DD52

2-hydroxynicotinic acid, DA93

nicotinic acid, DA90

5-nitro-2-furoic acid, DA55

uracil-6-carboxylic acid, DA58

TABLE V. Functional-Group Index

ACID SALTS, see ACIDS

ACIDS, SULFONIC, see SULFUR COMPOUNDS

ALCOHOLS, see HYDROXY COMPOUNDS

ALDEHYDES (see also UNSATURATED COMPOUNDS)

Aliphatic and Alicyclic

acrolein, DA15

cinnamaldehyde, DC72

crotonaldehyde, DA38

3,3-diphenylprop-2-eneal, DE89

formaldehyde sodium sulfoxylate, DA04

3-formylcyclohexene, DB86

D-glucose, DB37

2-methyl-2-nitratopropanal, DA42

phenylacetaldehyde, DC28

vitamin A aldehyde, DG17

Aromatic

benzaldehyde, DB66

4-chlorobenzaldehyde, DB60

3-nitrobenzaldehyde, DB62

4-nitrobenzaldehyde, DB63

salicylaldehyde, DB70

Heterocyclic

7-chloro-2,3-dihydro-1-methanoyl-5-phenyl-1H-1,4-benzodiazepine, DF06

5-nitro-2-furaldehyde, DA54

5-nitro-2-furaldehyde diacetate, DB82

4-nitro-2-thienylcarboxaldehyde, DA52

5-nitro-2-thienylcarboxaldehyde, DA53

pyridoxal, DC33

pyridoxal 5-phosphate, DC36

AMIDES

Aliphatic, non Aryl-substituted

acrylamide, DA18

N-benzylchloroacetamide, DC81

N-nitrososarcosinamide, DA24

oxamide, DA06

Aliphatic, Aryl-substituted

6-acetamido-2-phenylchromanone, DF60

7-acetamido-2,3-dihydro-5-phenyl-1H-1,4-benzodiazepin-2-one, DF61

3'-bromo-2-chloroacetanilide, DC16

4'-bromo-2-chloroacetanilide, DC17

chloroacetanilide, DC27

2,2-dichloroacetanilide, DC19

2,3'-dichloroacetanilide, DC20

2',3'-dimethyl-2-chloroacetanilide, DD23

3',4'-dimethyl-2-chloroacetanilide, DD24

4'-iodo-2-chloroacetanilide, DC18

3'-methyl-2,2-dichloroacetanilide, DC74

4'-methyl-2,2-dichloroacetanilide, DC75

3'-methyl-2,2,2-trichloroacetanilide, DC66

4'-methyl-2,2,2-trichloroacetanilide, DC67

s-tetramethylguanidine, DA80

trichloroacetanilide, DC10

Aliphatic, Heterocyclic-substituted

12-acetylbenzo[a]phenothiazine 7-dioxide, DF79

7-acetylbenzo[c]phenothiazine 12-dioxide, DF78

12-acetylbenzo[a]phenothiazine 7-oxide, DF77

1-acetyl-7-chloro-2,3-dihydro-5-phenyl-1H-1,4-benzodiazepine, DF56

7-chloro-1-acetamido-2,3-dihydro-5-phenyl-1H-benzodiazepine, DF64

7-chloro-2-(N-methylacetamido)-5-phenyl-3H-1,4-benzodiazepine 4-N-oxide, DF86

AMIDES (cont.)

Heterocyclic, benzodiazepines

7-acetamido-2,3-dihydro-5-phenyl 1H-1,4-benzodiazepin-2-one, DF61

7-acetyl-1,3-dihydro-5-phenyl-2H-1,4-benzodiazepin-2-one, DF55

7-bromo-1,3-dihydro-5-phenyl-2H-1,4-benzodiazepin-2-one, DE79

2-carbamoyl-7-chloro-2,3-dihydro-1-methyl-5-phenyl-1H-1,4-benzodiazepine, DF63

Chlorazepam, DF12

7-chloro-5-(2-chlorophenyl)-2,3-dihydro-1H-1,4-diazepin-2-one, DE73

7-chloro-5-(2-chlorophenyl)-2,3-dihydro-3-hydroxy-2H-1,4-benzodiazepin-2-one, DE74

7-chloro-1,3-dihydro-5-(2-fluorophenyl)-2H-1,4-benzodiazepin-2-one, DE72

7-chloro-1,3-dihydro-5-(1H-indol-1-yl)-2H-1,4-benzodiazepin-2-one, DF51

7-chloro-1,3-dihydro-5-(2-methoxyphenyl)-2H-1,4-benzodiazepin-2-one, DF10

7-chloro-2,3-dihydro-1-(N-methylcarbamoyl)-5-phenyl-1H-1,4-benzodiazepine, DF65

7-chloro-1,3-dihydro-5-(2-methylphenyl)-2H-1,4-benzodiazepin-2-one, DF07

7-chloro-1,3-dihydro-5-(1-methyl-1H-pyrazol-5-yl)-2H-1,4-benzodiazepin-2-one, DE03

7-chloro-1,3-dihydro-5-phenyl-2H-1,4-benzodiazepin-2-one, DE80

7-chloro-1,3-dihydro-5-(2-pyridinyl)-2H-1,4-benzodiazepin-2-one, DE33

7-chloro-1,3-dihydro-5-(2-thiazolyl)-2H-1,4-benzodiazepin-2-one, DD62

Diazepam, DF09

1,3-dihydro-7-fluoro-5-phenyl-2H-1,4-benzodiazepin-2-one, DE83

1,3-dihydro-7-methoxy-5-phenyl-2H-1,4-benzodiazepin-2-one, DF28

1,3-dihydro-7-methyl-5-phenyl-2H-1,4-benzodiazepin-2-one, DF26

1,3-dihydro-7-methylthio-5-phenyl-2H-1,4-benzodiazepin-2-one, DF27

1,3-dihydro-5-phenyl-2H-1,4-benzodiazepin-2-one, DE87

1,3-dihydro-5-phenyl-7-trifluoromethyl-2H-1,4-benzodiazepin-2-one, DF04

7,9-diiodo-1,3-dihydro-5-phenyl-2H-1,4-benzodiazepin-2-one, DE76

Nitrazepam, DE84

Oxazepam, DE81

Heterocyclic, others

allopurinol, DA60

5,5-bis(3-nitrophenyl)-2,4-imidazolidinedione, DE77

cytidine, DC88

cytidine monosphosphate, DC89

cytidylyl (3'→5') cytidine, DF98

cytidylyl (3'→5') guanidine, DG05

cytidylyl (3'→5') uridine, DF97

cytosine, DA36

3,3-diethyl-2,4-dioxopiperidine, DC90

3,3-diethyl-5-methyl-2,4-dioxopiperidine, DD37

3,3-diethyl-2,4-(1H,3H)pyridinedione, DC87

3,4-dihydro-2-methyl-4-oxo-3-o-tolylquinazoline, DF25

1,3-dimethylbenzimidazolone-4,7-quinone, DC68

1,3-dimethylbenzimidazolone-5,6-quinone, DC69

5-ethyl-5-(1-methylpropyl)-2-thiobarbituric acid, DD33

5-ethyl-5-(3-nitrophenyl)barbituric acid, DD80

3-ethyl-3-phenyl-2,6-dioxopiperidine, DE13

2-hydroxynicotinic acid, DA93

2-hydroxy-3-nitropyridine, DA56

2-hydroxy-5-nitropyridine, DA57

isatin 3-(O-ethyloxime), DD15

nicotinamide, DB04

nicotinamide adenine dinucleotide phosphate, reduced, DG27

TABLE V. Functional-Group Index

AMIDES (cont.)
  Heterocyclic, others (cont.)
    nicotinamide adenine dinucleotide, reduced, DG26
    pyridoxamine, DC42
    Thalidomide, DE01
    uracil-6-carboxylic acid, DA58
    uric acid, DA61
    1-vinyl-2-pyrrolidone, DB24
  Others
    6-chloro-3-chloromethyl-3,4-dihydro-2-methyl-7-sulfamoyl-1,2,4-benzothiadiazine 1,1-dioxide, DC84
    6-chloro-3,4-dihydro-7-sulfamyl-2H-1,2,4-benzothiadiazine 1,1-dioxide, DB74

AMINES (see also NITROSO COMPOUNDS)
  Aliphatic and Alicyclic
    1-aminobutane, DA49
    benzyloxycarbonylmethyl(dimethyl)-(2-dimethylaminoethyl)ammonium chloride, DE97
    cyclohexylhydroxylamine, DB38
    cysteine, DA23
    L-cystine, DB35
    diethylaminoethyl methacrylate, DD38
    N,N,N',N'-tetramethylethylenediamine, DB46
    N,N,N',N'-tetramethylmethylenediamine, DA81
    triethylamine, DB43

  Anilines
    N-Unsubstituted
      3-amino-4-hydroxyphenylarsonic acid, DB20
      2-amino-4-methylbenzenesulfonic acid, DB79
      2-amino-5-methylbenzenesulfonic acid, DB80
      5-amino-2-methylbenzenesulfonic acid, DB81
      4-aminophenol, DB15
      2-aminophenylarsonic acid, DB18
      4-aminophenylarsonic acid, DB19

      5-(2-aminophenyl)-7-chloro-2,3-dihydro-1-methyl-1H-1,4-benzodiazepine, DF35
      4-chloroaniline, DB03
      2-mercaptoaniline, DB16
      4-mercaptoaniline, DB17
      4-methyl-2-nitroaniline, DB76
      2-naphthylamine, DD07
      3-nitroaniline, DB05
      4-nitroaniline, DB06

    N-Substituted
      x-dimethylaminoazobenzene, DE45
      4-(dimethylamino)azobenzene, DE46
      3-[4-(dimethylamino)phenyl]-5-(4-methoxyphenyl)-1,2-dithiol-1-ium perchlorate, DF93
      N,N-dimethyl-4-toluidine, DC86
      1,3-diphenylguanidine, DE11
      4-trimethylammonioazobenzene perchlorate, DE95
      Wurster's Blue, DD32, DH32, DH33, DH34, DH35

    Aromatic, others
      1,6-bis(dimethylamino)pyrene, DG12

  Cyclic
    1-azabicyclo[2.2.2]octane, DB95
    1,4-dimethylhexahydropyrazine, DB39
    1,2-dimethylhexahydropyridazine, DB40
    2-(2'-ethyl-6'-methylpiperidyl)-ethyl methacrylate, DE67
    morpholine, DA48
    1,2,4,5-tetramethylhexahydro-s-tetrazine, DB47

  Heterocyclic, Benzodiazepines
    7-acetoxy-2-methylamino-5-phenyl-3H-1,4-benzodiazepine 4-N-oxide, DF92
    7-acetyl-5-(2-fluorophenyl)-2-methylamino-3H-1,4-benzodiazepine 4-N-oxide, DF87

AMINES (see also NITROSO COMPOUNDS) (cont.)
Heterocyclic, Benzodiazepines, (cont.)

    1-amino-7-chloro-2,3-dihydro-5-phenyl-1H-1,4-benzodiazepine, DE92

    2-amino-7-chloro-5-phenyl-3H-1,4-benzodiazepine 4-N-oxide, DE85

    7-bromo-2-methylamino-5-phenyl-3H-1,4-benzodiazepine 4-N-oxide, DF20

    2-butylamino-7-chloro-5-phenyl-3H-1,4-benzodiazepine 4-N-oxide, DG04

    7-chloro-2-dimethylamino-5-phenyl-3H-1,4-benzodiazepine 4-N-oxide, DF66

    7-chloro-2-ethylamino-5-phenyl-2H-1,4-benzodiazepine 4-N-oxide, DF67

    7-chloro-5-(4-hydroxyphenyl)-2-methylamino-3H-1,4-benzodiazepine 4-N-oxide, DF24

    7-chloro-2-methylamino-5-phenyl-3H-1,4-benzodiazepine 4-N-oxide, DF22

    7-chloro-2(N-methylpropionamido)-5-phenyl-3H-1,4-benzodiazepine 4-N-oxide, DG03

    5-(2-fluorophenyl)-7-iodo-2-methylamino-3H-1,4-benzodiazepine 4-N-oxide, DF17

    2-methylamino-7-methylmercapto-5-phenyl-3H-1,4-benzodiazepine 4-N-oxide, DF72

    2-methylamino-5-phenyl-3H-1,4-benzodiazepine 4-N-oxide, DF33

    7-methyl-2-methylamino-5-phenyl-3H-1,4-benzodiazepine 4-N-oxide, DF71

Heterocyclic, others

    2-amino-3-nitropyridine, DA66

    2-amino-5-nitropyridine, DA67

    cytidine, DC88

    cytidine monophosphate, DC89

    cytidylyl (3'→5') cytidine, DF98

    cytidylyl (3'→5') guanidine, DG05

    cytidylyl (3'→5') uridine, DF97

    cytosine, DA36

    1-methyl-2-dimethylaminobenzimidazole-4,7-quinone, DD19

    1-methyl-2-dimethylaminobenzimidazole-5,6-quinone, DD20

    1-methyl-2-dimethylaminobenzimidazole-6,7-quinone, DD21

    Methylene Blue, DF38

    nicotinamide adenine dinucleotide phosphate, reduced, DG27

    nicotinamide adenine dinucleotide, reduced, DG26

Quaternary

    benzyloxycarbonylmethyl(dimethyl)-(2-dimethylaminoethyl)ammonium chloride, DE97

    [3-(4-chlorophenyl)-3-(2-pyridyl)-1-propyl]dimethylammonium maleate, DG16

    4-trimethylammonioazobenzene perchlorate, DE95

AMINE SALTS, see AMINES

AMINO ACIDS, see ACIDS, CARBOXYLIC

ANHYDRIDES, CARBOXYLIC

    didecanoyl peroxide, DG20

    didodecanoyl peroxide, DG42

    dihexadecanoyl peroxide, DG70

    dinonanoyl peroxide, DG00

    dioctadecanoyl peroxide, DH09

    ditetradecanoyl peroxide, DG62

ARSENIC, see PHOSPHORUS

AZINES, see IMINES

AZO COMPOUNDS

    azobenzene, DD72

    azobenzene-4-sulfonic acid, DD75

    x-dimethylaminoazobenzene, DE45

    4-(dimethylamino)azobenzene, DE46

    diphenylthiocarbazone, DE10

    4-hydroxyazobenzene, DD73

    4-methoxyazobenzene, DE07

    naphthyl-1,1'-azo-2-naphthol-4',6-disulfonic acid, disodium salt, DG09

    4-nitroazobenzene, DD69

TABLE V. Functional-Group Index

AZO COMPOUNDS (cont.)

    1-o-tolylazo-2-naphthol, DF54

    4-trimethylammonioazobenzene perchlorate, DE95

BENZOQUINONES, see QUINONES

CARBAMATES, see ESTERS, NON-CARBOXYLIC

CARBOHYDRATES and SUGARS

    ascorbic acid, DB23

    cytidine, DC88

    cytidine monophosphate, DC89

    cytidylyl (3'→5') cytidine, DF98

    cytidylyl (3'→5') guanidine, DG05

    cytidylyl (3'→5') uridine, DF97

    D-glucose, DB37

    N-(2-hydroxymethyl-3,4-dihydroxy-5-tetrahydrofuryl)-3-carboxypyridine-(1H)2-one, DD52

    nicotinamide adenine dinucleotide phosphate, reduced, DG27

    nicotinamide adenine dinucleotide, reduced, DG26

    riboflavin, DF74

    sodium flavin mononucleotide, DF73

    D-xylose, DA77

CARBOXYLIC ACIDS, see ACIDS, CARBOXYLIC

DIAZO COMPOUNDS

    4-biphenylyl(phenyl)diazomethane, DG02

    3-trifluoromethyl-ω-diazoacetophenone, DC57

DYES

    Methylene Blue, DF38

    Wurster's Blue, DD32

EPOXIDES see ETHERS

ESTERS, CARBOXYLIC

    7-acetoxy-2-methylamino-5-phenyl-3H-1,4-benzodiazepine 4-N-oxide, DF92

    benzyl acetate, DC82

    benzyloxycarbonylmethyl(dimethyl)-(2-dimethylaminoethyl)ammonium chloride, DE97

    bis(α-carbethoxybenzyl)mercury, DG14

    bis(methoxycarbonyl)mercury(II), DA37

    t-butyl peroxymyristate, DG01

    tert-butyl peroxynonanoate, DE20

    cobalt(II)protoporphyrin(IX) dimethyl ester, DG74

    copper(II)protoporphyrin(IX) dimethyl ester, DG75

    2,2-dichloroethyl benzoate, DC65

    diethylaminoethyl methacrylate, DD38

    diethyl isopropylidenemalonate, DD35

    trans-di(ethylnicotinate)hematoporphyrincobalt(III), DH26

    dimethyl fumarate, DB21

    dimethyl maleate, DB22

    ethyl 2-bromomercurio-2-(4-isopropylphenyl)acetate,

    ethyl α-bromomercury-(2-bromophenyl)acetate, DD11

    ethyl α-bromomercury-(3-bromophenyl)acetate, DD12

    ethyl α-bromomercury-(4-bromophenyl)acetate, DD13

    ethyl 2-bromomercury-2-(3-tert-butylphenyl)acetate, DE59

    ethyl α-bromomercury-(4-chlorophenyl)acetate, DD08

    ethyl α-bromomercury-(4-ethylphenyl)acetate, DD84

    ethyl α-bromomercury-(4-fluorophenyl)acetate, DD09

    ethyl α-bromomercury-(4-iodophenyl)acetate, DD10

    ethyl α-bromomercury-(2-methylphenyl)acetate, DD50

    ethyl α-bromomercury-(3-methylphenyl)acetate, DD51

    ethyl α-bromomercuryphenylacetate, DD18

    ethyl 4-tert-butycyclohexylideneacetate, DE65

    ethyl 2-cyano-2,4-hexadienoate, DC85

ESTERS, CARBOXYLIC (cont.)

    2-(2'-ethyl-6'-methylpiperidyl)-ethyl methacrylate, DE67

    ethyl 5-nitro-2-furoate DB72

    O-ethyl thioacetothioacetate, DB28

    glycol dimercaptopropionate, DC45

    2-iodoethyl benzoate, DC76

    manganese(III) chloride protoporphyrin(IX) dimethyl ester monohydrate, DG78

    mesoporphyrin(IX) dimethyl ester, DG80

    methyl 4-bromo-5-oxohexanoate, DB91

    methyl 3-butenoate, DA72

    methyl 4-tert-butyl-1-cyclohex-1-enoate, DD92

    methyl cinnamate, DD16

    methyl trans-cinnamate, DD17

    methyl crotonate, DA73

    methyl 2-hexynoate, DB90

    methyl 5-(mercaptomethyl)-2-furoate, DB77

    methyl 2-methyl-5-(thiocyanatomethyl)-3-furoate, DC77

    methyl 4-pentenoate, DB29

    methyl 2-propenoate, DA40

    methyl propiolate, DA34

    methyl 5-(thiocyanatomethyl)-2-furoate, DC23

    nickel(II)protoporphyrin(IX) dimethyl ester, DG76

    5-nitrofuraldehyde diacetate, DB82

    4-oxo-1-pentyl acetate, DB94

    1-phenylethyl benzoate, DE94

    protoporphyrin(IX) dimethyl ester, DG79

    retinyl acetate, DG33

    retinyl palmitate, DH08

    tetraethyl ethenetetracarboxylic acid, DE60

    2,2,2-tribromoethyl benzoate, DC62

    2,2,2-trichloroethyl benzoate, DC63

    zinc(II)protoporphyrin(IX) dimethyl ester, DG77

ESTERS, NONCARBOXYLIC

  Carbamates, Thiocarbamates and Xanthates

    ammonium pyrrolidine-1-dithiocarboxylate, DA79

    diammonium proline-N-dithiocarboxylate, DB45

    ethyl N-methyl-N-nitrosocarbamate, DA44

    potassium ethylxanthate, DA17

    sodium dibutyldithiocarbamate, DC95

    sodium diethyldithiocarbamate, DA74

    sodium dimethyldithiocarbamate, DA20

    sodium dipropyldithiocarbamate, DB96

    sodium 1-piperidinedithiocarboxylate, DB26

    sodium pyrrolidine-1-dithiocarboxylate, DA70

  Nitrates

    2-methyl-2-nitratopropanal, DA42

    2-methyl-2-nitratopropanoic acid, DA43

    2-nitratopropanoic acid, DA19

  Phosphates

    cytidine monosphosphate, DC89

    cytidylyl (3'→5') cytidine, DF98

    cytidylyl (3'→5') guanidine, DG05

    cytidylyl (3'→5') uridine, DF97

    nicotinamide adenine dinucleotide phosphate, reduced, DG27

    nicotinamide adenine dinucleotide, reduced, DG26

    pyridoxal 5-phosphate, DC36

    sodium flavin mononucleotide, DF73

  Thioacetates

    bis(O-ethyl thioacetothioacetato)-cadminum(II), DD86

    bis(O-ethyl thioacetothioacetato)-mercury(II), DD87

    bis(O-ethyl thioacetothioacetato)-zinc(II), DD88

TABLE V. Functional-Group Index

ETHERS and EPOXIDES

    benzhydryl dodecyl ether, DG46

    benzhydryl methyl ether, DE43

    benzyl tert-butyl ether, DD53

    benzyl cyclohexyl ether, DE17

    benzyl ether, DE44

    benzyl methyl ether, DC37

    3,5-bis(4-methoxyphenyl)-1,2-dithiol-1-ium perchlorate, DF59

    2-butoxy-2-methoxyethylmercuric acetate, DC93

    7-chloro-1,3-dihydro-5-(2-methoxyphenyl)-2H-1,4-benzodiazepin-2-one, DF10

    cis-2-chloromercury-3-ethoxycyclononene, DD55

    cis-2-chloromercury-3-methoxycyclodecene, DD56

    trans-2-chloromercury-3-methoxycyclotridecene, DE66

    trans-2-chloromercury-3-methoxycycloundecene, DD93

    7-chloro-2-methoxy-5-phenyl-3H-1,4-benzodiazepine 4-N-oxide, DF11

    [2,3-dichloro-4-(2-methylene-2-oxobutyl)phenoxy]acetic acid, DE06

    1,3-dihydro-7-methoxy-5-phenyl-2H-1,4-benzodiazepin-2-one, DF28

    1,2-dimethoxybenzene, DC39

    3-[4-(dimethylamino)phenyl]-5-(4-methoxyphenyl)-1,2-dithiol-1-ium perchlorate, DF93

    3,8-dimethyl-2-methoxyazocine, DD25

    trans-2,3-diphenyloxirane, DE40

    2-ethoxy-2-cyclopentenone, DB89

    ethyl 1-phenylethyl ether, DD27

    ethyl vinyl ether, DA45

    2,3,6,7,10,11-hexamethoxytriphenylene, DG40

    2,3,6,7,10,11-hexamethoxytriphenylene radical cation, DG41

    4-methoxyazobenzene, DE07

    2-methoxybenz[b]azocine, DD78

    4-methoxybenz[d]azocine, DD79

    4-methoxy-2,2'-bipyridyl 1-oxide, DD49

    6-methoxydibenz[b,f]azocine, DF18

    6-methoxy-11,12-dihydrodibenz[b,f]azocine, DF32

    6-methoxy-1-hydroxyphenazine-5,10-dioxide, DE00

    3-(2-methoxyphenoxy)-1,2-propanediol, DD31

    3-(4-methoxyphenyl)-5-phenyl-1,2-dithiol-1-ium perchlorate, DF14

    6-methoxytribenz[b,d,f]azocine, DG10

    2-methoxy-3,5,5-trimethyl-2-cyclohexen-1-one, DD34

    3(2-methylpropoxy)-5,5-dimethyl-2-cyclohexen-1-one, DD91

    2-(2-methylpropoxy)-2-methoxyethylmercuric acetate, DC94

    2-octyl benzyl ether, DE96

    16α,17α-oxidoprogesterone, DG25

    4,4'-oxybis[benzenesulfonyl(isopropylidenehydrazide)], DF96

    retinyl methyl ether, DG28

    tetraphenylethylene oxide, DG49

    ubiquinone-10, DH30

GUANIDINES

    1,3-diphenylguanidine, DE11

    N-methyl-N-nitroso-N'-nitroguanidine, DA09

    s-tetramethyl guanidine, DA80

HALOGEN COMPOUNDS
  Aliphatic and Alicyclic
    Bromo

    4-bromobornane, DD36

    1,2-dibromoacenaphthene, DD61

    1,2-dibromobenzocyclobutene, DC09

    2,4-dibromo-2,4-dimethyl-3-pentanone, DB92

    1,2-dibromo-1,2-diphenylethane, DE38, DH42, DH44, DH45, DH46, DH49, DH51, DH52, DH60

    (1,2-dibromoethyl)benzene, DC26

    ethyl 2-bromomercurio-2-(4-isopropylphenyl)acetate, DE16

    methyl 4-bromo-5-oxohexanoate, DB91

    2,2,2-tribromoethyl benzoate, DC62

    Chloro

    carbon tetrachloride, DA00

    N-Benzylchloroacetamide, DC81

    bis(trichlorovinyl)mercury(II), DA28

(HALOGEN COMPOUNDS)(cont.)
  Aliphatic and Alicyclic (cont.)
    Chloro (cont.)

      chloroacetanilide, DC27

      2,2-dichloroacetanilide, DC19

      2,3-dichloroacetanilide, DC20

      2,2-dichloroethyl benzoate, DC65

      2,2-dichloronorbornane, DB85

      2',3'-dimethyl-2-chloroacetanilide, DD23

      3',4'-dimethyl-2-chloroacetanilide, DD24

      hexachlorobuta-1,3-diene, DA27

      3'-methyl-2,2-dichloroacetanilide, DC74

      4'-methyl-2,2-dichloroacetanilide, DC75

      3'-methyl-2,2,2-trichloroacetanilide, DC66

      4'-methyl-2,2,2-trichloroacetanilide, DC67

      4-nitrobenzyl chloride, DB65

      trichloroacetanilide, DC10

      2,2,2-trichloroethyl benzoate, DC63

    Fluoro

      bis[(tricarbonyl-η-cyclopentadienyl-trifluoromethyl)molybdenum], DF75

      η-cyclopentadienylperfluorodimethylethylene-1,2-disulfido-nickel (III), DC60

      η-cyclopentadienylperfluorodimethylethylene-1,2-disulfido-rhodium(III), DC61

      1,3-dihydro-5-phenyl-7-trifluoromethyl-2H-1,4-benzodiazepin-2-one, DF04

      2-nitro(trifluoromethyl)benzene, DB54

      3-nitro(trifluoromethyl)benzene, DB55

      4-nitro(trifluoromethyl)benzene, DB56

      3-trifluoromethyl-ω-diazoacetophenone, DC57

    Iodo

      2-cyanoethylmercuric iodide, DA13

      2-iodoethyl benzoate, DC76

      3-iodopropanonitrile, DA14

    Mixed Halogens

      3'-bromo-2-chloroacetanilide, DC16

      4'-bromo-2-chloroacetanilide, DC17

      2-exo-bromo-2-endo-chloro-norbornane, DB83

      3-bromo-2-chloro-3-phenylpropanoic acid, DC64

      2-exo-chloro-2-endo-bromo-norbornane, DB84

      1,3-dibromo-1-chloropropane, DA16

      4'-iodo-2-chloroacetanilide, DC18

Aromatic
  Bromo

    2-bromobenzonitrile, DB48

    3-bromobenzonitrile, DB49

    4-bromobenzonitrile, DB50

    3'-bromo-2-chloroacetanilide, DC16

    4'-bromo-2-chloroacetanilide, DC17

    5-bromo-3,6-dinitro-1,2,4-trimethylbenzene, DC73

    3-(4-bromophenyl)-5-phenyl-1,2-dithiolylium perchlorate, DE71

    ethyl α-bromomercury-(2-bromophenyl)acetate, DD11

    ethyl α-bromomercury-(3-bromophenyl)acetate, DD12

    ethyl α-bromomercury-(4-bromophenyl)acetate, DD13

  Chloro

    2,5-bis(4-chlorophenyl)-3,6-diphenylpyrazolo[4,3-d]pyrazolium anion, DG56

    bis(3,4,6-trichloro-2-hydroxyphenyl)methane, DD97

    4-chloroaniline, DB03

    4-chlorobenzaldehyde, DB60

    2-chlorobenzonitrile, DB51

    3-chlorobenzonitrile, DB52

    4-chlorobenzonitrile, DB53

TABLE V. Functional-Group Index

(HALOGEN COMPOUNDS)(cont.)
Aromatic (cont.)
Chloro (cont.)

    7-chloro-5-(2-chlorophenyl)-2,3-dihydro-1H-1,4-benzodiazepin-2-one, DE73

    7-chloro-5-(2-chlorophenyl)-2,3-dihydro-3-hydroxy-1H-1,4-benzodiazepin-2-one, DE74

    7-chloro-5-(2,3-dichlorophenyl)-2,3-dihydro-1-methyl-1H-1,4-benzodiazepine, DF15

    7-chloro-5-(2,6-dichlorophenyl)-2,3-dihydro-1-methyl-1H-1,4-benzodiazepine, DF16

    4-chlorophenylarsonic acid, DA97

    5-(2-chlorophenyl)-2,3-dihydro-7-iodo-1-methyl-1H-1,4-benzodiazepine, DF21

    5-(4-chlorophenyl)-2-phenyl-1,2,3-triazolo[4,5-d]triazolium anion, DE30

    [3-(4-chlorophenyl)-3-(2-pyridyl)-1-propyl]dimethylammonium maleate, DG16

    2,3'-dichloroacetanilide, DC20

    9,10-dichloro-9,10-dihydro-9,10-diphenylanthracene, DG48, DH47, DH50, DH54, DH55, DH61

    [2,3-dichloro-4-(2-methylene-1-oxobutyl)phenoxy]acetic acid, DE06

    ethyl α-bromomercury-(4-chlorophenyl)acetate, DD08

    pentachlorothiophenol, DA82

Fluoro

    7-acetyl-5-(2-fluorophenyl)-2-methylamino-3H-1,4-benzodiazepine 4-N-oxide, DF87

    7-chloro-1,3-dihydro-5-(2-fluorophenyl)-2H-1,4-benzodiazepin-2-one, DE72

    ethyl α-bromomercury-(4-fluorophenyl)acetate, DD09

    5-(2-fluorophenyl)-7-iodo-2-methylamino-3H-1,4-benzodiazepine 4-N-oxide, DF17

Iodo

    ethyl α-bromomercury-(4-iodophenyl)-acetate, DD10

    4-iodobenzonitrile, DB57

    4'-iodo-2-chloroacetanilide, DC18

Heterocyclic, Benzodiazepines
Bromo

    7-bromo-1,3-dihydro-5-phenyl-2H-1,4-benzodiazepin-2-one, DE79

    7-bromo-2-methylamino-5-phenyl-3H-1,4-benzodiazepine 4-N-oxide, DF20

Chloro

    1-acetyl-7-chloro-2,3-dihydro-5-phenyl-1H-1,4-benzodiazepine, DF56

    1-amino-7-chloro-2,3-dihydro-5-phenyl-1H-1,4-benzodiazepine, DE92

    2-amino-7-chloro-5-phenyl-3H-1,4-benzodiazepine 4-N-oxide, DE85

    5-(2-aminophenyl)-7-chloro-2,3-dihydro-1-methyl-1H-1,4-benzodiazepine, DF35

    2-butylamino-7-chloro-5-phenyl-3H-1,4-benzodiazepine 4-N-oxide, DG04

    2-carbamoyl-7-chloro-2,3-dihydro-1-methyl-5-phenyl-1H-1,4-benzodiazepine, DF63

Chlorazepam, DF12

    7-chloro-1-acetamido-2,3-dihydro-5-phenyl-1H-benzodiazepine, DF64

    7-chloro-5-(2-chlorophenyl)-2,3-dihydro-1H-1,4-diazepin-2-one, DE73

    7-chloro-5-(2-chlorophenyl)-2,3-dihydro-3-hydroxy-2H-1,4-benzodiazepin-2-one, DE74

    7-chloro-2-cyano-2,3-dihydro-1-methyl-5-phenyl-1H-1,4-benzodiazepine, DF53

    7-chloro-5-(2,3-dichlorophenyl-2,3-dihydro-1-methyl-1H-1,4-benzodiazepine, DF15

    7-chloro-5-(2,6-dichlorophenyl)-2,3-dihydro-1-methyl-1H-1,4-benzodiazepine, DF16

    7-chloro-1,3-dihydro-5-(2-fluorophenyl)-2H-1,4-benzodiazepin-2-one, DE72

    7-chloro-1,3-dihydro-2-hydrazino-5-phenyl-2H-1,4-benzodiazepine 4-N-oxide, DE91

(HALOGEN COMPOUNDS)(cont.)
  Heterocyclic, Benzodiazepines (cont.)
    Chloro

    7-chloro-2,3-dihydro-2-hydroxy-1-methyl-5-phenyl-1H-1,4-benzodiazepine, DF30

    7-chloro-1,3-dihydro-5-(1H-indol-1-yl)-2H-1,4-benzodiazepin-2-one, DF51

    7-chloro-2,3-dihydro-1-methanoyl-5-phenyl-1H-1,4-benzodiazepine, DF06

    7-chloro-1,3-dihydro-5-(2-methoxyphenyl)-2H-1,4-benzodiazepin-2-one, DF10

    7-chloro-2,3-dihydro-1-(N-methylcarbamoyl)-5-phenyl-1H-1,4-benzodiazepine, DF65

    7-chloro-2,3-dihydro-1-methyl-5(2-methylmercaptophenyl)-1H-1,4-benzodiazepine, DF70

    7-chloro-2,3-dihydro-1-methyl-5-(2-methylsulfinylphenyl)-1H-1,4-benzodiazepine, DF69

    7-chloro-2,3-dihydro-1-methyl-5-(2-nitrophenyl)-1H-1,4-benzodiazepine, DF23

    7-chloro-2,3-dihydro-1-methyl-5-phenyl-1H-1,4-benzodiazepin-2-carboxylic acid, DF57

    7-chloro-2,3-dihydro-1-methyl-5-phenyl-1H-1,4-benzodiazepine, DF29

    7-chloro-1,3-dihydro-5-(2-methylphenyl)-2H-1,4-benzodiazepin-2-one, DF07

    7-chloro-1,3-dihydro-5-(1-methyl-1H-pyrazol-5-yl)-2H-1,4-benzodiazepin-2-one, DE03

    7-chloro-2,3-dihydro-1-methylsulfonyl-5-phenyl-1H-1,4-benzodiazepine, DF31

    7-chloro-2,3-dihydro-5-phenyl-1H-1,4-benzodiazepine, DE90

    7-chloro-1,3-dihydro-5-phenyl-2H-1,4-benzodiazepin-2-one, DE80

    7-chloro-1,3-dihydro-5-(2-pyridinyl)-2H-1,4-benzodiazepin-2-one, DE33

    7-chloro-1,3-dihydro-5-(2-thiazolyl)-2H-1,4-benzodiazepin-2-one, DD62

    7-chloro-2-dimethylamino-5-phenyl-3H-1,4-benzodiazepine 4-N-oxide, DF66

    7-chloro-2-ethylamino-5-phenyl-2H-1,4-benzodiazepine 4-N-oxide, DF67

    7-chloro-5-(4-hydroxyphenyl)-2-methylamino-3H-1,4-benzodiazepine 4-N-oxide, DF24

    7-chloro-2-methoxy-5-phenyl-3H-1,4-benzodiazepine 4-N-oxide, DF11

    7-chloro-2-(N-methylacetamido)-5-phenyl-3H-1,4-benzodiazepine 4-N-oxide, DF86

    7-chloro-2-methylamino-5-phenyl-3H-1,4-benzodiazepine 4-N-oxide, DF22

    7-chloro-2-methyl-5-phenyl-3H-1,4-benzodiazepine 4-N-oxide, DF08

    7-chloro-2(N-methylpropionamido)-5-phenyl-3H-1,4-benzodiazepine 4-N-oxide, DG03

    Diazepam, DF09

    Oxazepam, DE81

    Fluoro

    1,3-dihydro-7-fluoro-5-phenyl-2H-1,4-benzodiazepin-2-one, DE83

    Iodo

    5-(2-chlorophenyl)-2,3-dihydro-7-iodo-1-methyl-1H-1,4-benzodiazepine, DF21

    7,9-diiodo-1,3-dihydro-5-phenyl-2H-1,4-benzodiazepin-2-one, DE76

    5-(2-fluorophenyl)-7-iodo-2-methylamino-3H-1,4-benzodiazepine 4-N-oxide, DF17

  Heterocyclic, Others

    9-(2-carboxyphenyl)-4,5-dibromo-6-hydroxy-3H-xanthen-2-one, DG06

    9-(2-carboxyphenyl)-4,5-dichloro-6-hydroxy-3H-xanthen-3-one, DG07

    9-(2-carboxyphenyl)-4,5-diiodo-6-hydroxy-3H-xanthen-3-one, DG08

    6-chloro-3-chloromethyl-3,4-dihydro-2-methyl-7-sulfamoyl-1,2,4-benzothiadiazine 1,1-dioxide, DC84

    6-chloro-3,4-dihydro-7-sulfamoyl-2H-12,4-benzothiadiazine 1,1-dioxide, DB74

TABLE V. Functional-Group Index

(HALOGEN COMPOUNDS)(cont.)
Heterocyclic, Others (cont.)

2,3-dibromothiophene, DA29

1-methyl-2-chlorobenzimidazole-4,7-quinone, DC04

1-methyl-2-chlorobenzimidazole-6,7-quinone, DC05

Steroids

16α-bromo-3-hydroxyestra-1,3,5(10)-trien-17-one, DF94

16β-bromo-3-hydroxyestra-1,3,5(10)-trien-17-one, DF95

9α-fluoro-17α,21-dihydroxy-5-pregnene-3,11,20-trione-3-ethyl-eneketal-17,20-21-bismethylene-dioxolane, DG45

HETEROCYCLIC COMPOUNDS
Nitrogen
Pyridines

2-amino-3-nitropyridine, DA66

2-amino-5-nitropyridine, DA67

2,2'-bipyridine, DD05

carbonyl(nitrosyl)-2,2'-bipyridinecobalt, DD45

[3-(4-chlorophenyl)-3-(2-pyridyl)-1-propyl]dimethylammonium maleate, DG16

5,9-diazonia-5,6,7-trihydrodibenzo[a,c]azepine bromide, DE12

1,1'-dimethyl-4,4'-bipyridylium bromide, DD83

2,7-dimethyl-9,10-dihydro-8a,10a-diazoniaphenanthrene bromide, DE49

3,6-dimethyl-9,10-dihydro-8a,10a-diazoniaphenanthrene bromide, DE50

dinitrosyl-2,2'-bipyridineiron, DD04

dinitrosyl[di-(2-pyridyl) ketone]-iron, DD46

di-2-pyridyl ketone, DD47

3-hydroxy-5-(hydroxymethyl)-2-methyl-4-pyridinecarboxylic acid, DC34

N-(2-hydroxymethyl-3,4-dihydroxy-5-tetrahydrofuryl)-3-carboxypyridine-(1H)2-one, DD52

2-hydroxynicotinic acid, DA93

2-hydroxy-3-nitropyridine, DA56

2-hydroxy-5-nitropyridine, DA57

4-methoxy-2,2'-bipyridyl 1-oxide, DD49

nicotinamide, DB04

nicotinic acid, DA90

pyridine, DA63

pyridoxal, DC33

pyridoxal 5-phosphate, DC36

pyridoxamine, DC42

pyridoxine, DC40

trisbipyridineiron(II)sulfate, DG65

trisbipyridineruthenium(II) dichloride, DG64, DH40, DH56

One Ring-Nitrogen Atom, Monocyclic (except Pyridines)

ammonium pyrrolidine-1-dithiocarboxylate, DA79

1,1'-bipyrrolidine, DC47

diammonium proline-N-dithiocarboxylate, DB45

3,3-diethyl-2,4-dioxopiperidine, DC90

3,3-diethyl-5-methyl-2,4-dioxopiperidine, DD37

3,3-diethyl-2,4-(1H,3H)pyradinedione, DC87

3,8-dimethyl-2-methoxyazocine, DD25

2-(2'-ethyl-6'-methylpiperidyl)-ethyl methacrylate, DE67

3-ethyl-3-phenyl-2,6-dioxopiperidine, DE13

morpholine, DA48

N-nitrosopiperidine, DA75

sodium 1-piperidinedithiocarboxylate, DB26

sodium 1-pyrrolidinedithiocarboxylate, DA70

1-vinyl-2-pyrrolidone, DB24

One Ring-Nitrogen Atom, Polycyclic

12-acetylbenzo[a]phenothiazine 7-dioxide, DF79

7-acetylbenzo[c]phenothiazine 12-dioxide, DF78

(HETEROCYCLIC COMPOUNDS)(cont.)
Nitrogen (cont.)
One Ring-Nitrogen Atom, Polycyclic, (cont.)

12-acetylbenzo[a]phenothiazine 7-oxide, DF77

1-azabicyclo[2.2.2]octane, DB95

bis(2-quinolyl)disulfide di-N-oxide, DF76

7-(1,3-dihydro-2-isoindolyl)-7-azabicyclo[4.1.0]heptane, DE57

isatin 3-(O-ethyloxime), DD15

2-methoxybenz[b]azocine, DD78

4-methoxybenz[d]azocine, DD79

6-methoxydibenz[b,f]azocine, DF18

6-methoxy-11,12-dihydrodibenz-[b,f]azacine, DF32

6-methoxytribenz[b,d,f]azocine, DG10

Methylene Blue, DF38

10-methylphenothiazine, DE04, DH38, DH39, DH40, DH41

phenoxazine, DD68

10-phenylphenothiazine, DF80, DH48

Thalidomide, DE01

Two Ring-Nitrogen Atoms, Monocyclic

5,5-bis(3-nitrophenyl)-2,4-imidazolidinedione, DE77

cytidine, DC88

cytidine monophosphate, DC89

cytidylyl (3'→5') cytidine, DF98

cytidylyl (3'→5') guanidine, DG05

cytidylyl (3'→5') uridine, DF97

cytosine, DA36

1,2-diethylpyrazolidine, DC01

1,2-diisopropylpyrazolidine, DC97

1,4-dimethylhexahydropyrazine, DB39

1,2-dimethylhexahydropyridazine, DB40

1,2-dimethyl-3-n-hexyl-1,2-diaziridine, DC98

1,2-dimethylpyrazolidine, DA78

1,2-dimethyl-1,2,3,6-tetrahydropyridazine, DB34

2,5-diphenyl-1,3,4-oxadiazole, DE35, DH37, DH38, DH43

pyrazine, DA31

pyridazine, DA32

pyrimidine, DA33

1,2-cis-3,6-tetramethylperhydropyridazine, DC52

1,2-trans-3,6-tetramethylperhydropyridazine, DC53

1,2,3-trimethylperhydropyridazine, DC02

uracil-6-carboxylic acid, DA58

Benzodiazepines

7-acetoxy-2-methylamino-5-phenyl-3H-1,4-benzodiazepine 4-N-oxide, DF92

7-acetamido-2,3-dihydro-5-phenyl-1H-1,4-benzodiazepin-2-one, DF61

1-acetyl-7-chloro-2,3-dihydro-5-phenyl-1H-1,4-benzodiazepine, DF56

7-acetyl-1,3-dihydro-5-phenyl-2H-1,4-benzodiazepin-2-one, DF55

7-acetyl-5-(2-fluorophenyl)-2-methylamino-3H-1,4-benzodiazepine 4-N-oxide, DF87

1-amino-7-chloro-2,3-dihydro-5-phenyl-1H-1,4-benzodiazepine, DE92

2-amino-7-chloro-5-phenyl-3H-1,4-benzodiazepine 4-N-oxide, DE85

5-(2-aminophenyl)-7-chloro-2,3-dihydro-1-methyl-1H-1,4-benzodiazepine, DF35

7-bromo-1,3-dihydro-5-phenyl-2H-1,4-benzodiazepin-2-one, DE79

7-bromo-2-methylamino-5-phenyl-3H-1,4-benzodiazepine 4-N-oxide, DF20

2-butylamino-7-chloro-5-phenyl-3H-1,4-benzodiazepine 4-N-oxide, DG04

2-carbamoyl-7-chloro-2,3-dihydro-1-methyl-5-phenyl-1H-1,4-benzodiazepine, DF63

Chloroazepam, DF12

7-chloro-1-acetamideo-2,3-dihydro-5-phenyl-1H-benzodiazepine, DF64

7-chloro-5-(2-chlorophenyl)-2,3-dihydro-1H-1,4-diazepin-2-one, DE73

TABLE V. Functional-Group Index

(HETEROCYCLIC COMPOUNDS (cont.)
  Nitrogen (cont.)
    Benzodiazepines (cont.)

  7-chloro-5-(2-chlorophenyl)-2,3-dihydro-3-hydroxy-1H-1,4-benzodiazepin-2-one, DE74

  7-chloro-2-cyano-2,3-dihydro-1-methyl-5-phenyl-1H-1,4-benzodiazepine, DF53

  7-chloro-5-(2,3-dichlorophenyl-2,3-dihydro-1-methyl-1H-1,4-benzodiazepine, DF15

  7-chloro-5-(2,6-dichlorophenyl)-2,3-dihydro-1-methyl-1H-1,4-benzodiazepine, DF16

  7-chloro-1,3-dihydro-5-(2-fluorophenyl)-2H-1,4-benzodiazepin-2-one, DE72

  7-chloro-1,3-dihydro-2-hydrazino-5-phenyl-2H-1,4-benzodiazepine 4-N-oxide, DE91

  7-chloro-2,3-dihydro-2-hydroxy-1-methyl-5-phenyl-1H-1,4-benzodiazepine, DF30

  7-chloro-1,3-dihydro-5-(1H-indol-1-yl)-2H-1,4-benzodiazepin-2-one, DF51

  7-chloro-2,3-dihydro-1-methanoyl-5-phenyl-1H-1,4-benzodiazepine, DF06

  7-chloro-1,3-dihydro-5-(2-methoxyphenyl)-2H-1,4-benzodiazepin-2-one, DF10

  7-chloro-2,3-dihydro-1-(N-methylcarbamoyl)-5-phenyl-1H-1,4-benzodiazepine, DF65

  7-chloro-2,3-dihydro-1-methyl-5-(2-methylmercaptophenyl)-1H-1,4-benzodiazepine, DF70

  7-chloro-2,3-dihydro-1-methyl-5-(2-methylsulfinylphenyl)-1H-1,4-benzodiazepine, DF69

  7-chloro-2,3-dihydro-1-methyl-5-(2-nitrophenyl)-1H-1,4-benzodiazepine, DF23

  7-chloro-2,3-dihydro-1-methyl-5-phenyl-1H-1,4-benzodiazepin-2-carboxylic acid, DF57

  7-chloro-2,3-dihydro-1-methyl-5-phenyl-1H-1,4-benzodiazepine, DF29

  7-chloro-1,3-dihydro-5-(2-methylphenyl)-2H-1,4-benzodiazepin-2-one, DF07

  7-chloro-1,3-dihydro-5-(1-methyl-1H-pyrazol-5-yl)-2H-1,4-benzodiazepin-2-one, DE03

  7-chloro-2,3-dihydro-1-methylsulfonyl-5-phenyl-1H-1,4-benzodiazepine, DF31

  7-chloro-2,3-dihydro-5-phenyl-1H-1,4-benzodiazepine, DE90

  7-chloro-1,3-dihydro-5-phenyl-2H-1,4-benzodiazepin-2-one, DE80

  7-chloro-1,3-dihydro-5-(2-pyridinyl)-2H-1,4-benzodiazepin-2-one, DE33

  7-chloro-1,3-dihydro-5-(2-thiazolyl)-2H-1,4-benzodiazepin-2-one, DD62

  7-chloro-2-dimethylamino-5-phenyl-3H-1,4-benzodiazepine 4-N-oxide, DF66

  7-chloro-2-ethylamino-5-phenyl-2H-1,4-benzodiazepine 4-N-oxide, DF67

  7-chloro-5-(4-hydroxyphenyl)-2-methylamino-3H-1,4-benzodiazepine 4-N-oxide, DF24

  7-chloro-2-methoxy-5-phenyl-3H-1,4-benzodiazepine 4-N-oxide, DF11

  7-chloro-2-(N-methylacetamido)-5-phenyl-3H-1,4-benzodiazepine 4-N-oxide, DF86

  7-chloro-2-methylamino-5-phenyl-3H-1,4-benzodiazepine 4-N-oxide, DF22

  7-chloro-2-methyl-5-phenyl-3H-1,4-benzodiazepine 4-N-oxide, DF08

  7-chloro-2(N-methylpropionamido)-5-phenyl-3H-1,4-benzodiazepine 4-N-oxide, DG03

  5-(2-chlorophenyl)-2,3-dihydro-7-iodo-1-methyl-1H-1,4-benzodiazepine, DF21

Diazepam, DF09

1,3-dihydro-7-fluoro-5-phenyl-2H-1,4-benzodiazepin-2-one, DE83

2,3-dihydro-5-(5-isothiazolyl)-1-methyl-7-nitro-1H-1,4-benzodiazepine, DE08

1,3-dihydro-7-methoxy-5-phenyl-2H-1,4-benzodiazepin-2-one, DF28

2,3-dihydro-1-methyl-5-(1-methyl-1H-imidazol-2-yl)-7-nitro-1H-1,4-benzodiazepine, DE47

(HETEROCYCLIC COMPOUNDS)(cont.)
  Nitrogen (cont.)
    Benzodiazepines (cont.)

  2,3-dihydro-1-methyl-5-(1-methyl-1H-pyrazol-5-yl)-7-nitro-1H-1,4-benzodiazepine, DE48

  2,3-dihydro-1-methyl-7-nitro-5-phenyl-1H-1,4-benzodiazepine, DF34

  2,3-dihydro-1-methyl-7-nitro-5-(4-pyridyl)-1H-1,4-benzodiazepine, DE93

  2,3-dihydro-1-methyl-7-nitro-5-(2-pyrimidinyl)-1H-1,4-benzodiazepine, DE41

  2,3-dihydro-1-methyl-7-nitro-5-(2-thiazolyl)-1H-1,4-benzodiazepine, DE09

  2,3-dihydro-1-methyl-5-phenyl-1H-1,4-benzodiazepine, DF37

  1,3-dihydro-7-methyl-5-phenyl-2H-1,4-benzodiazepin-2-one, DF26

  1,3-dihydro-7-methylthio-5-phenyl-2H-1,4-benzodiazepin-2-one, DF27

  1,3-dihydro-5-phenyl-2H-1,4-benzodiazepin-2-one, DE87

  1,3-dihydro-5-phenyl-7-trifluoromethyl-2H-1,4-benzodiazepin-2-one, DF04

  7,9-diiodo-1,3-dihydro-5-phenyl-2H-1,4-benzodiazepin-2-one, DE76

  5-(2-fluorophenyl)-7-iodo-2-methylamino-3H-1,4-benzodiazepine 4-N-oxide, DF17

  2-methylamino-7-methylmercapto-5-phenyl-3H-1,4-benzodiazepine 4-N-oxide, DF72

  2-methylamino-5-phenyl-3H-1,4-benzodiazepine 4-N-oxide, DF33

  7-methyl-2-methylamino-5-phenyl-3H-1,4-benzodiazepine 4-N-oxide, DF71

  Nitrazepam, DE84

  Oxazepam, DE81

Two Ring-Nitrogen Atoms Polycyclic (except Benzodiazepines)

  benzimidazole-4,7-quinone, DB58

  2-tert-butyl-2-azonia-3-aza-2-norbornene ion, DC92

  2-tert-butyl-2,3-diazanorbornane, DC96

  2-tert-butyl-3-methyl-2,3-diazanorbornane, DD39

  carbonyl(nitrosyl)-1,10-phenanthrolinecobalt, DD98

  6-chloro-3-chloromethyl-3,4-dihydro-2-methyl-7-sulfamoyl-1,2,4-benzothiadiazine 1,1-dioxide, DC84

  6-chloro-3,4-dihydro-7-sulfamoyl-2H-1,2,4-benzothiadiazine 1,1-dioxide, DB74

  6,6-cyclopentamethylene-1,5-diazabicyclo[3.1.0]hexane, DC91

  9,9-cyclopentamethylene-1,5-diazabicyclo[3.3.1]nonane, DD94

  8,8-cyclopentamethylene-1,5-diazabicyclo[3.2.1]octane, DD58

  1,6-diazabicyclo[4.4.0]deca-3,8-diene, DC41

  1,6-diazabicyclo[4.4.0]decane, DC48

  1,5-diazabicyclo[3.2.2]nonane, DB97

  1,5-diazabicyclo[3.3.1]nonane, DB98

  1,4-diazabicyclo[2.2.2]octane, DB31

  1,5-diazabicyclo[3.2.1]octane, DB32

  1,5-diazabicyclo[3.3.0]octane, DB33

  1,3-diazatricyclo[3.3.1.1^{3,7}]decane DC43

  3,4-di-tert-butyl-1,3,4-oxadiazolidine, DD41

  dicarbonyl-1,10-phenanthrolinenickel(0), DE24

  2,2-diethyl-5,8-ethanoperhydropyrazolo[1,2-a]pyridazine, DE19

  3,4-diethyl-1,3,4-oxadiazolidine, DB41

  3,4-dihydro-2-methyl-4-oxo-3-o-tolylquinazoline, DF25

  3,4-diisopropyl-1,3,4-oxadiazolidine, DC54

  1,3-dimethylbenzimidazolone-4,7-quinone, DC68

  1,3-dimethylbenzimidazolone-5,6-quinone, DC69

TABLE V. Functional-Group Index

(HETEROCYCLIC COMPOUNDS)(cont.)
  Nitrogen (cont.)
    Two Ring-Nitrogen Atoms Polycyclic (except Benzodiazepines)(cont.)

    2,3-dimethyl-2,3-diazabicyclo-[2.2.1]heptane, DB99

    2,3-dimethyl-2,3-diazabicyclo-[2.2.1]hept-5-ene, DB93

    2,3-dimethyl-2,3-diazabicyclo-[2.2.2]octa-5-ene, DC44

    2,3-dimethyl-2,3-diazabicyclo-[2.2.2]octane, DC49

    1,4-dimethyl-1,4-diazaspiro[4.5]-decane, DD40

    dinitrosyl-1,10-phenanthroline-iron, DD63

    5,7-diphenyl-1,3-diazatricyclo-[3.3.1.1^{3,7}]decane, DG15

    5,7-diphenyl-1,3-diazatricyclo-[3.3.1.1^{3,7}]decan-6-one, DG13

    5-ethyl-5-(1-methylpropyl)-2-thiobarbituric acid, DD33

    5-ethyl-5-(3-nitrophenyl)barbituric acid, DD80

    6-methoxy-1-hydroxyphenazine-5,10-dioxide, DE00

    1-methylbenzimidazole-4,7-quinone, DC11

    1-methylbenzimidazole-6,7-quinone, DC12

    2-methyl-7,8-benzophenazine, DF52

    1-methyl-2-chlorobenzimidazole-4,7-quinone, DC04

    1-methyl-2-chlorobenzimidazole-6,7-quinone, DC05

    1-methyl-2-dimethylaminobenzimidazole-4,7-quinone, DD19

    1-methyl-2-dimethylaminobenzimidazole-5,6-quinone, DD20

    1-methyl-2-dimethylaminobenzimidazole-6,7-quinone, DD21

    1-methyl(2,3-d)naphthoimidazole-4,9-quinone, DD65

    9,9-pentamethylene-1,5-diazatricyclo[3.3.1.1^{3,7}]decane, DE18

    1,10-phenanthroline, DD64

    strychnine, DG24

    tetrahydro-1,3-dimethyl-1H,3H-[1,3,4]-oxadiazolo[3,4-a]pyridazine, DE50

    1,2,3-trimethyl-1,2,3,6-tetrahydropyridazine, DC00

  Three-Ring Nitrogen Atoms

    1-methyl-3,5,7-triazatricyclo-[3.3.1.1^{3,7}]decane, DC46

    1,3,5-triazine, DA12

    1,3,5-trimethylperhydro-1,3,5-triazine, DB44

  Four-Ring Nitrogens, Tetraazacyclotetradecanes

    [b,i]dibenz-6,13-diacetyl-5,12-dimethyl-1,4,8,11-tetrazacyclotetradeca-2,4,6,9,11,13-hexaenenickel(II), DG39

    [b,i]dibenz-5,12,-dimethyl-1,4,8,11-tetrazacyclotetradeca-2,4,6,9,11,13-hexaenenickel(II), DG11

    dibenz(b,i]-1,4,8,11-tetraaza-2,5,7,9,12,14-cyclotetradecahexaenenickel(II), DF81

    7,14-diethylbis(1,2-cyclohexylenyl[c,j]-1,2,5,6,8,9,12,13-octaaza-2,4,7,9,11,14-cyclotetradecahexaenenickel(II), DF99

    2,3-dimethyl-1,4,8,11-tetraazacyclotetradecaneiron(II) dication, DD96

    5,7,7,12,14,14-hexamethyl-1,4,8,11-tetraaza-4,11-cyclotetradecadieneiron(II) dication, DF47

    5,5,7,12,12,14-hexamethyl-1,4,8,11-tetraazacyclotetradecaneiron(II) dication bis[tetrafluoroborate(1-)], DF49

    5,5,7,12,12,14-hexamethyl-1,4,8,11-tetraaza-1,3,7,10-cyclotetradecatetraeneiron(II) dication, DF41

    5,5,7,12,12,14-hexamethyl-1,4,8,11-tetraaza-1,3,8,10-cyclotetradecatetraeneiron(II) dication, DF42

    5,5,7,12,14,14-hexamethyl-1,4,8,11-tetraaza-1,3,7,11-cyclotetradecatetraeneiron(II) dication, DF43

    5,7,7,12,14,14-hexamethyl-1,4,8,11-tetraaza-1,4,8,11-cyclotetradecatetraeneiron(II) dication, DF44

(HETEROCYCLIC COMPOUNDS)(cont.)
Nitrogen(cont.)
Four-Ring Nitrogens, Tetraazacyclotetradecanes (cont.)

5,7,7,12,14,14-hexamethyl-1,4,8,11-tetraaza-1,3,8-cyclotetradecatrieneiron(II) dication, DF45

5,7,7,12,14,14-hexamethyl-1,4,8,11-tetraaza-1,4,11-cyclotetradecatrieneiron(II) dication, DF46

5,7,12,14-tetramethyl-1,4,8,11-tetraazacyclotetradeca-2,5,7,9,12,14-hexaenecopper(II), DE55

5,7,12,14-tetramethyl-1,4,8,11-tetraazacyclotetradeca-2,5,7,9,12,14-hexaenecopper(I) tetrafluoroborate, DE53

5,7,12,14-tetramethyl-1,4,8,11-tetraazacyclotetradeca-2,5,7,9,12,14-hexaenenickel(II), DE58

5,7,12,14-tetramethyl-1,4,8,11-tetraazacyclotetradeca-2,5,7,9,12,14-hexaenenickel(II) tetrafluoroborate, DE54

5,7,12,14-tetramethyl-1,4,8,11-tetraazacyclotetradeca-5,7,12,14-tetraenecopper(II), DE61

5,7,12,14-tetramethyl-1,4,8,11-tetraazacyclotetradeca-5,7,12,14-tetraenenickel(II), DE62

5,7,12,14-tetramethyl-1,4,8,11-tetraazacyclotetradeca-5,7,12,14-tetraenezinc(II), DE63

Four-Ring Nitrogens, Porphines

cobalt(II)protoporphyrin(IX) dimethyl ester, DG74

copper(II)protoporphyrin(IX) dimethyl ester, DG75

trans-di(4-acetylpyridine)hematoporphyrincobalt(III), DH25

trans-di(4,4'-bipyridine)hematoporphyrincobalt(III), DH28

trans-dicyanoprotoporphyrin(IX)iron(III), DG72

trans-di(ethylnicotinate)hematoporphyrincobalt(III), DH26

trans-di(2-methylpyridine)hematoporphyrincobalt(III), DH23

trans-di(4-methylpyridine)hematoporphyrincobalt(III), DH24

trans-di(3-methylpyridine)protoporphyriniron(III) chloride, DH21

trans-di(4-methylpyridine)protoporphyriniron(III) chloride, DH22

trans-di(4-phenylpyridine)hematoporphyrincobalt(III), DH29

diprotonated 2,3,7,8,12,13,17,18,-octaethylporphyrin perchlorate, DH07

trans-dipyridinehematoporphyrincobalt(III) chloride, DH20

trans-dipyridylprotoporphyrin(IX)-iron(III) chloride, DH19

hematoporphyrincobalt(III), DG73

manganese(III)chloride protoporphyrin(IX) dimethyl ester monohydrate, DG78

mesoporphyrin(IX) dimethyl ester, DG80

N-methyl-2,3,7,8,12,13,17,18,-octaethylporphyrin, DH10

nickel(II)protoporphyrin(IX) dimethyl ester, DG76

2,3,7,8,12,13,17,18-octaethylporphinatocalcium(II), DG82

2,3,7,8,12,13,17,18-octaethylporphinatocadmium(II), DG83

2,3,7,8,12,13,17,18-octaethylporphinatocobalt(II), DG84

2,3,7,8,12,13,17,18-octaethylporphinatocopper(II), DG85

2,3,7,8,12,13,17,18-octaethylporphinatodihydroxogermanium(IV), DH03

2,3,7,8,12,13,17,18-octaethylporphinatodihydroxosilicon(IV), DH05

2,3,7,8,12,13,17,18-octaethylporphinatodihydroxotin(IV), DH06

2,3,7,8,12,13,17,18-octaethylporphinatohydroxoaluminum(III), DG93

2,3,7,8,12,13,17,18-octaethylporphinatohydroxoantimony(III), DH00

2,3,7,8,12,13,17,18-octaethylporphinatohydroxochromium(III), DG94

2,3,7,8,12,13,17,18-octaethylporphinatohydroxogallium(III), DG96

TABLE V. Functional-Group Index

(HETEROCYCLIC COMPOUNDS)(cont.)
Nitrogen (cont.)
Four-Ring Nitrogens, Porphines (cont.)

2,3,7,8,12,13,17,18-octaethylporphinatohydroxoindium(III), DG97

2,3,7,8,12,13,17,18-octaethylporphinatohydroxoiron(III), DG95

2,3,7,8,12,13,17,18-octaethylporphinatohydroxomanganese(III), DG98

2,3,7,8,12,13,17,18-octaethylporphinatohydroxoooxomolybdenum(V), DG99

2,3,7,8,12,13,17,18-octaethylporphinatohydroxoscandium(III), DH01

2,3,7,8,12,13,17,18-octaethylporphinatohydroxothallium(III), DH02

2,3,7,8,12,13,17,18-octaethylporphinatolead(II), DG90

2,3,7,8,12,13,17,18-octaethylporphinatomagnesium(II), DG86

2,3,7,8,12,13,17,18-octaethylporphinatonickel(II), DG87

2,3,7,8,12,13,17,18-octaethylporphinatooxotitanium(IV), DG88

2,3,7,8,12,13,17,18-oxooctaethylporphinatooxovanadium(IV), DG89

2,3,7,8,12,13,17,18-octaethylporphinatopalladium(II), DG91

2,3,7,8,12,13,17,18-octaethylporphinatosilver(III) perchlorate, DG81

2,3,7,8,12,13,17,18-octaethylporphinatozinc(II), DG92

2,3,7,8,12,13,17,18-octaethylporphine, DH04

protoporphyrin(IX) dimethyl ester, DG79

protoporphyrin(IX) trans-cyanopyridine iron(III), DH12

protoporphyrin(IX) iron(III)-chloride, DG71

sodium phthalocyaninecopper(II)-sulfonate, DG68

sodium phthalocyanine-2-sulfonate, DG69

tetrabutylammonium N-methyl-2,3,7,8,12,13,17,18-octaethylporphyrin, DH27

5,10,15,20-tetraphenylporphinatocobalt(II), DH17

5,10,15,20-tetraphenylporphine, DH18, DH41, DH55, DH63, DH64

tetrasodium phthalocyanineiron-2,9,16,23-tetrasulfonate, DG67

tetrasodium 5,10,15,20-tetrakis(4-sulfonatophenyl)porphine, DH16

tetrasodium 5,10,15,20-tetra(4-sulfonatophenyl)porphinatocobalt(II), DH15

trisodium 5,10,15,20-tetrakis(4-sulfonatophenyl)porphinatocobalt(III), DH14

zinc(II)protoporphyrin(IX) dimethyl ester, DG77

Four-Ring Nitrogens, Others

allopurinol, DA60

2,5-bis(4-chlorophenyl)-3,6-diphenylpyrazolo[3,4-d]pyrazolium anion, DG56

1,4-dimethyl-3,6-diphenylpyrazolo[4,5-d]pyrazole, DF88

2,4-dimethyl-3,6-diphenylpyrazolo[3,4-d]pyrazole, DF89

4-methyl-2,3,6-triphenypyrazolo[3,4-d]pyrazole, DG34

nicotinamide adenine dinucleotide phosphate, reduced, DG27

nicotinamide adenine dinucleotide, reduce, DG26

purine, DA59

riboflavin, DF74

sodium flavin mononucleotide, DF73

1,3,5,7-tetraazatricyclo-[3.3.1.1^{3,7}]decane, DB36

1,3,6,8-tetraazatricyclo-[4.4.1.1^{3,8}]dodecane, DC51

1,2,4,5-tetraethylperhydro-s-tetrazine, DD42

1,2,4,5-tetraisopropylperhydro-s-tetrazine, DE68

1,2,4,5-tetramethylperhydro-s-tetrazine, DB47

2,4,6,8-tetramethyl-2,4,6,8-tetraazabicyclo[3.3.0]octane, DC55

uric acid, DA61

(HETEROCYCLIC COMPOUNDS)(cont.)
Nitrogen(cont.)
Six or More Ring Nitrogens

- 5-(4-chlorophenyl)-2-phenyl-1,2,3-triazolo[4,5-d]triazolium anion, DE30
- 2-cyclohexyl-5-phenyl-1,2,3-triazolo[4,5-d]-1,2,3-triazolium anion, DE51
- 2,5-dicyclohexyl-1,2,3-triazolo-[4,5-d]-1,2,3-triazolium anion, DE64
- 7,14-dimethyl-bis(1,2-cyclohexylenyl[c,j]-1,2,5,6,8,9,12,13-octaaza-2,4,7,9,11,14-cyclotetradecahexaene nickel(II), DF40
- 2,5-dimethyl-1,2,3,4,5,6-hexaazabicyclo[2.2.1]heptane, DA26
- 2,5-diphenyl-1,2,3,4,5,6-hexaazapentalene, DE36
- 1-methyl-5-phenylazonia-1,2,3-triazolo-[4,5-d]-1,2,3-triazole, DC70
- 2-methyl-5-phenylazonia-1,2,3-triazolo[4,5-d]-1,2,3-triazole anion, DC71

Oxygen

- 6-acetamido-2-phenylchromanone, DF60
- 2-acetyl-5-nitrofuran, DA95
- ascorbic acid, DB23
- 9-(2-carboxyphenyl)-4,5-dibromo-6-hydroxy-3H-xanthen-2-one, DG06
- 9-(2-carboxyphenyl)-4,5-dichloro-6-hydroxy-3H-xanthen-3-one, DG07
- 9-(2-carboxyphenyl)-4,5-diiodo-6-hydroxy-3H-xanthen-3-one, DG08
- 2-cyano-5-nitrofuran, DA51
- cytidine, DC88
- cytidine monophosphate, DC89
- cytidylyl (3'→5') cytidine, DF98
- cytidylyl (3'→5') guanidine, DG05
- cytidylyl (3'→5') uridine, DF97
- 3,4-di-tert-butyl-1,3,4-oxadiazolidine, DD41
- 3,4-diethyl-1,3,4-oxadiazolidine, DB41
- 3,4-diisopropyl-1,3,4-oxadiazolidine, DC54
- 2,6-diphenyl-4-methylpyrylium tetrachloroaluminate, DF82
- 2,5-diphenyl-1,3,4-oxadiazole, DE35, DH37, DH38, DH43
- trans-2,3-diphenyloxirane, DE40
- ethyl 5-nitro-2-furoate, DB72
- 9α-fluoro-17α,21-dihydroxy-5-pregnene-3,11,20-trione-3-ethyleneketal-17,20-21-bis-methylenedioxolane, DG45
- D-glucose, DB37
- N-(2-hydroxymethyl-3,4-dihydroxy-5-tetrahydrofuryl)-3-carboxy-pyridine-(1H)2-one, DD52
- 2-methyl-4,6-diphenylpyrylium iodide, DF85
- methyl 5-(mercaptomethyl)-2-furoate, DB77
- methyl 2-methyl-5-(thiocyanatomethyl)-3-furoate, DC77
- 2-methyl-5-nitrofuran, DA64
- methyl 5-(thiocyanatomethyl)-2-furoate, DC23
- 2-methyl-2-(4,8,12-trimethyltridecyl)-6-chromanol, DG50
- morpholine, DA48
- nicotinamide adenine dinucleotide phosphate, reduced, DG27
- nicotinamide adenine dinucleotide, reduced, DG26
- 5-nitro-2-furaldehyde, DA54
- 2-nitrofuran, DA30
- 5-nitrofuraldehyde diacetate, DB82
- 5-nitrofurfuryl alcohol, DA65
- 5-nitro-2-furoic acid, DA55
- 5-nitro-2-(thiocyanatomethyl)furan, DA83
- phenoxazine, DD68
- riboflavin, DF74
- sodium flavin mononucleotide, DF73
- strychnine, DG24
- tetrahydro-1,3-dimethyl-1H,3H-[1,3,4]-oxadiazolo[3,4-a]-pyridazine, DC50
- tetraphenlethylene oxide, DG49
- α-tocopherol, DG63
- β-tocopherol, DG60
- γ-tocopherol, DG61
- δ-tocopherol, DG54
- D-xylose, DA77

TABLE V. Functional-Group Index

(HETEROCYCLIC COMPOUNDS)(cont.)
Sulfur

12-acetylbenzo[a]phenothiazine 7-dioxide, DF79

7-acetylbenzo[c]phenothiazine 12-dioxide, DF78

12-acetylbenzo[a]phenothiazine 7-oxide, DF77

3,5-bis(4-methoxyphenyl)-1,2-dithiol-1-ium perchlorate, DF59

3,5-bis(4-methylphenyl)-1,2-dithiol-1-ium perchlorate, DF58

3-(4-bromophenyl)-5-phenyl-1,2-dithiolylium perchlorate, DE71

6-chloro-3-chloromethyl-3,4-dihydro-2-methyl-7-sulfamoyl-1,2,4-benzothiadiazine 1,1-dioxide, DC84

6-chloro-3,4-dihydro-7-sulfamoyl-2H-1,2,4-benzothiadiazine 1,1-dioxide, DB74

7-chloro-1,3-dihydro-5-(2-thiazolyl)-2H-1,4-benzodiazepin-2-one, DD62

dibenzothiophene, DD66

2,3-dibromothiophene, DA29

3,5-di(tert-butyl)-1,2-dithiolylium perchlorate, DD57

2,3-dihydro-5-(5-isothiazolyl)-1-methyl-7-nitro-1H-1,4-benzodiazepine, DE08

2,3-dihydro-1-methyl-7-nitro-5-(2-thiazolyl)-1H-1,4-benzodiazepine, DE09

2,6-dimercapto-1,4-thiopyrone-3-carboxylic acid, DA87

3-[4-(dimethylamino)phenyl]-5-(4-methoxyphenyl)-1,2-dithiol-1-ium perchlorate, DF93

3,5-dimethyl-1,2-dithiolylium perchlorate, DA69

3,5-diphenyl-1,2-dithiolylium perchlorate, DE82

3-(4-methoxyphenyl)-5-phenyl-1,2-dithiol-1-ium perchlorate, DF14

Methylene Blue, DF38

10-methylphenothiazine, DE04, DH38, DH39, DH40, DH41

3-methyl-5-phenyl-1,2-dithiolylium perchlorate, DD06

3-(4-methylphenyl-5-phenyl-1,2-dithiol-1-ium perchlorate, DF13

3-(N-nitrosomethylamino)thiolane, DA76

4-nitro-2-thienylcarboxaldehyde, DA52

5-nitro-2-thienylcarboxaldehyde, DA53

10-phenylphenothiazine, DF80, DH48

thianthrene, DD67, DH36, DH37

HYDRAZINES and HYDRAZIDES

1-amino-7-chloro-2,3-dihydro-5-phenyl-1H-1,4-benzodiazepine, DE92

1,1'-bipyrrolidine, DC47

2,5-bis(4-chlorophenyl)-3,6-diphenylpyrazolo[3,4-d]pyrazolium anion, DG56

2-tert-butyl-2-azonia-3-aza-2-norbornene ion, DC92

2-tert-butyl-2,3-diazanorbornane, DC96

2-tert-butyl-3-methyl-2,3-diazanorbornane, DD39

7-chloro-1-acetamido-2,3-dihydro-5-phenyl-1H-benzodiazepine, DF64

7-chloro-1,3-dihydro-2-hydrazino-5-phenyl-2H-1,4-benzodiazepine 4-N-oxide, DE91

6,6-cyclopentamethylene-1,5-diazabicyclo[3.1.0]hexane, DC91

1,6-diazabicyclo[4.4.0]deca-3,8 diene, DC41

1,6-diazabicyclo[4.4.0]decane, DC48

1,5-diazabicyclo[3.3.0]octane, DB33

3,4-di-tert-butyl-1,3,4-oxadiazolidine, DD41

2,2-diethyl-5,8-ethanoperhydropyrazolo[1,2-a]pyridazine, DE19

3,4-diethyl-1,3,4-oxadiazolidine, DB41

1,2-diethylpyrazolidine, DC01

7-(1,3-dihydro-2-isoindolyl)-7-azabicyclo[4.1.0]heptane, DE57

2,3-dihydro-1-methyl-5-(1-methyl-1H-pyrazol-5-yl)-7-nitro-1H-1,4-benzodiazepine, DE48

3,4-diisopropyl-1,3,4-oxadiazolidine, DC54

(HYDRAZINES and HYDRAZIDES)(cont.)

   1,2-diisopropylpyrazolidine, DC97

   2,5-trans-dimethyl-3,4-cyclotetramethyleneoxa-3,4-diazolidine, DC50

   2,3-dimethyl-2,3-diazabicyclo[2.2.1]heptane, DB99

   2,3-dimethyl-2,3-diazabicyclo[2.2.2]octa-5-ene, DC44

   2,3-dimethyl-2,3-diazabicyclo[2.2.2]octane, DC49

   2,5-dimethyl-1,2,3,4,5,6-hexaazabicyclo[2.2.1]heptane, DA26

   1,2-dimethylhexahydropyridazine, DB40

   1,2-dimethyl-3-n-hexyldiaziridine, DC98

   1,2-dimethylpyrazolidine, DA78

   1,2-dimethyl-1,2,3,6-tetrahydropyridazine, DB34

   diphenylthiocarbazone, DE10

   4-methyl-2,3,6-triphenylpyrazolo[3,4-d]pyrazole, DG34

   4,4'-oxybis[benzenesulfonyl(isopropylidenehydrazide)], DF96

   1,2,4,5-tetraethylperhydro-s-tetrazine, DD42

   1,2,4,5-tetraisopropylperhydro-s-tetrazine, DE68

   1,2,4,5-tetramethylperhydro-s-tetrazine, DB47

   tetramethylhydrazine, DA50

   1,2-cis-3,6-tetramethylperhydropyridazine, DC52

   1,2-trans-3,6-tetramethylperhydropyridazine, DC53

   tetraphenylhydrazine, DG38

   1,2,3-trimethylperhydropyridazine, DC02

   1,2,3-trimethyl-1,2,3,6-tetrahydropyridazine, DC00

HYDROCARBONS
  Aromatic
    One or More Rings, Uncondensed

     benzene, DA96

     1,4-diphenyl-1,3-butadiene, DF19

     diphenylbutadiyne, DE98

     α-methylstyrene, DC80

     phenylacetylene, DC08

     styrene, DC25

     tetraphenylethylene oxide, DG49

    Two Condensed Rings

     benzo[b]bicyclo[6.1.0]nona-2,4,6-triene, DE05

     benzocyclooctatetraene, DD70

     naphthalene, DD03

    Three or more Condensed Rings

     acenaphthylene, DD60

     anthracene, DE32, DH32, DH42

     (anthracene)tricarbonyliron(0), DF50

     benzo[g,h,i]perylene, DG29

     1,6-bis(dimethylamino)pyrene, DG12

     dibenzo[a,e]cyclooctatetraene, DF05

     dibenzotriptycene, DG55

     9,16-dichloro-9,10-dihydro-9,10-diphenylanthracene, DG48, DH47, DH50, DH54, DH55, DH61

     5,12-dihydro-5,12-o-benzenonaphthacene, DG35

     9,10-diphenylanthracene, DG47, DH33, DH52, DH53

     fluoranthene, DE99, DH38, DH48

     2,3,6,7,10,11-hexamethoxytriphenylene, DG40

     2,3,6,7,10,11-hexamethoxytriphenylene radical cation, DG41

     rubrene, DH13, DH35, DH59, DH60, DH61, DH62, DH63

     1,3,6,8-tetraphenylpyrene, DH11, DH34, DH58

  Aromatic Compounds in Complexes (see also ORGANOMETALLICS)

    acetyl-η-cyclopentadienyltricarbonylmanganese(II), DD02

    acetonyltricarbonyl-η-cyclopentadienylmolybdenum, DG32

    bis-[π-(η-cyclopentadienylnickel(I))]acetylene, DD81

    bis[dicarbonyl-(η-cyclopentadienyl)iron(I)], DE34

    bis[dicarbonyl-η-cyclopentadienyliron(I)]dimethyltin(II), DF36

(HYDROCARBONS)(cont.)
Aromatic Compounds in Complexes (cont.)

bis[dicarbonyl-(η-cyclopentadienyl)-ruthenium(I)], DE37

bis[(tricarbonyl)(η-cyclopentadienyl)chromium(I)], DF01

bis[(tricarbonyl)(η-cyclopentadienyl)chromium(I)]mercury(0), DF00

bis[(tricarbonyl)(η-cyclopentadienyl)molybdo(I)], DF03

bis[(tricarbonyl)(η-cyclopentadienyl)molybdo(I)mercury(0), DF02

bis[(tricarbonyl-η-cyclopentadienyltrifluoromethyl)molybdenum], DF75

η-cyclopentadienylperfluorodimethyl-ethylene-1,2-disulfidoiridium(III), DC59

η-cyclopentadienylperfluorodimethyl-ethylene-1,2-disulfidoiron(III), DC58

η-cyclopentadienylperfluorodimethyl-ethylene-1,2-disulfidonickel(III), DC60

η-cyclopentadienylperfluorodimethyl-ethylene-1,2-disulfidorhodium(III), DC61

η-cyclopentadienyltrifluoromethyl-tricarbonylmolybdenum(0), DC56

[dicarbonyl-η-cyclopentadienylmolybdenum(I)]tricarbonyl-η-cyclopentadienylmolybdenum(I)]dimethyltin(II), DF68

dicarbonyl-(η-cyclopentadienyl)trimethylstannio(III)molybdenum(I), DD26

dicarbonyl-(η-cyclopentadienyl)[triphenylplumbio(III)]iron(I), DG43

di-μ-carbonyl-(η-cyclopentadienyl)[triphenylstannio(III)iron(I), DG44

ferrocene, DD14

μ-hydrido-μ-dimethylphosphinobis-[tricarbonyl-(η-cyclopentadienyl)molybdenum(I)], DF91

1-hydroxybutylferrocene, DE56

pentacarbonylmanganese(0)(dicarbonyl)-η-cyclopentadienyliron(II), DD59

sodium cyclopentadienyldicarbonyl-ruthenate(0), DB64

sodium (cyclopentadienyl)tricarbonylchromate(0), DC06

sodium cyclopentadienyltricarbonylmolybdate(0), DC07

tetrakis[carbonyl-(η-cyclopentadienyl)iron(I)], DG37

tetrakis[carbonyl-(η-cyclopentadienyl)iron]hexafluorophosphate, DG36

α-tolylcyclopentadienyltricarbonylmolybdenum(II), DE86

tricarbonyl-η-cyclopentadienylmolybdenum(I)dicarbonyl-η-cyclopentadienyliron(II), DE75

1,2,2-trimethyl-1-propyl)ferrocene, DF39

tris(carbonyl-η-cyclopentadienylcobalt), DF84

tris(η-cyclopentadienyl)dicarbonyltrinickel, DF62

HYDROXY COMPOUNDS (see also QUINONES)
Aliphatic and Alicyclic

16α-bromo-3-hydroxyestra-1,3,5(10)-trien-17-one, DF94

16β-bromo-3-hydroxyestra-1,3,5(10)-trien-17-one, DF95

calciferol, DG57

cholesta-5,7-dien-3β-ol, DG52

trans-di(4-acetylpyridine)hematoporphyrincobalt(III), DH25

trans-di(4,4'-bipyridine)hematoporphyrincobalt(III), DH28

trans-di(ethylnicotinate)hematoporphyrincobalt(III), DH26

trans-di(2-methylpyridine)hematoporphyrincobalt(III), DH23

trans-di(4-methylpyridine)hematoporphyrincobalt(III), DH24

trans-di(4-phenylpyridine)hematoporphyrincobalt(III), DH29

trans-dipyridinehematoporphyrincobalt(III) chloride, DH20

ergosta-5,7,22-trien-3β-ol, DG58

formaldehyde sodium sulfoxylate, DA04

hematoporphyrincobalt(III), DG73

1-hydroxybutylferrocene, DE56

3-hydroxy-5-(hydroxymethyl)-2-methyl-4-pyridinecarboxylic acid, DC34

(HYDROXY COMPOUNDS)(see also QUINONES) (cont.)
  Aliphatic and Alicyclic (cont.)

    4-[2-(5-hydroxy-2-methylcyclohexen-1-yl)-ethenyl]-7a-methyl-3a,6,7,7a-tetrahydroindan, DG59

    3-(2-methoxyphenoxy)-1,2-propanediol, DD31

    5-nitrofurfuryl alcohol, DA65

    1-phenyl-2-nitroethanol, DC32

    pyridoxamine, DC42

    pyridoxine, DC40

    retinol, DG19

    vitamin D_3, DG53

  Aromatic
    Phenols

    3-amino-4-hydroxyphenylarsonic acid, DB20

    4-aminophenol, DB15

    bis(3,4,6-trichloro-2-hydroxyphenyl)methane, DD97

    7-chloro-5-(4-hydroxyphenyl)-2-methylamino-3H-1,4-benzodiazepine 4-N-oxide, DF24

    2,4-dihydroxy-5-nitrophenylarsonic acid, DB02

    2,4-dihydroxyphenylarsonic acid, DB13

    2,6-dimethylphenol, DC38

    hydroquinone, DB07

    4-hydroxyazobenzene, DD73

    4-hydroxynitrobenzene, DA94

    4-hydroxy-3-nitrophenylarsonic acid, DB01

    2-hydroxy-5-nitropyridine, DA57

    4-hydroxyphenylarsonic acid, DB12

    2-hydroxyphenylmercury chloride, DA88

    4-nitrosophenol, DA92

    quinhydrone, DD76

    resorcinol, DB08

    salicylaldehyde, DB70

    sodium 2-methylphenoxide, DB73

    2,3,5,6-tetramethylhydroquinone, DD30

  Two or More Rings

    Alizarine Red S, DE22

    naphthyl-1,1'-azo-2-naphthol-4',6-disulfonic acid, disodium salt, DG09

    1-o-tolylazo-2-naphthol, DF54

  Heterocyclic
  Carbohydrates

    ascorbic acid, DB23

    cytidine, DC88

    cytidine monosphosphate, DC89

    cytidylyl (3'→5') cytidine, DF98

    cytidylyl (3'→5') guanidine, DG05

    cytidylyl (3'→5') uridine, DF97

    D-glucose, DB37

    N-(2-hydroxymethyl-3,4-dihydroxy-5-tetrahydrofuryl)-3-carboxypyridine-(1H)2-one, DD52

    nicotinamide adenine dinucleotide phosphate, reduced, DG27

    nicotinamide adenine dinucleotide, reduced, DG26

    riboflavin, DF74

    sodium flavin mononucleotide, DF73

    D-xylose, DA77

  Others

    allopurinol, DA60

    9-(2-carboxyphenyl)-4,5-dibromo-6-hydroxy-3H-xanthen-2-one, DG06

    9-(2-carboxyphenyl)-4,5-dichloro-6-hydroxy-3H-xanthen-3-one, DG07

    9-(2-carboxyphenyl)-4,5-diiodo-6-hydroxy-3H-xanthen-3-one, DG08

    Chlorazepam, DF12

    7-chloro-5-(2-chlorophenyl)-2,3-dihydro-3-hydroxy-1H-1,4-benzodiazepine-2-one, DE74

    7-chloro-2,3-dihydro-2-hydroxy-1-methyl-5-phenyl-1H-1,4-benzodiazepine, DF30

    3,3-diethyl-2,4-dioxopiperidine, DC90

TABLE V. Functional-Group Index

(HYDROXY COMPOUNDS)(see also QUINONES) (cont.)
  Heterocyclic (cont.)
    Others (cont.)

    2-hydroxynicotinic acid, DA93

    2-hydroxy-3-nitropyridine, DA56

    6-methoxy-1-hydroxy phenazine-5,10-dioxide, DE00

    2-methyl-2-(4,8,12-trimethyltridecyl)-6-chromanol, DG50

    Oxazepam, DE81

    pyridoxal, DC33

    pyridoxal 5-phosphate, DC36

    pyridixamine, DC42

    4-pyridoxic acid, DC34

    pyridoxine, DC40

    riboflavin, DF74

    α-tocopherol, DG63

    β-tocopherol, DG60

    γ-tocopherol, DG61

    δ-tocopherol, DG54

    uric acid, DA61

  Hydroxylamines

    cyclohexylhydroxylamine, DB38

HYDROXY SALTS, see HYDROXY COMPOUNDS

IMIDES, see HETEROCYCLIC COMPOUNDS

IMINES AND OTHER AZOMETHINES (see also Oximes)

  Aliphatic

    4,4'-oxybis[benzenesulfonyl(isopropylidenehydrazide)], DF96

    s-tetramethylguanidine, DA80

  Aromatic

    1-naphthaldehydeazine, DG30

    2-naphthaldehydeazine, DG31

  Cyclic

    7-acetoxy-2-methylamino-5-phenyl-3H-1,4-benzodiazepine 4-N-oxide, DF92

    7-acetamido-2,3-dihydro-5-phenyl-1H-1,4-benzodiazepin-2-one, DF61

    1-acetyl-7-chloro-2,3-dihydro-5-phenyl-1H-1,4-benzodiazepine, DF56

    7-acetyl-1,3-dihydro-5-phenyl-2H-1,4-benzodiazepin-2-one, DF55

    7-acetyl-5-(2-fluorophenyl)-2-methylamino-3H-1,4-benzodiazepine 4-N-oxide, DF87

    1-amino-7-chloro-2,3-dihydro-5-pheny-1H-1,4-benzodiazepine, DE92

    2-amino-7-chloro-5-phenyl-3H-1,4-benzodiazepine 4-N-oxide, DE85

    5-(2-aminophenyl)-7-chloro-2,3-dihydro-1-methyl-1H-1,4-benzodiazepine, DF35

    7-bromo-1,3-dihydro-5-phenyl-2H-1,4-benzodiazepin-2-one, DE79

    7-bromo-2-methylamino-5-phenyl-3H-1,4-benzodiazepine 4-N-oxide, DF20

    2-butylamino-7-chloro-5-phenyl-3H-1,4-benzodiazepine 4-N-oxide, DG04

    2-carbamoyl-7-chloro-2,3-dihydro-1-methyl-5-phenyl-1H-1,4-benzodiazepine, DF63

    Chlorazepam, DF12

    7-chloro-1-acetamido-2,3-dihydro-5-phenyl-1H-benzodiazepine, DF64

    7-chloro-5-(2-chlorophenyl)-2,3-dihydro-1H-1,4-diazepin-2-one, DE73

    7-chloro-5-(2-chlorophenyl)-2,3-dihydro-3-hydroxy-1H-1,4-benzodiazepin-2-one, DE74

    7-chloro-2-cyano-2,3-dihydro-1-methyl-5-phenyl-1H-1,4-benzodiazepine, DF53

    7-chloro-5-(2,3-dichlorophenyl-2,3-dihydro-1-methyl-1H-1,4-benzodiazepine, DF15

    7-chloro-5-(2,6-dichlorophenyl)-2,3-dihydro-1-methyl-1H-1,4-benzodiazepine, DF16

    7-chloro-1,3-dihydro-5-(2-fluorophenyl)-2H-1,4-benzodiazepin-2-one, DE72

    7-chloro-1,3-dihydro-2-hydrazino-5-phenyl-2H-1,4-benzodiazepine 4-N-oxide, DE91

(IMINES and OTHER AZOMETHINES)(see also Oximes)(cont.)
Cyclic (cont.)

7-chloro-2,3-dihydro-2-hydroxy-1-methyl-5-phenyl-1H-1,4-benzodiazepine, DF30

7-chloro-1,3-dihydro-5-(1H-indol-1-yl)-2H-1,4-benzodiazepin-2-one, DF51

7-chloro-2,3-dihydro-1-methanoyl-5-phenyl-1H-1,4-benzodiazepine, DF06

7-chloro-1,3-dihydro-5-(2-methoxyphenyl)-2H-1,4-benzodiazepin-2-one, DF10

7-chloro-2,3-dihydro-1-(N-methylcarbamoyl)-5-phenyl-1H-1,4-benzodiazepine, DF65

7-chloro-2,3-dihydro-1-methyl-5-(2-methylmercaptophenyl)-1H-1,4-benzodiazepine, DF70

7-chloro-2,3-dihydro-1-methyl-5-(2-methylsulfinylphenyl)-1H-1,4-benzodiazepine, DF69

7-chloro-2,3-dihydro-1-methyl-5-(2-nitrophenyl)-1H-1,4-benzodiazepine, DF23

7-chloro-2,3-dihydro-1-methyl-5-phenyl-1H-1,4-benzodiazepin-2-carboxylic acid, DF57

7-chloro-2,3-dihydro-1-methyl-5-phenyl-1H-1,4-benzodiazepine, DF29

7-chloro-1,3-dihydro-5-(2-methylphenyl)-2H-1,4-benzodiazepin-2-one, DF07

7-chloro-1,3-dihydro-5-(1-methyl-1H-pyrazol-5-yl)-2H-1,4-benzodiazepin-2-one, DE03

7-chloro-2,3-dihydro-1-methylsulfonyl-5-phenyl-1H-1,4-benzodiazepine, DF31

7-chloro-2,3-dihydro-5-phenyl-1H-1,4-benzodiazepine, DE90

7-chloro-1,3-dihydro-5-phenyl-2H-1,4-benzodiazepin-2-one, DE80

7-chloro-1,3-dihydro-5-(2-pyridinyl)-2H-1,4-benzodiazepin-2-one, DE33

7-chloro-1,3-dihydro-5-(2-thiazolyl)-2H-1,4-benzodiazepin-2-one, DD62

7-chloro-2-dimethylamino-5-phenyl-3H-1,4-benzodiazepine 4-N-oxide, DF66

7-chloro-2-ethylamino-5-phenyl-2H-1,4-benzodiazepine 4-N-oxide, DF67

7-chloro-5-(4-hydroxyphenyl)-2-methylamino-3H-1,4-benzodiazepine 4-N-oxide, DF24

7-chloro-2-methoxy-5-phenyl-3H-1,4-benzodiazepine 4-N-oxide, DF11

7-chloro-2-(N-methylacetamido)-5-phenyl-3H-1,4-benzodiazepine 4-N-oxide, DF86

7-chloro-2-methylamino-5-phenyl-3H-1,4-benzodiazepine 4-N-oxide, DF22

7-chloro-2-methyl-5-phenyl-3H-1,4-benzodiazepine 4-N-oxide, DF08

7-chloro-2(N-methylpropionamido)-5-phenyl-3H-1,4-benzodiazepine 4-N-oxide, DG03

5-(2-chlorophenyl)-2,3-dihydro-7-iodo-1-methyl-1H-1,4-benzodiazepine, DF21

cytidylyl (3'→5') cytidine, DF98

cytidylyl (3'→5') guanidine, DG05

cytidylyl (3'→5') uridine, DF97

Diazepam, DF09

1,3-dihydro-7-fluoro-5-phenyl-2H-1,4-benzodiazepin-2-one, DE83

2,3-dihydro-5-(5-isothiazolyl)-1-methyl-7-nitro-1H-1,4-benzodiazepine, DE08

1,3-dihydro-7-methoxy-5-phenyl-2H-1,4-benzodiazepin-2-one, DF28

2,3-dihydro-1-methyl-5-(1-methyl-1H-imidazol-2-yl)-7-nitro-2H-1,4-benzodiazepine, DE47

2,3-dihydro-1-methyl-5-(1-methyl-1H-pyrazoly-5-yl)-7-nitro-2H-1,4-benzodiazepine, DE48

2,3-dihydro-1-methyl-7-nitro-5-phenyl-1H-1,4-benzodiazepine, DF34

2,3-dihydro-1-methyl-7-nitro-5-(4-pyridyl)-1H-1,4-benzodiazepine, DE93

2,3-dihydro-1-methyl-7-nitro-5-(2-pyrimidinyl)-1H-1,4-benzodiazepine, DE41

2,3-dihydro-1-methyl-7-nitro-5-(2-thiazolyl)-1H-1,4-benzodiazepine, DE09

3,4-dihydro-2-methyl-4-oxo-3-o-tolyl-quinazoline, DF25

(IMINES AND OTHER AZOMETHINES)(see also Oximes)(cont.)
Cyclic (cont.)

    2,3-dihydro-1-methyl-5-phenyl-1H-1,4-benzodiazepine, DF37

    1,3-dihydro-7-methyl-5-phenyl-2H-1,4-benzodiazepin-2-one, DF26

    1,3-dihydro-7-methylthio-5-phenyl-2H-1,4-benzodiazepin-2-one, DF27

    1,3-dihydro-5-phenyl-2H-1,4-benzodiazepin-2-one, DE87

    1,3-dihydro-5-phenyl-7-trifluoromethyl-2H-1,4-benzodiazepin-2-one, DF04

    7,9-diiodo-1,3-dihydro-5-phenyl-2H-1,4-benzodiazepin-2-one, DE76

    1,4-dimethyl-3,6-diphenylpyrazolo-[4,5-d]pyrazole, DF88

    2,4-dimethyl-3,6-diphenylpyrazolo-[3,4-d]pyrazole, DF89

    5-(2-fluorophenyl)-7-iodo-2-methylamino-3H-1,4-benzodiazepine 4-N-oxide, DF17

    2-methylamino-7-methylmercapto-5-phenyl-3H-1,4-benzodiazepine 4-N-oxide, DF72

    2-methylamino-5-phenyl-3H-1,4-benzodiazepine 4-N-oxide, DF33

    7-methyl-2-methylamino-5-phenyl-3H-1,4-benzodiazepine 4-N-oxide, DF71

    nicotinamide adenine dinucleotide phosphate, reduced, DG27

    nicotinamide adenine dinucleotide, reduced, DG26

    Nitrazepam, DE84

    Oxazepam, DE81

ISOTHIOCYANATES and THIOCYANATES

    methyl 2-methyl-5-(thiocyanatomethyl)-3-furoate, DC77

    methyl 5-(thiocyanatomethyl)-2-furoate, DC23

    4-nitrobenzyl thiocyanate, DC14

    5-nitro-2-(thiocyanatomethyl)furan, DA83

KETONES (see also HETEROCYCLIC COMPOUNDS; IMIDES; QUINONES and HYDROQUINONES; and UNSATURATED COMPOUNDS

Dialkyl Ketones

    1-acetyl-4-tert-butyl-1-cyclohexene, DD90

    diacetonylmercury, DB25

    2,4-dibromo-2,4-dimethyl-3-pentanone, DB92

    trans-4,4-dimethyl-1-phenyl-1-penten-3-one, DE14

    3-heptyn-2-one, DB87

    methyl 4-bromo-5-oxohexanoate, DB91

    3-methyl-3-penten-2-one, DB27

    4-oxo-1-pentyl acetate, DB94

    trans-3-penten-2-one, DA71

    3-phenyl-3-hepten-2-one, DE15

    D-xylose, DA77

Aryl-Alkyl Ketones

    acetyl-η-cyclopentadienyltricarbonylmanganese(II), DD02

    chalcone, DE88

    [2,3-dichloro-4-(2-methylene-1-oxobutyl)phenoxy]acetic acid, DE06

    phenylglyoxalic acid, DC15

    1,2,3,3-tetraphenylpropenone, DG51

    3-trifluoromethyl-ω-diazoacetophenone, DC57

    1,1,3-triphenylpropenone, DG21

    cis-1,2,3-triphenyl-2-propenone, DG22

    trans-1,2,3-triphenyl-2-propenone, DG23

Alkyl-Heterocyclic Ring Ketones

    7-acetylbenzo[c]phenothiazine 12-dioxide, DF78

    12-acetylbenzo[a]phenothiazine 7-dioxide, DF79

    12-acetylbenzo[a]phenothiazine 7-oxide, DF77

    1-acetyl-7-chloro-2,3-dihydro-5-phenyl-1H-1,4-benzodiazepine, DF56

    7-acetyl-1,3-dihydro-5-phenyl-2H-1,4-benzodiazepin-2-one, DF55

    2-acetyl-5-nitrofuran, DA95

(KETONES (see also HETEROCYCLIC COMPOUNDS; IMIDES; QUINONES and HYDROQUINONES; and UNSATURATED COMPOUNDS) (cont.)

Diaryl Ketones

benzophenone, DE02

9-fluorenone, DD99

Diheterocyclic-Ring Ketones

dinitrosyl[di-(2-pyridyl) ketone]-iron, DD46

di-2-pyridyl ketone, DD47

Alicyclic Ketones

16α-bromo-3-hydroxyestra-1,3,5(10)-trien-17-one, DF94

16β-bromo-3-hydroxyestra-1,3,5(10)-trien-17-one, DF95

2-cyclopentenone, DA68

5,5-dimethyl-3-phenyl-2-cyclohexenone, DE52

2-ethoxy-2-cyclopentenone, DB89

9α-fluoro-17α,21-dihydroxy-5-pregnene-3,11,20-trione-3-ethyleneketal-17,20,21-bis-methylenedioxolane, DG45

2,4a,5,6,7,8-hexahydro-4a-8-dimethylnaphthalen-2-one, DD85

3-isobutoxy-5,5-dimethyl-2-cyclohexen-1-one, DD91

2-methoxy-3,5,5-trimethyl-2-cyclohexen-1-one, DD34

5-methyl-2-cyclohexenone, DB88

1,2,3,4,4a,5,6,7-octahydronaphthalen-1-one, DD28

1,2,3,4,5,6,7,8-octahydronaphthalen-1-one, DD29

$16\alpha,17\alpha$-oxidoprogesterone, DG25

phenindione, DE78

2-phenyl-2-cyclohexenone, DD82

Heterocyclic Ketones

6-acetamido-2-phenylchromanone, DF60

9-(2-carboxyphenyl)-4,5-dibromo-6-hydroxy-3H-xanthen-2-one, DG06

9-(2-carboxyphenyl)-4,5-dichloro-6-hydroxy-3H-xanthen-3-one, DG07

9-(2-carboxyphenyl)-4,5-diiodo-6-hydroxy-3H-xanthen-3-one, DG08

cytidylyl (3'→5') cytidine, DF98

cytidylyl (3'→5') guanidine, DG05

cytidylyl (3'→5') uridine, DF97

3,3-diethyl-2,4-dioxopiperidine, DC90

3,3-diethyl-5-methyl-2,4-dioxopiperidine, DD37

3,3-diethyl-2,4-(1H,3H)pyridinedione, DC87

2,6-dimercapto-1,4-thiopyrone-3-carboxylic acid, DA87

5,7-diphenyl-1,3-diazatricyclo-[3.3.1.1^{3,7}]decan-6-one, DG13

N-(2-hydroxymethyl-3,4-dihydroxy-5-tetrahydrofuryl)-3-carboxy-pyridine-(1H)2-one, DD52

strychnine, DG24

LACTAMS, see LACTONES

LACTONES and LACTAMS (see also HETEROCYCLIC COMPOUNDS)

ascorbic acid, DB23

9-(2-carboxyphenyl)-4,5-dibromo-6-hydroxy-3H-xanthen-2-one, DG06

9-(2-carboxyphenyl)-4,5-dichloro-6-hydroxy-3H-xanthen-3-one, DG07

9-(2-carboxyphenyl)-4,5-diiodo-6-hydroxy-3H-xanthen-3-one, DG08

cytidylyl (3'→ 5') cytidine, DF98

cytidylyl (3'→ 5') guanidine, DG05

cytidylyl (3'→ 5') uridine, DF97

orotic acid, DA58

uric acid, DA61

1-vinyl-2-pyrrolidone, DB24

METHOXIMES, see OXIMES

NITRILES

Aliphatic and Alicyclic

acrylonitrile, DA11

4-(tert-butyl)-1-cyanocyclohexene, DD54

crotonitrile, DA35

2-cyanoethylmercuric iodide, DA13

ethyl 2-cyano-2,4-hexadienoate, DC85

TABLE V. Functional-Group Index

(NITRILES)(cont.)
Aliphatic and Alicyclic (cont.)

3-iodopropanonitrile, DA14

2,4-pentadienenitrile, DA62

Benzonitriles

benzonitrile, DB61

2-bromobenzonitrile, DB48

3-bromobenzonitrile, DB49

4-bromobenzonitrile, DB50

2-chlorobenzonitrile, DB51

3-chlorobenzonitrile, DB52

4-chlorobenzonitrile, DB53

1,4-dicyanobenzene, DC03

4,4'-dicyanobiphenyl, DE23

4-iodobenzonitrile, DB57

4-methylbenzonitrile, DC21

4-nitrobenzonitrile, DB59

4-nitrophenylacetonitrile, DC13

Heterocyclic

7-chloro-2-cyano-2,3-dihydro-1-methyl-5-phenyl-1H-1,4-benzodiazepine, DF53

2-cyano-5-nitrofuran, DA51

protoporphyrin(IX)-trans-cyanopyridineiron(III), DH12

NITRO COMPOUNDS
Aliphatic and Alicyclic

N-methyl-N-nitroso-N'-nitroguanidine, DA09

nitrocyclohexane, DB30

β-nitrostyrene, DC22

1-phenyl-2-nitroethane, DC31

1-phenyl-2-nitroethanol, DC32

tetranitromethane, DA01

trinitromethane, DA02

Nitrobenzenes

1,2-bis(4-nitrophenyl)ethane, DE39

5,5-bis(3-nitrophenyl)-2,4-imidazolidinedione, DE77

5-bromo-3,6-dinitro-1,2,4-trimethylbenzene, DC73

7-chloro-2,3-dihydro-1-methyl-5-(2-nitrophenyl)-1H-1,4-benzodiazepine, DF23

2,4-dihydroxy-5-nitrophenylarsonic acid, DB02

1,3-dinitrobenzene, DA84

5-ethyl-5-(3-nitrophenyl)barbituric acid, DD80

4-hydroxynitrobenzene, DA94

4-hydroxy-3-nitrophenylarsonic acid, DB01

4-methyl-2-nitroaniline, DB76

3-nitroaniline, DB05

4-nitroaniline, DB06

4-nitroazobenzene, DD69

3-nitrobenzaldehyde, DB62

4-nitrobenzaldehyde, DB63

nitrobenzene, DA91

4-nitrobenzonitrile, DB59

4-nitrobenzyl chloride, DB65

4-nitrobenzyl thiocyanate, DC14

4-nitrophenylacetonitrile, DC13

2-nitrophenylarsonic acid, DA98

3-nitrophenylarsonic acid, DA99

4-nitrophenylarsonic acid, DB00

4-nitrotoluene, DB71

2-nitro(trifluoromethyl)benzene, DB54

3-nitro(trifluoromethyl)benzene, DB55

4-nitro(trifluoromethyl)benzene, DB56

Nitroaromatics with Condensed Rings

9,10-dinitroanthracene, DE25

9-nitroanthracene, DE31

Nitro Derivatives of N-Heterocycles

2-amino-3-nitropyridine, DA66

2-amino-5-nitropyridine, DA67

2,3-dihydro-5-(5-isothiazolyl)-1-methyl-7-nitro-1H-1,4-benzodiazepine, DE08

2,3-dihydro-1-methyl-5-(1-methyl-1H-imidazol-2-yl)-7-nitro-1H-1,4-benzodiazepine, DE47

(NITRO COMPOUNDS)(cont.)
  Nitro Derivatives of N-Heterocycles (cont.)

  2,3-dihydro-1-methyl-5-(1-methyl-1H-pyrazol-5-yl)-7-nitro-1H-1,4-benzodiazepine, DE48

  2,3-dihydro-1-methyl-7-nitro-5-phenyl-1H-1,4-benzodiazepine, DF34

  2,3-dihydro-1-methyl-7-nitro-5-(4-pyridyl)-1H-1,4-benzodiazepine, DE93

  2,3-dihydro-1-methyl-7-nitro-5-(2-pyrimidinyl)-1H-1,4-benzodiazepine, DE41

  2,3-dihydro-1-methyl-7-nitro-5-(2-thiazolyl)-1H-1,4-benzodiazepine, DE09

  2-hydroxy-3-nitropyridine, DA56

  2-hydroxy-5-nitropyridine, DA57

  Nitrazepam, DE84

Nitro Derivatives of O-Heterocycles

  2-acetyl-5-nitrofuran, DA95

  2-cyano-5-nitrofuran, DA51

  ethyl-5-nitro-2-furoate, DB72

  2-methyl-5-nitrofuran, DA64

  5-nitro-2-furaldehyde, DA54

  5-nitrofuraldehyde diacetate, DB82

  2-nitrofuran, DA30

  5-nitrofurfuryl alcohol, DA65

  5-nitro-2-(thiocyanatomethyl)furan, DA83

  5-nitro-2-furoic acid, DA55

Nitro Derivatives of S-Heterocycles

  4-nitro-2-thienylcarboxaldehyde, DA52

  5-nitro-2-thienylcarboxaldehyde, DA53

NITROSO and NITROSYL COMPOUNDS

  carbonyl(nitrosyl)-2,2'-bipyridinecobalt, DD45

  carbonyl(nitrosyl)-1,10-phenanthrolinecobalt, DD98

  dibenzyl-N-nitrosoamine, DE42

  dicyclohexyl-N-nitrosoamine, DD95

  diisopropyl-N-nitrosoamine, DB42

  dimethyl-N-nitrosoamine, DA10

  dinitrosyl-2,2'-bipyridineiron, DD04

  dinitrosyl[di-(2-pyridyl) ketone]-iron, DD46

  dinitrosyl-1,10-phenanthroline-iron, DD63

  ethyl N-methyl-N-nitrosocarbamate, DA44

  N-methyl-N-nitrosoaniline, DB75

  N-methyl-N-nitroso-N'-nitroguanidine, DA09

  N-methyl-N-nitrosourea, DA08

  nitrosobenzene, DA89

  N-nitrosodiphenylamine, DD74

  3-(N-nitrosomethylamino)thiolane, DA76

  4-nitrosophenol, DA92

  N-nitrosopiperidine, DA75

  N-nitrososarcosinamide, DA24

ORGANOMETALLICS
  Al

  2,3,7,8,12,13,17,18-octaethylporphinatohydroxoaluminum(III), DG93

  Ag

  2,3,7,8,12,13,17,18-octaethylporphinatosilver(III) perchlorate, DG81

  B

  bis[π-μ-1,3-trimethylene-1,2-dicarbaundecaborane(9)]nickel(IV) (orange form), DD43

  bis[π-μ-1,2-trimethylene-1,2-dicarbaundecaborane(9)]nickel(IV) (yellow form), DD44

  tetramethylammonium bis[π-μ-1,2-trimethylene(3)-1,2-dicarbaundecaboranyl(9)]cobaltate(III), DE69

  tetramethylammonium bis[π-μ-1,2-trimethylene(3)-1,2-dicarbaundecaboranyl(9)]nickelate(III), DE70

  Ca

  2,3,7,8,12,13,17,18-octaethylporphinatocalcium(II), DG82

TABLE V. Functional-Group Index

(ORGANOMETALLICS)(cont.)

Cd

bis(0-ethyl thioacetothioacetato)-cadmium(II), DD86

2,3,7,8,12,13,17,18-octaethylporphinatocadmium(II), DG83

Cr

bis[(tricarbonyl)(η-cyclopentadienyl)chromium(I)], DF01

bis[(tricarbonyl)(η-cyclopentadienyl)chromium(I)]mercury(0), DF00

2,3,7,8,12,13,17,18-octaethylporphinatohydroxochromium(III), DG94

sodium (cyclopentadienyl)tricarbonylchromate(0), DC06

Co

carbonyl(nitrosyl)-2,2'-bipyridinecobalt, DD45

carbonyl(nitrosyl)-1,10-phenanthrolinecobalt, DD98

cobalt(II)protoporphyrin(IX) dimethyl ester, DG74

trans-di(4-acetylpyridine)hematoporphyrincobalt(III), DH25

trans-di(4,4'-bipyridine)hematoporphyrincobalt(III), DH28

trans-di(ethylnicotinate)hematoporphyrincobalt(III), DH26

trans-di(2-methylpyridine)hematoporphyrincobalt(III), DH23

trans-di(4-methylpyridine)hematoporphyrincobalt(III), DH24

trans-di(4-phenylpyridine)hematoporphyrincobalt(III), DH29

trans-dipyridinehematoporphyrincobalt(III) chloride, DH20

hematoporphyrincobalt(III), DG73

2,3,7,8,12,13,17,18-octaethylporphinatocobalt(II), DG84

tetramethylammonium bis[π-μ-1,2-trimethylene-(3)-1,2-dicarbaundecaboranyl(9)]cobaltate(III), DE69

5,10,15,20-tetraphenylporphinatocobalt(II), DH17

tetrasodium 5,10,15,20-tetrakis(4-sulfonatophenyl)porphinatocobalt(II), DH15

tris(carbonyl-η-cyclopentadienylcobalt), DF84

trisodium 5,10,15,20-tetrakis(4-sulfonatophenyl)porphinatocobaltate(III), DH14

tris[tricarbonylcobalt(I)]methyl chloride, DC99

Cu

copper(II)protoporphyrin(IX) dimethyl ester, DG75

2,3,7,8,12,13,17,18-octaethylporphinatocopper(II), DG85

sodium phthalocyaninecopper(II)-2-sulfonate, DG68

5,7,12,14-tetramethyl-1,4,8,11-tetraazacyclotetradeca-2,5,7,9,12,14-hexaenecopper(II), DE55

5,7,12,14-tetramethyl-1,4,8,11-tetraazacyclotetradeca-2,5,7,9,12,14-hexaenecopper(I) tetrafluoroborate, DE53

5,7,12,14-tetramethyl-1,4 8,11-tetraazacyclotetradeca-5,7,12,14-tetraenecopper(II), DE61

Fe

(anthracene)tricarbonyliron(0), DF50

bis[dicarbonyl-(η-cyclopentadienyl)iron(I)], DE34

bis(dicarbonyl-η-cyclopentadienyliron(I))dimethyltin(II), DF36

η-cyclopentadienylperfluorodimethylethylene-1,2-disulfidoiron(III), DC58

dicarbonyl-(η-cyclopentadienyl)-[triphenylplumbio(III)]iron(I), DG43

dicarbonyl-(η-cyclopentadienyl)-[triphenylstannio(III)]iron(I), DG44

trans-dicyanoprotoporphyrin(IX)-iron(III), DG72

trans-di(3-methylpyridine)protoporphyriniron(III) chloride, DH21

trans-di(4-methylpyridine)protoporphyriniron(III) chloride, DH22

(ORGANOMETALLICS)(cont.)

Fe (cont.)

2,3-dimethyl-1,4,8,11-tetraazacyclotetradecaneiron(II) dication, DD96

dinitrosyl-2,2'-bipyridineiron, DD04

dinitrosyl[di-(2-pyridyl)ketone]iron, DD46

dinitrosyl-1,10-phenanthrolineiron, DD63

trans-dipyridylprotoporphyrin(IX)iron(III) chloride, DH19

ferrocene, DD14

5,7,7,12,14,14-hexamethyl-1,4,8,11-tetraaza-4,11-cyclotetradecadieneiron(II) dication, DF47

5,5,7,12,12,14-hexamethyl-1,4,8,11-tetraazacyclotetradecaneiron(II) dication bis[tetrafluoroborate(1-)], DF49

5,5,7,12,12,14-hexamethyl-1,4,8,11-tetraaza-1,3,7,10-cyclotetradecatetraeneiron(II) dication, DF41

5,5,7,12,12,14-hexamethyl-1,4,8,11-tetraaza-1,3,8,10-cyclotetradecatetraeneiron(II) dication, DF42

5,5,7,12,14,14-hexamethyl-1,4,8,11-tetraaza-1,3,7,11-cyclotetradecatetraeneiron(II) dication, DF43

5,7,7,12,14,14-hexamethyl-1,4,8,11-tetraaza-1,4,8,11-cyclotetradecatetraeneiron(II) dication, DF44

5,7,7,12,14,14-hexamethyl-1,4,8,11-tetraaza-1,3,8-cyclotetradecatrieneiron(II) dication, DF45

5,7,7,12,14,14-hexamethyl-1,4,8,11-tetra-1,4,11-cyclotetradecatrieneiron(II) dication, DF46

1-hydroxybutylferrocene, DE56

2,3,7,8,12,13,17,18-octaethylporphinatohydroxoiron(III), DG95

pentacarbonylmanganese(0)(dicarbonyl)-η-cyclopentadienyliron(I), DD59

protoporphyrin(IX)-trans-cyanopyridineiron(III), DH12

protoporphyrin(IX) iron(III) chloride, DG71

tetrakis[carbonyl-(η-cyclopentadienyl)iron(I)], DG37

tetrakis[carbonyl-(η-cyclopentadienyl)iron] hexafluorophosphate, DG36

tetrasodium phthalocyanineiron-2,9,16,23-tetrasulfonate, DG67

tricarbonyl-η-cyclopentadienylmolybdenum(I)dicarbonyl-η-cyclopentadienyliron(II), DE75

1,2,2-trimethyl-1-propyl)ferrocene, DF39

tris(bipyridineiron(II) sulfate, DG65

Ga

2,3,7,8,12,13,17,18-octaethylporphinatohydroxogallium(III), DG96

Ge

2,3,7,8,12,13,17,18-octaethylporphinatodihydroxogermanium(IV), DH03

Hg

bis(α-carbethoxybenzyl)mercury, DG14

bis(O-ethyl thioacetothioacetato)mercury(II), DD87

bis(methoxycarbonyl)mercury(II), DA37

bis[(tricarbonyl)(η-cyclopentadienyl)chromium(I)]mercury(0), DF00

bis[(tricarbonyl)(η-cyclopentadienyl)molybdo(I)mercury(0), DF02

bis(trichlorovinyl)mercury(II), DA28

2-butoxy-2-methoxyethylmercuric acetate, DC93

cis-2-chloromercury-3-ethoxycyclononene, DD55

cis-2-chloromercury-3-methoxycyclodecene, DD56

trans-2-chloromercury-3-methoxycyclotridecene, DE66

trans-2-chloromercury-3-methoxycycloundecene, DD93

2-cyanoethylmercuric iodide, DA13

diacetonylmercury, DB25

ethyl 2-bromomercurio-2-(4-isopropylphenyl)acetate, DE16

ethyl α-bromomercury-(2-bromophenyl)acetate, DD11

ethyl α-bromomercury-(3-bromophenyl)acetate, DD12

TABLE V. Functional-Group Index

(ORGANOMETALLICS)(cont.)
Hg (cont.)

ethyl α-bromomercury-(4-bromophenyl)acetate, DD13

ethyl α-bromomercury-(3-tert-butylphenyl)acetate, DE59

ethyl α-bromomercury-(4-chlorophenyl)acetate, DD08

ethyl α-bromomercury-(4-ethylphenyl)acetate, DD84

ethyl α-bromomercury-(4-fluorophenyl)acetate, DD09

ethyl α-bromomercury-(4-iodophenyl)acetate, DD10

ethyl α-bromomercury-(2-methylphenyl)acetate, DD50

ethyl α-bromomercury-(3-methylphenyl)acetate, DD51

ethyl α-bromomercuryphenylacetate, DD18

2-hydroxyphenylmercury chloride, DA88

2-isobutoxy-2-methoxyethylmercuric acetate, DC94

mercury phenylmercaptide, DD71

methylmercuric chloride, DA03

In

2,3,7,8,12,13,17,18-octaethylporphinatohydroxoindium(III), DG97

Ir

η-cyclopentadienylperfluorodimethylethylene-1,2-disulfido-iridium, DC59

Mg

isopropylmagnesium bromide, DA47

1-methylethylmagnesium bromide, DA22

2,3,7,8,12,13,17,18-octaethylporphinatomagnesium(II), DG86

Mn

acetyl-η-cyclopentadienyltricarbonylmanganese(II), DD02

manganese(III)chloride protoporphyrin(IX) dimethyl ester monohydrate, DG78

2,3,7,8,12,13,17,18-octaethylporphinatohydroxomanganese(III), DG98

pentacarbonylmanganese(0)(dicarbonyl)-η-cyclopentadienyliron(I), DD59

Mo

acetonyltricarbonyl-η-cyclopentadienylmolybdenum, DG32

bis[(tricarbonyl)(η-cyclopentadienyl)molybdo(I)], DF03

bis[(tricarbonyl)(η-cyclopentadienyl)molybdo(I)]mercury(0), DF02

bis[(tricarbonyl-η-cyclopentadienyltrifluoromethyl)molybdenum], DF75

η-cyclopentadienyltrifluoromethyltricarbonylmolybdenum(0), DC56

[dicarbonyl-η-cyclopentadienylmolybdenum(I)][tricarbonyl-η-cyclopentadienylmolybdenum(I)]-dimethyltin(II), DF68

dicarbonyl-(η-cyclopentadienyl)trimethylstannio(III)molybdenum(I), DD26

μ-hydrido-μ-dimethylphosphinobis-[tricarbonyl-(η-cyclopentadienyl)molybdenum(I)], DF91

2,3,7,8,12,13,17,18-octaethylporphinatohydroxooxomolybdenum(V), DG99

sodium cyclopentadienetricarbonylmolybdate(0), DC07

α-tolylcyclopentadienyltricarbonylmolybdenum(II), DE86

tricarbonyl-η-cyclopentadienylmolybdenum(I)dicarbonyl-η-cyclopentadienyliron(II), DE75

Ni

bis[π-(η-cyclopentadienylnickel(I)]-acetylene, DD81

bis[π-μ-1,2-trimethylene-1,2-dicarbaundecaborane(9)]nickel(IV), (orange form), DD43

bis[π-μ-1,-trimethylene-1,2-dicarbaundecaborane(9)]nickel(IV) (yellow form), DD44

η-cyclopentadienylperfluorodimethylethylene-1,2-disulfido-nickel(III), DC60

(ORGANOMETALLICS)(cont.)
Ni (cont.)

[b,i]dibenz-6,13-diacetyl-5,12-dimethyl-1,4,8,11-tetraazacyclotetradeca-2,4,6,9,11,13-hexaenenickel(II), DG39

[b,i]dibenz-5,12-dimethyl-1,4,8,11-tetraazacyclotetradeca-2,4,6,9,11,13-hexaenenickel(II), DG11

dibenzo[b,i]-1,4,8,11-tetraaza-2,5,7,9,12,14-cyclotetradecahexaenenickel(II), DF81

dicarbonyl-1,10-phenanthrolinenickel(0), DE24

7,14-diethylbis(1,2-cyclohexylenyl[c,j]-1,2,5,6,8,9,12,13-octaaza-2,4,7,9,11,14-cyclotetradecahexaenenickel(II), DF99

7,14-dimethylbis(1,2-cyclohexylenyl[c,j]-1,2,5,6,8,9,12,13-octaaza-2,4,7,9,11,14-cyclotetradecahexaenenickel(II), DF40

nickel(II)protoporphyrin(IX) dimethyl ester, DG76

2,3,7,8,12,13,17,18-octaethylporphinatonickel(II), DG87

tetramethylammonium bis[π-μ-1,2-trimethylene(3)-1,2-dicarbaundecaboranyl(9)nickelate(III), DE70

5,7,12,14-tetramethyl-1,4,8,11-tetraazacyclotetradeca-2,5,7,9,12,14-hexaenenickelate(II), DE58

5,7,12,14-tetramethyl-1,4,8,11-tetraazacyclotetradeca-2,5,7,9,12,14-hexaenenickel(I)tetrafluoroborate, DE54

5,7,12,14-tetramethyl-1,4,8,11-tetraazacyclotetradeca-5,7,12,-14-tetraenenickel(II), DE62

tris(η-cyclopentadienyl)dicarbonyltrinickel, DF62

Pb

dicarbonyl-(η-cyclopentadienyl)(triphenylplumbio(III))iron(I), DG43

2,3,7,8,12,13,17,18-octaethylporphinatolead(II), DG90

Pd

2,3,7,8,12,13,17,18-octaethylporphinatopalladium(II), DG91

Rh

η-cyclopentadienylperfluorodimethylethylene-1,2-disulfidorhodium(III), DC61

Ru

bis[dicarbonyl-(η-cyclopentadienyl)ruthenium(I)], DE37

sodium cyclopentadienyldicarbonylruthenate(0), DB64

trisbipyridineruthenium(II) dichloride, DG64, DH40, DH56

Sb

2,3,7,8,12,13,17,18-octaethylporphinatohydroxoantimony(III), DH00

Sc

2,3,7,8,12,13,17,18-octaethylporphinatohydroxoscandium(III), DH01

Si

2,3,7,8,12,13,17,18-octaethylporphinatodihydroxosilicon(IV), DH05

Sn

dicarbonyl-(η-cyclopentadienyl)-trimethylstannio(III)molybdenum-(I), DD26

dicarbonyl-(η-cyclopentadienyl)-[triphenylstannio(III)]iron(I), DG44

2,3,7,8,12,13,17,18-octaethylporphinatodihydroxotin(IV), DH06

triphenyltin hydroxide, DF90

Ti

2,3,7,8,12,13,17,18-oxooctaethylporphinatooxotitanium(IV), DG88

Tl

2,3,7,8,12,13,17,18-octaethylporphinatohydroxothallium(III), DH02

V

2,3,7,8,12,13,17,18-oxooctaethylporphinatooxovanadium(IV), DG89

TABLE V. Functional-Group Index

(ORGANOMETALLICS)(cont.)

Zn

bis(O-ethyl thioacetothioacetato)zinc(II), DD88

2,3,7,8,12,13,17,18-octaethylporphinatozinc(II), DG92

5,7,12,14-tetramethyl-1,4,8,11-tetraazacyclotetradeca-5,7,12,14-tetraenezinc, DE63

zinc(II)protoporphyrin(IX) dimethyl ester, DG77

N-OXIDES

7-acetoxy-2-methylamino-5-phenyl-3H-1,4-benzodiazepine 4-N-oxide, DF92

7-acetyl-5-(2-fluorophenyl)-2-methylamino-3H-1,4-benzodiazepine 4-N-oxide, DF87

2-amino-7-chloro-5-phenyl-3H-1,4-benzodiazepine 4-N-oxide, DE85

bis(2-quinolyl)disulfide di-N-oxide, DF76

7-bromo-2-methylamino-5-phenyl-3H-1,4-benzodiazepine 4-N-oxide, DF20

2-butylamino-7-chloro-5-phenyl-3H 1,4-benzodiazepine 4-N-oxide, DG04

7-chloro-1,3-dihydro-2-hydrazino-5-phenyl-2H-1,4-benzodiazepine 4-N-oxide, DE91

7-chloro-2-dimethylamino-5-phenyl-3H-1,4-benzodiazepine 4-N-oxide, DF66

7-chloro-2-ethylamino-5-phenyl-2H-1,4-benzodiazepine 4-N-oxide, DF67

7-chloro-5-(4-hydroxyphenyl)-2-methylamino-3H-1,4-benzodiazepine 4-N-oxide, DF24

7-chloro-2-methoxy-5-phenyl-3H-1,4-benzodiazepine 4-N-oxide, DF11

7-chloro-2-(N-methylacetamido)-5-phenyl-3H-1,4-benzodiazepine 4-N-oxide, DF86

7-chloro-2-methylamino-5-phenyl-3H-1,4-benzodiazepine 4-N-oxide, DF22

7-chloro-2-methyl-5-phenyl-3H-1,4-benzodiazepine 4-N-oxide, DF08

7-chloro-2-[N-methylpropionamido)-5-phenyl-3H-1,4-4-N-oxide, DG03

5-(2-fluorophenyl)-7-iodo-2-methylamino-3H-1,4-benzodiazepine 4-N-oxide, DF17

4-methoxy-2,2'-bipyridyl 1-oxide, DD49

6-methoxy-1-hydroxyphenazine-5,10-dioxide, DE00

2-methylamino-7-methylmercapto-5-phenyl-3H-1,4-benzodiazepine 4-N-oxide, DF72

2-methylamino-5-phenyl-3H-1,4-benzodiazepine 4-N-oxide, DF33

7-methyl-2-methylamino-5-phenyl-3H-1,4-benzodiazepine 4-N-oxide, DF71

OXIMES and ALKOXIMES

isatin 3-(O-ethyloxime), DD15

phenylacetaldoxime, DC30

PHENOLS, see HYDROXY COMPOUNDS

PEROXIDES, HYDROPEROXIDES and PER-ACIDS

tert-butyl peroxydodecanoate, DF48

tert-butyl peroxynonanoate, DE20

t-butyl peroxytetradecanoate, DG01

didecanoyl peroxide, DG20

didodecanoyl peroxide, DG42

dihexadecanoyl peroxide, DG70

dinonanoyl peroxide, DG00

dioctadecanoyl peroxide, DH09

ditetradecanoyl peroxide, DG62

PHOSPHORUS and ARSENIC COMPOUNDS (see also ESTERS, NONCARBOXYLIC and SULFUR COMPOUNDS)

3-amino-4-hydroxyphenylarsonic acid, DB20

2-aminophenylarsonic acid, DB18

4-aminophenylarsonic acid, DB19

4-chlorophenylarsonic acid, DA97

cytidine monosphosphate, DC89

cytidylyl (3'→5') cytidine, DF98

cytidylyl (3'→5') guanidine, DG05

(PHOSPHORUS and ARSENIC COMPOUNDS)
(see also ESTERS, NONCARBOXYLIC
and SULFUR COMPOUNDS)(cont.)

    cytidylyl (3'→5') uridine, DF97

    2,4-dihydroxy-5-nitrophenylarsonic acid, DB02

    2,4-dihydroxyphenylarsonic acid, DB13

    μ-hydrido-μ-dimethylphosphinobis-[tricarbonyl-(η-cyclopentadienyl)-molybdenum(I), DF91

    4-hydroxy-3-nitrophenylarsonic acid, DB01

    4-hydroxyphenylarsonic acid, DB12

    4-methylphenylarsonic acid, DB78

    nicotinamide adenine dinucleotide phosphate, reduced, DG27

    nicotinamide adenine dinucleotide reduced, DG26

    2-nitrophenylarsonic acid, DA98

    3-nitrophenylarsonic acid, DA99

    4-nitrophenylarsonic acid, DB00

    phenylarsonic acid, DB11

    phenylarsonous acid, DB10

    1-propanearsonic acid, DA25

    pyridoxal-5-phosphate, DC36

    riboflavin, DF74

    sodium flavin mononucleotide, DF73

    triphenylarsine oxide, DF83

QUARTERNARY COMPOUNDS, see AMINES

QUINONES and HYDROQUINONES (see also HYDROXY COMPOUNDS)
Benzoquinones

    1,2-benzoquinone, DB85

    1,4-benzoquinone, DA86, DH59

    hydroquinone, DB07

    methyl-1,4-benzoquinone, DB69

    quinhydrone, DD76

    Ubiquinone-10, DH30

Polycyclic Quinones

    Alizarine Red S, DE22

    anthraquinone-1,5-disulfonic acid, DE28

    anthraquinone-2,6-disulfonic acid, DE29

    anthraquinone-1-sulfonic acid, DE26

    anthraquinone-2-sulfonic acid, DE27

    2-methyl-1,4-naphthoquinone, DD48

    2-methyl-3-(3,7,11,15-tetramethyl-2-hexadecenyl)-1,4-naphthoquinone, DG66

    1,2-naphthoquinone, DD00

    1,4-naphthoquinone, DD01

    sodium 1-anthraquinonesulfonate, DE21

Heterocyclic Quinones

    benzimidazole-4,7-quinone, DB58

    1,3-dimethylbenzimidazolone-4,7-quinone, DC68

    1,3-dimethylbenzimidazolone-5,6-quinone, DC69

    1-methylbenzimidazole-4,7-quinone, DC11

    1-methylbenzimidazole-6,7-quinone, DC12

    1-methyl-2-chlorobenzimidazole-4,7-quinone, DC04

    1-methyl-2-chlorobenzimidazole-6,7-quinone, DC05

    1-methyl-2-dimethylaminobenzimidazole-4,7-quinone, DD19

    1-methyl-2-dimethylaminobenzimidazole-5,6-quinone, DD20

    1-methyl-2-dimethylaminobenzimidazole-6,7-quinone, DD21

    1-methyl-(2,3-d)naphthoimidazole-4,9-quinone, DD65

SUGARS, see CARBOHYDRATES

SULFATES, see ESTERS, NONCARBOXYLIC

SULFIDES, see SULFUR COMPOUNDS

SULFONAMIDES, see SULFUR COMPOUNDS

SULFONATES, see SULFUR COMPOUNDS

SULFONES, see SULFUR COMPOUNDS

SULFONIC ACIDS, see SULFUR COMPOUNDS

SULFUR COMPOUNDS (see also HETEROCYCLIC COMPOUNDS; ORGANOMETALLICS; and ESTERS, NON-CARBOXYLIC)
Thiols

    benzenethiol, DB09

    cysteine, DA23

TABLE V. Functional-Group Index

(SULFUR COMPOUNDS)(see also HETERO-
CYCLIC COMPOUNDS; ORGANOMETALLICS;
and ESTERS, NON-CARBOXYLIC)(cont.)

Thiols (cont.)

- 2,6-dimercapto-1,4-thiopyrone-3-carboxylic acid, DA87
- diphenylthiocarbazone, DE10
- o-ethyl thioacetothioacetate, DB28
- glycol dimercaptopropionate, DC45
- 2-mercaptoaniline, DB16
- 4-mercaptoaniline, DB17
- methyl 5-(mercaptomethyl)-2-furoate, DB77
- pentachlorothiophenol, DA82
- thiobenzoic acid, DB67
- thioglycolic acid, DA07

Sulfides, Thioethers and Sulfonium Ions

- 7-chloro-2,3-dihydro-1-methyl-5-(2-methylmercaptophenyl)-1H-1,4-benzodiazepine, DF70
- 1,3-dihydro-7-methylthio-5-phenyl-2H-1,4-benzodiazepin-2-one, DF27
- diphenyl sulfide, DD77
- 2-methylamino-7-methylmercapto-5-phenyl-3H-1,4-benzodiazepine 4-N-oxide, DF72

Disulfides

- bis(2-quinolyl)disulfide di-N-oxide, DF76
- L-cystine, DB35

Dithiolium Compounds

- 3,5-bis(4-methoxyphenyl)-1,2-dithiol-1-ium perchlorate, DF59
- 3,5-bis(4-methylphenyl-1,2-dithiol-1-ium perchlorate, DF58
- 3-(4-bromophenyl)-5-phenyl-1,2-dithiolylium perchlorate, DE71
- 3,5-di(tert-butyl)-1,2-dithiolylium perchlorate, DD57
- 3-[4-(dimethylamino)phenyl]-5-(4-methoxyphenyl)-1,2-dithiol-1-ium perchlorate, DF93
- 3,5-dimethyl-1,2-dithiolylium perchlorate, DA69
- 3,5-diphenyl-1,2-dithiolylium perchlorate, DE82
- 3-(4-methoxyphenyl)-5-phenyl-1,2-dithiol-1-ium perchlorate, DF14
- 3-methyl-5-phenyl-1,2-dithiolylium perchlorate, DD06
- 3-(4-methylphenyl)-5-phenyl-1,2-dithiol-1-ium perchlorate, DF13

Thiocarbamates, Dithiocarbamates, and Xanthates

- ammonium pyrrolidine-1-dithiocarboxylate, DA79
- diammonium proline-N-dithiocarboxylate, DB45
- potassium ethylxanthate, DA17
- sodium dibutyldithiocarbamate, DC95
- sodium diethyldithiocarbamate, DA74
- sodium dimethyldithiocarbamate, DA20
- sodium dipropyldithiocarbamate, DB96
- sodium 1-piperidinedithiocarboxylate, DB26
- sodium 1-pyrrolidinedithiocarboxylate, DA70

Thiones, Thioamides, Thiobarbiturates, Thiopyrimidines, and Thiopurines

- 5-ethyl-5-(1-methylpropyl)-2-thiobarbituric acid, DD33

Sulfoxides, Sulfones, and S-dioxides

- 7-acetylbenzo[c]phenothiazine 12-dioxide, DF78
- 12-acetylbenzo[a]phenothiazine 7-dioxide, DF79
- 12-acetylbenzo[a]phenothiazine 7-oxide,
- benzyl vinyl sulfone, DC83
- 6-chloro-3-chloromethyl-3,4-dihydro-2-methyl-7-sulfamoyl-1,2,4-benzothiadiazine 1,1-dioxide, DC84
- 7-chloro-2,3-dihydro-1-methyl-5-(2-methylsulfinylphenyl)-1H-1,4-benzodiazepine, DF69
- 7-chloro-2,3-dihydro-1-methylsulfonyl-5-phenyl-1H-1,4-benzodiazepine, DF31

(SULFUR COMPOUNDS)(see also HETERO-
CYCLIC COMPOUNDS; ORGANOMETALLICS;
and ESTERS, NON-CARBOXYLIC)(cont.)

Sulfoxides, Sulfones, and S-dioxides (cont.)

  6-chloro-3,4-dihydro-7-sulfamyl-2H-1,2,4-benzothiadiazine 1,1-dioxide, DB74

  divinyl sulfone, DA41

  divinyl sulfoxide, DA39

  ethyl vinyl sulfone, DA46

  methyl vinyl sulfone, DA21

  3-(N-Nitrosomethylamino)thiolane, DA76

  phenyl vinyl sulfone, DC29

Sulfonamides

  6-chloro-3,4-dihydro-7-sulfamoyl-2H-1,2,4-benzothiadiazine 1,1-dioxide, DB74

  4,4'-oxybis[benzenesulfonyl(isopropylidenehydrazide)], DF96

Sulfuric and Thiosulfuric Acid Derivatives, Sulfonates, and Sulfinates

  Alizarine Red S, DE22

  2-amino-4-methylbenzenesulfonic acid, DB79

  2-amino-5-methylbenzenesulfonic acid, DB80

  5-amino-2-methylbenzenesulfonic acid, DB81

  anthraquinone-1,5-disulfonic acid, DE28

  anthraquinone-2,6-disulfonic acid, DE29

  anthraquinone-1-sulfonic acid, DE26

  anthraquinone-2-sulfonic acid, DE27

  azobenzene-4-sulphonic acid, DD75

  formaldehyde sodium sulfoxylate, DA04

  naphthyl-1,1'-azo-2-naphthol-4',6-disulfonic acid disodium salt, DG09

  sodium 1-anthraquinonesulfonate, DE21

  sodium phthalocyaninecopper(II)-2-sulfonate, DG68

  sodium phthalocyanine-2-sulfonate, DG69

  tetrasodium phthalocyanineiron-2,9,16,23-tetrasulfonate, DG67

  tetrasodium 5,10,15,20-tetrakis-(4-sulfonatophenyl)porphine, DH16

  tetrasodium 5,10,15,20-tetrakis(4-sulfonatophenyl)porphinatocobalt(II), DH15

  trisodium 5,10,15,20-tetrakis(4-sulfonatophenyl)porphinatocobalt(III), DH14

Organometallic Complexes

  bis(O-ethyl thioacetothioacetato)-cadmium(II), DD86

  bis(O-ethyl thioacetothioacetato)-mercury(II), DD87

  bis(O-ethyl thioacetothioacetato)-zinc(II), DD88

  η-cyclopentadienylperfluorodimethylethylene-1,2-disulfido-iridium(III), DC59

  η-cyclopentadienylperfluorodimethylethylene-1,2-disulfido-iron(III), DC58

  η-cyclopentadienylperfluorodimethylethylene-1,2-disulfido-nickel(III), DC60

  η-cyclopentadienylperfluorodimethylethylene-1,2-disulfido-rhodium(III), DC61

  mercury phenylmercaptide, DD71

THIOCYANATES, see ISOTHIOCYANATES

THIOLS, see SULFUR COMPOUNDS

THIOUREAS, see UREAS

UNSATURATED COMPOUNDS (see also HYDROCARBONS, and QUINONES and HYDROQUINONES)

Ethylenic

Double Bond Isolated From Nonhydrocarbon Functional Groups

  cis-2-chloromercury-3-methoxy-cyclodecene, DD56

  cis-2-chloromercury-3-ethoxy-cyclonene, DD55

  trans-2-chloromercury-3-methoxy-cyclotridecene, DE66

  trans-2-chloromercury-3-methoxy-cycloundecene, DD93

TABLE V. Functional-Group Index

(UNSATURATED COMPOUNDS)(see also HYDRO-
CARBONS, and QUINONES and HYDRO-
QUINONES)(cont.)

Ethylenic

Double Bond Isolated From Nonhydro-
Carbon Functional Groups (cont.)

1,4-diphenyl-1,3-butadiene, DF19

methyl 3-butenoate, DA72

methyl 4-pentenoate, DB29

2-methyl-3-(3,7,11,15-tetramethyl-
2-hexadecenyl)-1,4-naphthoquinone,
DG66

retinyl methyl ether, DG28

ubiquinone-10, DH30

1-vinyl-2-pyrrolidone, DB24

α,β-Unsaturated Aldehydes

cinnamaldehyde, DC72

crotonaldehyde, DA38

3,3-diphenylprop-2-eneal, DE89

3-formylcyclohexene, DB86

vitamin A aldehyde, DG17

α,β-Unsaturated Ketones

1-acetyl-4-tert-butyl-1-cyclo-
hexene, DD90

acrolein, DA15

chalcone, DE88

2-cyclopentenone, DA68

[2,3-dichloro-4-(2-methylene-1-
oxobutyl)phenoxy]acetic acid,
DE06

5,5-dimethyl-3-phenyl-2-cyclo-
hexenone, DE52

trans-4,4-dimethyl-1-phenyl-1-
penten-3-one, DE14

2-ethoxy-2-cyclopentenone, DB89

2,4a,5,6,7,8-hexahydro-4a,8-di-
methylnaphthalen-2-one, DD85

2-methoxy-3,5,5-trimethyl-2-cyclo-
hexen-1-one, DD34

5-methyl-2-cyclohexenone, DB88

3-methyl-3-penten-2-one, DB27

3-(2-methylpropoxy)-5,5-dimethyl-
2-cyclohexen-1-one, DD91

1,2,3,4,4a,5,6,7-octahydronaphtha-
len-1-one, DD28

1,2,3,4,5,6,7,8-octahydronaphtha-
len-1-one, DD29

trans-3-penten-2-one, DA71

2-phenyl-2-cyclohexenone, DD82

3-phenyl-3-hepten-2-one, DE15

1,2,3,3-tetraphenylpropenone,
DG51

1,1,3-triphenylpropenone, DG21

cis-1,2,3-triphenyl-2-propenone,
DG22

trans-1,2,3-triphenyl-2-propen-
one, DG23

α,β-Unsaturated Sulfones and Sulfoxides

benzyl vinyl sulfone, DC83

divinyl sulfone, DA41

divinyl sulfoxide, DA39

ethyl vinyl sulfone, DA46

methyl vinyl sulfone, DA21

phenyl vinyl sulfone, DC29

α,β-Unsaturated Acids, Thioacids,
Esters and Amides

acrylamide, DA18

bis(O-ethyl thioacetothioacetato)-
cadmium(II), DD86

bis(O-ethyl thioacetothioacetato)-
mercury(II), DD87

bis(O-ethyl thioacetothioacetato)-
zinc(II), DD88

[3-(4-chlorophenyl)-3-(2-pyridyl-
1-propyl]dimethylammonium
maleate, DG16

diethylaminoethyl methacrylate,
DD38

diethyl isopropylidenemalonate,
DD35

dimethyl fumarate, DB21

dimethyl maleate, DB22

ethyl 4-tert-butylcyclohexylidene
acetate, DE65

2-(2'-ethyl-6'-methylpiperidyl)-
ethyl methacrylate, DE67

lithium 4-tert-butylcyclohexyli-
dene acetate, DD89

methyl 4-tert-butyl-1-cyclohexen-
oate, DD92

(UNSATURATED COMPOUNDS)(see also HYDRO-
CARBONS, and QUINONES and HYDRO-
QUINONES)(cont.)

Ethylenic(cont.)

α,β-Unsaturated Acids, Thioacids, Esters and Amides (cont.)

methyl cinnamate, DD16

methyl trans-cinnamate, DD17

methyl crotonate, DA73

methyl 2-propenoate, DA40

retinoic acid, DG18

retinyl acetate, DG33

retinyl palmitate, DH08

tetraethylethenetetracarboxylic acid, DE60

α,β-Unsaturated Nitriles

acrylonitrile, DA11

4-(tert-butyl)-1-cyanocyclohex-ene, DD54

crotonitrile, DA35

ethyl 2-cyano-2,4-hexadienoate, DC85

2,4-pentadienenitrile, DA62

Unsaturated Amines, Halogens, Alcohols, Ethers and Nitro-compounds (see also α,β-Unsaturated Imines)

bis(trichlorovinyl)mercury(II), DA28

1,2-dibromobenzocyclobutene, DC09

3,3-diethyl-2,4-(1H,3H)pyridine-dione, DC87

2-ethoxy-2-cyclopentenone, DB89

ethyl vinyl ether, DA45

hexachlorobuta-1,3-diene, DA27

3-isobutoxy-5,5-dimethyl-2-cyclo-hexen-1-one, DD91

2-methoxy-3,5,5-trimethyl-2-cyclo-hexen-1-one, DD34

2-(2-methylpropoxy)-2-methoxy-ethylmercuric acetate, DC94

β-nitrostyrene, DC22

retinol, DG19

Alicyclic

1-acetyl-4-tert-butyl-1-cyclo-hexene, DD90

cis-bicyclo[6.2.0]deca-2,4,6-triene, DD22

cis-bicyclo[6.1.0]nona-2,4,6-triene, DC78

trans-bicyclo[6.1.0]nona-2,4,6-triene, DC79

4-(tert-butyl)-1-cyanocyclo-hexene, DD54

cis-2-chloromercury-3-ethoxy-cyclononene, DD55

cis-2-chloromercury-3-methoxy-cyclodecene, DD56

trans-2-chloromercury-3-methoxy-cyclotridecene, DE66

trans-2-chloromercury-3-methoxy-cycloundecene, DD93

cyclooctatetraene, DC24

1,3,5-cyclooctatriene, DC35

2-cyclopentenone, DA68

dibenzo[a,e]cyclooctatetraene, DF05

1,2-dibromobenzocyclobutene, DC09

5,5-dimethyl-3-phenyl-2-cyclo-hexenone, DE52

2-ethoxy-2-cyclopentenone, DB89

ethyl 4-tert-butylcyclohexylidene acetate, DE65

3-formylcyclohexene, DB86

2,4a,5,6,7,8-hexahydro-4a,8-di-methylnaphthalen-2-one, DD85

3-isobutoxy-5,5-dimethyl-2-cyclo-hexen-1-one, DD91

lithium 4-tert-butylcyclohexyli-deneacetate, DD89

2-methoxy-3,5,5-trimethyl-2-cyclo-hexen-1-one, DD34

methyl 4-tert-butyl-1-cyclohexen-oate, DD92

5-methyl-2-cyclohexenone, DB88

1,2,3,4,4a,5,6,7-octahydronaphtha-len-1-one, DD28

1,2,3,4,5,6,7,8-octahydronapha-len-1-one, DD29

2-phenyl-2-cyclohexenone, DD82

retinyl methyl ether, DG28

retinoic acid, DG18

TABLE V. Functional-Group Index

(UNSATURATED COMPOUNDS)(see also HYDRO-
CARBONS, and QUINONES and HYDRO-
QUINONES)(cont.)

Alicyclic (cont.)

retinol, DG19

retinyl acetate, DG33

retinyl palmitate, DH08

vitamin A aldehyde, DG17

Steroids

16α-bromo-3-hydroxyestra-1,3,5(10)-trien-17-one, DF94

16β-bromo-3-hydroxyestra-1,3,5(10)-trien-17-one, DF95

calciferol, DG57

cholestra-5,7-dien-3β-ol, DG52

ergosta-5,7,22-trien-3β-ol, DG58

9α-fluoro-17α,21-dihydroxy-5-pregnene-3,11,20-trione-3-ethyleneketal-17,20,21-bis-methylenedioxolane, DG45

4-[2-(5-hydroxy-2-methylcyclohexen-1-yl)ethenyl]-7a-methyl-3a,6,7,7a-tetrahydroindan, DG59

16α,17α-oxidoprogesterone, DG25

vitamin D_3, DG53

Heterocyclic Compounds (nonaromatic)

1,6-diazabicyclo[4.4.0]deca-3,8-diene, DC41

2,6-dimercapto-1,4-thiopyrone-3-carboxylic acid, DA87

2,3-dimethyl-2,3-diazabicyclo-[2.2.1]hept-5-ene, DB93

2,3-dimethyl-2,3-diazabicyclo-[2.2.2]octa-5-ene, DC44

1,2-dimethyl-1,2,3,6-tetrahydro-pyridazine, DB34

1,2,3-trimethyl-1,2,3,6-tetrahydropyridazine, DC00

Porphines

cobalt(II)protoporphyrin(IX) dimethyl ester, DG74

copper(II)protoporphyrin(IX) dimethyl ester, DG75

trans-dicyanoprotoporphyrin(IX)-iron(III), DG72

trans-di(3-methylpyridine)protoporphyriniron(III) chloride, DH21

trans-di(4-methylpyridine)protoporphyriniron(II) chloride, DH22

trans-dipyridylprotoporphyrin(IX)-iron(III) chloride, DH19

hematoporphyrincobalt(III), DG73

manganese(III)chloride protoporphyrin(IX) dimethyl ester monohydrate, DG78

nickel(II)protoporphyrin(IX) dimethyl ester, DG76

protoporphyrin(IX)-trans-cyanopyridineiron(III), DH12

protoporphyrin(IX) dimethyl ester, DG79

protoporphyrin(IX) iron(III)-chloride, DG71

zinc(II)protoporphyrin(IX) dimethyl ester, DG77

Acetylenic

diphenylbutadiyne, DE98

3-heptyn-2-one, DB87

lithium 2-hexynoate, DB14

methyl 2-hexynoate, DB90

methyl propiolate, DA34

UREAS and THIOUREAS (see also HETERO-
CYCLIC COMPOUNDS and KETONES, Heterocyclic)

diphenylthiocarbazone, DE10

N-methyl-N-nitrosourea, DA08

TABLE VI.
CHEMICAL ABSTRACTS REGISTRY NUMBERS

This is a list of the Chemical Abstracts Registry Numbers given under the names of the compounds appearing in Table I. Registry Numbers have not been assigned to some of the compounds in Table I, and these compounds are not included in this table. The Registry Numbers are listed in numerical order, and each is followed by the code number assigned to the corresponding compound in Table I.

TABLE VI. Chemical Abstracts Registry Numbers

50-14-6, DG57	77-21-4, DE13	100-42-5, DC25
50-35-1, DE01	77-77-0, DA41	100-47-0, DB61
50-99-7, DB37	79-06-1, DA18	100-52-7, DB66
51-46-7, DC51	79-81-2, DH08	100-75-4, DA75
51-80-9, DA81	82-82-6, DC34	100-76-5, DB95
52-90-4, DA23	83-12-5, DE78	100-97-0, DB36
53-57-6, DG27	83-88-5, DF74	102-06-7, DE11
54-47-7, DC36	84-48-0, DE27	102-54-5, DD14
56-23-5, DA00	84-50-4, DE29	102-96-5, DC22
56-89-3, DB35	84-80-0, DG66	103-26-4, DD16
57-24-9, DG24	85-88-0, DC42	103-33-3, DD72
57-87-4, DG58	86-30-6, DD74	103-50-4, DE44
58-25-3, DF22	87-68-3, DA27	104-85-8, DC21
58-27-5, DD48	88-12-0, DB24	104-88-1, DB60
58-54-8, DE06	88-44-8, DB80	104-89-6, DA17
58-68-4, DG26	89-62-3, DB76	104-91-6, DA92
58-86-6, DA77	90-02-8, DB70	105-16-8, DD38
58-93-5, DB74	90-03-9, DA88	105-74-8, DG42
59-02-9, DG63	91-16-7, DC39	106-34-3, DD76
59-67-6, DA90	91-20-3, DD03	106-47-8, DB03
59-82-5, DA51	91-59-8, DD07	106-51-4, DA85
60-10-6, DE10	92-85-3, DD67	106-58-1, DB39
60-11-7, DE46	93-14-1, DD31	107-02-8, DA15
62-75-9, DA10	93-52-7, DC26	107-13-1, DA11
65-23-6, DC40	96-33-3, DA40	107-34-6, DA25
65-46-3, DC88	98-05-5, DB11	108-46-3, DB08
65-85-0, DB68	98-14-6, DB12	108-74-7, DB44
65-86-1, DA58	98-46-4, DB55	108-98-5, DB09
66-71-7, DD64	98-50-0, DB19	109-73-9, DA49
66-72-8, DC33	98-72-6, DB00	109-92-2, DA45
67-97-0, DG53	98-91-9, DB67	110-18-9, DB46
68-11-1, DA07	98-92-0, DB04	110-86-1, DA63
68-26-8, DG19	98-95-3, DA91	110-91-8, DA48
69-93-2, DA61	99-09-2, DB05	113-92-8, DG16
70-25-7, DA09	99-61-6, DB62	115-09-3, DA03
70-30-4, DD97	99-65-0, DA84	116-31-4, DG17
71-30-7, DA36	99-97-8, DC86	117-14-6, DE28
71-43-2, DA96	99-99-0, DB71	118-88-7, DB81
72-44-6, DF25	100-01-6, DB06	119-13-1, DG54
76-87-9, DF90	100-02-7, DA94	119-61-9, DE02
77-03-2, DC90	100-14-1, DB65	119-98-2, DG50
77-04-3, DC87	100-22-1, DD32	120-12-7, DE32

120-73-0, DA59	434-16-2, DG52	629-20-9, DC24
121-19-7, DB01	439-14-5, DF09	632-51-0, DG38
121-44-8, DB43	469-06-7, DG59	645-12-5, DA55
122-78-1, DC28	470-35-9, DG49	645-49-7, DB21
123-30-8, DB15	471-46-5, DA06	684-93-5, DA08
123-31-9, DB07	486-25-9, DD99	698-63-5, DA54
123-73-9, DA38	509-14-8, DA01	716-54-1, DD09
125-64-4, DD37	509-85-3, DD80	725-12-2, DE35
128-04-1, DA20	517-25-9, DA02	736-30-1, DE39
130-15-4, DD01	517-51-1, DH13	762-12-9, DG20
130-22-3, DE22	514-42-5, DD00	762-13-0, DG00
130-40-5, DF73	527-18-4, DD30	766-84-7, DB52
132-65-0, DD66	536-74-3, DC08	779-01-1, DD51
133-49-3, DA82	538-86-3, DC37	781-99-7, DD84
135-07-9, DC84	553-97-9, DB69	808-57-1, DG40
136-30-1, DC95	555-16-8, DB63	818-57-5, DB29
139-66-2, DD77	555-21-5, DC13	823-74-5, DA64
144-62-7, DA05	565-62-8, DB27	831-08-3, DD18
146-22-5, DE84	576-26-1, DC38	833-29-4, DD12
148-03-8, DG60	583-63-1, DA86	849-01-4, DG21
148-18-5, DA74	586-96-9, DA89	872-71-9, DA70
149-44-0, DA04	587-65-5, DC27	873-32-5, DB51
191-24-2, DG29	596-03-2, DG06	873-57-4, DB26
206-44-0, DE99	601-77-4, DB42	883-17-0, DD50
208-96-8, DD60	602-60-8, DE31	886-65-7, DF19
262-89-5, DF05	604-75-1, DE81	886-66-8, DE98
265-49-6, DD70	609-39-2, DA30	917-23-7, DH18
280-28-4, DB32	609-71-2, DA93	920-39-8, DA22
280-57-9, DB31	611-73-4, DC15	922-67-8, DA34
281-17-4, DB98	614-00-6, DB75	930-30-3, DA68
281-29-8, DC43	614-47-1, DE88	943-37-3, DB72
283-47-6, DB97	615-53-2, DA44	947-69-3, DD11
289-80-5, DA32	618-07-5, DA99	947-92-2, DD95
289-95-2, DA33	619-72-7, DB59	1016-09-7, DE43
290-37-9, DA31	623-00-7, DB50	1088-11-5, DE80
290-87-9, DA12	623-03-0, DB53	1115-15-7, DA39
302-79-4, DG18	623-26-7, DC03	1122-60-7, DB30
315-30-0, DA60	623-43-8, DA73	1153-05-5, DF83
366-18-7, DD05	624-48-6, DB22	1193-02-8, DB17
384-22-5, DB54	625-33-2, DA71	1207-72-3, DE04
402-54-0, DB56	627-26-9, DA35	1210-39-5, DE89
		1270-66-2, DE82

TABLE VI. Chemical Abstracts Registry Numbers

1321-16-0, DB86	2564-09-2, DC67	5358-96-3, DF28
1333-46-6, DC82	2646-17-5, DF54	5410-29-7, DA98
1439-07-2, DE40	2648-00-2, DE83	5522-66-7, DG79
1499-10-1, DG47	2683-82-1, DH04	5535-48-8, DC29
1591-30-6, DE23	2697-96-3, DG70	5540-04-0, DA97
1615-70-9, DA62	2842-11-7, DC75	5571-63-1, DF26
1617-35-2, DA43	2886-65-9, DE72	5682-76-8, DD34
1689-82-3, DD73	2891-12-5, DF27	5789-30-0, DE38
1694-78-6, DE90	2894-61-3, DE79	6174-95-4, DE60
1694-79-7, DF06	2894-67-9, DE73	6191-11-3, DA57
1738-10-9, DA87	2898-08-0, DE87	6269-96-1, DB13
1754-62-7, DD17	2898-12-6, DF29	6332-56-5, DA56
1803-95-8, DF56	2898-19-3, DF34	6333-11-5, DG51
1812-32-4, DE33	2898-21-7, DF37	6400-42-6, DC54
1829-28-3, DC76	3023-44-7, DF10	6415-12-9, DA50
1860-18-0, DA19	3058-39-7, DB57	6486-01-7, DG48
1871-52-9, DC35	3140-93-0, DA29	6704-33-2, DB25
1889-59-4, DA46	3240-78-6, DD83	6802-75-1, DD35
2042-37-7, DB48	3273-75-4, DH09	6952-59-6, DB49
2045-00-3, DB18	3289-76-7, DC19	7028-48-0, DC30
2095-57-0, DD33	3299-05-6, DD27	7152-42-3, DF80
2123-88-8, DF48	3451-65-8, DD85	7214-50-8, DB88
2144-00-5, DG30	3459-80-1, DD53	7370-14-1, DD54
2144-02-7, DG31	3530-28-7, DG62	7474-65-9, DG23
2163-77-1, DB20	3680-02-2, DA21	7512-67-6, DG22
2211-64-5, DB38	3693-14-9, DF66	7616-22-0, DG61
2285-16-7, DF04	3724-55-8, DA72	7711-39-9, DB58
2382-64-1, DF97	3897-18-5, DF11	7711-63-9, DC11
2382-65-2, DG05	3969-54-8, DB78	7713-21-5, DF08
2396-60-3, DE07	4063-38-1, DA83	7722-15-8, DE85
2484-88-0, DD75	4143-50-4, DB96	7773-19-5, DG54
2491-52-3, DD69	4214-75-9, DA66	10075-93-1, DG12
2493-04-1, DA65	4214-76-0, DA67	10191-41-0, DG63
2517-76-2, DA14	4521-33-9, DA53	10338-77-9, DD08
2517-78-4, DA13	4549-72-8, DB73	10338-78-0, DD13
2536-99-4, DF98	4556-09-6, DD82	10338-79-1, DD10
2563-06-4, DC66	4928-03-4, DF61	10504-35-5, DB23
2563-97-5, DC10	4998-93-0, DC62	10507-36-5, DG14
2563-98-6, DC74	5185-97-7, DB94	10507-38-7, DA28
2564-00-3, DC18	5275-69-4, DA95	10507-39-8, DA37
2564-02-5, DC17	5336-53-8, DE42	12088-73-2, DD59
2564-05-8, DC20	5358-35-0, DF07	12091-64-4, DF03
2564-06-9, DC81		

12091-98-4, DF36	14323-06-9, DG64	22825-58-7, DC01
12094-57-4, DF50	14371-10-9, DC72	22913-02-6, DE20
12107-35-6, DC07	14409-63-3, DG85	22960-08-3, DF65
12116-28-8, DD02	14882-62-3, DD65	23193-88-6, DF72
12130-13-1, DE75	14932-10-6, DG74	23211-28-1, DE19
12131-42-9, DF91	15304-09-3, DG77	23388-71-8, DH12
12132-08-0, DG43	15304-60-6, DG75	24037-79-4, DD28
12132-09-1, DG44	15304-70-8, DG76	24803-99-4, DG87
12132-87-5, DE37	15466-96-3, DD91	24804-00-0, DG91
12152-60-2, DC56	15489-47-1, DG71	24804-01-1, DG95
12154-95-9, DE34	15753-89-6, DC83	25013-15-4, DC80
12169-93-6, DC59	15990-45-1, DC32	25087-66-5, DG88
12169-95-8, DC60	16224-09-2, DE17	25226-58-8, DD61
12169-96-9, DC61	16290-29-2, DF31	25400-22-0, DB10
12193-46-3, DD26	16904-05-5, DG73	26059-43-8, DB87
12194-07-9, DE86	17043-14-0, DC98	26105-53-3, DF85
12194-11-5, DF00	17263-70-6, DC57	26132-66-1, DC78
12194-12-6, DF01	17346-16-6, DB92	26163-36-0, DB34
12194-13-7, DF02	17632-18-7, DG92	26163-37-1, DB40
12194-20-6, DF84	17632-19-8, DG84	26171-64-2, DC52
12194-69-3, DF62	18012-46-9, DC79	26440-32-4, DE41
12203-12-2, DC06	18091-88-8, DE91	26868-58-6, DF20
12203-87-1, DG37	18389-95-2, DC47	27214-06-8, DC89
12261-16-4, DC58	18470-19-4, DF76	27600-99-3, DE21
12275-31-9, DA69	18582-60-0, DH19	27860-55-5, DG89
12791-65-0, DG36	18631-96-4, DD29	28705-30-8, DE36
13256-21-8, DA76	18920-63-3, DG63	28739-21-1, DF30
13279-35-1, DF96	18937-79-6, DB90	29008-64-8, DH06
13287-49-5, DC14	19066-35-4, DG13	29387-92-6, DE45
13358-49-1, DE94	19437-26-4, DD47	29497-66-3, DG72
13395-89-6, DG35	19529-55-6, DG93	29569-91-3, DE14
13638-82-9, DH11	19531-64-7, DE18	29776-28-1, DF14
13682-02-5, DC99	19916-65-5, DB85	29776-30-5, DF59
13733-50-1, DE65	20205-53-2, DD25	30116-09-7, DH10
13925-12-7, DE00	20717-38-8, DB47	30179-51-2, DC31
14163-05-4, DD49	21910-35-4, DG86	30597-85-4, DE30
14172-90-8, DH17	21690-94-8, DB83	30597-87-6, DC70
14263-81-1, DG65	21690-95-9, DB84	30597-88-7, DC71
14287-89-9, DB99	22173-19-9, DD99	30597-89-8, DE51
14287-91-3, DC44	22250-72-2, DC09	30597-90-1, DE64
14287-92-4, DC49	22357-85-3, DG78	30637-95-7, DE26
14288-15-4, DB93	22504-50-3, DC45	30905-13-6, DG02

TABLE VI. Chemical Abstracts Registry Numbers

31075-97-5, DE63	38577-97-8, DG08	41661-29-1, DH14
31863-29-3, DE57	38704-82-4, DD40	41954-69-2, DG69
32393-41-2, DD56	38704-87-9, DC97	42414-57-3, DC93
32393-42-3, DD93	38704-89-1, DA78	42576-34-1, DD52
32450-56-9, DG28	38704-91-5, DC53	42802-20-0, DB64
32462-45-6, DE66	38704-92-6, DC02	42842-99-9, DD39
33176-67-9, DF93	38704-93-7, DE68	42877-41-8, DE71
33217-24-2, DD55	38704-94-8, DC00	42877-42-9, DF13
33242-40-9, DG46	38704-98-2, DC50	42877-43-0, DF58
33269-25-9, DG90	38705-00-9, DB41	48223-98-9, DG60
33685-60-8, DE25	38705-05-4, DD58	49540-66-1, DB45
34784-43-5, DD22	38705-06-5, DD94	49606-58-8, DG82
34827-66-2, DB28	38705-08-7, DG15	49661-61-2, DG83
34867-01-1, DD88	38705-09-8, DC55	50732-69-9, DA79
34867-02-2, DD86	38705-10-1, DC46	50733-37-4, DH00
34867-03-3, DD87	38786-33-3, DD41	50733-38-5, DG98
35335-01-4, DE05	38903-46-7, DD44	50733-39-6, DG97
35610-83-4, DD57	38904-47-1, DE69	50733-40-9, DG96
36047-17-3, DE52	38904-48-2, DD43	50733-41-0, DG94
36093-53-5, DF55	39011-19-3, DE53	50733-45-4, DG99
36454-18-9, DD04	39018-12-7, DE54	50820-15-0, DH05
36454-19-0, DD63	39018-13-8, DE58	50854-51-8, DA80
36454-20-3, DD46	39018-17-2, DE55	50936-73-7, DH03
36454-21-4, DD45	39018-20-7, DG11	50938-43-7, DH01
36454-22-5, DD98	39018-22-9, DG39	51102-49-9, DF16
36454-23-6, DE24	39050-26-5, DH16	51319-02-9, DH07
36683-55-3, DE70	39060-35-0, DE62	51481-19-7, DG68
36869-44-0, DF40	39060-36-1, DE61	52445-55-3, DD81
36869-45-1, DF99	39251-81-5, DF81	52677-47-1, DH22
37881-11-1, DB14	39254-69-8, DD62	52887-21-5, DF89
37881-12-2, DD89	39254-76-7, DE08	52887-25-9, DF88
37882-95-4, DD42	39264-08-9, DE03	52887-27-1, DG34
37908-48-8, DF18	39264-20-5, DE09	55098-43-6, DE92
37908-49-9, DG10	39264-24-9, DE48	55098-44-7, DF64
37908-50-2, DD79	39264-38-5, DE47	55098-46-9, DF63
37908-51-3, DD78	39936-46-1, DA24	55098-47-0, DF57
37908-52-4, DF32	40006-62-0, DE15	55098-48-1, DF15
37934-98-8, DC65	40006-64-2, DB89	55098-49-2, DF23
37934-99-9, DC63	40953-63-7, DC96	55098-51-6, DF70
38032-87-0, DG41	41206-74-0, DG81	55098-52-7, DF69
38469-75-9, DD15	41322-56-9, DC92	55098-53-8, DG04
38523-68-1, DE96	41467-62-3, DD06	55098-54-9, DF67

55098-55-0, DG03
55098-56-1, DF86
55098-57-2, DF33
55098-58-3, DF92
55098-59-4, DF17
55098-60-7, DF87
55098-61-8, DF71
55098-62-9, DE76
55098-63-0, DF51
55098-64-1, DE93
55153-72-5, DF21
57178-74-2, DB02
57500-53-5, DA52
58482-63-6, DH02
59710-71-3, DG01
60160-48-7, DG80
60684-33-5, DH30
61004-83-9, DH15

TABLE VII.
INDEX OF SOLVENTS EMPLOYED

This table is an index that lists the solvents and mixtures of solvents that appear in Table I. It enables the user to identify all of the entries in that table that contain information pertaining to any particular solvent or mixture of solvents.

The entry

Propylene carbonate, DH64

signifies that data are given in Table I for the compound designated by code number DH64 in nominally pure propylene carbonate as the solvent. We have considered a solvent to be nominally pure whenever there was no deliberate addition of a second solvent or whenever the stated purity of the solvent was at least 99%. This criterion does not enable the user of this table to distinguish between entries for, say, reagent-grade acetonitrile and those for acetonitrile that had been rigorously purified to remove as much as possible of the water and other proton donors.

The entry

Benzene-methanol (20:80), DC26

provides references to data obtained in binary mixtures containing 20% of benzene and 80% of methanol by volume. Binary (and other) mixtures are listed in the alphabetical order of the names of their constituents, so that mixtures of benzene with benzonitrile precede those of benzene with ethanol. Binary mixtures containing the same constituents in different proportions are listed in the order of decreasing percentage, which is always by volume unless otherwise stated, of the constituent whose name is given first. Thus 2-propanol-water (75:25) precedes 2-propanol-water (40:60). Occasionally in this series we decided to include data even though the information given in the original literature left us more or less uncertain about the proportions of the components of a binary mixture. These cases are represented by entries of the form

Acetone-toluene-water (?:?:?), DF96

which denotes a mixture of acetone, tuolene, and water that contain unknown amounts of each solvent.

The citations for binary mixtures are given only once, in the entry for the constituent whose name occurs first in the alphabet. A cross-reference is always given under the name of the other constituent and in the following form:

Methanol

-benzene, see benzene-methanol

Ternary mixtures are treated similarly. For a mixture of ethanol, methanol, and water the citation is given in the entry for ethanol, and in that entry it appears as part of the subentry for mixtures of ethanol with methanol, so that the three constituents are named in alphabetical order:

Ethanol

-methanol

-water (85:5:10), DF94, DF95, DG45

All ternary mixtures are fully cross-referenced so that they can be located in the entries for all of their constituents.

TABLE VII. Index of Solvents Employed

Acetic acid-water (80:20), DD66, DD77
Acetone, DB28, DD86, DD87, DD88
    -toluene-water (? : ? : ?), DF96
    -water (60:40), DA89
          (40:60), DA89
          (20:80), DA89
          (10:90), DA89
Acetonitrile, DA12, DA26, DA31, DA32, DA33, DA34, DA40, DA45, DA50, DA59, DA63,
            DA72, DA78, DA81, DA84, DA86, DB03, DB05, DB06, DB21, DB22, DB29,
            DB31, DB32, DB33, DB34, DB36, DB39, DB40, DB41, DB43, DB44, DB46,
            DB47, DB65, DB68, DB69, DB71, DB76, DB92, DB93, DB95, DB97, DB98,
            DB99, DC00, DC01, DC02, DC08, DC09, DC13, DC14, DC24, DC25, DC37,
            DC38, DC41, DC43, DC44, DC46, DC47, DC48, DC49, DC50, DC51, DC52,
            DC53, DC54, DC55, DC78, DC80, DC82, DC91, DC92, DC96, DC97, DC98,
            DD00, DD16, DD22, DD27, DD30, DD39, DD40, DD41, DD42, DD43, DD44,
            DD53, DD57, DD58, DD60, DD61, DD67, DD68, DD69, DD72, DD73, DD75,
            DD76, DD94, DD96, DE05, DE07, DE17, DE18, DE19, DE35, DE39, DE40,
            DE43, DE44, DE46, DE53, DE54, DE55, DE56, DE57, DE58, DE63, DE68,
            DE69, DE70, DE71, DE82, DE94, DE95, DE96, DE98, DF13, DF14, DF19,
            DF39, DF41, DF42, DF43, DF44, DF45, DF46, DF47, DF49, DF59, DG02,
            DG10, DG12, DG13, DG15, DG29, DG36, DG38, DG39, DG46, DG48, DG49,
            DG64, DH36, DH37, DH38, DH40, DH43, DH56
    -water (50:50), DA06
Ammonia, DA96, DD03, DE02, DE03

Benzene-methanol (50:50), DE20, DF48, DG00, DG01, DG20, DG42, DG62, DG70, DH09
               (33:67), DG52, DG53, DG57, DG58, DG59, DH08
               (20:80), DC26
Benzonitrile, DG81, DG87, DG90, DH01, DH13
Butanol-water (96:04), DB60
Butyronitrile, DD60, DD61, DG36, DG82, DG83, DG84, DG85, DG86, DG88, DG89, DG91,
             DG92, DG93, DG94, DG95, DG96, DG97, DG98, DG99, DH00, DH02, DH03,
             DH04, DH05, DH06, DH07, DH10, DH27

m-Cresol, DA48, DA49, DA80, DB43, DB73, DE11, DE45, DG24

Dichloromethane, DA69, DD06, DD57, DE04, DE71, DE82, DF13, DF14, DF40, DF58, DF59,
               DF93, DF99, DG36, DG37, DH13, DH18, DH41, DH55, DH57, DH63, DH64
    -trifluoroacetic acid (90:10), DC39
1,2-Dimethoxyethane, DA22, DB64, DC06, DC07, DC56, DC58, DC59, DC60, DC61, DC99,
                 DD04, DD05, DD26, DD45, DD46, DD47, DD59, DD63, DD64, DD81,
                 DD98, DD99, DE24, DE34, DE37, DE75, DE86, DF00, DF01, DF02,
                 DF03, DF36, DF50, DF62, DF68, DF75, DF84, DF91, DG33, DG37,
                 DG43, DG44
Dimethylformamide, DA11, DA35, DA38, DA59, DA62, DA68, DA71, DA73, DA85, DB07,
              DB14, DB24, DB27, DB48, DB49, DB50, DB51, DB52, DB53, DB54,
              DB55, DB56, DB57, DB58, DB59, DB61, DB62, DB63, DB83, DB84,
              DB85, DB86, DB87, DB88, DB89, DB90, DB92, DC03, DC04, DC05,
              DC11, DC12, DC21, DC63, DC68, DC69, DC70, DC71, DC85, DD01,
              DD07, DD17, DD19, DD20, DD21, DD28, DD29, DD32, DD34, DD35,
              DD36, DD38, DD54, DD65, DD82, DD85, DD89, DD90, DD91, DD92,
              DE04, DE14, DE15, DE23, DE25, DE30, DE31, DE32, DE36, DE38,
              DE51, DE52, DE60, DE61, DE62, DE64, DE65, DE88, DE89, DE99,
              DF52, DF80, DF88, DF89, DG12, DG21, DG22, DG23, DG25, DG28,
              DG32, DG35, DG36, DG47, DG48, DG51, DG55, DG56, DG71, DG74,
              DG75, DG76, DG77, DG78, DG79, DG80, DG91, DH11, DH13, DH17,
              DH18, DH32, DH33, DH34, DH35, DH39, DH41, DH42, DH43, DH44,
              DH45, DH46, DH47, DH48, DH49, DH50, DH51, DH52, DH53, DH54,
              DH58, DH59, DH60, DH61, DH62, DH64
    -water (95:05), DA62, DG28
          (80:20), DA62

Dimethyl sulfoxide, DD76, DF81, DG11, DG36, DG81, DG82, DG83, DG84, DG85, DG86,
DG88, DG89, DG90, DG92, DG93, DG94, DG95, DG96, DG97, DG98,
DG99, DH00, DH01, DH02, DH03, DH04, DH05, DH06, DH10, DH27
Dioxane-water (75:25), DH13
(70:30), DA89
(50:50), DD55, DD56, DD93, DE66
(40:60), DA89
(20:80), DD05, DD49
(10:90), DA89

Ethanol, DA29, DB09, DD71, DF77, DF78, DF79
   -methanol-water (85:05:10), DF94, DF95, DG45
   -water (96.0:04.0), DB60
       (96 :04), DB66, DB70, DC28
       (90 :10), DB66, DB70, DC28
       (86 :14), DA83, DB77, DC23, DC77
       (80 :20), DB66, DB70, DC28, DG30, DH30
       (75 :25), DA00, DA27, DB69, DG17, DG18, DG19, DG34, DG50, DG54,
                  DG60, DG61, DG63, DH08
       (70 :30), DA89
       (80 :40), DA89, DB66, DB70, DC28, DE06
       (50 :50), DA03, DA14, DA30, DA51, DA64, DA95, DB67, DC10, DC16,
                  DC17, DC18, DC19, DC20, DC22, DC27, DC30, DC66, DC67,
                  DC74, DC75, DC81, DD23, DD24, DF60, DF76, DF90
       (48 :52), DA15
       (~40 :~60), DA93, DD52
       (40 :60), DA16, DA89, DB66, DB70, DC28, DD02, DF54
       (30 :70), DA30, DA51, DA64, DA95, DG73, DH12, DH19, DH21, DH22
       (25 :75), DC72
       (20 :80), DA30, DA51, DA64, DA89, DA95, DB66, DB70, DE10
       (10 :90), DA30, DA51, DA54, DA55, DA64, DA65, DA89, DA95, DB66,
                  DB70, DB72, DB82, DC22, DC23, DC30, DC31, DC45
       (05 :95), DE78
       (02 :98), DE06
       (01 :99), DC57

Fluorosulfonic acid-trifluoroacetic acid (50:50), DG41

Isopropyl alcohol, see 2-propanol

Methanol, DA45, DC25, DC62, DC63, DC65, DC68, DC69, DC76, DC80, DD14, DD19, DD20,
    DD76
Methanol-benzene, see benzene-methanol
Methanol-ethanol-water, see Ethanol-methanol-water
    -tetrahydrofurfurylalcohol (50:50), DE01
    -water (98.8:01.2), DB60
       (92:08), DA40, DD38, DE67, DG31
       (80:20), DA16
       (75:25), DA28, DG52, DG53, DG57, DG58
       (70:30), DA16
       (60:40), DA16, DC73
       (50:50), DA37, DB25, DC93, DC94, DD08, DD09, DD10, DD11, DD12,
                  DD13, DD18, DD50, DD51, DD84, DE16, DE59, DG14
       (40:60), DA16, DE01
       (33:67), DD62
       (25:75), DD48
       (20:80), DA21, DA39, DA41, DA46, DC29, DC83
       (9.1:90.9), DC32
       (3.3:96.7), DE03, DE08, DE09, DE33, DE41, DE47, DE48, DE72, DE73,
                   DE76, DE79, DE80, DE83, DE84, DE85, DE87, DE90, DE91,
                   DE92, DE93, DF04, DF06, DF07, DF08, DF10, DF11, DF15,
                   DF16, DF17, DF20, DF21, DF22, DF23, DF24, DF26, DF27,

TABLE VII. Index of Solvents Employed

                DF28, DF29, DF30, DF31, DF33, DF34, DF35, DF37, DF51,
                DF53, DF55, DF56, DF57, DF61, DF63, DF64, DF65, DF66,
                DF67, DF69, DF70, DF71, DF72, DF86, DF87, DF92, DG03,
                DG04

Pentanonitrile, DC09, DD61
Propanol-water (96:04), DB60
2-Propanol-water (75:25), DG66
        (69.4w/v:30.6), DA91
           (65:35), DA01, DA02
           (40:60), DB69, DE26, DE27, DE28, DE29
      (22.6w/v:77.4), DA91
      (6.5w/v:93.5), DA91
Propylene carbonate, DH64
Pyridine, DD76
Pyridine-water (20:80), DH20

Tetrahydrofuran, DC24, DC35, DC78, DC79, DD22, DD25, DD70, DD78, DD79, DF05, DF18,
        DF32, DG10, DG48
Tetrahydrofurfuryl alcohol-methanol, see methanol-tetrahydrofurfuryl alcohol
Trifluoroacetic acid, DG40
Trifluoroacetic acid-dichloromethane, see dichloromethane-trifluoroacetic acid
Trifluoroacetic acid-fluorosulfonic acid, see fluorosulfonic acid-trifluoroacetic
  acid

Water, DA04, DA05, DA06, DA07, DA08, DA09, DA10, DA13, DA14, DA17, DA18, DA19,
      DA20, DA23, DA24, DA25, DA30, DA31, DA36, DA42, DA43, DA44, DA51, DA52,
      DA53, DA54, DA55, DA56, DA57, DA58, DA60, DA61, DA64, DA66, DA67, DA70,
      DA74, DA75, DA76, DA77, DA79, DA82, DA87, DA88, DA90, DA91, DA92, DA94,
      DA97, DA98, DA99, DB00, DB01, **DB02**, DB04, DB08, DB09, DB10, DB11, DB12,
      DB13, DB15, DB16, DB17, DB18, DB19, DB20, DB23, DB26, DB35, DB37, DB38,
      DB42, DB45, **DB74**, DB75, DB78, DB79, DB80, DB81, DB91, DB94, DB96, DC15,
      DC33, DC34, DC36, DC40, DC42, DC64, DC84, DC86, DC87, DC88, DC89, DC90,
      DC92, DC95, DD05, DD15, DD31, DD33, DD37, DD74, DD80, DD83, DD95, DD97,
      DE00, DE12, DE13, DE21, DE22, DE42, DE49, DE50, DE74, DE77, DE81, DE84,
      DE97, DF09, DF12, DF25, DF38, DF54, DF73, DF74, DF82, DF83, DF85, DF96,
      DF97, DF98, DG05, DG06, DG07, DG08, DG09, DG16, DG26, DG27, **DG65**, DG67,
      DG68, DG69, DG72, DG73, DH14, DH15, DH16, **DH23**, DH24, DH25, DH26, DH28,
      DH29, DH30, DH31
    -acetic acid, see acetic acid-water
    -acetone, see acetone-water
    -acetone-toluene, see acetone-toluene-water
    -acetonitrile, see acetonitrile-water
    -butanol, see butanol-water
    -dimethylformamide, see dimethylformamide-water
    -dioxane, see dioxane-water
    -ethanol, see ethanol-water
    -ethanol-methanol, see ethanol-methanol-water
    -isopropanol, see 2-propanol-water
    -methanol, see methanol-water
    -methanol-ethanol-see ethanol-methanol-water
    -propanol, see propanol-water
    -pyridine, see pyridine-water
    -toluene-acetone, see acetone-toluene-water

TABLE VIII.
INDEX OF TECHNIQUES EMPLOYED

This is an index that lists the techniques employed in obtaining the data that appear in Table I. It enables the user to identify all of the entries in that table that contain information obtained by any particular technique.

In the typical entry

Chronoamperometry (IR), DB53, DB57, DC21, DE25, DE31

the name is the one recommended by the International Union of Pure and Applied Chemistry (*Pure and Applied Chemistry,* **45**, 81, 1976). Cross-references are provided liberally, so that, for example, the entries for polarographic coulometry can be found by users who seek them under "dropping-electrode coulometry", "millicoulometry", or "microcoulometry", and these cross-references also serve to notify the user that the name followed by the citations is recommended in preference to one cross-referenced to it.

Following each recommended name there appears, in parentheses, the two-letter symbol by which the technique is denoted in Column 6 (or, occasionally, in Column 21 or 26) of Table I. This enables the user to tell, without having to consult the list of abbreviations repeatedly, what part of the information contained in a long entry in Table I has been obtained by the technique in which he is interested.

Finally, for every technique except polarography itself, there appear in alphanumeric order the code numbers for all of the compounds about which information obtained by that technique appears in Table I. There are a few entries of the form

Alternating-current chronopotentiometry (EF)

in which the name of a technique is followed by a two-letter symbol that identifies it as a recommended name, but in which no code numbers follow the symbol. Such an entry signifies that the technique is included in our sphere of activity but has produced no information that is given in this volume. At the other extreme, the entry for polarography is of this same form but for a very different reason: as was stated previously, a list of the code numbers of the compounds for which polarographic data are given would be too long to be of any use. We estimate that it would have contained over 600 citations. If this figure is compared with the total of 569 citations given below, it becomes apparent that, during the period covered by these volumes, the contribution made by polarography to our fundamental knowledge of the electrochemical behaviors of organic, biochemical, and organometallic substances appreciably exceeded that of all the other techniques combined.

TABLE VIII. Index of Techniques Employed

Ac polarography (PV), DA36, DA79, DA87, DB04, DB09, DB45, DB69, DC88, DC89, DD86,
                DD87, DD88, DE25, DE31, DF97, DF98, DG05, DG16, DG74, DG79
    higher-harmonic, see Higher-harmonic ac polarography
    in-phase (PVI), DA36, DC88, DC89, DF98
Ac voltammetry (VV), DA61, DE22
Alternating-current chronopotentiometry (EF)
Alternating-voltage chronopotentiometry (EK)
Amperometry (IO)
    differential (ID)
    with two indicator electrodes (IB)
Alternating-voltage polarography (PI)

Cathode-ray polarography, see Single-sweep polarography
Chronoamperometry (IR), DB53, DB57, DC21, DE25, DE31
    convective, see Convective chronoamperometry
    double potential-step, see Double potential-step chronoamperometry
    dropping-electrode, with linear potential sweep, see Single-sweep polarography
    polarographic, see Polarographic chronoamperometry
    rotating-disc-electrode, see Convective chronoamperometry
    stirred-pool-electrode, see Convective chronoamperometry
    thin-layer (TI)
    with linear potential sweep, (IL), DA05, DA20, DA60, DA61, DA70, DA74, DB23,
                                    DB26, DB65, DB68, DB71, DB96, DC33, DC34,
                                    DC36, DC38, DC40, DC42, DC86, DC95, DD66,
                                    DD68, DD77, DE22, DE39, DE77, DE78, DG17,
                                    DG18, DG19, DG26, DG27, DG34, DG50, DG52,
                                    DG53, DG54, DG57, DG58, DG59, DG60, DG61,
                                    DG63, DH08
        second-derivative (XT)
    with non-linear potential sweep (IM)
Chronocoulometry (QR)
    convective, see Convective chronocoulometry
    double potential-step, see Double potential-step chronocoulometry
    potential-step, see Chronocoulometry
    rotating-disc-electrode, see Convective chronocoulometry
    stirred-pool-electrode, see Convective chronocoulometry
    thin-layer (TQ)
Chronopotentiometry (ER), DA49, DF39, DG73, DH12, DH19, DH21, DH22
    alternating-current (EF)
    alternating-voltage (EK)
    current-cessation, see Current-step chronopotentiometry
    current-reversal, see Current-step chronopotentiometry
    current-step, see Current-step chronopotentiometry
    cyclic, see Cyclic chronopotentiometry
    derivative, see Derivative chronopotentiometry
    programmed-current (EC)
    with linear current sweep (EL)
    with superimposed ac (EV)
Controlled-current coulometry (QP)
Controlled-potential coulometry (QE), DA31, DB92, DC37, DC82, DD27, DD53, DE17,
                                    DE40, DE43, DE44, DE82, DE94, DE96, DG36,
                                    DG37, DG46, DG49
Controlled-potential electrolysis (CP), DC63
Convective chronoamperometry (CA)
    controlled-current (with a reagent precursor), see Coulometric titration
                      (without a reagent precursor)(QP)
    controlled-potential, see Controlled-potential coulometry
    polarographic, see Polarographic coulometry
Coulometry
    controlled-current, see Controlled-current coulometry
    controlled-potential, see Controlled-potential coulometry
    dropping-electrode, see Dropping-electrode coulometry
    polarographic, see Polarographic coulometry

Current-cessation chronopotentiometry, see Current-step chronopotentiometry
Current-reversal chronopotentiometry, see Current-step chronopotentiometry
Current-scanning polarography (PC)
Current-step chronopotentiometry (EE), DE56, DG73
Cyclic chronopotentiometry (EY)
Cyclic triangular-wave polarography (PR), DA31
Cyclic triangular-wave voltammetry (VR), DA26, DA36, DA50, DA59, DA63, DA69, DA78,
　　　DA81, DA84, DA85, DA96, DB03, DB05, DB06, DB31, DB32, DB33, DB34, DB36,
　　　DB39, DB40, DB41, DB43, DB44, DB46, DB47, DB48, DB49, DB50, DB51, DB52,
　　　DB53, DB54, DB55, DB56, DB57, DB59, DB61, DB76, DB93, DB95, DB97, DB98,
　　　DB99, DC00, DC01, DC02, DC03, DC09, DC14, DC21, DC41, DC43, DC44, DC46,
　　　DC47, DC48, DC49, DC50, DC51, DC52, DC53, DC54, DC55, DC79, DC89, DC91,
　　　DC92, DC96, DC97, DC98, DD03, DD06, DD22, DD30, DD32, DD39, DD40, DD41,
　　　DD42, DD43, DD44, DD57, DD58, DD60, DD61, DD67, DD76, DD94, DD96, DE02,
　　　DE04, DE05, DE18, DE19, DE23, DE32, DE35, DE38, DE55, DE57, DE58, DE68,
　　　DE69, DE70, DE71, DE82, DE84, DE99, DF13, DF14, DF41, DF42, DF43, DF44,
　　　DF45, DF46, DF47, DF49, DF58, DF59, DF80, DF93, DF97, DF98, DG10, DG13,
　　　DG15, DG16, DG36, DG38, DG41, DG47, DG48, DG64, DG81, DG82, DG83, DG84,
　　　DG85, DG86, DG87, DG88, DG89, DG90, DG91, DG92, DG93, DG94, DG95, DG96,
　　　DG97, DG98, DG99, DH00, DH01, DH02, DH03, DH04, DH05, DH06, DH07, DH10,
　　　DH11, DH13, DH18, DH20, DH27

Dc polarography, see Polarography
Demodulation polarography (FP)
Derivative chronopotentiometry (EM)
Derivative polarography (PD), DA00, DA27, DE22
Derivative pulse polarography (DP)
Derivative voltammetry (VD)
Differential amperometry, see Amperometry, differential
Differential linear-sweep polarography, see Polarography, linear-sweep, differential
Differential polarography (PF)
Differential pulse polarography (DI), DA08, DA09, DA10, DA24, DA75, DA90, DA92, DA93,
　　　　　　　　　　　　　　　　DA94, DB04, DB75, DC87, DD33, DD48, DD52, DD62,
　　　　　　　　　　　　　　　　DD74, DD80, DE03, DE08, DE09, DE33, DE41, DE42,
　　　　　　　　　　　　　　　　DE47, DE48, DE72, DE73, DE76, DE77, DE79, DE80,
　　　　　　　　　　　　　　　　DE83, DE84, DE85, DE87, DE90, DE91, DE92, DE93,
　　　　　　　　　　　　　　　　DF04, DF06, DF07, DF08, DF09, DF10, DF11, DF15,
　　　　　　　　　　　　　　　　DF16, DF17, DF20, DF21, DF22, DF23, DF24, DF25,
　　　　　　　　　　　　　　　　DF26, DF27, DF28, DF29, DF30, DF31, DF33, DF34,
　　　　　　　　　　　　　　　　DF35, DF37, DF51, DF53, DF55, DF56, DF57, DF61,
　　　　　　　　　　　　　　　　DF63, DF64, DF65, DF66, DF67, DF69, DF70, DF71,
　　　　　　　　　　　　　　　　DF72, DF74, DF86, DF87, DF92, DG03, DG04, DG66,
Differential voltammetry (VF)
Double potential-step chronoamperometry (IU), DA31
Double potential-step chronocoulometry, (QU)
Double-tone polarography (PX)
Dropping-electrode coulometry, see Polarographic coulometry
Dynamic capacity, measurement of (NF)

Electrochemiluminescence (HN), DH32 - DH64
Electrogravimetry (QW)

Faradaic rectification, high-level (HL), DG74, DG75, DG76, DG77, DG78, DG79, DG80,

High-level faradaic rectification, see Faradaic rectification, high-level
Higher-harmonic ac polarography (PH)
Higher-harmonic ac polarography with phase-sensitive rectification (PB)
Hydrodynamic voltammetry (VY), DA04, DA45, DA49, DB07, DB08, DB15, DB64, DC06,
　　　　　　　　　　　　　DC07, DC08, DC25, DC80, DD14, DE53, DE54, DE55,
　　　　　　　　　　　　　DE58, DE61, DE62, DE63, DE98, DF19, DF40, DF81,
　　　　　　　　　　　　　DF99, DG02, DG11, DG12, DG29, DG39, DG73, DH12,
　　　　　　　　　　　　　DH19, DH21, DH22

TABLE VIII. Index of Techniques Employed

Incremental-charge polarography (DQ)
Intermodulation polarography (PL)

Kalousek polarography (PK)

Linear-sweep differential polarography (XI)
Linear-sweep voltammetry, see Chronoamperometry with linear potential sweep

Microcoulometry, see Polarographic coulometry
Millicoulometry, see Polarographic coulometry
Modulation polarography (MM)
Multisweep polarography (PM)
Multisweep voltammetry (VM)

Non-faradaic admittance, see Dynamic capacity

Oscillographic polarography, see Oscillopolarography
    single-sweep, see Single-sweep polarography
Oscillopolarography (PO), DA18, DD31, DE01, DF60
Oscillovoltammetry (VO)

Polarographic chronoamperometry (IV), DA25, DA97, DA98, DA99, DB00, DB01, DB02, DB10, DB11, DB12, DB13, DB19, DB20, DB78
Polarographic coulometry (PQ), DC23
Polarography (PY)
    ac, see Ac polarography
    alternating-voltage, see alternating-voltage polarography
    cathode-ray, see Single-sweep polarography
    current scanning, see Current-scanning polarography
    cyclic triangular wave, see Cyclic triangular-wave polarography
    demodulation, see Demodulation polarography
    derivative, see Derivative polarography
    derivative pulse, see Derivative pulse polarography
    differential, see Differential polarography
    differential pulse, see Differential pulse polarography
    double-tone, see Double-tone polarography
    higher-harmonic ac, see Higher-harmonic ac polarography
    incremental-charge, see Incremental-charge polarography
    in-phase ac, see Ac polarography, in-phase
    intermodulation, see Intermodulation polarography
    kalousek, see Kalousek polarography
    linear-sweep differential, see Linear-sweep differential polarography
    modulation, see Modulation polarography
    multisweep, see Multisweep polarography
    oscillographic, see Oscillopolarography
    pulse, see Pulse polarography
    rf, see, Rf polarography
    single-sweep, see Single-sweep polarography
    square-wave, see Square-wave polarography
    staircase, see Staircase polarography
    Tast, see Tast polarography
    triangular-wave, see Triangular-wave polarography
        cyclic, see Cyclic triangular-wave polarography
Programmed-current chronopotentiometry (EC)
Pulse polarography (PP)
Pulse voltammetry (XT)

Rf polarography (RP), DG80

Single-sweep polarography (PW)
Square-wave polarography (PG), DG80
Staircase polarography (PS)
Stirred-pool-electrode chronoamperometry, see Convective chronoamperometry

Tast polarography (PT), DA08, DA10, DA24, DA44, DA75, DA76, DB42, DB75, DC87, DC90, DD37, DD74, DD95, DE13, DE42, DF12, DF76
Thin-layer chronoamperometry, see Chronoamperometry, thin-layer
Thin-layer chronocoulometry, see Chronocoulometry, thin-layer
Titration, controlled-potential coulometric, see Controlled-potential coulometry
    coulometric, see Coulometric titration
Triangular-wave polarography (PA), DC56, DC58, DC59, DC60, DC61, DC99, DF18, DF91, DG10, DG74, DG75, DG79, DH19
    cyclic, see Cyclic triangular-wave polarography
Triangular-wave voltammetry (VA), DA06, DA17, DA31, DB04, DB53, DC13, DC39, DD25, DD66, DD70, DD76, DD77, DD78, DD79, DD97, DE10, DE25, DE31, DE70, DG10, DG40, DG53, DH08, DH13
    cyclic, see Cyclic triangular-wave voltammetry

Voltammetry (V)
    ac, see Ac voltammetry
    cyclic triangular-wave, see Cyclic Triangular-wave voltammetry
    derivative, see Derivative voltammetry
    differential, see Differential voltammetry
    hydrodynamic, see Hydrodynamic voltammetry
    linear-sweep, see Chronoamperometry with linear potential sweep
    multisweep, see Multisweep voltammetry
    oscillographic, see Oscillovoltammetry
    pulse, see Pulse voltammetry
    triangular-wave, see Triangular-wave voltammetry
        cyclic, see Cyclic triangular-wave voltammetry

TABLE IX.
INDEX OF INDICATOR ELECTRODES EMPLOYED

This is an index that lists the indicator and working electrodes employed in obtaining the data that appear in Table I. It enables the user to identify all of the entries in that table that contain information obtained with any particular indicator or working electrode.

In the typical entry

Carbon, cylindrical, rotating (RCCE), DA45, DC25, DC80, DD14, DF19

the initial portion identifies the electrode used and gives its configuration whenever possible and, in parentheses, the abbreviation used to denote the electrode in Column 11 of Table I. This enables the user to tell, without having to consult the list of abbreviations repeatedly, what part of the information contained in a long entry in Table I has been obtained with the electrode and configuration in which he is interested. Following the abbreviation, there appear the code numbers of the compounds that have been studied with the electrode. Finally, liberal cross-references are provided to ensure that no entry that might be of interest will be overlooked.

Precious though space in the original literature certainly is, it does not seem to us to be valuable enough to justify the frequency with which sentences like "The graphite indicator electrode was described previously" are used by authors meticulous in specifying the other circumstances under which their data were obtained. As it is well known that electrochemical properties depend on the configuration of the indicator electrode as well as on the material from which it is constructed, our original intention was to provide a complete index of both configurations and materials. A few cases in which the reference number appended to a sentence like the one quoted led only to an exactly identical sentence in another article convinced us, however, that this intention was unlikely to be achievable, and we have therefore confined ourselves to transcribing the information provided. We recognize that this complicates the use of this table, even with the cross-references provided, and urge the authors of future papers to describe their electrodes as carefully as they do the other facets of their work.

There is one null entry in this table, and that is the one for the dropping mercury electrode. This reflects the same fact that led to the omission of polarography from Table VIII. There are some compounds in Table I for which no information obtained with a dropping mercury electrode is given in this volume. The great majority of these are compounds that undergo oxidation at relatively positive potentials and that were therefore studied with platinum, graphite, carbon paste, or other indicator electrodes. Many such compounds are not amenable to investigation with dropping mercury electrodes, but there are many others that are. Some of these can also be reduced within the range of potentials accessible with a dropping mercury electrode while some undergo oxidation within that range under certain conditions but not under others; for some there may be polarographic data in the prior literature, while for others the possibility of obtaining a polarographic response may not have been investigated. In any event, an entry documenting the use of the dropping mercury electrode in obtaining the data included in this volume would have contained approximately 650 entries and would have been too long to be of any practical use.

TABLE IX. Index of Indicated Electrodes Employed

Bead, see Platinum, spherical bead

Carbon
    cylindrical, rotating (RCCE), DA45, DC25, DC80, DD14, DF19
    glassy (GCE), DG27, DG34, DG52, DG53, DG57, DG58, DG59, DH08
        cylindrical, rotating (RCGE), DC25
    paste (CPE)
        containing dissolved organic substances (CPDO), DG17, DG18, DG19, DG34, DG50, DG53, DC54, DG57, DG58, DG60, DG61, DG63, DH08
        with wax and silicone (XEI), DB23, DC33, DC34, DC36, DC40, DC42
    See also Graphite and Ring-disc

Disc
    mercury-plated (MPDE), DA31, DA63
    See also Gold disc and Platinum disc
Drop, see Mercury, hanging drop

Gauze, see Platinum-gauze
Gold (Au)
    bead (AuBE), DA26, DA50, DA78, DA81, DB31, DB32, DB33, DB34, DB36, DB39, DB40, DB41, DB43, DB44, DB46, DB47, DB93, DB95, DB97, DB98, DB99, DC00, DC01, DC02, DC41, DC43, DC44, DC46, DC47, DC48, DC49, DC50, DC51, DC52, DC53, DC54, DC55, DC91, DC97, DC98, DD40, DD41, DD42, DD58, DD94, DE18, DE19, DE57, DE68, DG13, DG15, DG38
    disc (AuDE), DA96, DD03, DE02
    rotating mercury-plated (XEI), DA18
Graphite (GE),
    pyrolytic (PGE), DA05, DA60, DA61, DD76
    See also Carbon

Lead disc (PbDE), DA17
Lead sulfide (PbSE), DA17

Mercury
    hanging drop (HMDE), DA06, DA20, DA36, DA59, DA70, DA74, DA84, DB03, DB04, DB05, DB06, DB26, DB76, DB96, DC09, DC56, DC58, DC59, DC60, DC61, DC89, DC95, DC99, DD60, DD61, DD97, DE10, DE22, DE31, DE77, DE78, DE84, DF91, DF97, DF98, DG16, DH19
    pool (MP), DA49, DB92
    streaming (SME), DF60

Platinum (Pt), DB07, DB65, DB68, DB71, DC14, DC37, DC38, DC82, DC86, DD27, DD30, DD53, DD66, DD67, DD68, DD76, DD77, DE04, DE17, DE35, DE38, DE39, DE43, DE44, DE55, DE56, DE58, DE94, DE96, DF39, DG36, DG37, DG46, DG48, DG49, DG64, DH13, DH18, DH32, DH33, DH34, DH35, DH36, DH37, DH38, DH39, DH40, DH41, DH42, DH43, DH44, DH45, DH46, DH47, DH49, DH50, DH51, DH52, DH53, DH54, DH55, DH56, DH57, DH58, DH59, DH60, DH61, DH62, DH63, DH64

disc (PDE), DA85, DB54, DB55, DB56, DD32, DE04, DE31, DE32, DE99, DF80,
            DG47, DG73, DH11, DH12, DH13, DH19, DH20, DH21, DH22, DH48
gauze (Pt Gauze), DC63, DE82
rotating (RPE), DA04, DA45, DA64, DC06, DC07, DC08, DC25, DC80, DD14, DE53,
            DE54, DE55, DE58, DE61, DE62, DE63, DE98, DF19, DF40, DF81,
            DF99, DG11, DG12, DG29, DG39, DG73, DH12, DH19, DH21, DH22
    disc (RPDE), DA49, DB08, DG02
spherical bead (PBE), DA69, DB48, DB49, DB50, DB51, DB52, DB53, DB57, DB59,
            DB61, DC03, DC09, DC13, DC21, DC79, DD06, DD22, DD25,
            DD43, DD44, DD57, DD60, DD61, DD70, DD78, DD79, DE05,
            DE23, DE25, DE69, DE70, DE71, DE82, DF58, DF59, DF93,
            DG10, DG36, DG37, DG81, DG82, DG83, DG84, DG85, DG86,
            DG87, DG88, DG89, DG90, DG91, DG92, DG93, DG94, DG95,
            DG96, DG97, DG98, DG99, DH00, DH01, DH02, DH03, DH04,
            DH05, DH06, DH07, DH10, DH27

Pool, see Mercury pool

Ring-disc, rotating (RRDE)   DB15

Silver (Ag), DD96
Sphere; see Mercury, hanging drop and Platinum, spherical bead

TABLE X.
KEY TO LITERATURE CITATIONS

This table is provided to permit decoding the literature references that appear in Column 14 of Table I and obtaining full citations from them.

The form and significance of those references are described in the introduction to Table I. To illustrate the use of this table we shall suppose that a full citation is wanted for the information given in this volume for cytidine monophosphate. In Table I this compound appears under code number DC89, and the literature reference is given as JA095-0991 in Column 14 on the first of the lines dealing with this compound.

As was discussed in the introduction to Table I, the letters "JA" denote the journal. The next three digits give the volume, in this case 95. As that number contains only two digits, it is preceded by a zero (if the volume number contained only one digit, it would be preceded by two zeros). The last four digits give the page number. Here this is 991 and is preceded by one zero. Occasionally two or three zeros are needed to give the total of four digits. Hence the reference is to volume 95, page 991, of the journal denoted by the two-letter code "JA".

To identify the journal that is denoted by the abbreviation JA, this table must be inspected. The journals are arranged in the alphabetical order of the two-letter codes assigned to them, and each such code is followed in parentheses by the full title of the journal to which it corresponds. It can be found that the two-letter code "JA" denotes the *Journal of the American Chemical Society*. Hence the reference is to the *Journal of the American Chemical Society*, volume 95, page 991, and this table provides the further information that the paper was published in 1972.

Names of some journals published in other languages than English are given in transliteration of the original title. When data were obtained from English translations (e.g., of Russian journals) the English translation of the name of the journal is given.

Beneath the abbreviation and the name of the journal, volume numbers and years are given, followed by references to all of the publications from which data for Table I in this volume were abstracted. These references are given in exactly the same form as is used in Table I, and are followed by the names of the authors. The references to each volume of each journal are listed in the order of increasing page numbers.

For example, the reference "BF63O-2252" is the third one which appears underneath volume 1963 of the *Bulletin de la Société chimique de France* and gives as authors J. Tirouflet and P. I. Cheng. Consequently the full citation in the usual form is J. Tirouflet and P. I. Cheng, *Bull. Soc. Chim. Fr.*, 2252, 1963.

Following the author's name there appear the code numbers of all of the compounds appearing in this volume for which data were taken from the reference given. For the reference just mentioned, the code number DF60 is the only one given, and it means that data on the compound denoted by that code number are the only ones taken from that paper in this volume.

Data for more than one compound were often obtained from a single publication. In such a case the code numbers of all the compounds are given in alphanumeric order. Usually these additional code numbers denote compounds that are closely related to the one that was originally of interest, or represent data obtained by similar techniques under similar conditions, and in either case provide cross-references that may further illuminate the information already obtained. It should not be inferred that these are complete lists of the compounds about which information is given in the original references, for some of these may have been omitted from this volume as having been studied in insufficient detail or for other reasons.

TABLE X. Key to Literature Citations

AA (ANALYTICA CHIMICA ACTA)

Vol. 30 (1964)
AA030-0313 Hetman, J.S., DE01

Vol. 58 (1971)
AA058-0183 Dryhurst, G., and De, P.K., DA60, DA61

Vol. 59 (1972)
AA059-0127 Halvorsen, S., and Jacobsen, E., DE84

Vol. 60 (1972)
AA060-0472 Jacobsen, E., and Jacobsen, T.V., DF09

Vol. 61 (1972)
AA061-0320 Jacobsen, E., and Rojahn, T., DD97

Vol. 62 (1972)
AA062-0405 Jacobsen, E., and Klevan, K.H., DE78

Vol. 63 (1972)
AA063-0175 Berge, H., and Brügmann, L. DE22
AA063-0415 Donahue, J.J., and Oliveri-Vigh, S., DE00

Vol. 64 (1973)
AA064-0055 Davis, D.G., and Truxillo, L.A., DG72, DH20, DH23, DH24, DH25, DH26, DH27, DH28, DH29
AA064-0165 Brooks, M.A., DeSilva, J.A.F., and Hackman, M.R., DD80, DE77

Vol. 66 (1973)
AA066-0023 Arthur, J., Silva, F. de, Strojny, N., and Munno, N., DA93, DD52
AA066-0427 Goldsmith, J.A., Jenkins, H.A., Grant, J., and Smyth, W.F., DE74, DE81

Vol. 67 (1973)
AA067-0415 McAllister, D., Pinson, J.P., and Dryhurst, G., DA06

Vol. 70 (1974)
AA070-0411 Tomcsány, L., DE10

Vol. 71 (1974)
AA071-0157 Jacobsen, E., and Høgberg, K., DG16

AA071-0175 Jacobsen, E., and Thorgersen, K.B., DB04, DG16
AA071-0433 Dickerson, R.L., and Rogers, J.W., DB54, DB55, DB56

Vol. 72 (1974)
AA072-0169 Bos, M., and Dahmen, E.A.M.F., DA48, DA49, DA80, DB43, DB73, DE11, DE45, DG24
AA072-0209 Dryhurst, G., and McAllister, D.L., DA05

Vol. 73 (1974)
AA073-0337 Deshler, L., and Zuman, P., DE06

Vol. 74 (1955)
AA074-0367 Brooks, M.A., Bel Bruno, J.J., De Silva, J.A.F. and Hackman, M.R., DD62, DE03, DE08, DE09, DE33, DE41, DE47, DE48, DE72, DE73, DE76, DE79, DE80, DE83, DE84, DE85, DE87, DE90, DE91, DE92, DE93, DF04, DF06, DF07, DF08, DF10, DF11, DF15, DF16, DF17, DF20, DF21, DF22, DF23, DF24, DF26, DF27, DF28, DF29, DF30, DF31, DF33, DF34, DF35, DF37, DF51, DF53, DF55, DF56, DF57, DF61, DF63, DF64, DF65, DF66, DF67, DF69, DF70, DF71, DF72, DF86, DF87, DF92, DG03, DG04

Vol. 76 (1975)
AA076-0289 Smyth, W.F., and Leo, B., DF12

Vol. 78 (1975)
AA078-0081 Smyth, W.F., Watkiss, D., Burmicz, J.S., and Hamley, H.O., DA08, DA09, DA10, DA24, DA44, DA75, DA76, DA92, DA94, DB42, DB75, DD74, DD95, DE42
AA078-0271 Thomas, L.C., and Christian, G.D., DG26, DG27

Vol. 80 (1975)
AA080-0017 Kkolos, E., and Walker, J. DB74, DC84
AA080-0233 Smyth, W.F., Jenkins, T., Siekiera, J., and Baydar, A., DC87, DC90, DD33, DD37, DE13, DF25

AC (ANALYTICAL CHEMISTRY)

Vol. 34 (1962)
AC034-1440 Kabasakalian, P., and McGlotten, J., DF94, DF95, DG45

Vol. 35 (1963)
AC035-0128 Cohen, A.I., DG25
AC035-0880 Swern, D., and Silbert, L.S., DE20, DF48, DG00, DG01, DG20, DG42, DG62, DG70, DH09

Vol. 36 (1964)
AC036-0523 Budke, C.C., Banerjee, D.K., and Miller, F.D., DF96

Vol. 42 (1970)
AC042-0825 Booth, M.D., and Fleet, B., DF90

AJ (AUSTRALIAN JOURNAL OF CHEMISTRY)

Vol. 25 (1972)
AJ025-2329 Woods, R., DA17

Vol. 26 (1973)
AJ026-1251 Boto, K.G., and Thomas, F.G., DD69, DD72
AJ026-1669 Boto, K.G., and Thomas, F.G., DD69, DD72, DE07, DE46, DE95

Vol. 27 (1974)
AJ027-1215 Boto, K.G., and Thomas, F.G., DD73, DD75
AJ027-2495 Mohamad, M., Khan, A.Y., Afzal, M., Nisa, A., and Ahmed, R., DA84, DB03, DB05, DB06, DB76
AJ028-0021 Randle, T.H., Cardwell, T.J., and Magee, R.J., DA20, DA70, DA74, DB26, DB96, DC95

AL (ANALYTICAL LETTERS)

Vol. 2 (1969)
AL002-0123 Badoz-Lambling, J., and Demange-Guerin, G., DB07

AN (THE ANALYST)

Vol. 95 (1970)
AN095-0387 Brand, M.J.D., Fleet, B., and Weaver, M.R.H., DC64

Vol. 98 (1973)
AN098-0886 Atuma, S.S., and Lindquist, J., DG50, DG54, DG60, DG61, DG63

Vol. 99 (1974)
AN099-0683 Atuma, S.S., Lindquist, J., and Lundström, K., DG17, DG18, DG19, DG34, DH08

Vol. 100 (1975)
AN100-0339 Lindquist, J., DB23
AN100-0349 Söderhjelm, P., and Lindquist, J., DC33, DC34, DC36, DC40, DC42
AN100-0377 Lindquist, J., and Farroha, S.M., DA90, DD48, DF74, DG66
AN100-0503 Powell, F.E. and Snowden, C.J., DG09
AN100-0489 Watson, A., and Svehla, G., DA25, DA97, DA98, DA99, DB00, DB01, DB02, DB11, DB12, DB13, DB18, DB19, DB20, DB78
AN100-0573 Watson, A., and Svehla, G., DB10
AN100-0584 Watson, A., and Svehla, G., DF83
AN100-0735 Edgar, J.S., DA04
AN100-0827 Atuma, S.S., Lundström, K., and Lindquist, J., DG52, DG53, DG57, DG58, DG59, DH08

BF (BULLETIN DE LA SOCIÉTÉ CHIMIQUE DE FRANCE)

1963
BF630-0479 Fournari, P., and Chane, J.P., DA52, DA53
BF530-1655 Tirouflet, J., Dabard, R., and Laviron, É., DD02
BF630-2252 Tirouflet, J., and Cheng, P.I., DF60

TABLE X. Key to Literature Citations

BJ (BULLETIN OF THE CHEMICAL SOCIETY OF JAPAN)

Vol. 45 (1972)
BJ045-0685 Katiyar, S.S., Lalithambika, M., and Shukig, R.P., DD05, DD49
BJ045-2898 Kageyama, H., Hidai, M., and Uchida, Y., DH17

Vol. 46 (1973)
BJ046-0147 Matsui, Y., Kurosaki, Y., and Date, Y., DD07
BJ046-0430 Matsuda, Y., Kimura, K., Iwakura, C., and Tamura, H., DB68
BJ046-2107 Ikeda, T., and Senda, M., DA77, DB37
BJ046-2129 Murayama, T., and Morioka, M., DA88
BJ046-2151 Kitagawa, T., and Tako, K., DA79, DB45
BJ046-3652 Kakutani, T., Totsuka, S., and Senda, M., DG71, DG74, DG75, DG76, DG77, DG78, DG79, DG80
BJ046-3720 Kakutani, T., and Senda, M., DG74, DG75, DG76, DG77, DG78, DG79, DG80
BJ046-3792 Kitagawa, T., and Ichimura, A., DE25, DE31

Vol. 47 (1974)
BJ047-1093 Suzuki, H., Shinozuka, N., and Hayano, S., DF54
BJ047-1490 Matsumoto, A., Lee, J.H., Yoshida, M., and Simamura, O., DC70, DC71, DE30, DE36, DE51, DE64, DF88, DF89, DG32, DG56
BJ047-2650 Yao, T., and Musha, S., DA59

Vol. 48 (1975)
BJ048-0416 Hayashi, T., Mataga, N., Sakata, Y., and Misumi, S., DG35, DG55
BJ048-0435 Yao, T., and Musha, S., DA58
BJ048-1354 Erabi, T., Hiura, H., and Tanaka, M., DH30
BJ048-2176 Wasa, T., and Musha, S., DA56, DA57, DA66, DA67

BP (BERICHTE DER BUNSENGESELLSCHAFT FUR PHYSKALISCHE CHEMIE)

Vol. 68 (1974)
BP068-0003 Berg, H., and Kramarczyk, K., DB69, DE26, DE27, DE28, DE29

CO (COLLECTION OF CZECHLOSLOVAK CHEMICAL COMMUNICATIONS)

Vol. 32 (1967)
CO032-0246 Zikán, J., and Kalous, V., DB35

Vol. 36 (1971)
CO036-1035 Mader, P., DA07, DA23, DA82, DB09, DB16, DB17

CZ (CHEMICKÉ ZVESTI)

Vol. 18 (1964)
CZ018-0422 Hetman, J.S., DE01
CA018-0427 Hynie, I., and Prokeš, J., DD31

DS (DOKLADY AKADEMII NAUK SSSR)

Vol. 171 (1966)
DS171-0860 Arbuzov, A., and Berdnikov, E.A., DA21, DA39, DA41, DA46, DC29, DC83

EA (ELECTROCHIMICA ACTA)

Vol. 11 (1966)
EA011-1189 Brook, P.A., and Crossley, J.A., DB79, DB80, DB81

Vol. 13 (1968)
EA013-1187 Visco, R.E., and Chandross, E.A., DA13
EA013-1197 Chang, J., Hercules, D.M., and Roe, D.K., DH13

EA (ELECTROCHIMICA ACTA)(cont.)
Vol. 13 (1968)(cont.)
   EA013-1209 Maricle, D.L., Zweig, A., Maurer, A.H., and Brinen, J.S., DH13

Vol. 15 (1969)
   EA015-1543 Pragst, F., and Jugelt, W., DG02

Vol. 17 (1971)
   EA017-0471 Vijayalakshamma, S.K., and Subrahmanya, R.S., DA89
   EA017-0511 Sadek, H., and Abd-El-Nabey, B.A., DA91
   EA017-1195 Issa, I.M., El-Samahy, A.A., Issa, R.M., and Ghoneim, M.M., DG06, DG07, DG08
   EA017-1391 Iwakura, C., Tsunage, M., and Tamura, H., DC38
   EA017-1421 Houghton, D.S., and Humffray, A.A., DD77
   EA017-1524 Andruzzi, R., Cardinali, M.E., and Trazza, A., DD15
   EA017-1595 Katz, M., Riemenschneider, P., and Wendt, H., DA45, DC08, DC25, DC80, DD14, DE98, DF19
   EA017-1615 Issa, I.M., El-Samahy, A.A., Issa, R.M., and Temerik, Y.M., DB35
   EA017-2009 Savena, R.S., and Chaturvedi, U.S., DC45
   EA017-2065 Sadek, H., and Abd-El-Nabey, B.A., DA91
   EA017-2077 Katiyar, S.S., Lalithambika, M., and Devaprabhakara, D., DD55, DD56, DD93, DE66
   EA017-2085 Tiwari, S.K., and Kumar, A., DB67
   EA017-2145 Houghton, D.S., and Humffray, A.A., DD66

ER (SOVIET ELECTROCHEMISTRY)

Vol. 1 (1965)
   ER001-0060 Mairanovskii, S.G., Barashova, N.V., and Vol'kenshtein, Y.B., DA29
   ER001-0759 Rusakova, M.S., Tur'yan, Y.I., and Ustavshchikov, B.F., DA19, DA42, DA43
   ER001-1345 Avrutskaya, I.A., and Fioshin, M. Ya, DB30, DB38

Vol. 2 (1966)
   ER002-0042 Bezuglyi, V.D., and Rapota, T.M., DB60, DB66, DB70, DC28
   ER002-0108 Mairanovskii, S.G., and Petrosyna, V.A., DA01, DA02
   ER002-0587 Butin, K.P., Beletskaya, I.P., and Reutov, O.A., DA28, DA37, DB25, DG14

Vol. 3 (1967)
   ER003-0170 Afanas'ev, B.N., DC15
   ER003-0263 Leibzon, V.N., Belikov, V.M., and Kozlov, L.M., DC32
   ER003-0494 Sevast'yanova, I.G., and Tomilov, A.P., DA11, DA62
   ER003-0538 Mairanovskii, V.G., Valashek, I.G., and Samokhvalov, G.I., DG28
   ER003-0623 Sister, Yu. D., Kiselev, B.A., Zhdanov, S.I., and Lyalikov, Yu. S., DA87, DB09, DD71

GA (GAZZETTA CHIMICA ITALIANA)

Vol. 96 (1966)
   GA096-0578 Cabani, S., Ceccanti, N., and Pagan, P., DA15

IV (IZVESTIYA VYSSHIKH UCHEBNYKH ZAVEDENII)

Vol. 8 (1965)
   IV008-0069 Komarova, G.E., and Lenskaya, V.N., DE22

JA (JOURNAL OF THE AMERICAN CHEMICAL SOCIETY)

Vol. 83 (1961)
   JA083-3949 Hoh, G.L.K., McEwen, W.E., and Kleinberg, J., DF39

Vol. 85 (1963)
   JA085-2124 Pysh, E.S., and Yang, N.C., DG29

JA (JOURNAL OF THE AMERICAN CHEMICAL SOCIETY)(cont.)

Vol. 86 (1964)
JA086-1382 Little, W.F., Reilley, C.N., Johnson, J.D., and Sanders, A.P., DE56
JA086-3155 Lambert, F.L., Albert, A.H., and Hardy, J.P., DD36

Vol. 88 (1966)
JA088-0471 Dessy, R.E., Stary, F.E., King, R.B., and Waldrop, M., DC56, DC60
JA088-1365 Davis, D.G., and Martin, R.F., DG73, DH12, DH19, DH21, DH22
JA088-5112 Dessy, R.E., King, R.B., and Waldrop, M., DC59, DC99, DF91
JA088-5117 Dessy, R.E., Weissman, P.M., and Pohl, R.L., DD26, DD59, DD81, DE34, DE37, DE75, DE86, DF00, DF01, DF02, DF03, DF36, DF62, DF68, DF75, DF84, DG33, DG37, DG43, DG44
JA088-5121 Dessy, R.E., Pohl, R.L., and King, R.B., DB64, DC06, DC07
JA088-5124 Dessy, R.E., and Weissman, P.M., DD59
JA088-5132 Psarras, T., and Dessy, R.E., DA22

Vol. 90 (1968)
JA090-1995 Dessy, R.E., and Pohl, R.L., DF50
JA090-2001 Dessy, R.E., Kornmann, R., Smith, C., and Haytor, R., DC58, DC61

Vol. 94 (1972)
JA094-0240 Dirlam, J.P., Eberson, L., and Casanova, J., DB92
JA094-0640 Bartak, D.E., and Hawley, M.D., DB65, DB71, DC13, DE39
JA094-0691 Faulkner, L.R., Tachikawa, H., and Bard, A.J., DA85, DD32, DE04, DE32, DE99, DG47, DH11, DH32, DH33, DH34, DH35, DH53, DH58, DH59, DH62
JA094-0738 Dessy, R.E., Charkoudian, J.C., and Rheingold, A.L., DD04, DD05, DD45, DD46, DD47, DD63, DD64, DD98, DE24
JA094-1522 Keszthelyi, C.P., Tachikawa, H., and Bard, A.J., DD67, DE35, DE36, DE37, DE38, DH43
JA094-2521 Ihielen, D.R., and Anderson, L.B., DC24
JA094-2862 Tokel, N.E., and Bard, A.J., DG64, DH40, DH56
JA094-3409 Ferguson, J.A., and Meyer, T.J., DG36
JA094-4529 Truex, T.J., and Holm, R.H., DE53, DE54, DE55, DE58, DE61, DE62, DE63, DF40, DF81, DF99, DG11, DG39
JA094-4749 Bechgaard, K., and Parker, V.D., DC39, DG40, DG41
JA094-4790 Freed, D.J., and Faulkner, L.R., DE04, DE99, DF80, DH48
JA094-4872 Tokel, N.E., Keszthelyi, C.P., and Bard, A.J., DE04, DH13, DH18, DH41, DH55, DH57, DH63, DH64
JA094-4882 Paxson, T.E., Kaloustian, M.K., Tom, G.M., Wiersema, R.J., and Hawthorne, M.F., DD43, DD44, DD69, DD70
JA094-4907 Paquette, L.A., Anderson, L.B., Hansen, J.F., Lang, S.A., Jr., and Berk, H., DD25, DD78, DD79, DF18, DF32, DG10
JA094-4915 Anderson, L.B., and Paquette, L.A., DC24, DD70, DF05
JA094-5139 Semmelhack, M.F., and Heinsohn, G.E., DC62, DC63, DC65, DC76
JA094-5502 Dabrowiak, J.C., Lovecchio, F.V., Goedken, V.L., and Busch, D.H., DD96, DF41, DF42, DF43, DF44, DF45, DF46, DF47, DF49
JA094-5538 Kowert, B.A., Marcoux, L., and Bard, A.J., DD68
JA094-6317 Bezman, R., and Faulkner, L.R., DG47, DH53
JA094-6324 Bezman, R., and Faulkner, L.R., DH62
JA094-6331 Bezman, R., and Faulkner, L.R., DH39
JA094-6812 Mayeda, F.A., Miller, L.L., and Wolf, J.F., DC37, DC82, DD27, DE17, DE40, DE43, DE44, DE94, DE96, DG46, DG49
JA094-7108 Nelsen, S.F., and Hintz, P.J., DA26, DA50, DA78, DB33, DB34, DB40, DB41, DB47, DB93, DB99, DC00, DC01, DC02, DC41, DC44, DC47, DC48, DC49, DC50, DC52, DC53, DC54, DC91, DC97, DC98, DD41, DD42, DE19, DE57, DE68, DG38

JA (JOURNAL OF THE AMERICAN CHEMICAL
   SOCIETY)(cont.)
Vol. 94 (1972)(cont.)
   JA094-7114 Nelsen, S.F., and Hintz,
      P.J., DA81, DB31, DB32,
      DB36, DB39, DB43, DB44,
      DB46, DB95, DB97, DB98,
      DC43, DC46, DC51, DC55,
      DD40, DD58, DD94, DE18,
      DG13, DG15
   JA094-7295 Klatt, L.N., and Rouseff,
      R.L., DA31
   JA094-7526 Bartak, D.E., Houser, K.J.,
      Rudy, B.C., and Hawley,
      M.D., DB48, DB49, DB50,
      DB51, DB52, DB53, DB57,
      DB59, DB61, DC03, DC21,
      DE23
   JA094-7941 O'Reilly, J.E., and Elving,
      P.J., DA12, DA31, DA32,
      DA33, DA63
   JA094-8197 Besto, S.R., Klapper, M.H.,
      and Anderson, L.B., DH31
   JA094-8471 House, H.O., Huber, L.E.,
      and Umen, M.J., DA35, DA38,
      DA68, DA71, DA73, DB14,
      DB27, DB86, DB87, DB88,
      DB89, DB90, DC72, DD17,
      DD28, DD29, DD34, DD35,
      DD54, DD82, DD85, DD89,
      DD90, DD91, DD92, DD99,
      DE14, DE52, DE60, DE65,
      DE88, DE89, DG21, DG22,
      DG23, DG51
   JA094-8475 Fry, A.J., and Reed, R.G.,
      DB83, DB84, DB85
   JA094-9020 Siegel, I.M., and Mark,
      H.B., Jr., DE38, DG48,
      DH42, DH44, DH45, DH46,
      DH47, DH49, DH50, DH51,
      DH52, DH54, DH60, DH61

Vol. 95 (1973)
   JA095-0991 Webb, J.W., Janik, B., and
      Elving, P.J., DA36, DC88,
      DC89, DF97, DF98, DG05
   JA095-1449 Bond, A.M., Hendrickson, A.
      R., and Martin, R.L.,
      DB28, DD86, DD87, DD88
   JA095-1703 Hurst, J.K., and Lane,
      R.H., DA34, DA40, DA72,
      DB21, DB22, DB29, DD16
   JA095-2198 Anderson, J.B., Broadhurst,
      M.J., and Paquette, L.A.,
      DC35, DC78, DC79, DD22,
      DE05
   JA095-2408 Chatt, J., Elson, C.M., and
      Leigh, G.S., DH14, DH15,
      DH16
   JA095-2646 Rieke, R.D., and Hudnall,
      P.M., DC09, DD60, DD61

JA095-3411 Ichikawa, M., and Meshituka,
   S., DD05, DD83, DE12,
   DE21, DE49, DE50, DF38,
   DF73, DG65, DG67, DG68,
   DG69
JA095-3495 Demortier, A., and Bard,
   A.J., DA96, DD03, DE02
JA095-4373 Bechgaard, K., Parker, V.
   D., and Pedersen, C.T.,
   DA69, DD06, DD57, DE71,
   DE82, DF13, DF14, DF58,
   DF59, DF93
JA095-5140 Fuhrhop, J., Kadish, K.M.,
   and Davis, D.G., DG81,
   DG82, DG83, DG84, DG85,
   DG86, DG87, DG88, DG89,
   DG90, DG91, DG92, DG93,
   DG94, DG95, DG96, DG97,
   DG98, DG99, DH00, DH01,
   DH03, DH04, DH05, DH06,
   DH07, DH10, DH27
JA095-5422 Nelsen, S.F., and Landis,
   R.T., II, DC92, DC96,
   DD39

JE (JOURNAL OF ELECTROANALYTICAL
   CHEMISTRY)

Vol. 4 (1962)
   JE004-0048 Gîrd, E., and Balaban,
      A.R., DF82, DF85
   JE004-0321 Holleck, L., and Becher,
      D., DB62, DB63

Vol. 6 (1963)
   JE006-0034 Hush, N.S., and Oldham,
      K.B., DA03

Vol. 16 (1968)
   JE016-0041 Malachesky, P.A., Prater,
      K.B., Petrue, G., and
      Adams, R.N., DB15

Vol. 24 (1970)
   JE024-0230 Kamel, L.A., and Shams El
      Din, A.M., DD76

Vol. 25 (1970)
   JE025-0397 Fleet, B., and Jee, R.D.,
      DC93, DC94

Vol. 30 (1971)
   JE030-0289 Bartak, D.E., Shields, T.,
      M., and Hawley, M.D.,
      DC14

TABLE X. Key to Literature Citations

JL (JOURNAL OF THE CHEMICAL SOCIETY [London])

1962
JL62A-4558 Lindsey, A.S., Peover, M.E., and Savill, N.G., DB69
JL62B-4540 Peover, M.E., DB69

1970
JL70B-0034 Bailes, M., and Leveson, L.L., DC57

JO (JOURNAL OF ORGANIC CHEMISTRY)

Vol. 25 (1960)
JO025-0611 Klemm, L.H., Lind, C.D., and Spence, J.T., DH13

Vol. 32 (1967)
JO032-1322 Zweig, A; Maurer, A.H., and Roberts, B.G., DG12

JP (JOURNAL OF PHYSICAL CHEMISTRY)

Vol. 67 (1963)
JP067-0862 Galus, Z., and Adams, R.N., DC86

JS (JOURNAL OF THE ELECTROCHEMICAL SOCIETY)

Vol. 108 (1961)
JS108-0980 Kolthoff, I.M., and Reddy, T.B., DD76

Vol. 112 (1965)
JS112-1215 Turner, W.R., and Elving, P.J., DD76

Vol. 116 (1969)
JS116-0743 Petrovich, J.P., Baizer, M.M., and Ort, M.R., DC84

Vol. 117 (1970)
JS117-0186 Eggins, B.R., and Chambers, J.Q., DD30, DD76

LU (VESTNIK LENINGRADSKOGO UNIVERSITETA, SERIYA FIZIKI I KHIMII)

Vol. 21 (1966)
LU021-0150 Reishahrit, L.S., and Batasova, L.V., DB08

LV (LATVIJAS PSR ZINATNU AKADEMIJAS VESTIS)

1967
LV670-0023 Reihmanis, G. and Stradins, J., DA30, DA51, DA54, DA55, DA64, DA65, DA95, DB72, DB82

RG (JOURNAL OF GENERAL CHEMISTRY OF THE USSR)

Vol. 35 (1965)
RG035-0011 Kudryavtseva, N.A., Pushkareva, Z.V., and Gryazev, V.F., DF77, DF78, DF79
RG035-0015 Bezuglyi, V.D., and Shimanskaya, N.P., DG30, DG31
RG035-0773 Stradyn', Ya. P., Yurashek, A., and Reikhmanis, G., DA83, DB77, DC23, DC77
RG035-0843 Svetin, Yu., V. and Andreeva, L., DC10, DC16, DC17, DC18, DC19, DC20, DC27, DC66, DC67, DC74, DC75, DC81, DD23, DD24

Vol. 36 (1966)
RG036-0622 Lopushanskii, A.I., and Shnarevich, A.I., DE97

RL (INDUSTRIAL LABORATORY [USSR])

Vol. 32 (1966)
RL032-0344 Kutanina, L.K., Berezina, K.G., and Gudzenko, Zh. D., DA00, DA27

RL (INDUSTRIAL LABORATORY [USSR])
   (cont.)

Vol. 33 (1967)
RL033-0029 Kheifets, L. Ya., Preobrazhenskaya, E.A., Bezuglyi, V.D., DC26

RR (JOURNAL OF ANALYTICAL CHEMISTRY OF THE USSR)

Vol. 20 (1965)
RR020-0469 Bezuglyi, V.D., and Ponomarev, Yu. P., DD38, DE67
RR020-1055 Tikhomirova, G.P., Belen'kaya, S.L., DC72
RR020-1167 Levinson, I.M., Skvortsova, A.E., DA16
RR020-1274 Bezuglyi, V.D., and Ponomarev, Yu. P., DB24

Vol. 21 (1966)
RR021-0322 Chursina, V.M., Litvin, E., F., and Freidlin, L. Kh. DC22, DC30, DC31

Vol. 22 (1967)
RR022-0263 Tikhomirova, G.P., Belen'kaya, S.L., and Madudina, N.F., DB91, DB94

RU (SOVIET PROGRESS IN CHEMISTRY)

Vol. 32 (1966)
RU032-0244 Shapoval, V.I., and Shapoval, G.S., DA18

RY (JOURNAL OF ORGANIC CHEMISTRY [USSR])

Vol. 2 (1966)
RY002-1096 Bezuglyi, V.D., Kheifets, L. Ya., Zakhs, É.R., and Éfros, L.S., DA85, DA86, DB58, DC04, DC05, DC11, DC12, DC68, DC69, DD00, DD01, DD19, DD20, DD21, DD65

Vol. 3 (1967)
RY003-0218 Beletskaya, I.P., Butin, K.P., and Reutov, O.A., DD08, DD09, DD10, DD11, DD12, DD13, DD18, DD50, DD51, DD84, DE16, DE59

TA (TALANTA)

Vol. 14 (1967)
TA014-0745 El Khiami, I., and Johnson, R.M., DF76

TF (TRANSACTIONS OF THE FARADAY SOCIETY)

Vol. 61 (1965)
TF061-1516 Davis, K.M.C., Hammond, P.R., and Peover, M.E., DB69

Vol. 65 (1969)
TF065-1668 Barnes, D., and Zuman, P., DC72

UK (UKRAINSKII KHIMICHESKII ZHURNAL)

Vol. 31 (1965)
UK031-0392 Bezuglyi, V.D., and Alekseeva, T.A., DA40

VZ (MENDELEEVA CHEMISTRY JOURNAL)

Vol. 11 (1966)
VZ011-0144 Brodskii, A.I., Gordienko, L.L., and Degtyarev, L. S., DF52

TABLE XI.
AUTHOR INDEX

This is an index to Table X. It lists, in alphabetical order, the names of all the authors cited in Table X and gives, for each author, the number of each page on which a citation of his name appears in Table X. Each reference given on the cited page should be inspected, for there may be two or more references on the page that includes the name sought, and such situations are not identified here.

TABLE XI. Author Index

Abd-El-Nabey, B.A., 466
Adams, R.N., 468, 469
Afanas'ev, B.N., 466
Afzal, M., 464
Ahmed, R., 464
Albert, A.H., 467
Alekseeva, T.A., 470
Anderson, J.B., 468
Anderson, L.B., 467, 468
Andreeva, L., 469
Andruzzi, R., 466
Arbuzov, A., 465
Arthur, J., 463
Atuma, S.S., 464
Avrutskaya, I.A., 466

Badoz-Lambling, J., 464
Bailes, M., 469
Baizer, M.M., 469
Balaban, A.R., 468
Banerjee, D.K., 464
Barashova, N.V., 466
Bard, A.J., 467, 468
Barnes, D., 470
Bartak, D.E., 467, 468
Batasova, L.V., 469
Baydar, A., 463
Becher, D., 468
Bechgaard, K., 467, 468
Bel Bruno, J.J., 463
Belen'kaya, S.L., 470
Beletskaya, I.P., 466, 470
Belikov, V.M., 466
Berdnikov, E.A., 465
Berezina, K.G., 469
Berg, H., 465
Berge, H., 463
Berk, H., 467
Besto, S.R., 468
Bezman, R., 467
Bezuglyi, V.D., 466, 469, 470
Bond, A.M., 468
Booth, M.S., 464
Bos, M., 463
Boto, K.G., 464
Brand, M.J.D., 464
Brinen, J.S., 466
Broadhurst, M.J., 468
Brodskii, A.I., 470
Brook, P.A., 465
Brooks, M.A., 463
Brügmann, L., 463
Budke, C.C., 464
Burmicz, J.S., 463
Busch, D.H., 467
Butin, K.P., 466, 470

Cabani, S., 466
Cardinali, M.E., 466

Cardwell, T.J., 464
Casanova, J., 467
Ceccanti, N., 466
Chambers, J.Q., 469
Chandross, E.A., 465
Chang, J., 465
Chane, J.P., 464
Charkoudian, J.C., 467
Chatt, J., 468
Chaturvedi, U.S., 466
Cheng, P.I., 464
Christian, G.D., 463
Chursina, V.M., 470
Cohen, A.I., 464
Crossley, J.A., 465

Dabard, R., 464
Dabrowiak, J.C., 467
Dahmen, E.A.M.F., 463
Date, Y., 465
Davis, D.G., 463, 467, 468
Davis, K.M.C., 470
De, P.K., 463
Degtyarev, L.S., 470
Demange-Guerin, G., 464
Demortier, A., 468
Deshler, L., 463
De Silva, J.A.F., 463
Dessy, R.E., 467
Devaprabhakara, D., 466
Dickerson, R.L., 463
Dirlam, J.P., 467
Donahue, J.J., 463
Dryhurst, G., 463

Eberson, L., 467
Edgar, J.S., 464
Efros, L.S., 470
Eggins, B.R., 469
El Khiami, I., 470
El-Samahy, A.A., 466
Elson, C.M., 468
Elving, P.J., 468, 469
Erabi, T., 465

Farroha, S.M., 464
Faulkner, L.R., 467
Ferguson, J.A., 467
Fioshin, M. Ya., 466
Fleet, B., 464, 468
Fournari, P., 464
Freed, D.J., 467
Freidlin, L. Kh., 470
Fry, A.J., 468
Fuhrhop, J., 468

Galus, Z., 469
Ghoneim, M.M., 466
Gîrd, E., 468
Goedken, V.L., 467
Goldsmith, J.A., 463
Gordienko, L.L., 470
Grant, J., 463
Gryazev, V.F., 469
Gudzenko, Zh.D., 469

Hackman, M.R., 463
Halvorsen, S., 463
Hamley, H.O., 463
Hammond, P.R., 470
Hansen, J.F., 467
Hardy, J.P., 467
Hawley, M.D., 467, 468
Hawthorne, M.F., 467
Hayano, S., 465
Hayashi, T., 465
Haytor, R., 467
Heinsohn, G.E., 467
Hendrickson, A.R., 468
Hercules, D.M., 465
Hetman, J.S., 463, 465
Hidai, M., 465
Hintz, P.J., 467, 468
Hiura, H., 465
Høgberg, K., 463
Hoh, G.L.K., 466
Holleck, L., 468
Holm, R.H., 467
Houghton, D.S., 466
House, H.O., 468
Houser, K.J., 468
Huber, L.E., 468
Hudnall, P.M., 468
Humffray, A.A., 466
Hurst, J.K., 468
Hush, N.S., 468
Hynie, I., 465

Ichikawa, M., 468
Ichimura, A., 465
Ihielen, D.R., 467
Ikeda, T., 465
Issa, I.M., 466
Issa, R.M., 466
Iwakura, C., 465, 466

Jacobsen, E., 463
Jacobsen, T.V., 463
Janík, B., 468
Jee, R.D., 468
Jenkins, H.A., 463
Jenkins, T., 463

Johnson, J.D., 467
Johnson, R.M., 470
Jugelt, W., 466

Kabasakalian, P., 464
Kadish, K.M., 468
Kageyama, H., 465
Kakutani, T., 465
Kalous, V., 465
Kaloustian, M.K., 467
Kamel, L.A., 468
Katiyar, S.S., 465, 466
Katz, M., 466
Keszthelyi, C.P., 467
Khan, A.Y., 464
Kheifets, L. Ya., 470
Kimura, K., 465
King, R.B., 467
Kiselev, B.A., 466
Kitagawa, T., 465
Kkolos, E., 463
Klapper, M.H., 468
Klatt, L.N., 468
Kleinberg, J., 466
Klemm, L.H., 469
Klevan, K.H., 463
Kolthoff, I.M., 469
Komarova, G.E., 466
Kornmann, R., 467
Kowert, B.A., 467
Kozlov, L.M., 466
Kramarczyk, K., 465
Kudryavtseva, N.A., 469
Kumar, A., 466
Kurosaki, Y., 465
Kutanina, L.K., 469

Lalithambika, M., 465, 466
Lambert, F.L., 467
Landis, R.T., 468
Lane, R.H., 468
Lang, S.A., Jr., 467
Laviron, E., 464
Lee, J.H., 465
Leibzon, V.N., 466
Leigh, G.S., 468
Lenskaya, V.N., 466
Leo, B., 463
Leveson, L.L., 469
Levinson, I.M., 470
Lind, C.D., 469
Lindquist, J., 464
Lindsey, A.S., 469
Little, W.F., 467
Litvin, E.F., 470
Lopushanskii, A.I., 469
Lovecchio, F.V., 467
Lundström, K., 464
Lyalikov, Yu. S., 466

TABLE XI. Author Index

McAllister, D., 463
McEwen, W.E., 466
McGlotten, J., 464
Mader, P., 465
Madudina, N.F., 470
Magee, R.J., 464
Mairanovskii, S.G., 466
Mairanovskii, V.G., 466
Malachesky, P.A., 468
Marcoux, L., 467
Maricle, D.L., 466
Mark, H.B. Jr., 468
Martin, R.F., 467
Martin, R.L., 468
Mataga, N., 465
Matsuda, Y., 465
Matsui, Y., 465
Matsumoto, A., 465
Maurer, A.H., 466, 469
Mayeda, F.A., 467
Meshituka, S., 468
Meyer, T.J., 467
Miller, F.D., 464
Miller, L.L., 467
Misumi, S., 465
Mohamad, M., 464
Morioka, M., 465
Munno, N., 463
Murayama, T., 465
Musha, S., 465

Nelson, S.F., 467, 468
Nisa, A., 464

Oldham, K.B., 468
Oliveri-Vigh, S., 463
O'Reilly, J.E., 468
Ort, M.R., 469

Pagan, P., 466
Paquette, L.A., 467, 468
Parker, V.D., 467, 468
Paxson, T.E., 467
Pedersen, C.T., 468
Peover, M.E., 469, 470
Petrosyan, V.A., 466
Petrovich, J.P., 469
Petrue, G., 468
Pinson, J.P., 463
Pohl, R.L., 467
Ponomarev, Yu. P., 470
Powell, F.E., 464
Pragst, F., 466
Prater, K.B., 468
Preobrazhenskaya, E.A., 470
Prokeš, J., 465

Psarras, T., 467
Pushkareva, Z.N., 469
Pysh, E.S., 466

Randle, T.H., 464
Rapota, T.M., 466
Reddy, T.B., 469
Reed, R.G., 468
Reihmanis, G., 469
Reilley, C.N., 467
Reingold, A.L., 467
Reishahrit, L.S., 469
Reutov, O.A., 466, 470
Rieke, R.D., 468
Riemenschneider, P., 466
Roberts, B.G., 469
Roe, D.K., 465
Rogers, J.W., 463
Rojahn, T., 463
Rouseff, R.L., 468
Rudy, B.C., 468
Rusakova, M.S., 466

Sadek, H., 466
Sakata, Y., 465
Samokhvalov, G.I., 466
Sanders, A.P., 467
Savena, R.S., 466
Savill, N.G., 469
Semmelhack, M.F., 467
Senda, M., 465
Sevast'yanova, I.G., 466
Shams El Din, A.M., 468
Shapoval, G.S., 470
Shapoval, V.I., 470
Shields, T.M., 468
Shimanskaya, N.P., 469
Shinozuka, N., 465
Shnarevich, A.I., 469
Shukig, R.P., 465
Siegel, I.M., 468
Siekiera, J., 463
Silbert, L.S., 464
Silva, F. de, 463
Simamura, O., 465
Sister, Yu. D., 466
Skvortsova, A.E., 470
Smith, C., 467
Smyth, W.F., 463
Snowden, C.J., 464
Söderhjelm, P., 464
Spence, J.T., 469
Stary, F.E., 467
Stradyn', Ya. P., 469
Stradins, J., 469
Strójny, N., 463
Subrahmanya, R.S., 466
Suzuki, H., 465

Svehla, G., 464
Svetin, Yu. V., 469
Swern, D., 464

Tachikawa, H., 467
Tako, K., 465
Tamura, H., 465, 466
Tanaka, M., 465
Temerik, Y.M., 466
Thomas, F.G., 464
Thomas, L.C., 463
Thorgersen, K.B., 463
Tikhomirova, G.P., 470
Tirouflet, J., 464
Tiwari, S.K., 466
Tokel, N.E., 467
Tom, G.M., 467
Tomcsány, L., 463
Tomilov, A.P., 466
Totsuka, S., 465
Trazza, A., 466
Truex, T.J., 467
Truxillo, L.A., 463
Tsunage, M., 466
Turner, W.R., 469
Tur'yan, Y.I., 466

Uchida, Y., 465
Umen, M.J., 468
Ustavshchinkov, R.E., 466

Valashek, I.G., 466
Vijayalakshamma, S.K., 466
Visco, R.E., 465
Vol'kenshtein, Y.B., 466

Waldrop, M., 467
Walker, J., 463
Wasa, T., 465
Watkiss, D., 463
Watson, A., 464
Weaver, M.R.H., 464
Webb, J.W., 468
Weissman, P.D., 467
Wendt, H., 466
Wiersema, F., 467
Wolf, J.F., 467
Woods, R., 464

Yang, N.C., 466
Yao, T., 465
Yoshida, M., 465
Yurashek, A., 469

Zakhs, É.R., 470
Zhdanov, S.I., 466
Zikán, J., 465
Zuman, P., 463, 470
Zweig, A., 466, 469

CORRIGENDA FOR PRECEDING VOLUMES OF THE CRC HANDBOOK SERIES IN ORGANIC ELECTROCHEMISTRY

VOLUME I
Table I

Page	Code number	Correction
38	AR51	The structural formula should read $C_6H_5CH{:}CHCOCH_3$.

VOLUME II
Table II

38	BH13	The correct structure is

[9-methylanthracene structure]

46	BK31	The correct structure is

[3,4,3',4'-tetramethoxybenzophenone structure]

63	BP65	The plus and minus signs should be reversed.

Table IX

587	Vol. 15 (1969) should read Vol. 15 (1970).
608	Under JS107-0537, Gendersen, A., should read Gundersen, A.
608	Under JS107-0616, Day, R. J., Jr., should read Day, R. A., Jr.
610	Under JS117-0485, Dunnig, J. S., should read Dunning, J. S.
616	Under TF064-1070, Parsons, R. and Symons, P. C., should read Edwards, T. G. and Grinter, R. The new names should be added to Table X, with references to page 616.

VOLUME III
Table I

Page	Code	Correction
42	CA96	The empirical formula should read $C_5H_5N_5O$ and the compound should appear on page 46 between CB05 and CB06.
186	CG18	The structural formula should read $[(CH_3)_2N(CH{:}CH)_3CH{:}N(CH_3)_2]^+$ ClO_4^-.
262	CJ39	The empirical formula should read $C_{15}H_{10}O_2$ and the compound should appear on the same page between CJ42 and CJ43.
	CJ43	The structural formula should read $3\text{-}BrC_6H_4COCHC_6H_5$ and the WLN code should read ER CV1U1R.
	CJ97	The WLN code should read QR D BX FX.
318	CM18	The empirical formula should read $C_{21}H_{29}N_7O_{14}P_2$.

The cross references given in parentheses should be added beneath the following code numbers:

CA24(ab40), CA25(ab44), CA49(ac15), CA74(ad01), CA94(ad97), CC27(ai47), CC82(ak87),
CD74(an87), CE01(ao48), CE30(ap24), CE42(ap58), CF39(ar87), CF59as34), CG98(ba91),
CH57(bc77), CH61(bc85), CJ43(bg90), CJ47bg98), CJ51(bh03), CJ60(bh13), CJ65(bh22),
CJ97(bh99), CK18(bi86), CK24(bi96), CK26(bi97), CK51(bj44), CK83(bk13), CK91bk31),
CL17(bk84), CL20(bk88), CL54(bl91), CL56(bm05), CL61(bm21), CL68(bm41), CL69(bm42),
CM09(bn71), CM10(bn74), CM41(bp13), CM50(bp65), CM98(bs21), CM99(bs29), CN00(bs31).

CORRIGENDA FOR PRECEDING VOLUMES OF THE CRC HANDBOOK SERIES IN ORGANIC ELECTROCHEMISTRY (continued)

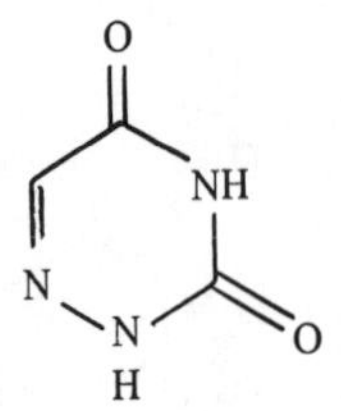

Page	Code number	Correction

Table III (continued)

357   CA24   The correct structure is

[Structure: 1,2,4-triazine-3,5(2H,4H)-dione ring with O at position 5, NH, N-H, and C=O at position 3]

370   CK91   The correct structure is

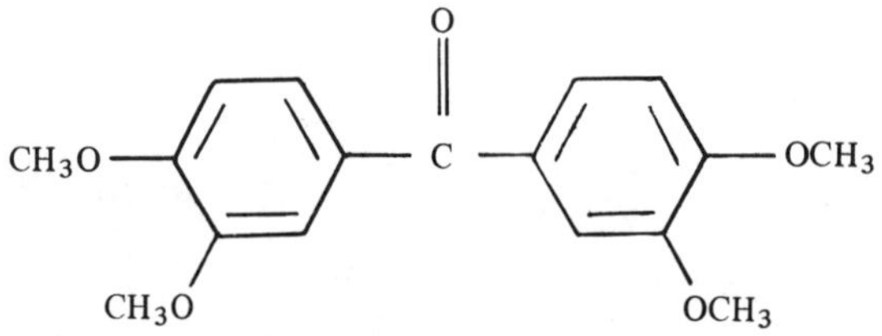

[Structure: 3,4,3',4'-tetramethoxybenzophenone — two aryl rings bearing CH$_3$O and OCH$_3$ groups connected by C=O]

Table IV

Page	Correction
515	5-*tert*-butyl-1,2,4-triazacyclohex-6-ene-3,5-dione, CC95 should read 6-*tert*-butyl-1,2,4-triazacyclohex-6-ene-3,5-dione, CC95.
	5-*tert*-butyl-1,2,4-triazine-3,5(2H,4H)dione, CC95 should read 6-*tert*-butyl-1,2,4-triazine-3,5(2H,4H)dione, CC95.
535	5-ethyl-1,2,4-triazacyclohex-6-ene-3,5-dione, CB12 should read 6-ethyl-1,2,4-triazacyclohex-6-ene-3,5-dione, CB12.
	5-ethyl-1,2,4-triazine-3,5(2H,4H)dione, CB12 should read 6-ethyl-1,2,4-triazine-3,5(2H,4H)dione, CB12.
556	protoporphineiron(III), CM86 should read protoporphine(IX) iron(III) chloride, CM86.

Ref
QD
272
E4
C17
v.4